M. Karimi / IUP 2004

INSTRUCTOR'S SOLUTIONS MANUAL

D1762742

INTRODUCTION to ELECTRODYNAMICS

Third Edition

David J. Griffiths

PRENTICE HALL, Upper Saddle River, NJ 07458

Executive Editor: Alison Reeves
Project Manager: Elizabeth Kell
Special Projects Manager: Barbara A. Murray
Production Editor: Barbara A. Till
Supplement Cover Manager: Paul Gourhan
Supplement Cover Designer: Liz Nemeth
Manufacturing Manager: Trudy Pisciotti

Upper Saddle River, NJ 07458

Printed in the United States of America

10 9 8 7 6 5 4 3

ISBN 0-13-859851-7

Prentice-Hall International (UK) Limited, London
Prentice-Hall of Australia Pty. Limited, Sydney
Prentice-Hall Canada, Inc., Toronto
Prentice-Hall Hispanoamericana, S.A., Mexico
Prentice-Hall of India Private Limited, New Delhi
Prentice-Hall (Singapore) Pte. Ltd.
Prentice-Hall of Japan, Inc., Tokyo
Editora Prentice-Hall do Brazil, Ltda., Rio de Janeiro

ACKNOWLEDGMENTS

Although I wrote these solutions, most of the typesetting was done by Jonah Gollub, Christopher Lee, and James Terwilliger (any errors are, of course, entirely their fault). Chris also did most of the figures, and I would like to thank him particularly for all his help. We have carefully proofread every page, but if you do find mistakes, please let me know (griffith@reed.edu).

TABLE OF CONTENTS

Chapter 1

Vector Analysis

Problem 1.1

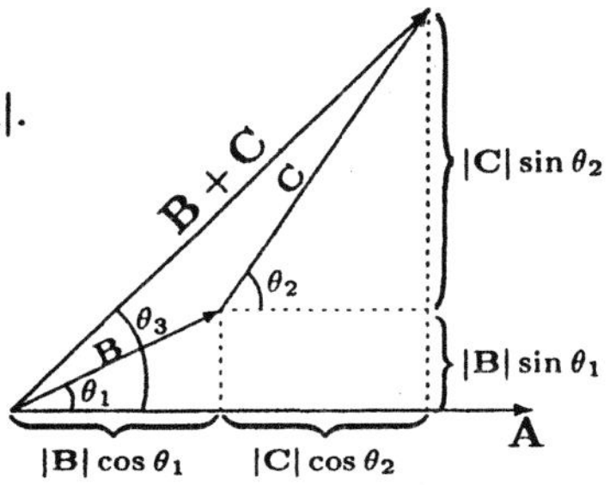

(a) From the diagram, $|\mathbf{B}+\mathbf{C}|\cos\theta_3 = |\mathbf{B}|\cos\theta_1 + |\mathbf{C}|\cos\theta_2$. Multiply by $|\mathbf{A}|$.
$|\mathbf{A}||\mathbf{B}+\mathbf{C}|\cos\theta_3 = |\mathbf{A}||\mathbf{B}|\cos\theta_1 + |\mathbf{A}||\mathbf{C}|\cos\theta_2$.
So: $\mathbf{A}\cdot(\mathbf{B}+\mathbf{C}) = \mathbf{A}\cdot\mathbf{B} + \mathbf{A}\cdot\mathbf{C}$. (Dot product is distributive.)

Similarly: $|\mathbf{B}+\mathbf{C}|\sin\theta_3 = |\mathbf{B}|\sin\theta_1 + |\mathbf{C}|\sin\theta_2$. Mulitply by $|\mathbf{A}|\,\hat{\mathbf{n}}$.
$|\mathbf{A}||\mathbf{B}+\mathbf{C}|\sin\theta_3\,\hat{\mathbf{n}} = |\mathbf{A}||\mathbf{B}|\sin\theta_1\,\hat{\mathbf{n}} + |\mathbf{A}||\mathbf{C}|\sin\theta_2\,\hat{\mathbf{n}}$.
If $\hat{\mathbf{n}}$ is the unit vector pointing out of the page, it follows that
$\mathbf{A}\times(\mathbf{B}+\mathbf{C}) = (\mathbf{A}\times\mathbf{B}) + (\mathbf{A}\times\mathbf{C})$. (Cross product is distributive.)

(b) For the general case, see G. E. Hay's *Vector and Tensor Analysis*, Chapter 1, Section 7 (dot product) and Section 8 (cross product).

Problem 1.2

The triple cross-product is *not* in general associative. For example, suppose $\mathbf{A} = \mathbf{B}$ and $\mathbf{C}$ is perpendicular to $\mathbf{A}$, as in the diagram. Then $(\mathbf{B}\times\mathbf{C})$ points out-of-the-page, and $\mathbf{A}\times(\mathbf{B}\times\mathbf{C})$ points *down*, and has magnitude ABC. But $(\mathbf{A}\times\mathbf{B}) = 0$, so $(\mathbf{A}\times\mathbf{B})\times\mathbf{C} = 0 \neq \mathbf{A}\times(\mathbf{B}\times\mathbf{C})$.

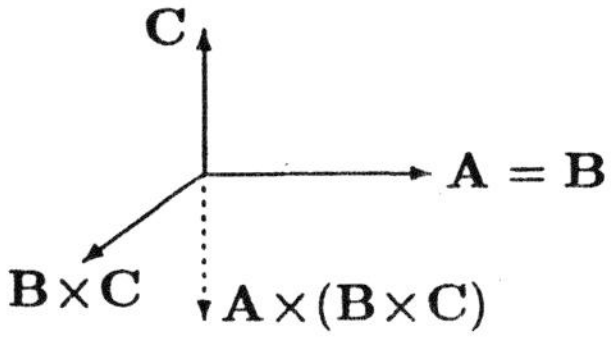

Problem 1.3

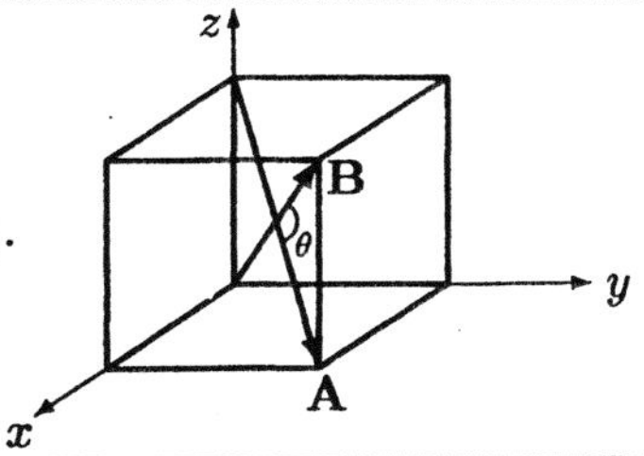

$\mathbf{A} = +1\,\hat{\mathbf{x}} + 1\,\hat{\mathbf{y}} - 1\,\hat{\mathbf{z}};\ A = \sqrt{3};\ \mathbf{B} = 1\,\hat{\mathbf{x}} + 1\,\hat{\mathbf{y}} + 1\,\hat{\mathbf{z}};\ B = \sqrt{3}.$

$\mathbf{A}\cdot\mathbf{B} = +1 + 1 - 1 = 1 = AB\cos\theta = \sqrt{3}\sqrt{3}\cos\theta \Rightarrow \cos\theta = \frac{1}{3}.$

$$\boxed{\theta = \cos^{-1}\left(\tfrac{1}{3}\right) \approx 70.5288^\circ}$$

Problem 1.4

The cross-product of any two vectors in the plane will give a vector perpendicular to the plane. For example, we might pick the base ($\mathbf{A}$) and the left side ($\mathbf{B}$):

$\mathbf{A} = -1\,\hat{\mathbf{x}} + 2\,\hat{\mathbf{y}} + 0\,\hat{\mathbf{z}};\ \mathbf{B} = -1\,\hat{\mathbf{x}} + 0\,\hat{\mathbf{y}} + 3\,\hat{\mathbf{z}}.$

$$\mathbf{A}\times\mathbf{B} = \begin{vmatrix} \hat{\mathbf{x}} & \hat{\mathbf{y}} & \hat{\mathbf{z}} \\ -1 & 2 & 0 \\ -1 & 0 & 3 \end{vmatrix} = 6\,\hat{\mathbf{x}} + 3\,\hat{\mathbf{y}} + 2\,\hat{\mathbf{z}}.$$

This has the right *direction*, but the wrong *magnitude*. To make a *unit* vector out of it, simply divide by its length:

$$|\mathbf{A}\times\mathbf{B}| = \sqrt{36+9+4} = 7. \qquad \hat{\mathbf{n}} = \frac{\mathbf{A}\times\mathbf{B}}{|\mathbf{A}\times\mathbf{B}|} = \boxed{\tfrac{6}{7}\hat{\mathbf{x}} + \tfrac{3}{7}\hat{\mathbf{y}} + \tfrac{2}{7}\hat{\mathbf{z}}}.$$

Problem 1.5

$$\mathbf{A}\times(\mathbf{B}\times\mathbf{C}) = \begin{vmatrix} \hat{\mathbf{x}} & \hat{\mathbf{y}} & \hat{\mathbf{z}} \\ A_x & A_y & A_z \\ (B_yC_z - B_zC_y) & (B_zC_x - B_xC_z) & (B_xC_y - B_yC_x) \end{vmatrix}$$

$$= \hat{\mathbf{x}}[A_y(B_xC_y - B_yC_x) - A_z(B_zC_x - B_xC_z)] + \hat{\mathbf{y}}() + \hat{\mathbf{z}}()$$

(I'll just check the x-component; the others go the same way.)

$$= \hat{\mathbf{x}}(A_yB_xC_y - A_yB_yC_x - A_zB_zC_x + A_zB_xC_z) + \hat{\mathbf{y}}() + \hat{\mathbf{z}}().$$

$$\mathbf{B}(\mathbf{A}\cdot\mathbf{C}) - \mathbf{C}(\mathbf{A}\cdot\mathbf{B}) = [B_x(A_xC_x + A_yC_y + A_zC_z) - C_x(A_xB_x + A_yB_y + A_zB_z)]\,\hat{\mathbf{x}} + ()\,\hat{\mathbf{y}} + ()\,\hat{\mathbf{z}}$$

$$= \hat{\mathbf{x}}(A_yB_xC_y + A_zB_xC_z - A_yB_yC_x - A_zB_zC_x) + \hat{\mathbf{y}}() + \hat{\mathbf{z}}().$$ They agree.

Problem 1.6

$$\mathbf{A}\times(\mathbf{B}\times\mathbf{C}) + \mathbf{B}\times(\mathbf{C}\times\mathbf{A}) + \mathbf{C}\times(\mathbf{A}\times\mathbf{B}) = \mathbf{B}(\mathbf{A}\cdot\mathbf{C}) - \mathbf{C}(\mathbf{A}\cdot\mathbf{B}) + \mathbf{C}(\mathbf{A}\cdot\mathbf{B}) - \mathbf{A}(\mathbf{C}\cdot\mathbf{B}) + \mathbf{A}(\mathbf{B}\cdot\mathbf{C}) - \mathbf{B}(\mathbf{C}\cdot\mathbf{A}) = 0.$$

So: $\mathbf{A}\times(\mathbf{B}\times\mathbf{C}) - (\mathbf{A}\times\mathbf{B})\times\mathbf{C} = -\mathbf{B}\times(\mathbf{C}\times\mathbf{A}) = \mathbf{A}(\mathbf{B}\cdot\mathbf{C}) - \mathbf{C}(\mathbf{A}\cdot\mathbf{B})$.

If this is zero, then either **A** is parallel to **C** (including the case in which they point in *opposite* directions, or one is zero), or else $\mathbf{B}\cdot\mathbf{C} = \mathbf{B}\cdot\mathbf{A} = 0$, in which case **B** is perpendicular to **A** and **C** (including the case $\mathbf{B} = 0$).

Conclusion: $\boxed{\mathbf{A}\times(\mathbf{B}\times\mathbf{C}) = (\mathbf{A}\times\mathbf{B})\times\mathbf{C} \iff \text{either } \mathbf{A} \text{ is parallel to } \mathbf{C}, \text{ or } \mathbf{B} \text{ is perpendicular to } \mathbf{A} \text{ and } \mathbf{C}.}$

Problem 1.7

$$\boldsymbol{\mathscr{r}} = (4\,\hat{\mathbf{x}} + 6\,\hat{\mathbf{y}} + 8\,\hat{\mathbf{z}}) - (2\,\hat{\mathbf{x}} + 8\,\hat{\mathbf{y}} + 7\,\hat{\mathbf{z}}) = \boxed{2\,\hat{\mathbf{x}} - 2\,\hat{\mathbf{y}} + \hat{\mathbf{z}}}$$

$$\mathscr{r} = \sqrt{4+4+1} = \boxed{3}$$

$$\hat{\boldsymbol{\mathscr{r}}} = \frac{\boldsymbol{\mathscr{r}}}{\mathscr{r}} = \boxed{\tfrac{2}{3}\hat{\mathbf{x}} - \tfrac{2}{3}\hat{\mathbf{y}} + \tfrac{1}{3}\hat{\mathbf{z}}}$$

Problem 1.8

(a) $\bar{A}_y\bar{B}_y + \bar{A}_z\bar{B}_z = (\cos\phi A_y + \sin\phi A_z)(\cos\phi B_y + \sin\phi B_z) + (-\sin\phi A_y + \cos\phi A_z)(-\sin\phi B_y + \cos\phi B_z)$
$= \cos^2\phi A_yB_y + \sin\phi\cos\phi(A_yB_z + A_zB_y) + \sin^2\phi A_zB_z + \sin^2\phi A_yB_y - \sin\phi\cos\phi(A_yB_z + A_zB_y) + \cos^2\phi A_zB_z$
$= (\cos^2\phi + \sin^2\phi)A_yB_y + (\sin^2\phi + \cos^2\phi)A_zB_z = A_yB_y + A_zB_z.$ ✓

(b) $(\overline{A}_x)^2 + (\overline{A}_y)^2 + (\overline{A}_z)^2 = \Sigma_{i=1}^3 \overline{A}_i\overline{A}_i = \Sigma_{i=1}^3\left(\Sigma_{j=1}^3 R_{ij}A_j\right)\left(\Sigma_{k=1}^3 R_{ik}A_k\right) = \Sigma_{j,k}\left(\Sigma_i R_{ij}R_{ik}\right)A_jA_k.$

This equals $A_x^2 + A_y^2 + A_z^2$ *provided* $\boxed{\Sigma_{i=1}^3 R_{ij}R_{ik} = \begin{Bmatrix} 1 & if & j = k \\ 0 & if & j \neq k \end{Bmatrix}}$

Moreover, if R is to preserve lengths for *all* vectors **A**, then this condition is not only *sufficient* but also *necessary*. For suppose $\mathbf{A} = (1,0,0)$. Then $\Sigma_{j,k}\left(\Sigma_i R_{ij}R_{ik}\right)A_jA_k = \Sigma_i R_{i1}R_{i1}$, and this must equal 1 (since we want $\overline{A}_x^2 + \overline{A}_y^2 + \overline{A}_z^2 = 1$). Likewise, $\Sigma_{i=1}^3 R_{i2}R_{i2} = \Sigma_{i=1}^3 R_{i3}R_{i3} = 1$. To check the case $j \neq k$, choose $\mathbf{A} = (1,1,0)$. Then we want $2 = \Sigma_{j,k}\left(\Sigma_i R_{ij}R_{ik}\right)A_jA_k = \Sigma_i R_{i1}R_{i1} + \Sigma_i R_{i2}R_{i2} + \Sigma_i R_{i1}R_{i2} + \Sigma_i R_{i2}R_{i1}$. But we already know that the first two sums are both 1; the third and fourth are *equal*, so $\Sigma_i R_{i1}R_{i2} = \Sigma_i R_{i2}R_{i1} = 0$, and so on for other unequal combinations of j, k. ✓ In matrix notation: $\tilde{R}R = 1$, where $\tilde{R}$ is the *transpose* of R.

120 clockwize

$$\begin{pmatrix} i\cdot i & i\cdot j & i\cdot k \\ j\cdot i & j\cdot j & j\cdot k \\ k\cdot i & k\cdot j & k\cdot k \end{pmatrix}$$

Problem 1.9

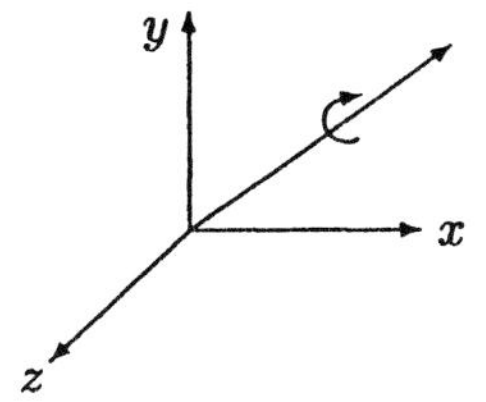

Looking down the axis:

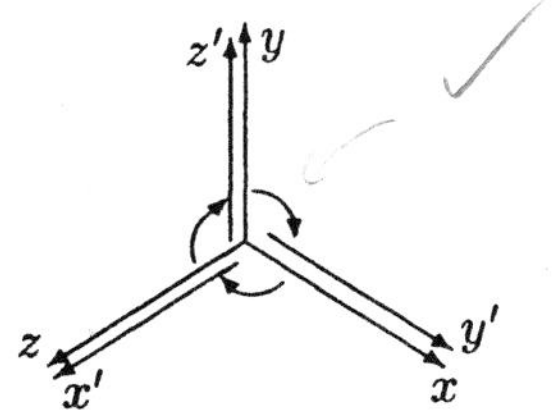

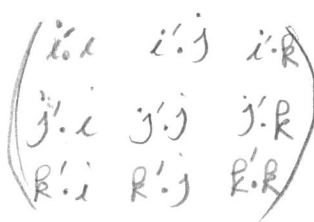

A 120° rotation carries the z axis into the y $(=\bar{z})$ axis, y into x $(=\bar{y})$, and x into z $(=\bar{x})$. So $\overline{A}_x = A_z$, $\overline{A}_y = A_x$, $\overline{A}_z = A_y$.

$$\boxed{R = \begin{pmatrix} 0 & 0 & 1 \\ 1 & 0 & 0 \\ 0 & 1 & 0 \end{pmatrix}}$$

Problem 1.10

(a) $\boxed{\text{No change.}}$ $(\overline{A}_x = A_x,\ \overline{A}_y = A_y,\ \overline{A}_z = A_z)$

(b) $\boxed{\mathbf{A} \longrightarrow -\mathbf{A},}$ in the sense $(\overline{A}_x = -A_x,\ \overline{A}_y = -A_y,\ \overline{A}_z = -A_z)$

(c) $(\mathbf{A}\times\mathbf{B}) \longrightarrow (-\mathbf{A})\times(-\mathbf{B}) = (\mathbf{A}\times\mathbf{B})$. That is, if $\mathbf{C} = \mathbf{A}\times\mathbf{B}$, $\boxed{\mathbf{C} \longrightarrow \mathbf{C}}$. *No* minus sign, in contrast to behavior of an "ordinary" vector, as given by (b). If **A** and **B** are *pseudovectors*, then $(\mathbf{A}\times\mathbf{B}) \longrightarrow (\mathbf{A})\times(\mathbf{B}) = (\mathbf{A}\times\mathbf{B})$. So the cross-product of two pseudovectors is again a *pseudovector.* In the cross-product of a vector and a pseudovector, one changes sign, the other doesn't, and therefore the cross-product is itself a *vector.* *Angular momentum* $(\mathbf{L} = \mathbf{r}\times\mathbf{p})$ and *torque* $(\mathbf{N} = \mathbf{r}\times\mathbf{F})$ are pseudovectors.

(d) $\mathbf{A}\cdot(\mathbf{B}\times\mathbf{C}) \longrightarrow (-\mathbf{A})\cdot((-\mathbf{B})\times(-\mathbf{C})) = -\mathbf{A}\cdot(\mathbf{B}\times\mathbf{C})$. So, if $a = \mathbf{A}\cdot(\mathbf{B}\times\mathbf{C})$, then $\boxed{a \longrightarrow -a;}$ a pseudoscalar *changes sign* under inversion of coordinates.

Problem 1.11

$$\boxed{\begin{aligned} &(a)\nabla f = 2x\,\hat{\mathbf{x}} + 3y^2\,\hat{\mathbf{y}} + 4z^3\,\hat{\mathbf{z}} \\ &(b)\nabla f = 2xy^3z^4\,\hat{\mathbf{x}} + 3x^2y^2z^4\,\hat{\mathbf{y}} + 4x^2y^3z^3\,\hat{\mathbf{z}} \\ &(c)\nabla f = e^x \sin y \ln z\,\hat{\mathbf{x}} + e^x \cos y \ln z\,\hat{\mathbf{y}} + e^x \sin y (1/z)\,\hat{\mathbf{z}} \end{aligned}}$$

Problem 1.12

(a) $\nabla h = 10[(2y - 6x - 18)\,\hat{\mathbf{x}} + (2x - 8y + 28)\,\hat{\mathbf{y}}]$. $\nabla h = 0$ at summit, so

$$\left.\begin{aligned} &2y - 6x - 18 = 0 \\ &2x - 8y + 28 = 0 \Longrightarrow 6x - 24y + 84 = 0 \end{aligned}\right\} 2y - 18 - 24y + 84 = 0.$$

$22y = 66 \Longrightarrow y = 3 \Longrightarrow 2x - 24 + 28 = 0 \Longrightarrow x = -2.$

Top is $\boxed{\text{3 miles north, 2 miles west, of South Hadley.}}$

(b) Putting in $x = -2$, $y = 3$:
$h = 10(-12 - 12 - 36 + 36 + 84 + 12) = \boxed{720 \text{ ft.}}$

(c) Putting in $x = 1$, $y = 1$: $\nabla h = 10[(2 - 6 - 18)\,\hat{\mathbf{x}} + (2 - 8 + 28)\,\hat{\mathbf{y}}] = 10(-22\,\hat{\mathbf{x}} + 22\,\hat{\mathbf{y}}) = 220(-\,\hat{\mathbf{x}} + \hat{\mathbf{y}})$.
$|\nabla h| = 220\sqrt{2} \approx \boxed{311 \text{ ft/mile}}$; direction: $\boxed{\text{northwest.}}$

Problem 1.13

$\boldsymbol{\imath} = (x - x')\,\hat{\mathbf{x}} + (y - y')\,\hat{\mathbf{y}} + (z - z')\,\hat{\mathbf{z}}; \quad \imath = \sqrt{(x - x')^2 + (y - y')^2 + (z - z')^2}.$

(a) $\boldsymbol{\nabla}(\imath^2) = \frac{\partial}{\partial x}[(x-x')^2 + (y-y')^2 + (z-z')^2]\,\hat{\mathbf{x}} + \frac{\partial}{\partial y}()\,\hat{\mathbf{y}} + \frac{\partial}{\partial z}()\,\hat{\mathbf{z}} = 2(x-x')\,\hat{\mathbf{x}} + 2(y-y')\,\hat{\mathbf{y}} + 2(z-z')\,\hat{\mathbf{z}} = 2\boldsymbol{\imath}.$

(b) $\boldsymbol{\nabla}(\frac{1}{\imath}) = \frac{\partial}{\partial x}[(x - x')^2 + (y - y')^2 + (z - z')^2]^{-\frac{1}{2}}\,\hat{\mathbf{x}} + \frac{\partial}{\partial y}()^{-\frac{1}{2}}\,\hat{\mathbf{y}} + \frac{\partial}{\partial z}()^{-\frac{1}{2}}\,\hat{\mathbf{z}}$
$= -\frac{1}{2}()^{-\frac{3}{2}}2(x - x')\,\hat{\mathbf{x}} - \frac{1}{2}()^{-\frac{3}{2}}2(y - y')\,\hat{\mathbf{y}} - \frac{1}{2}()^{-\frac{3}{2}}2(z - z')\,\hat{\mathbf{z}}$
$= -()^{-\frac{3}{2}}[(x - x')\,\hat{\mathbf{x}} + (y - y')\,\hat{\mathbf{y}} + (z - z')\,\hat{\mathbf{z}}] = -(1/\imath^3)\boldsymbol{\imath} = -(1/\imath^2)\hat{\boldsymbol{\imath}}.$

(c) $\frac{\partial}{\partial x}(\imath^n) = n\imath^{n-1}\frac{\partial \imath}{\partial x} = n\imath^{n-1}(\frac{1}{2}\frac{1}{\imath}2\imath_x) = n\imath^{n-1}\hat{\imath}_x$, so $\boxed{\boldsymbol{\nabla}(\imath^n) = n\imath^{n-1}\,\hat{\boldsymbol{\imath}}.}$

Problem 1.14

$\bar{y} = +y\,\cos\phi + z\,\sin\phi$; multiply by $\sin\phi$: $\bar{y}\sin\phi = +y\,\sin\phi\cos\phi + z\,\sin^2\phi$.
$\bar{z} = -y\,\sin\phi + z\,\cos\phi$; multiply by $\cos\phi$: $\bar{z}\cos\phi = -y\,\sin\phi\cos\phi + z\,\cos^2\phi$.
Add: $\bar{y}\sin\phi + \bar{z}\cos\phi = z(\sin^2\phi + \cos^2\phi) = z$. Likewise, $\bar{y}\cos\phi - \bar{z}\sin\phi = y$.
So $\frac{\partial y}{\partial \bar{y}} = \cos\phi$; $\frac{\partial y}{\partial \bar{z}} = -\sin\phi$; $\frac{\partial z}{\partial \bar{y}} = \sin\phi$; $\frac{\partial z}{\partial \bar{z}} = \cos\phi$. Therefore

$$\left.\begin{array}{l} \overline{(\boldsymbol{\nabla} f)}_y = \frac{\partial f}{\partial \bar{y}} = \frac{\partial f}{\partial y}\frac{\partial y}{\partial \bar{y}} + \frac{\partial f}{\partial z}\frac{\partial z}{\partial \bar{y}} = +\cos\phi(\boldsymbol{\nabla} f)_y + \sin\phi(\boldsymbol{\nabla} f)_z \\ \overline{(\boldsymbol{\nabla} f)}_z = \frac{\partial f}{\partial \bar{z}} = \frac{\partial f}{\partial y}\frac{\partial y}{\partial \bar{z}} + \frac{\partial f}{\partial z}\frac{\partial z}{\partial \bar{z}} = -\sin\phi(\boldsymbol{\nabla} f)_y + \cos\phi(\boldsymbol{\nabla} f)_z \end{array}\right\} \text{So } \boldsymbol{\nabla} f \text{ transforms as a vector.} \quad \text{qed}$$

Problem 1.15

(a) $\boldsymbol{\nabla}\cdot\mathbf{v}_a = \frac{\partial}{\partial x}(x^2) + \frac{\partial}{\partial y}(3xz^2) + \frac{\partial}{\partial z}(-2xz) = 2x + 0 - 2x = 0.$

(b) $\boldsymbol{\nabla}\cdot\mathbf{v}_b = \frac{\partial}{\partial x}(xy) + \frac{\partial}{\partial y}(2yz) + \frac{\partial}{\partial z}(3xz) = y + 2z + 3x.$

(c) $\boldsymbol{\nabla}\cdot\mathbf{v}_c = \frac{\partial}{\partial x}(y^2) + \frac{\partial}{\partial y}(2xy + z^2) + \frac{\partial}{\partial z}(2yz) = 0 + (2x) + (2y) = 2(x + y).$

Problem 1.16

$$\boldsymbol{\nabla}\cdot\mathbf{v} = \frac{\partial}{\partial x}(\frac{x}{r^3}) + \frac{\partial}{\partial y}(\frac{y}{r^3}) + \frac{\partial}{\partial z}(\frac{z}{r^3}) = \frac{\partial}{\partial x}\left[x(x^2 + y^2 + z^2)^{-\frac{3}{2}}\right] + \frac{\partial}{\partial y}\left[y(x^2 + y^2 + z^2)^{-\frac{3}{2}}\right] + \frac{\partial}{\partial z}\left[z(x^2 + y^2 + z^2)^{-\frac{3}{2}}\right]$$
$$= ()^{-\frac{3}{2}} + x(-3/2)()^{-\frac{5}{2}}2x + ()^{-\frac{3}{2}} + y(-3/2)()^{-\frac{5}{2}}2y + ()^{-\frac{3}{2}} + z(-3/2)()^{-\frac{5}{2}}2z$$
$$= 3r^{-3} - 3r^{-5}(x^2 + y^2 + z^2) = 3r^{-3} - 3r^{-3} = 0.$$

This conclusion is surprising, because, from the diagram, this vector field is obviously diverging away from the origin. How, then, can $\boldsymbol{\nabla}\cdot\mathbf{v} = 0$? The answer is that $\boldsymbol{\nabla}\cdot\mathbf{v} = 0$ everywhere *except* at the origin, but at the origin our calculation is no good, since $r = 0$, and the expression for $\mathbf{v}$ blows up. In fact, $\boldsymbol{\nabla}\cdot\mathbf{v}$ is ***infinite*** at that one point, and zero elsewhere, as we shall see in Sect. 1.5.

Problem 1.17

$\bar{v}_y = \cos\phi\,v_y + \sin\phi\,v_z$; $\bar{v}_z = -\sin\phi\,v_y + \cos\phi\,v_z$.

$\frac{\partial \bar{v}_y}{\partial \bar{y}} = \frac{\partial v_y}{\partial \bar{y}}\cos\phi + \frac{\partial v_z}{\partial \bar{y}}\sin\phi = \left(\frac{\partial v_y}{\partial y}\frac{\partial y}{\partial \bar{y}} + \frac{\partial v_y}{\partial z}\frac{\partial z}{\partial \bar{y}}\right)\cos\phi + \left(\frac{\partial v_z}{\partial y}\frac{\partial y}{\partial \bar{y}} + \frac{\partial v_z}{\partial z}\frac{\partial z}{\partial \bar{y}}\right)\sin\phi$. Use result in Prob. 1.14:
$= \left(\frac{\partial v_y}{\partial y}\cos\phi + \frac{\partial v_y}{\partial z}\sin\phi\right)\cos\phi + \left(\frac{\partial v_z}{\partial y}\cos\phi + \frac{\partial v_z}{\partial z}\sin\phi\right)\sin\phi.$

$\frac{\partial \bar{v}_z}{\partial \bar{z}} = -\frac{\partial v_y}{\partial \bar{z}}\sin\phi + \frac{\partial v_z}{\partial \bar{z}}\cos\phi = -\left(\frac{\partial v_y}{\partial y}\frac{\partial y}{\partial \bar{z}} + \frac{\partial v_y}{\partial z}\frac{\partial z}{\partial \bar{z}}\right)\sin\phi + \left(\frac{\partial v_z}{\partial y}\frac{\partial y}{\partial \bar{z}} + \frac{\partial v_z}{\partial z}\frac{\partial z}{\partial \bar{z}}\right)\cos\phi$
$= -\left(-\frac{\partial v_y}{\partial y}\sin\phi + \frac{\partial v_y}{\partial z}\cos\phi\right)\sin\phi + \left(-\frac{\partial v_z}{\partial y}\sin\phi + \frac{\partial v_z}{\partial z}\cos\phi\right)\cos\phi.$ So

$\frac{\partial \bar{v}_y}{\partial \bar{y}} + \frac{\partial \bar{v}_z}{\partial \bar{z}} = \frac{\partial v_y}{\partial y}\cos^2\phi + \frac{\partial v_y}{\partial z}\sin\phi\cos\phi + \frac{\partial v_z}{\partial y}\sin\phi\cos\phi + \frac{\partial v_z}{\partial z}\sin^2\phi + \frac{\partial v_y}{\partial y}\sin^2\phi - \frac{\partial v_y}{\partial z}\sin\phi\cos\phi$

$$-\frac{\partial v_z}{\partial y}\sin\phi\cos\phi + \frac{\partial v_z}{\partial z}\cos^2\phi$$
$$= \frac{\partial v_y}{\partial y}\left(\cos^2\phi + \sin^2\phi\right) + \frac{\partial v_z}{\partial z}\left(\sin^2\phi + \cos^2\phi\right) = \frac{\partial v_y}{\partial y} + \frac{\partial v_z}{\partial z}. \checkmark$$

Problem 1.18

(a) $$\boldsymbol{\nabla}\times\mathbf{v}_a = \begin{vmatrix} \hat{\mathbf{x}} & \hat{\mathbf{y}} & \hat{\mathbf{z}} \\ \frac{\partial}{\partial x} & \frac{\partial}{\partial y} & \frac{\partial}{\partial z} \\ x^2 & 3xz^2 & -2xz \end{vmatrix} = \hat{\mathbf{x}}(0-6xz) + \hat{\mathbf{y}}(0+2z) + \hat{\mathbf{z}}(3z^2-0) = \boxed{-6xz\,\hat{\mathbf{x}} + 2z\,\hat{\mathbf{y}} + 3z^2\,\hat{\mathbf{z}}.}$$

(b) $$\boldsymbol{\nabla}\times\mathbf{v}_b = \begin{vmatrix} \hat{\mathbf{x}} & \hat{\mathbf{y}} & \hat{\mathbf{z}} \\ \frac{\partial}{\partial x} & \frac{\partial}{\partial y} & \frac{\partial}{\partial z} \\ xy & 2yz & 3xz \end{vmatrix} = \hat{\mathbf{x}}(0-2y) + \hat{\mathbf{y}}(0-3z) + \hat{\mathbf{z}}(0-x) = \boxed{-2y\,\hat{\mathbf{x}} - 3z\,\hat{\mathbf{y}} - x\,\hat{\mathbf{z}}.}$$

(c) $$\boldsymbol{\nabla}\times\mathbf{v}_c = \begin{vmatrix} \hat{\mathbf{x}} & \hat{\mathbf{y}} & \hat{\mathbf{z}} \\ \frac{\partial}{\partial x} & \frac{\partial}{\partial y} & \frac{\partial}{\partial z} \\ y^2 & (2xy+z^2) & 2yz \end{vmatrix} = \hat{\mathbf{x}}(2z-2z) + \hat{\mathbf{y}}(0-0) + \hat{\mathbf{z}}(2y-2y) = \boxed{0.}$$

Problem 1.19

$\mathbf{v} = y\,\hat{\mathbf{x}} + x\,\hat{\mathbf{y}}$; or $\mathbf{v} = yz\,\hat{\mathbf{x}} + xz\,\hat{\mathbf{y}} + xy\,\hat{\mathbf{z}}$; or $\mathbf{v} = (3x^2z - z^3)\,\hat{\mathbf{x}} + 3\,\hat{\mathbf{y}} + (x^3 - 3xz^2)\,\hat{\mathbf{z}}$;
or $\mathbf{v} = (\sin x)(\cosh y)\,\hat{\mathbf{x}} - (\cos x)(\sinh y)\,\hat{\mathbf{y}}$; etc.

Problem 1.20

(i) $$\boldsymbol{\nabla}(fg) = \frac{\partial(fg)}{\partial x}\hat{\mathbf{x}} + \frac{\partial(fg)}{\partial y}\hat{\mathbf{y}} + \frac{\partial(fg)}{\partial z}\hat{\mathbf{z}} = \left(f\frac{\partial g}{\partial x} + g\frac{\partial f}{\partial x}\right)\hat{\mathbf{x}} + \left(f\frac{\partial g}{\partial y} + g\frac{\partial f}{\partial y}\right)\hat{\mathbf{y}} + \left(f\frac{\partial g}{\partial z} + g\frac{\partial f}{\partial z}\right)\hat{\mathbf{z}}$$
$$= f\left(\frac{\partial g}{\partial x}\hat{\mathbf{x}} + \frac{\partial g}{\partial y}\hat{\mathbf{y}} + \frac{\partial g}{\partial z}\hat{\mathbf{z}}\right) + g\left(\frac{\partial f}{\partial x}\hat{\mathbf{x}} + \frac{\partial f}{\partial y}\hat{\mathbf{y}} + \frac{\partial f}{\partial z}\hat{\mathbf{z}}\right) = f(\boldsymbol{\nabla}g) + g(\boldsymbol{\nabla}f). \quad \text{qed}$$

(iv) $$\boldsymbol{\nabla}\cdot(\mathbf{A}\times\mathbf{B}) = \frac{\partial}{\partial x}(A_yB_z - A_zB_y) + \frac{\partial}{\partial y}(A_zB_x - A_xB_z) + \frac{\partial}{\partial z}(A_xB_y - A_yB_x)$$
$$= A_y\frac{\partial B_z}{\partial x} + B_z\frac{\partial A_y}{\partial x} - A_z\frac{\partial B_y}{\partial x} - B_y\frac{\partial A_z}{\partial x} + A_z\frac{\partial B_x}{\partial y} + B_x\frac{\partial A_z}{\partial y} - A_x\frac{\partial B_z}{\partial y} - B_z\frac{\partial A_x}{\partial y}$$
$$+A_x\frac{\partial B_y}{\partial z} + B_y\frac{\partial A_x}{\partial z} - A_y\frac{\partial B_x}{\partial z} - B_x\frac{\partial A_y}{\partial z}$$
$$= B_x\left(\frac{\partial A_z}{\partial y} - \frac{\partial A_y}{\partial z}\right) + B_y\left(\frac{\partial A_x}{\partial z} - \frac{\partial A_z}{\partial x}\right) + B_z\left(\frac{\partial A_y}{\partial x} - \frac{\partial A_x}{\partial y}\right) - A_x\left(\frac{\partial B_z}{\partial y} - \frac{\partial B_y}{\partial z}\right)$$
$$-A_y\left(\frac{\partial B_x}{\partial z} - \frac{\partial B_z}{\partial x}\right) - A_z\left(\frac{\partial B_y}{\partial x} - \frac{\partial B_x}{\partial y}\right) = \mathbf{B}\cdot(\boldsymbol{\nabla}\times\mathbf{A}) - \mathbf{A}\cdot(\boldsymbol{\nabla}\times\mathbf{B}). \quad \text{qed}$$

(v) $$\boldsymbol{\nabla}\times(f\mathbf{A}) = \left(\frac{\partial(fA_z)}{\partial y} - \frac{\partial(fA_y)}{\partial z}\right)\hat{\mathbf{x}} + \left(\frac{\partial(fA_x)}{\partial z} - \frac{\partial(fA_z)}{\partial x}\right)\hat{\mathbf{y}} + \left(\frac{\partial(fA_y)}{\partial x} - \frac{\partial(fA_x)}{\partial y}\right)\hat{\mathbf{z}}$$
$$= \left(f\frac{\partial A_z}{\partial y} + A_z\frac{\partial f}{\partial y} - f\frac{\partial A_y}{\partial z} - A_y\frac{\partial f}{\partial z}\right)\hat{\mathbf{x}} + \left(f\frac{\partial A_x}{\partial z} + A_x\frac{\partial f}{\partial z} - f\frac{\partial A_z}{\partial x} - A_z\frac{\partial f}{\partial x}\right)\hat{\mathbf{y}}$$
$$+\left(f\frac{\partial A_y}{\partial x} + A_y\frac{\partial f}{\partial x} - f\frac{\partial A_x}{\partial y} - A_x\frac{\partial f}{\partial y}\right)\hat{\mathbf{z}}$$
$$= f\left[\left(\frac{\partial A_z}{\partial y} - \frac{\partial A_y}{\partial z}\right)\hat{\mathbf{x}} + \left(\frac{\partial A_x}{\partial z} - \frac{\partial A_z}{\partial x}\right)\hat{\mathbf{y}} + \left(\frac{\partial A_y}{\partial x} - \frac{\partial A_x}{\partial y}\right)\hat{\mathbf{z}}\right]$$
$$-\left[\left(A_y\frac{\partial f}{\partial z} - A_z\frac{\partial f}{\partial y}\right)\hat{\mathbf{x}} + \left(A_z\frac{\partial f}{\partial x} - A_x\frac{\partial f}{\partial z}\right)\hat{\mathbf{y}} + \left(A_x\frac{\partial f}{\partial y} - A_y\frac{\partial f}{\partial x}\right)\hat{\mathbf{z}}\right]$$
$$= f(\boldsymbol{\nabla}\times\mathbf{A}) - \mathbf{A}\times(\boldsymbol{\nabla}f). \quad \text{qed}$$

Problem 1.21

(a) $$(\mathbf{A}\cdot\boldsymbol{\nabla})\mathbf{B} = \left(A_x\frac{\partial B_x}{\partial x} + A_y\frac{\partial B_x}{\partial y} + A_z\frac{\partial B_x}{\partial z}\right)\hat{\mathbf{x}} + \left(A_x\frac{\partial B_y}{\partial x} + A_y\frac{\partial B_y}{\partial y} + A_z\frac{\partial B_y}{\partial z}\right)\hat{\mathbf{y}}$$
$$+\left(A_x\frac{\partial B_z}{\partial x} + A_y\frac{\partial B_z}{\partial y} + A_z\frac{\partial B_z}{\partial z}\right)\hat{\mathbf{z}}.$$

(b) $\hat{\mathbf{r}} = \frac{\mathbf{r}}{r} = \frac{x\,\hat{\mathbf{x}} + y\,\hat{\mathbf{y}} + z\,\hat{\mathbf{z}}}{\sqrt{x^2+y^2+z^2}}$. Let's just do the x component.

$$[(\hat{\mathbf{r}}\cdot\boldsymbol{\nabla})\hat{\mathbf{r}}]_x = \frac{1}{\sqrt{}}\left(x\frac{\partial}{\partial x} + y\frac{\partial}{\partial y} + z\frac{\partial}{\partial z}\right)\frac{x}{\sqrt{x^2+y^2+z^2}}$$

$$= \frac{1}{r}\left\{x\left[\frac{1}{\sqrt{\ }} + x\left(-\frac{1}{2}\right)\frac{1}{(\sqrt{\ })^3}2x\right] + yx\left[-\frac{1}{2}\frac{1}{(\sqrt{\ })^3}2y\right] + zx\left[-\frac{1}{2}\frac{1}{(\sqrt{\ })^3}2z\right]\right\}$$
$$= \frac{1}{r}\left\{\frac{x}{r} - \frac{1}{r^3}\left(x^3 + xy^2 + xz^2\right)\right\} = \frac{1}{r}\left\{\frac{x}{r} - \frac{x}{r^3}\left(x^2 + y^2 + z^2\right)\right\} = \frac{1}{r}\left(\frac{x}{r} - \frac{x}{r}\right) = 0.$$

Same goes for the other components. Hence: $\boxed{(\hat{\mathbf{r}}\cdot\nabla)\,\hat{\mathbf{r}} = 0}$.

(c) $(\mathbf{v}_a\cdot\nabla)\,\mathbf{v}_b = \left(x^2\frac{\partial}{\partial x} + 3xz^2\frac{\partial}{\partial y} - 2xz\frac{\partial}{\partial z}\right)(xy\,\hat{\mathbf{x}} + 2yz\,\hat{\mathbf{y}} + 3xz\,\hat{\mathbf{z}})$

$$= x^2\left(y\,\hat{\mathbf{x}} + 0\,\hat{\mathbf{y}} + 3z\,\hat{\mathbf{z}}\right) + 3xz^2\left(x\,\hat{\mathbf{x}} + 2z\,\hat{\mathbf{y}} + 0\,\hat{\mathbf{z}}\right) - 2xz\left(0\,\hat{\mathbf{x}} + 2y\,\hat{\mathbf{y}} + 3x\,\hat{\mathbf{z}}\right)$$
$$= \left(x^2y + 3x^2z^2\right)\hat{\mathbf{x}} + \left(6xz^3 - 4xyz\right)\hat{\mathbf{y}} + \left(3x^2z - 6x^2z\right)\hat{\mathbf{z}}$$
$$= \boxed{x^2\left(y + 3z^2\right)\hat{\mathbf{x}} + 2xz\left(3z^2 - 2y\right)\hat{\mathbf{y}} - 3x^2z\,\hat{\mathbf{z}}}$$

Problem 1.22

(ii) $[\nabla(\mathbf{A}\cdot\mathbf{B})]_x = \frac{\partial}{\partial x}(A_xB_x + A_yB_y + A_zB_z) = \frac{\partial A_x}{\partial x}B_x + A_x\frac{\partial B_x}{\partial x} + \frac{\partial A_y}{\partial x}B_y + A_y\frac{\partial B_y}{\partial x} + \frac{\partial A_z}{\partial x}B_z + A_z\frac{\partial B_z}{\partial x}$

$$[\mathbf{A}\times(\nabla\times\mathbf{B})]_x = A_y(\nabla\times\mathbf{B})_z - A_z(\nabla\times\mathbf{B})_y = A_y\left(\frac{\partial B_y}{\partial x} - \frac{\partial B_x}{\partial y}\right) - A_z\left(\frac{\partial B_x}{\partial z} - \frac{\partial B_z}{\partial x}\right)$$
$$[\mathbf{B}\times(\nabla\times\mathbf{A})]_x = B_y\left(\frac{\partial A_y}{\partial x} - \frac{\partial A_x}{\partial y}\right) - B_z\left(\frac{\partial A_x}{\partial z} - \frac{\partial A_z}{\partial x}\right)$$
$$[(\mathbf{A}\cdot\nabla)\mathbf{B}]_x = \left(A_x\frac{\partial}{\partial x} + A_y\frac{\partial}{\partial y} + A_z\frac{\partial}{\partial z}\right)B_x = A_x\frac{\partial B_x}{\partial x} + A_y\frac{\partial B_x}{\partial y} + A_z\frac{\partial B_x}{\partial z}$$
$$[(\mathbf{B}\cdot\nabla)\mathbf{A}]_x = B_x\frac{\partial A_x}{\partial x} + B_y\frac{\partial A_x}{\partial y} + B_z\frac{\partial A_x}{\partial z}$$

So $[\mathbf{A}\times(\nabla\times\mathbf{B}) + \mathbf{B}\times(\nabla\times\mathbf{A}) + (\mathbf{A}\cdot\nabla)\mathbf{B} + (\mathbf{B}\cdot\nabla)\mathbf{A}]_x$

$$\begin{aligned}
= \;& A_y\frac{\partial B_y}{\partial x} - A_y\frac{\partial B_x}{\partial y} - A_z\frac{\partial B_x}{\partial z} + A_z\frac{\partial B_z}{\partial x} + B_y\frac{\partial A_y}{\partial x} - B_y\frac{\partial A_x}{\partial y} - B_z\frac{\partial A_x}{\partial z} + B_z\frac{\partial A_z}{\partial x} \\
& + A_x\frac{\partial B_x}{\partial x} + A_y\frac{\partial B_x}{\partial y} + A_z\frac{\partial B_x}{\partial z} + B_x\frac{\partial A_x}{\partial x} + B_y\frac{\partial A_x}{\partial y} + B_z\frac{\partial A_x}{\partial z} \\
= \;& B_x\frac{\partial A_x}{\partial x} + A_x\frac{\partial B_x}{\partial x} + B_y\left(\frac{\partial A_y}{\partial x} - \cancel{\frac{\partial A_x}{\partial y}} + \cancel{\frac{\partial A_x}{\partial y}}\right) + A_y\left(\frac{\partial B_y}{\partial x} - \cancel{\frac{\partial B_x}{\partial y}} + \cancel{\frac{\partial B_x}{\partial y}}\right) \\
& + B_z\left(-\cancel{\frac{\partial A_x}{\partial z}} + \frac{\partial A_z}{\partial x} + \cancel{\frac{\partial A_x}{\partial z}}\right) + A_z\left(-\cancel{\frac{\partial B_x}{\partial z}} + \frac{\partial B_z}{\partial x} + \cancel{\frac{\partial B_x}{\partial z}}\right) \\
= \;& [\nabla(\mathbf{A}\cdot\mathbf{B})]_x \text{ (same for } y \text{ and } z)
\end{aligned}$$

(vi)
$$\begin{aligned}
[\nabla\times(\mathbf{A}\times\mathbf{B})]_x &= \frac{\partial}{\partial y}(\mathbf{A}\times\mathbf{B})_z - \frac{\partial}{\partial z}(\mathbf{A}\times\mathbf{B})_y = \frac{\partial}{\partial y}(A_xB_y - A_yB_x) - \frac{\partial}{\partial z}(A_zB_x - A_xB_z) \\
&= \frac{\partial A_x}{\partial y}B_y + A_x\frac{\partial B_y}{\partial y} - \frac{\partial A_y}{\partial y}B_x - A_y\frac{\partial B_x}{\partial y} - \frac{\partial A_z}{\partial z}B_x - A_z\frac{\partial B_x}{\partial z} + \frac{\partial A_x}{\partial z}B_z + A_x\frac{\partial B_z}{\partial z}
\end{aligned}$$

$[(\mathbf{B}\cdot\nabla)\mathbf{A} - (\mathbf{A}\cdot\nabla)\mathbf{B} + \mathbf{A}(\nabla\cdot\mathbf{B}) - \mathbf{B}(\nabla\cdot\mathbf{A})]_x$

$$\begin{aligned}
&= B_x\frac{\partial A_x}{\partial x} + B_y\frac{\partial A_x}{\partial y} + B_z\frac{\partial A_x}{\partial z} - A_x\frac{\partial B_x}{\partial x} - A_y\frac{\partial B_x}{\partial y} - A_z\frac{\partial B_x}{\partial z} + A_x\left(\frac{\partial B_x}{\partial x} + \frac{\partial B_y}{\partial y} + \frac{\partial B_z}{\partial z}\right) - B_x\left(\frac{\partial A_x}{\partial x} + \frac{\partial A_y}{\partial y} + \frac{\partial A_z}{\partial z}\right) \\
&= B_y\frac{\partial A_x}{\partial y} + A_x\left(-\cancel{\frac{\partial B_x}{\partial x}} + \cancel{\frac{\partial B_x}{\partial x}} + \frac{\partial B_y}{\partial y} + \frac{\partial B_z}{\partial z}\right) + B_x\left(\cancel{\frac{\partial A_x}{\partial x}} - \cancel{\frac{\partial A_x}{\partial x}} - \frac{\partial A_y}{\partial y} - \frac{\partial A_z}{\partial z}\right) \\
&\quad + A_y\left(-\frac{\partial B_x}{\partial y}\right) + A_z\left(-\frac{\partial B_x}{\partial z}\right) + B_z\left(\frac{\partial A_x}{\partial z}\right) \\
&= [\nabla\times(\mathbf{A}\times\mathbf{B})]_x \text{ (same for } y \text{ and } z)
\end{aligned}$$

Problem 1.23

$$\begin{aligned}
\nabla(f/g) &= \frac{\partial}{\partial x}(f/g)\,\hat{\mathbf{x}} + \frac{\partial}{\partial y}(f/g)\,\hat{\mathbf{y}} + \frac{\partial}{\partial z}(f/g)\,\hat{\mathbf{z}} \\
&= \frac{g\frac{\partial f}{\partial x} - f\frac{\partial g}{\partial x}}{g^2}\,\hat{\mathbf{x}} + \frac{g\frac{\partial f}{\partial y} - f\frac{\partial g}{\partial y}}{g^2}\,\hat{\mathbf{y}} + \frac{g\frac{\partial f}{\partial z} - f\frac{\partial g}{\partial z}}{g^2}\,\hat{\mathbf{z}} \\
&= \frac{1}{g^2}\left[g\left(\frac{\partial f}{\partial x}\hat{\mathbf{x}} + \frac{\partial f}{\partial y}\hat{\mathbf{y}} + \frac{\partial f}{\partial z}\hat{\mathbf{z}}\right) - f\left(\frac{\partial g}{\partial x}\hat{\mathbf{x}} + \frac{\partial g}{\partial y}\hat{\mathbf{y}} + \frac{\partial g}{\partial z}\hat{\mathbf{z}}\right)\right] = \frac{g\nabla f - f\nabla g}{g^2}. \quad \text{qed}
\end{aligned}$$

$$\begin{aligned}
\nabla\cdot(\mathbf{A}/g) &= \frac{\partial}{\partial x}(A_x/g) + \frac{\partial}{\partial y}(A_y/g) + \frac{\partial}{\partial z}(A_z/g) \\
&= \frac{g\frac{\partial A_x}{\partial x} - A_x\frac{\partial g}{\partial x}}{g^2} + \frac{g\frac{\partial A_y}{\partial y} - A_y\frac{\partial g}{\partial y}}{g^2} + \frac{g\frac{\partial A_z}{\partial z} - A_z\frac{\partial g}{\partial z}}{g^2} \\
&= \frac{1}{g^2}\left[g\left(\frac{\partial A_x}{\partial x} + \frac{\partial A_y}{\partial y} + \frac{\partial A_z}{\partial z}\right) - \left(A_x\frac{\partial g}{\partial x} + A_y\frac{\partial g}{\partial y} + A_z\frac{\partial g}{\partial z}\right)\right] = \frac{g\nabla\cdot\mathbf{A} - \mathbf{A}\cdot\nabla g}{g^2}. \quad \text{qed}
\end{aligned}$$

$$
\begin{aligned}
[\nabla\times(\mathbf{A}/g)]_x &= \frac{\partial}{\partial y}(A_z/g) - \frac{\partial}{\partial z}(A_y/g) \\
&= \frac{g\frac{\partial A_z}{\partial y} - A_z\frac{\partial g}{\partial y}}{g^2} - \frac{g\frac{\partial A_y}{\partial z} - A_y\frac{\partial g}{\partial z}}{g^2} \\
&= \frac{1}{g^2}\left[g\left(\frac{\partial A_z}{\partial y} - \frac{\partial A_y}{\partial z}\right) - \left(A_z\frac{\partial g}{\partial y} - A_y\frac{\partial g}{\partial z}\right)\right] \\
&= \frac{g(\nabla\times\mathbf{A})_x + (\mathbf{A}\times\nabla g)_x}{g^2} \text{ (same for } y \text{ and } z\text{). qed}
\end{aligned}
$$

Problem 1.24

(a) $\mathbf{A}\times\mathbf{B} = \begin{vmatrix} \hat{\mathbf{x}} & \hat{\mathbf{y}} & \hat{\mathbf{z}} \\ x & 2y & 3z \\ 3y & -2x & 0 \end{vmatrix} = \hat{\mathbf{x}}(6xz) + \hat{\mathbf{y}}(9zy) + \hat{\mathbf{z}}(-2x^2 - 6y^2)$

$\nabla\cdot(\mathbf{A}\times\mathbf{B}) = \frac{\partial}{\partial x}(6xz) + \frac{\partial}{\partial y}(9zy) + \frac{\partial}{\partial z}(-2x^2 - 6y^2) = 6z + 9z + 0 = 15z$

$\nabla\times\mathbf{A} = \hat{\mathbf{x}}\left(\frac{\partial}{\partial y}(3z) - \frac{\partial}{\partial z}(2y)\right) + \hat{\mathbf{y}}\left(\frac{\partial}{\partial z}(x) - \frac{\partial}{\partial x}(3z)\right) + \hat{\mathbf{z}}\left(\frac{\partial}{\partial x}(2y) - \frac{\partial}{\partial y}(x)\right) = 0$; $\mathbf{B}\cdot(\nabla\times\mathbf{A}) = 0$

$\nabla\times\mathbf{B} = \hat{\mathbf{x}}\left(\frac{\partial}{\partial y}(0) - \frac{\partial}{\partial z}(-2x)\right) + \hat{\mathbf{y}}\left(\frac{\partial}{\partial z}(3y) - \frac{\partial}{\partial x}(0)\right) + \hat{\mathbf{z}}\left(\frac{\partial}{\partial x}(-2x) - \frac{\partial}{\partial y}(3y)\right) = -5\,\hat{\mathbf{z}}$; $\mathbf{A}\cdot(\nabla\times\mathbf{B}) = -15z$

$\nabla\cdot(\mathbf{A}\times\mathbf{B}) \overset{?}{=} \mathbf{B}\cdot(\nabla\times\mathbf{A}) - \mathbf{A}\cdot(\nabla\times\mathbf{B}) = 0 - (-15z) = 15z.$ ✓

(b) $\mathbf{A}\cdot\mathbf{B} = 3xy - 4xy = -xy$; $\nabla(\mathbf{A}\cdot\mathbf{B}) = \nabla(-xy) = \hat{\mathbf{x}}\frac{\partial}{\partial x}(-xy) + \hat{\mathbf{y}}\frac{\partial}{\partial y}(-xy) = -y\,\hat{\mathbf{x}} - x\,\hat{\mathbf{y}}$

$\mathbf{A}\times(\nabla\times\mathbf{B}) = \begin{vmatrix} \hat{\mathbf{x}} & \hat{\mathbf{y}} & \hat{\mathbf{z}} \\ x & 2y & 3z \\ 0 & 0 & -5 \end{vmatrix} = \hat{\mathbf{x}}(-10y) + \hat{\mathbf{y}}(5x)$; $\mathbf{B}\times(\nabla\times\mathbf{A}) = 0$

$(\mathbf{A}\cdot\nabla)\mathbf{B} = \left(x\frac{\partial}{\partial x} + 2y\frac{\partial}{\partial y} + 3z\frac{\partial}{\partial z}\right)(3y\,\hat{\mathbf{x}} - 2x\,\hat{\mathbf{y}}) = \hat{\mathbf{x}}(6y) + \hat{\mathbf{y}}(-2x)$

$(\mathbf{B}\cdot\nabla)\mathbf{A} = \left(3y\frac{\partial}{\partial x} - 2x\frac{\partial}{\partial y}\right)(x\,\hat{\mathbf{x}} + 2y\,\hat{\mathbf{y}} + 3z\,\hat{\mathbf{z}}) = \hat{\mathbf{x}}(3y) + \hat{\mathbf{y}}(-4x)$

$\mathbf{A}\times(\nabla\times\mathbf{B}) + \mathbf{B}\times(\nabla\times\mathbf{A}) + (\mathbf{A}\cdot\nabla)\mathbf{B} + (\mathbf{B}\cdot\nabla)\mathbf{A}$
$= -10y\,\hat{\mathbf{x}} + 5x\,\hat{\mathbf{y}} + 6y\,\hat{\mathbf{x}} - 2x\,\hat{\mathbf{y}} + 3y\,\hat{\mathbf{x}} - 4x\,\hat{\mathbf{y}} = -y\,\hat{\mathbf{x}} - x\,\hat{\mathbf{y}} = \nabla\cdot(\mathbf{A}\cdot\mathbf{B}).$ ✓

(c) $\nabla\times(\mathbf{A}\times\mathbf{B}) = \hat{\mathbf{x}}\left(\frac{\partial}{\partial y}(-2x^2 - 6y^2) - \frac{\partial}{\partial z}(9zy)\right) + \hat{\mathbf{y}}\left(\frac{\partial}{\partial z}(6xz) - \frac{\partial}{\partial x}(-2x^2 - 6y^2)\right) + \hat{\mathbf{z}}\left(\frac{\partial}{\partial x}(9zy) - \frac{\partial}{\partial y}(6xz)\right)$
$= \hat{\mathbf{x}}(-12y - 9y) + \hat{\mathbf{y}}(6x + 4x) + \hat{\mathbf{z}}(0) = -21y\,\hat{\mathbf{x}} + 10x\,\hat{\mathbf{y}}$

$\nabla\cdot\mathbf{A} = \frac{\partial}{\partial x}(x) + \frac{\partial}{\partial y}(2y) + \frac{\partial}{\partial z}(3z) = 1 + 2 + 3 = 6$; $\nabla\cdot\mathbf{B} = \frac{\partial}{\partial x}(3y) + \frac{\partial}{\partial y}(-2x) = 0$

$(\mathbf{B}\cdot\nabla)\mathbf{A} - (\mathbf{A}\cdot\nabla)\mathbf{B} + \mathbf{A}(\nabla\cdot\mathbf{B}) - \mathbf{B}(\nabla\cdot\mathbf{A}) = 3y\,\hat{\mathbf{x}} - 4x\,\hat{\mathbf{y}} - 6y\,\hat{\mathbf{x}} + 2x\,\hat{\mathbf{y}} - 18y\,\hat{\mathbf{x}} + 12x\,\hat{\mathbf{y}} = -21y\,\hat{\mathbf{x}} + 10x\,\hat{\mathbf{y}}$
$= \nabla\times(\mathbf{A}\times\mathbf{B}).$ ✓

Problem 1.25

(a) $\frac{\partial^2 T_a}{\partial x^2} = 2$; $\frac{\partial^2 T_a}{\partial y^2} = \frac{\partial^2 T_a}{\partial z^2} = 0 \Rightarrow \boxed{\nabla^2 T_a = 2.}$

(b) $\frac{\partial^2 T_b}{\partial x^2} = \frac{\partial^2 T_b}{\partial y^2} = \frac{\partial^2 T_b}{\partial z^2} = -T_b \Rightarrow \boxed{\nabla^2 T_b = -3T_b = -3\sin x\sin y\sin z.}$

(c) $\frac{\partial^2 T_c}{\partial x^2} = 25T_c$; $\frac{\partial^2 T_c}{\partial y^2} = -16T_c$; $\frac{\partial^2 T_c}{\partial z^2} = -9T_c \Rightarrow \boxed{\nabla^2 T_c = 0.}$

(d)
$$
\left.\begin{array}{l}
\frac{\partial^2 v_x}{\partial x^2} = 2 \; ; \; \frac{\partial^2 v_x}{\partial y^2} = \frac{\partial^2 v_x}{\partial z^2} = 0 \Rightarrow \nabla^2 v_x = 2 \\
\frac{\partial^2 v_y}{\partial x^2} = \frac{\partial^2 v_y}{\partial y^2} = 0 \; ; \; \frac{\partial^2 v_y}{\partial z^2} = 6x \Rightarrow \nabla^2 v_y = 6x \\
\frac{\partial^2 v_z}{\partial x^2} = \frac{\partial^2 v_z}{\partial y^2} = \frac{\partial^2 v_z}{\partial z^2} = 0 \Rightarrow \nabla^2 v_z = 0
\end{array}\right\} \boxed{\nabla^2 \mathbf{v} = 2\,\hat{\mathbf{x}} + 6x\,\hat{\mathbf{y}}.}
$$

Problem 1.26

$$\nabla\cdot(\nabla\times\mathbf{v}) = \frac{\partial}{\partial x}\left(\frac{\partial v_z}{\partial y} - \frac{\partial v_y}{\partial z}\right) + \frac{\partial}{\partial y}\left(\frac{\partial v_x}{\partial z} - \frac{\partial v_z}{\partial x}\right) + \frac{\partial}{\partial z}\left(\frac{\partial v_y}{\partial x} - \frac{\partial v_x}{\partial y}\right)$$
$$= \left(\frac{\partial^2 v_z}{\partial x\,\partial y} - \frac{\partial^2 v_z}{\partial y\,\partial x}\right) + \left(\frac{\partial^2 v_x}{\partial y\,\partial z} - \frac{\partial^2 v_x}{\partial z\,\partial y}\right) + \left(\frac{\partial^2 v_y}{\partial z\,\partial x} - \frac{\partial^2 v_y}{\partial x\,\partial z}\right) = 0, \text{ by equality of cross-derivatives.}$$

From Prob. 1.18: $\nabla\times\mathbf{v}_b = -2y\,\hat{\mathbf{x}} - 3z\,\hat{\mathbf{y}} - x\,\hat{\mathbf{z}} \Rightarrow \nabla\cdot(\nabla\times\mathbf{v}_b) = \frac{\partial}{\partial x}(-2y) + \frac{\partial}{\partial y}(-3z) + \frac{\partial}{\partial z}(-x) = 0.$ ✓

Problem 1.27

$$\nabla\times(\nabla t) = \begin{vmatrix} \hat{\mathbf{x}} & \hat{\mathbf{y}} & \hat{\mathbf{z}} \\ \frac{\partial}{\partial x} & \frac{\partial}{\partial y} & \frac{\partial}{\partial z} \\ \frac{\partial t}{\partial x} & \frac{\partial t}{\partial y} & \frac{\partial t}{\partial z} \end{vmatrix} = \hat{\mathbf{x}}\left(\frac{\partial^2 t}{\partial y\,\partial z} - \frac{\partial^2 t}{\partial z\,\partial y}\right) + \hat{\mathbf{y}}\left(\frac{\partial^2 t}{\partial z\,\partial x} - \frac{\partial^2 t}{\partial x\,\partial z}\right) + \hat{\mathbf{z}}\left(\frac{\partial^2 t}{\partial x\,\partial y} - \frac{\partial^2 t}{\partial y\,\partial x}\right)$$

$= 0$, by equality of cross-derivatives.

In Prob. 1.11(b), $\nabla f = 2xy^3z^4\,\hat{\mathbf{x}} + 3x^2y^2z^4\,\hat{\mathbf{y}} + 4x^2y^3z^3\,\hat{\mathbf{z}}$, so

$$\nabla\times(\nabla f) = \begin{vmatrix} \hat{\mathbf{x}} & \hat{\mathbf{y}} & \hat{\mathbf{z}} \\ \frac{\partial}{\partial x} & \frac{\partial}{\partial y} & \frac{\partial}{\partial z} \\ 2xy^3z^4 & 3x^2y^3z^4 & 4x^2y^3z^3 \end{vmatrix}$$

$= \hat{\mathbf{x}}(3\cdot 4x^2y^2z^3 - 4\cdot 3x^2y^2z^3) + \hat{\mathbf{y}}(4\cdot 2xy^3z^3 - 2\cdot 4xy^3z^3) + \hat{\mathbf{z}}(2\cdot 3xy^2z^4 - 3\cdot 2xy^2z^4) = 0.$ ✓

Problem 1.28

(a) $(0,0,0) \longrightarrow (1,0,0)$. $x: 0\to 1, y = z = 0; d\mathbf{l} = dx\,\hat{\mathbf{x}}; \mathbf{v}\cdot d\mathbf{l} = x^2\,dx; \int \mathbf{v}\cdot d\mathbf{l} = \int_0^1 x^2\,dx = (x^3/3)|_0^1 = 1/3.$
$(1,0,0) \longrightarrow (1,1,0)$. $x = 1, y: 0\to 1, z = 0; d\mathbf{l} = dy\,\hat{\mathbf{y}}; \mathbf{v}\cdot d\mathbf{l} = 2yz\,dy = 0; \int \mathbf{v}\cdot d\mathbf{l} = 0.$
$(1,1,0) \longrightarrow (1,1,1)$. $x = y = 1, z: 0\to 1; d\mathbf{l} = dz\,\hat{\mathbf{z}}; \mathbf{v}\cdot d\mathbf{l} = y^2\,dz = dz; \int \mathbf{v}\cdot d\mathbf{l} = \int_0^1 dz = z|_0^1 = 1.$
Total: $\int \mathbf{v}\cdot d\mathbf{l} = (1/3) + 0 + 1 = \boxed{4/3.}$

(b) $(0,0,0) \longrightarrow (0,0,1)$. $x = y = 0, z: 0\to 1; d\mathbf{l} = dz\,\hat{\mathbf{z}}; \mathbf{v}\cdot d\mathbf{l} = y^2\,dz = 0; \int \mathbf{v}\cdot d\mathbf{l} = 0.$
$(0,0,1) \longrightarrow (0,1,1)$. $x = 0, y: 0\to 1, z = 1; d\mathbf{l} = dy\,\hat{\mathbf{y}}; \mathbf{v}\cdot d\mathbf{l} = 2yz\,dy = 2y\,dy; \int \mathbf{v}\cdot d\mathbf{l} = \int_0^1 2y\,dy = y^2|_0^1 = 1.$
$(0,1,1) \longrightarrow (1,1,1)$. $x: 0\to 1, y = z = 1; d\mathbf{l} = dx\,\hat{\mathbf{x}}; \mathbf{v}\cdot d\mathbf{l} = x^2\,dx; \int \mathbf{v}\cdot d\mathbf{l} = \int_0^1 x^2\,dx = (x^3/3)|_0^1 = 1/3.$
Total: $\int \mathbf{v}\cdot d\mathbf{l} = 0 + 1 + (1/3) = \boxed{4/3.}$

(c) $x = y = z: 0\to 1; dx = dy = dz; \mathbf{v}\cdot d\mathbf{l} = x^2\,dx + 2yz\,dy + y^2\,dz = x^2\,dx + 2x^2\,dx + x^2\,dx = 4x^2\,dx;$
$\int \mathbf{v}\cdot d\mathbf{l} = \int_0^1 4x^2\,dx = (4x^3/3)|_0^1 = \boxed{4/3.}$

(d) $\oint \mathbf{v}\cdot d\mathbf{l} = (4/3) - (4/3) = \boxed{0.}$

Problem 1.29

$x, y: 0\to 1, z = 0; d\mathbf{a} = dx\,dy\,\hat{\mathbf{z}}; \mathbf{v}\cdot d\mathbf{a} = y(z^2 - 3)\,dx\,dy = -3y\,dx\,dy; \int \mathbf{v}\cdot d\mathbf{a} = -3\int_0^2 dx \int_0^2 y\,dy = -3(x|_0^2)(\frac{y^2}{2}|_0^2) = -3(2)(2) = \boxed{12.}$ In Ex. 1.7 we got 20, for the same boundary line (the square in the *xy*-plane), so the answer is $\boxed{\text{no:}}$ the surface integral does *not* depend only on the boundary line. The *total* flux for the cube is $20 + 12 = \boxed{32.}$

Problem 1.30

$\int T\,d\tau = \int z^2\,dx\,dy\,dz$. You can do the integrals in any order—here it is simplest to save z for last:

$$\int z^2\left[\int\left(\int dx\right)dy\right]dz.$$

The sloping surface is $x + y + z = 1$, so the x integral is $\int_0^{(1-y-z)} dx = 1 - y - z$. For a given z, y ranges from 0 to $1 - z$, so the y integral is $\int_0^{(1-z)}(1 - y - z)\,dy = [(1-z)y - (y^2/2)]|_0^{(1-z)} = (1-z)^2 - [(1-z)^2/2] = (1-z)^2/2 =$

$(1/2) - z + (z^2/2)$. Finally, the z integral is $\int_0^1 z^2(\frac{1}{2} - z + \frac{z^2}{2})\,dz = \int_0^1(\frac{z^2}{2} - z^3 + \frac{z^4}{2})\,dz = (\frac{z^3}{6} - \frac{z^4}{4} + \frac{z^5}{10})|_0^1 = \frac{1}{6} - \frac{1}{4} + \frac{1}{10} = \boxed{1/60.}$

Problem 1.31

$T(\mathbf{b}) = 1 + 4 + 2 = 7;\; T(\mathbf{a}) = 0. \Rightarrow \boxed{T(\mathbf{b}) - T(\mathbf{a}) = 7.}$

$\nabla T = (2x + 4y)\hat{\mathbf{x}} + (4x + 2z^3)\hat{\mathbf{y}} + (6yz^2)\hat{\mathbf{z}};\; \nabla T\cdot d\mathbf{l} = (2x + 4y)dx + (4x + 2x^3)dy + (6yz^2)dz$

(a) Segment 1: $x: 0 \to 1,\ y = z = dy = dz = 0.\ \int \nabla T\cdot d\mathbf{l} = \int_0^1 (2x)\,dx = x^2\big|_0^1 = 1.$
Segment 2: $y: 0 \to 1,\ x = 1,\ z = 0,\ dx = dz = 0.\ \int \nabla T\cdot d\mathbf{l} = \int_0^1 (4)\,dy = 4y|_0^1 = 4.$
Segment 3: $z: 0 \to 1,\ x = y = 1,\ dx = dy = 0.\ \int \nabla T\cdot d\mathbf{l} = \int_0^1 (6z^2)\,dz = 2z^3\big|_0^1 = 2.$
$\Big\}\ \int_{\mathbf{a}}^{\mathbf{b}} \nabla T\cdot d\mathbf{l} = 7.$ ✓

(b) Segment 1: $z: 0 \to 1,\ x = y = dx = dy = 0.\ \int \nabla T\cdot d\mathbf{l} = \int_0^1 (0)\,dz = 0.$
Segment 2: $y: 0 \to 1,\ x = 0,\ z = 1,\ dx = dz = 0.\ \int \nabla T\cdot d\mathbf{l} = \int_0^1 (2)\,dy = 2y|_0^1 = 2.$
Segment 3: $x: 0 \to 1,\ y = z = 1,\ dy = dz = 0.\ \int \nabla T\cdot d\mathbf{l} = \int_0^1 (2x + 4)\,dx$
$= (x^2 + 4x)\big|_0^1 = 1 + 4 = 5.$
$\Big\}\ \int_{\mathbf{a}}^{\mathbf{b}} \nabla T\cdot d\mathbf{l} = 7.$ ✓

(c) $x: 0 \to 1,\ y = x,\ z = x^2,\ dy = dx\ dz = 2x\,dx.$

$\nabla T\cdot d\mathbf{l} = (2x + 4x)dx + (4x + 2x^6)dx + (6xx^4)2x\,dx = (10x + 14x^6)dx.$

$\int_{\mathbf{a}}^{\mathbf{b}} \nabla T\cdot d\mathbf{l} = \int_0^1 (10x + 14x^6)dx = (5x^2 + 2x^7)\big|_0^1 = 5 + 2 = 7.$ ✓

Problem 1.32

$\nabla\cdot\mathbf{v} = y + 2z + 3x$

$$\int(\nabla\cdot\mathbf{v})d\tau = \int(y + 2z + 3x)\,dx\,dy\,dz = \iint\left\{\int_0^2(y + 2z + 3x)\,dx\right\}dy\,dz$$

$$\hookrightarrow \left[(y + 2z)x + \tfrac{3}{2}x^2\right]_0^2 = 2(y + 2z) + 6$$

$$= \int\left\{\int_0^2(2y + 4z + 6)dy\right\}dz$$

$$\hookrightarrow \left[y^2 + (4z + 6)y\right]_0^2 = 4 + 2(4z + 6) = 8z + 16$$

$$= \int_0^2(8z + 16)dz = (4z^2 + 16z)\big|_0^2 = 16 + 32 = \boxed{48.}$$

Numbering the surfaces as in Fig. 1.29:

(i) $d\mathbf{a} = dy\,dz\,\hat{\mathbf{x}}, x = 2.\ \mathbf{v}\cdot d\mathbf{a} = 2y\,dy\,dz.\ \int \mathbf{v}\cdot d\mathbf{a} = \iint 2y\,dy\,dz = 2y^2\big|_0^2 = 8.$
(ii) $d\mathbf{a} = -dy\,dz\,\hat{\mathbf{x}}, x = 0.\ \mathbf{v}\cdot d\mathbf{a} = 0.\ \int \mathbf{v}\cdot d\mathbf{a} = 0.$
(iii) $d\mathbf{a} = dx\,dz\,\hat{\mathbf{y}}, y = 2.\ \mathbf{v}\cdot d\mathbf{a} = 4z\,dx\,dz.\ \int \mathbf{v}\cdot d\mathbf{a} = \iint 4z\,dx\,dz = 16.$
(iv) $d\mathbf{a} = -dx\,dz\,\hat{\mathbf{y}}, y = 0.\ \mathbf{v}\cdot d\mathbf{a} = 0.\ \int \mathbf{v}\cdot d\mathbf{a} = 0.$
(v) $d\mathbf{a} = dx\,dy\,\hat{\mathbf{z}}, z = 2.\ \mathbf{v}\cdot d\mathbf{a} = 6x\,dx\,dy.\ \int \mathbf{v}\cdot d\mathbf{a} = 24.$
(vi) $d\mathbf{a} = -dx\,dy\,\hat{\mathbf{z}}, z = 0.\ \mathbf{v}\cdot d\mathbf{a} = 0.\ \int \mathbf{v}\cdot d\mathbf{a} = 0.$
$\Rightarrow \int \mathbf{v}\cdot d\mathbf{a} = 8 + 16 + 24 = 48$ ✓

Problem 1.33

$\nabla\times\mathbf{v} = \hat{\mathbf{x}}(0 - 2y) + \hat{\mathbf{y}}(0 - 3z) + \hat{\mathbf{z}}(0 - x) = -2y\,\hat{\mathbf{x}} - 3z\,\hat{\mathbf{y}} - x\,\hat{\mathbf{z}}.$
$d\mathbf{a} = dy\,dz\,\hat{\mathbf{x}}$, if we agree that the path integral shall run counterclockwise. So
$(\nabla\times\mathbf{v})\cdot d\mathbf{a} = -2y\,dy\,dz.$

$$\begin{aligned}\int(\nabla\times\mathbf{v})\cdot d\mathbf{a} &= \int\left\{\int_0^{2-z}(-2y)dy\right\}dz \\ &\hookrightarrow y^2\Big|_0^{2-z} = -(2-z)^2 \\ &= -\int_0^2(4-4z+z^2)dz = -\left(4z-2z^2+\frac{z^3}{3}\right)\Big|_0^2 \\ &= -\left(8-8+\frac{8}{3}\right) = \boxed{-\frac{8}{3}}\end{aligned}$$

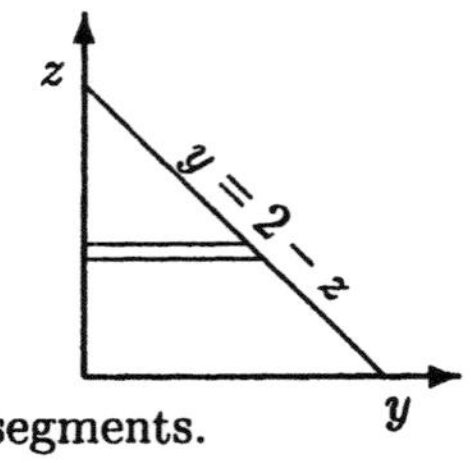

Meanwhile, $\mathbf{v}\cdot d\mathbf{l} = (xy)dx + (2yz)dy + (3zx)dz$. There are three segments.

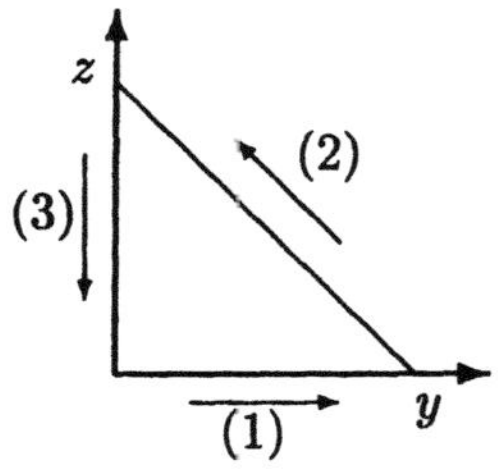

(1) $x = z = 0$; $dx = dz = 0$. $y: 0 \to 2$. $\int\mathbf{v}\cdot d\mathbf{l} = 0$.
(2) $x = 0$; $z = 2 - y$; $dx = 0$, $dz = -dy$, $y: 2 \to 0$. $\mathbf{v}\cdot d\mathbf{l} = 2yz\,dy$.
$\int\mathbf{v}\cdot d\mathbf{l} = \int_2^0 2y(2-y)dy = -\int_0^2(4y-2y^2)dy = -\left(2y^2-\frac{2}{3}y^3\right)\Big|_0^2 = -\left(8-\frac{2}{3}\cdot 8\right) = -\frac{8}{3}$.
(3) $x = y = 0$; $dx = dy = 0$; $z: 2 \to 0$. $\mathbf{v}\cdot d\mathbf{l} = 0$. $\int\mathbf{v}\cdot d\mathbf{l} = 0$. So $\oint\mathbf{v}\cdot d\mathbf{l} = -\frac{8}{3}$. ✓

Problem 1.34

By Corollary 1, $\int(\nabla\times\mathbf{v})\cdot d\mathbf{a}$ should equal $\frac{4}{3}$. $\nabla\times\mathbf{v} = (4z^2-2x)\hat{\mathbf{x}} + 2z\,\hat{\mathbf{z}}$.

(i) $d\mathbf{a} = dy\,dz\,\hat{\mathbf{x}}$, $x = 1$; $y, z: 0 \to 1$. $(\nabla\times\mathbf{v})\cdot d\mathbf{a} = (4z^2-2)dy\,dz$; $\int(\nabla\times\mathbf{v})\cdot d\mathbf{a} = \int_0^1(4z^2-2)dz$ $= \left(\frac{4}{3}z^3-2z\right)\Big|_0^1 = \frac{4}{3}-2 = -\frac{2}{3}$.
(ii) $d\mathbf{a} = -dx\,dy\,\hat{\mathbf{z}}$, $z = 0$; $x, y: 0 \to 1$. $(\nabla\times\mathbf{v})\cdot d\mathbf{a} = 0$; $\int(\nabla\times\mathbf{v})\cdot d\mathbf{a} = 0$.
(iii) $d\mathbf{a} = dx\,dz\,\hat{\mathbf{y}}$, $y = 1$; $x, z: 0 \to 1$. $(\nabla\times\mathbf{v})\cdot d\mathbf{a} = 0$; $\int(\nabla\times\mathbf{v})\cdot d\mathbf{a} = 0$.
(iv) $d\mathbf{a} = -dx\,dz\,\hat{\mathbf{y}}$, $y = 0$; $x, z: 0 \to 1$. $(\nabla\times\mathbf{v})\cdot d\mathbf{a} = 0$; $\int(\nabla\times\mathbf{v})\cdot d\mathbf{a} = 0$.
(v) $d\mathbf{a} = dx\,dy\,\hat{\mathbf{z}}$, $z = 1$; $x, y: 0 \to 1$. $(\nabla\times\mathbf{v})\cdot d\mathbf{a} = 2\,dx\,dy$; $\int(\nabla\times\mathbf{v})\cdot d\mathbf{a} = 2$.

$\Rightarrow \int(\nabla\times\mathbf{v})\cdot d\mathbf{a} = -\frac{2}{3}+2 = \frac{4}{3}$. ✓

Problem 1.35

(a) Use the product rule $\nabla\times(f\mathbf{A}) = f(\nabla\times\mathbf{A}) - \mathbf{A}\times(\nabla f)$:

$$\int_S f(\nabla\times\mathbf{A})\cdot d\mathbf{a} = \int_S \nabla\times(f\mathbf{A})\cdot d\mathbf{a} + \int_S[\mathbf{A}\times(\nabla f)]\cdot d\mathbf{a} = \oint_{\mathcal{P}} f\mathbf{A}\cdot d\mathbf{l} + \int_S[\mathbf{A}\times(\nabla f)]\cdot d\mathbf{a}. \quad \text{qed.}$$

(I used Stokes' theorem in the last step.)

(b) Use the product rule $\nabla\cdot(\mathbf{A}\times\mathbf{B}) = \mathbf{B}\cdot(\nabla\times\mathbf{A}) - \mathbf{A}\cdot(\nabla\times\mathbf{B})$:

$$\int_{\mathcal{V}} \mathbf{B}\cdot(\nabla\times\mathbf{A})d\tau = \int_{\mathcal{V}} \nabla\cdot(\mathbf{A}\times\mathbf{B})\,d\tau + \int_{\mathcal{V}} \mathbf{A}\cdot(\nabla\times\mathbf{B})\,d\tau = \oint_S(\mathbf{A}\times\mathbf{B})\cdot d\mathbf{a} + \int_{\mathcal{V}} \mathbf{A}\cdot(\nabla\times\mathbf{B})\,d\tau. \quad \text{qed.}$$

(I used the divergence theorem in the last step.)

Problem 1.36 $\boxed{r = \sqrt{x^2+y^2+z^2}; \quad \theta = \cos^{-1}\left(\frac{z}{\sqrt{x^2+y^2+z^2}}\right); \quad \phi = \tan^{-1}\left(\frac{y}{x}\right).}$

Problem 1.37

There are many ways to do this one—probably the most illuminating way is to work it out by trigonometry from Fig. 1.36. The most systematic approach is to study the expression:

$$\mathbf{r} = x\,\hat{\mathbf{x}} + y\,\hat{\mathbf{y}} + z\,\hat{\mathbf{z}} = r\sin\theta\cos\phi\,\hat{\mathbf{x}} + r\sin\theta\sin\phi\,\hat{\mathbf{y}} + r\cos\theta\,\hat{\mathbf{z}}.$$

If I only vary r slightly, then $d\mathbf{r} = \frac{\partial}{\partial r}(\mathbf{r})dr$ is a short vector pointing in the direction of increase in r. To make it a unit vector, I must divide by its length. Thus:

$$\hat{\mathbf{r}} = \frac{\frac{\partial \mathbf{r}}{\partial r}}{\left|\frac{\partial \mathbf{r}}{\partial r}\right|}; \quad \hat{\boldsymbol{\theta}} = \frac{\frac{\partial \mathbf{r}}{\partial \theta}}{\left|\frac{\partial \mathbf{r}}{\partial \theta}\right|}; \quad \hat{\boldsymbol{\phi}} = \frac{\frac{\partial \mathbf{r}}{\partial \phi}}{\left|\frac{\partial \mathbf{r}}{\partial \phi}\right|}.$$

$$\frac{\partial \mathbf{r}}{\partial r} = \sin\theta\cos\phi\,\hat{\mathbf{x}} + \sin\theta\sin\phi\,\hat{\mathbf{y}} + \cos\theta\,\hat{\mathbf{z}}; \quad \left|\frac{\partial \mathbf{r}}{\partial r}\right|^2 = \sin^2\theta\cos^2\phi + \sin^2\theta\sin^2\phi + \cos^2\theta = 1.$$
$$\frac{\partial \mathbf{r}}{\partial \theta} = r\cos\theta\cos\phi\,\hat{\mathbf{x}} + r\cos\theta\sin\phi\,\hat{\mathbf{y}} - r\sin\theta\,\hat{\mathbf{z}}; \quad \left|\frac{\partial \mathbf{r}}{\partial \theta}\right|^2 = r^2\cos^2\theta\cos^2\phi + r^2\cos^2\theta\sin^2\phi + r^2\sin^2\theta = r^2.$$
$$\frac{\partial \mathbf{r}}{\partial \phi} = -r\sin\theta\sin\phi\,\hat{\mathbf{x}} + r\sin\theta\cos\phi\,\hat{\mathbf{y}}; \quad \left|\frac{\partial \mathbf{r}}{\partial \phi}\right|^2 = r^2\sin^2\theta\sin^2\phi + r^2\sin^2\theta\cos^2\phi = r^2\sin^2\theta.$$

$$\Rightarrow \boxed{\begin{array}{l} \hat{\mathbf{r}} = \sin\theta\cos\phi\,\hat{\mathbf{x}} + \sin\theta\sin\phi\,\hat{\mathbf{y}} + \cos\theta\,\hat{\mathbf{z}}. \\ \hat{\boldsymbol{\theta}} = \cos\theta\cos\phi\,\hat{\mathbf{x}} + \cos\theta\sin\phi\,\hat{\mathbf{y}} - \sin\theta\,\hat{\mathbf{z}}. \\ \hat{\boldsymbol{\phi}} = -\sin\phi\,\hat{\mathbf{x}} + \cos\phi\,\hat{\mathbf{y}}. \end{array}}$$

Check: $\hat{\mathbf{r}}\cdot\hat{\mathbf{r}} = \sin^2\theta(\cos^2\phi + \sin^2\phi) + \cos^2\theta = \sin^2\theta + \cos^2\theta = 1, \checkmark$

$\hat{\boldsymbol{\theta}}\cdot\hat{\boldsymbol{\phi}} = -\cos\theta\sin\phi\cos\phi + \cos\theta\sin\phi\cos\phi = 0, \checkmark$ etc.

$$\sin\theta\,\hat{\mathbf{r}} = \sin^2\theta\cos\phi\,\hat{\mathbf{x}} + \sin^2\theta\sin\phi\,\hat{\mathbf{y}} + \sin\theta\cos\theta\,\hat{\mathbf{z}}.$$
$$\cos\theta\,\hat{\boldsymbol{\theta}} = \cos^2\theta\cos\phi\,\hat{\mathbf{x}} + \cos^2\theta\sin\phi\,\hat{\mathbf{y}} - \sin\theta\cos\theta\,\hat{\mathbf{z}}.$$

Add these:

(1) $\sin\theta\,\hat{\mathbf{r}} + \cos\theta\,\hat{\boldsymbol{\theta}} = +\cos\phi\,\hat{\mathbf{x}} + \sin\phi\,\hat{\mathbf{y}};$

(2) $\hat{\boldsymbol{\phi}} = -\sin\phi\,\hat{\mathbf{x}} + \cos\phi\,\hat{\mathbf{y}}.$

Multiply (1) by $\cos\phi$, (2) by $\sin\phi$, and subtract:

$$\boxed{\hat{\mathbf{x}} = \sin\theta\cos\phi\,\hat{\mathbf{r}} + \cos\theta\cos\phi\,\hat{\boldsymbol{\theta}} - \sin\phi\,\hat{\boldsymbol{\phi}}.}$$

Multiply (1) by $\sin\phi$, (2) by $\cos\phi$, and add:

$$\boxed{\hat{\mathbf{y}} = \sin\theta\sin\phi\,\hat{\mathbf{r}} + \cos\theta\sin\phi\,\hat{\boldsymbol{\theta}} + \cos\phi\,\hat{\boldsymbol{\phi}}.}$$

$$\cos\theta\,\hat{\mathbf{r}} = \sin\theta\cos\theta\cos\phi\,\hat{\mathbf{x}} + \sin\theta\cos\theta\sin\phi\,\hat{\mathbf{y}} + \cos^2\theta\,\hat{\mathbf{z}}.$$
$$\sin\theta\,\hat{\boldsymbol{\theta}} = \sin\theta\cos\theta\cos\phi\,\hat{\mathbf{x}} + \sin\theta\cos\theta\sin\phi\,\hat{\mathbf{y}} - \sin^2\theta\,\hat{\mathbf{z}}.$$

Subtract these:

$$\boxed{\hat{\mathbf{z}} = \cos\theta\,\hat{\mathbf{r}} - \sin\theta\,\hat{\boldsymbol{\theta}}.}$$

Problem 1.38

(a) $\nabla\cdot\mathbf{v}_1 = \frac{1}{r^2}\frac{\partial}{\partial r}(r^2 r^2) = \frac{1}{r^2}4r^3 = 4r$

$\int(\nabla\cdot\mathbf{v}_1)d\tau = \int(4r)(r^2\sin\theta\, dr\, d\theta\, d\phi) = (4)\int_0^R r^3 dr \int_0^\pi \sin\theta\, d\theta \int_0^{2\pi} d\phi = (4)\left(\frac{R^4}{4}\right)(2)(2\pi) = \boxed{4\pi R^4}$

$\int \mathbf{v}_1\cdot d\mathbf{a} = \int(r^2\hat{\mathbf{r}})\cdot(r^2\sin\theta\, d\theta\, d\phi\,\hat{\mathbf{r}}) = r^4\int_0^\pi \sin\theta\, d\theta \int_0^{2\pi} d\phi = 4\pi R^4$ ✓ (*Note:* at surface of sphere $r = R$.)

(b) $\nabla\cdot\mathbf{v}_2 = \frac{1}{r^2}\frac{\partial}{\partial r}\left(r^2\frac{1}{r^2}\right) = 0 \Rightarrow \boxed{\int(\nabla\cdot\mathbf{v}_2)d\tau = 0}$

$\int \mathbf{v}_2\cdot d\mathbf{a} = \int\left(\frac{1}{r^2}\hat{\mathbf{r}}\right)(r^2\sin\theta\, d\theta\, d\phi\,\hat{\mathbf{r}}) = \int\sin\theta\, d\theta\, d\phi = \boxed{4\pi.}$

They *don't* agree! The point is that this divergence is zero ***except at the origin***, where it blows up, so our calculation of $\int(\nabla\cdot\mathbf{v}_2)$ is *incorrect.* The right answer is 4π.

Problem 1.39

$$\begin{aligned}\nabla\cdot\mathbf{v} &= \frac{1}{r^2}\frac{\partial}{\partial r}(r^2\, r\cos\theta) + \frac{1}{r\sin\theta}\frac{\partial}{\partial\theta}(\sin\theta\, r\sin\theta) + \frac{1}{r\sin\theta}\frac{\partial}{\partial\phi}(r\sin\theta\cos\phi)\\ &= \frac{1}{r^2}3r^2\cos\theta + \frac{1}{r\sin\theta}r\,2\sin\theta\cos\theta + \frac{1}{r\sin\theta}r\sin\theta(-\sin\phi)\\ &= 3\cos\theta + 2\cos\theta - \sin\phi = 5\cos\theta - \sin\phi\end{aligned}$$

$\int(\nabla\cdot\mathbf{v})d\tau = \int(5\cos\theta - \sin\phi)\, r^2\sin\theta\, dr\, d\theta\, d\phi = \int_0^R r^2\, dr \int_0^{\frac{\pi}{2}}\left[\int_0^{2\pi}(5\cos\theta - \sin\phi)\, d\phi\right] d\theta\sin\theta$

$\hookrightarrow 2\pi(5\cos\theta)$

$= \left(\frac{R^3}{3}\right)(10\pi)\int_0^{\frac{\pi}{2}}\sin\theta\cos\theta\, , d\theta$

$\hookrightarrow \left.\frac{\sin^2\theta}{2}\right|_0^{\frac{\pi}{2}} = \frac{1}{2}$

$= \boxed{\frac{5\pi}{3}R^3.}$

Two surfaces—one the hemisphere: $d\mathbf{a} = R^2\sin\theta\, d\theta\, d\phi\,\hat{\mathbf{r}}$; $r = R$; $\phi : 0\to 2\pi$, $\theta : 0\to\frac{\pi}{2}$.

$\int\mathbf{v}\cdot d\mathbf{a} = \int(r\cos\theta)R^2\sin\theta\, d\theta\, d\phi = R^3\int_0^{\frac{\pi}{2}}\sin\theta\cos\theta\, d\theta\int_0^{2\pi} d\phi = R^3\left(\frac{1}{2}\right)(2\pi) = \pi R^3.$

other the flat bottom: $d\mathbf{a} = (dr)(r\sin\theta\, d\phi)(+\hat{\boldsymbol{\theta}}) = r\, dr\, d\phi\,\hat{\boldsymbol{\theta}}$ (here $\theta = \frac{\pi}{2}$). $r : 0\to R$, $\phi : 0\to 2\pi$.

$\int\mathbf{v}\cdot d\mathbf{a} = \int(r\sin\theta)(r\, dr\, d\phi) = \int_0^R r^2\, dr\int_0^{2\pi} d\phi = 2\pi\frac{R^3}{3}.$

Total: $\int\mathbf{v}\cdot d\mathbf{a} = \pi R^3 + \frac{2}{3}\pi R^3 = \frac{5}{3}\pi R^3.$ ✓

Problem 1.40 $\boxed{\nabla t = (\cos\theta + \sin\theta\cos\phi)\hat{\mathbf{r}} + (-\sin\theta + \cos\theta\cos\phi)\hat{\boldsymbol{\theta}} + \frac{1}{\sin\theta}(-\sin\theta\sin\phi)\hat{\boldsymbol{\phi}}}$

$$\begin{aligned}\nabla^2 t &= \nabla\cdot(\nabla t)\\ &= \frac{1}{r^2}\frac{\partial}{\partial r}\left(r^2(\cos\theta + \sin\theta\cos\phi)\right) + \frac{1}{r\sin\theta}\frac{\partial}{\partial\theta}\left(\sin\theta(-\sin\theta + \cos\theta\cos\phi)\right) + \frac{1}{r\sin\theta}\frac{\partial}{\partial\phi}(-\sin\phi)\\ &= \frac{1}{r^2}2r(\cos\theta + \sin\theta\cos\phi) + \frac{1}{r\sin\theta}(-2\sin\theta\cos\theta + \cos^2\theta\cos\phi - \sin^2\theta\cos\phi) - \frac{1}{r\sin\theta}\cos\phi\\ &= \frac{1}{r\sin\theta}[2\sin\theta\cos\theta + 2\sin^2\theta\cos\phi - 2\sin\theta\cos\theta + \cos^2\theta\cos\phi - \sin^2\theta\cos\phi - \cos\phi]\\ &= \frac{1}{r\sin\theta}\left[(\sin^2\theta + \cos^2\theta)\cos\phi - \cos\phi\right] = 0.\end{aligned}$$

$\Rightarrow \boxed{\nabla^2 t = 0}$

Check: $r\cos\theta = z$, $r\sin\theta\cos\phi = x$ $\Rightarrow$ in Cartesian coordinates $t = x + z$. Obviously, Laplacian is zero.

Gradient Theorem: $\int_{\mathbf{a}}^{\mathbf{b}}\nabla t\cdot d\mathbf{l} = t(\mathbf{b}) - t(\mathbf{a})$

Segment 1: $\theta = \frac{\pi}{2}$, $\phi = 0$, $r : 0\to 2$. $d\mathbf{l} = dr\,\hat{\mathbf{r}}$; $\nabla t\cdot d\mathbf{l} = (\cos\theta + \sin\theta\cos\phi)dr = (0 + 1)dr = dr.$

$\int\nabla t\cdot d\mathbf{l} = \int_0^2 dr = 2.$

Segment 2: $\theta = \frac{\pi}{2}$, $r = 2$, $\phi : 0\to\frac{\pi}{2}$. $d\mathbf{l} = r\sin\theta\, d\phi\,\hat{\boldsymbol{\phi}} = 2\, d\phi\,\hat{\boldsymbol{\phi}}.$

$\nabla t\cdot d\mathbf{l} = (-\sin\phi)(2\, d\phi) = -2\sin\phi\, d\phi$. $\int\nabla t\cdot d\mathbf{l} = -\int_0^{\frac{\pi}{2}} 2\sin\phi\, d\phi = 2\cos\phi|_0^{\frac{\pi}{2}} = -2.$

Segment 3: $r = 2,\ \phi = \frac{\pi}{2};\ \theta : \frac{\pi}{2} \to 0.$

$d\mathbf{l} = r\,d\theta\,\hat{\boldsymbol{\theta}} = 2\,d\theta\,\hat{\boldsymbol{\theta}};\ \nabla t\cdot d\mathbf{l} = (-\sin\theta + \cos\theta\cos\phi)(2\,d\theta) = -2\sin\theta\,d\theta.$

$\int \nabla t\cdot d\mathbf{l} = -\int_{\frac{\pi}{2}}^{0} 2\sin\theta\,d\theta = 2\cos\theta\big|_{\frac{\pi}{2}}^{0} = 2.$

Total: $\int_{\mathbf{a}}^{\mathbf{b}} \nabla t\cdot d\mathbf{l} = 2 - 2 + 2 = \boxed{2}$. Meanwhile, $t(\mathbf{b}) - t(\mathbf{a}) = [2(1+0)] - [0(\ )] = 2.$ ✓

Problem 1.41 From Fig. 1.42, $\boxed{\hat{\mathbf{s}} = \cos\phi\,\hat{\mathbf{x}} + \sin\phi\,\hat{\mathbf{y}};\ \hat{\boldsymbol{\phi}} = -\sin\phi\,\hat{\mathbf{x}} + \cos\phi\,\hat{\mathbf{y}};\ \hat{\mathbf{z}} = \hat{\mathbf{z}}}$

Multiply first by $\cos\phi$, second by $\sin\phi$, and subtract:

$\hat{\mathbf{s}}\cos\phi - \hat{\boldsymbol{\phi}}\sin\phi = \cos^2\phi\,\hat{\mathbf{x}} + \cos\phi\sin\phi\,\hat{\mathbf{y}} + \sin^2\phi\,\hat{\mathbf{x}} - \sin\phi\cos\phi\,\hat{\mathbf{y}} = \hat{\mathbf{x}}(\sin^2\phi + \cos^2\phi) = \hat{\mathbf{x}}.$

So $\boxed{\hat{\mathbf{x}} = \cos\phi\,\hat{\mathbf{s}} - \sin\phi\,\hat{\boldsymbol{\phi}}.}$

Multiply first by $\sin\phi$, second by $\cos\phi$, and add:

$\hat{\mathbf{s}}\sin\phi + \hat{\boldsymbol{\phi}}\cos\phi = \sin\phi\cos\phi\,\hat{\mathbf{x}} + \sin^2\phi\,\hat{\mathbf{y}} - \sin\phi\cos\phi\,\hat{\mathbf{x}} + \cos^2\phi\,\hat{\mathbf{y}} = \hat{\mathbf{y}}(\sin^2\phi + \cos^2\phi) = \hat{\mathbf{y}}.$

So $\boxed{\hat{\mathbf{y}} = \sin\phi\,\hat{\mathbf{s}} + \cos\phi\,\hat{\boldsymbol{\phi}}.}$ $\boxed{\hat{\mathbf{z}} = \hat{\mathbf{z}}.}$

Problem 1.42

(a)
$$\begin{aligned}\nabla\cdot\mathbf{v} &= \tfrac{1}{s}\tfrac{\partial}{\partial s}\left(s\,s(2+\sin^2\phi)\right) + \tfrac{1}{s}\tfrac{\partial}{\partial\phi}(s\sin\phi\cos\phi) + \tfrac{\partial}{\partial z}(3z)\\ &= \tfrac{1}{s}\,2s(2+\sin^2\phi) + \tfrac{1}{s}\,s(\cos^2\phi - \sin^2\phi) + 3\\ &= 4 + 2\sin^2\phi + \cos^2\phi - \sin^2\phi + 3\\ &= 4 + \sin^2\phi + \cos^2\phi + 3 = \boxed{8.}\end{aligned}$$

(b) $\int(\nabla\cdot\mathbf{v})d\tau = \int(8)s\,ds\,d\phi\,dz = 8\int_0^2 s\,ds\int_0^{\frac{\pi}{2}} d\phi\int_0^5 dz = 8(2)\left(\frac{\pi}{2}\right)(5) = \boxed{40\pi.}$

Meanwhile, the surface integral has five parts:

top: $z = 5,\ d\mathbf{a} = s\,ds\,d\phi\,\hat{\mathbf{z}};\ \mathbf{v}\cdot d\mathbf{a} = 3z\,s\,ds\,d\phi = 15s\,ds\,d\phi.\ \int\mathbf{v}\cdot d\mathbf{a} = 15\int_0^2 s\,ds\int_0^{\frac{\pi}{2}} d\phi = 15\pi.$

bottom: $z = 0,\ d\mathbf{a} = -s\,ds\,d\phi\,\hat{\mathbf{z}};\ \mathbf{v}\cdot d\mathbf{a} = -3z\,s\,ds\,d\phi = 0.\ \int\mathbf{v}\cdot d\mathbf{a} = 0.$

back: $\phi = \frac{\pi}{2},\ d\mathbf{a} = ds\,dz\,\hat{\boldsymbol{\phi}};\ \mathbf{v}\cdot d\mathbf{a} = s\sin\phi\cos\phi\,ds\,dz = 0.\ \int\mathbf{v}\cdot d\mathbf{a} = 0.$

left: $\phi = 0,\ d\mathbf{a} = -ds\,dz\,\hat{\boldsymbol{\phi}};\ \mathbf{v}\cdot d\mathbf{a} = -s\sin\phi\cos\phi\,ds\,dz = 0.\ \int\mathbf{v}\cdot d\mathbf{a} = 0.$

front: $s = 2,\ d\mathbf{a} = s\,d\phi\,dz\,\hat{\mathbf{s}};\ \mathbf{v}\cdot d\mathbf{a} = s(2+\sin^2\phi)s\,d\phi\,dz = 4(2+\sin^2\phi)d\phi\,dz.$

$\int\mathbf{v}\cdot d\mathbf{a} = 4\int_0^{\frac{\pi}{2}}(2+\sin^2\phi)d\phi\int_0^5 dz = (4)(\pi + \frac{\pi}{4})(5) = 25\pi.$

So $\oint\mathbf{v}\cdot d\mathbf{a} = 15\pi + 25\pi = 40\pi.$ ✓

(c)
$$\begin{aligned}\nabla\times\mathbf{v} &= \left(\tfrac{1}{s}\tfrac{\partial}{\partial\phi}(3z) - \tfrac{\partial}{\partial z}(s\sin\phi\cos\phi)\right)\hat{\mathbf{s}} + \left(\tfrac{\partial}{\partial z}\left(s(2+\sin^2\phi)\right) - \tfrac{\partial}{\partial s}(3z)\right)\hat{\boldsymbol{\phi}}\\ &\quad + \tfrac{1}{s}\left(\tfrac{\partial}{\partial s}(s^2\sin\phi\cos\phi) - \tfrac{\partial}{\partial\phi}\left(s(2+\sin^2\phi)\right)\right)\hat{\mathbf{z}}\\ &= \tfrac{1}{s}(2s\sin\phi\cos\phi - s\,2\sin\phi\cos\phi) = \boxed{0.}\end{aligned}$$

Problem 1.43

(a) $3(3^2) - 2(3) - 1 = 27 - 6 - 1 = \boxed{20.}$

(b) $\cos\pi = \boxed{-1.}$

(c) $\boxed{\text{zero.}}$

(d) $\ln(-2+3) = \ln 1 = \boxed{\text{zero.}}$

Problem 1.44

(a) $\int_{-2}^{2}(2x+3)\frac{1}{3}\delta(x)\,dx = \frac{1}{3}(0+3) = \boxed{1.}$

(b) By Eq. 1.94, $\delta(1-x) = \delta(x-1)$, so $1 + 3 + 2 = \boxed{6.}$

(c) $\int_{-1}^{1} 9x^2 \frac{1}{3}\,\delta(x+\frac{1}{3})\,dx = 9\left(-\frac{1}{3}\right)^2 \frac{1}{3} = \boxed{\frac{1}{3}.}$

(d) $\boxed{1 \text{ (if } a > b\text{)}, 0 \text{ (if } a < b\text{)}.}$

Problem 1.45

(a) $\int_{-\infty}^{\infty} f(x)\left[x\frac{d}{dx}\delta(x)\right]dx = x\,f(x)\delta(x)\big|_{-\infty}^{\infty} - \int_{-\infty}^{\infty}\frac{d}{dx}\left(x\,f(x)\right)\delta(x)\,dx.$

The first term is zero, since $\delta(x) = 0$ at $\pm\infty$; $\frac{d}{dx}\left(x\,f(x)\right) = x\frac{df}{dx} + \frac{dx}{dx}f = x\frac{df}{dx} + f.$

So the integral is $-\int_{-\infty}^{\infty}\left(x\frac{df}{dx} + f\right)\delta(x)\,dx = 0 - f(0) = -f(0) = -\int_{-\infty}^{\infty} f(x)\delta(x)\,dx.$

So, $x\frac{d}{dx}\delta(x) = -\delta(x).$ qed

(b) $\int_{-\infty}^{\infty} f(x)\frac{d\theta}{dx}dx = f(x)\theta(x)\big|_{-\infty}^{\infty} - \int_{-\infty}^{\infty}\frac{df}{dx}\theta(x)dx = f(\infty) - \int_0^{\infty}\frac{df}{dx}dx = f(\infty) - (f(\infty) - f(0))$
$= f(0) = \int_{-\infty}^{\infty} f(x)\delta(x)\,dx.$ So $\frac{d\theta}{dx} = \delta(x).$ qed

Problem 1.46

(a) $\boxed{\rho(\mathbf{r}) = q\delta^3(\mathbf{r}-\mathbf{r}').}$ Check: $\int\rho(\mathbf{r})d\tau = q\int\delta^3(\mathbf{r}-\mathbf{r}')\,d\tau = q.$ ✓

(b) $\boxed{\rho(\mathbf{r}) = q\delta^3(\mathbf{r}-\mathbf{r}') - q\delta^3(\mathbf{r}).}$

(c) Evidently $\rho(r) = A\delta(r-R)$. To determine the constant A, we require

$Q = \int\rho\,d\tau = \int A\delta(r-R)4\pi r^2\,dr = A\,4\pi R^2.$ So $A = \frac{Q}{4\pi R^2}.$ $\boxed{\rho(r) = \frac{Q}{4\pi R^2}\delta(r-R).}$

Problem 1.47

(a) $a^2 + \mathbf{a}\cdot\mathbf{a} + a^2 = \boxed{3a^2.}$

(b) $\int(\mathbf{r}-\mathbf{b})^2\frac{1}{5^3}\delta^3(\mathbf{r})\,d\tau = \frac{1}{125}b^2 = \frac{1}{125}(4^2+3^2) = \boxed{\frac{1}{5}.}$

(c) $c^2 = 25 + 9 + 4 = 38 > 36 = 6^2$, so $\mathbf{c}$ is outside $\mathcal{V}$, so the integral is $\boxed{\text{zero.}}$

(d) $(\mathbf{e} - (2\,\hat{\mathbf{x}} + 2\,\hat{\mathbf{y}} + 2\,\hat{\mathbf{z}}))^2 = (1\,\hat{\mathbf{x}} + 0\,\hat{\mathbf{y}} + (-1)\,\hat{\mathbf{z}})^2 = 1 + 1 = 2 < (1.5)^2 = 2.25$, so $\mathbf{e}$ is inside $\mathcal{V}$, and hence the integral is $\mathbf{e}\cdot(\mathbf{d}-\mathbf{e}) = (3,2,1)\cdot(-2,0,2) = -6 + 0 + 2 = \boxed{-4.}$

Problem 1.48

First method: use Eq. 1.99 to write $J = \int e^{-r}\left(4\pi\delta^3(\mathbf{r})\right)d\tau = 4\pi e^{-0} = \boxed{4\pi.}$

Second method: integrating by parts (use Eq. 1.59).

$$\begin{aligned}
J &= -\int_{\mathcal{V}}\frac{\hat{\mathbf{r}}}{r^2}\cdot\nabla(e^{-r})\,d\tau + \oint_{\mathcal{S}} e^{-r}\frac{\hat{\mathbf{r}}}{r^2}\cdot d\mathbf{a}. \quad \text{But} \quad \nabla\left(e^{-r}\right) = \left(\frac{\partial}{\partial r}e^{-r}\right)\hat{\mathbf{r}} = -e^{-r}\hat{\mathbf{r}}. \\
&= \int\frac{1}{r^2}e^{-r}4\pi r^2\,dr + \int e^{-r}\frac{\hat{\mathbf{r}}}{r^2}\cdot r^2\sin\theta\,d\theta\,d\phi\,\hat{\mathbf{r}} = 4\pi\int_0^{\infty} e^{-r}\,dr + e^{-R}\int\sin\theta\,d\theta\,d\phi \\
&= 4\pi\left(-e^{-r}\right)\Big|_0^{\infty} + 4\pi e^{-R} = 4\pi\left(-e^{-\infty} + e^{-0}\right) = 4\pi.✓ \quad (\text{Here } R = \infty, \text{ so } e^{-R} = 0.)
\end{aligned}$$

Problem 1.49 (a) $\nabla\cdot\mathbf{F}_1 = \frac{\partial}{\partial x}(0) + \frac{\partial}{\partial y}(0) + \frac{\partial}{\partial z}\left(x^2\right) = \boxed{0}$; $\nabla\cdot\mathbf{F}_2 = \frac{\partial x}{\partial x} + \frac{\partial y}{\partial y} + \frac{\partial z}{\partial z} = 1 + 1 + 1 = \boxed{3}$

$$\nabla\times\mathbf{F}_1 = \begin{vmatrix}\hat{\mathbf{x}} & \hat{\mathbf{y}} & \hat{\mathbf{z}} \\ \frac{\partial}{\partial x} & \frac{\partial}{\partial y} & \frac{\partial}{\partial z} \\ 0 & 0 & x^2\end{vmatrix} = -\hat{\mathbf{y}}\frac{\partial}{\partial x}\left(x^2\right) = \boxed{-2x\hat{\mathbf{y}}}; \quad \nabla\times\mathbf{F}_2 = \begin{vmatrix}\hat{\mathbf{x}} & \hat{\mathbf{y}} & \hat{\mathbf{z}} \\ \frac{\partial}{\partial x} & \frac{\partial}{\partial y} & \frac{\partial}{\partial z} \\ x & y & z\end{vmatrix} = \boxed{0}$$

$\nabla\cdot(\nabla\times V) = 0$

$\nabla\times(\nabla T) = 0$

$\boxed{\mathbf{F_2} \text{ is a gradient; } \mathbf{F_1} \text{ is a curl}}$ $\boxed{U_2 = \frac{1}{2}\left(x^2 + y^2 + z^2\right)}$ would do ($\mathbf{F_2} = \boldsymbol{\nabla} U_2$).

For $\mathbf{A_1}$, we want $\left(\frac{\partial A_y}{\partial z} - \frac{\partial A_z}{\partial y}\right) = \left(\frac{\partial A_x}{\partial z} - \frac{\partial A_z}{\partial x}\right) = 0;$ $\frac{\partial A_y}{\partial x} - \frac{\partial A_x}{\partial y} = x^2.$ $A_y = \frac{x^3}{3}$, $A_x = A_z = 0$ would do it.

$\boxed{\mathbf{A_1} = \frac{1}{3}x^2\,\hat{\mathbf{y}}}$ $(\mathbf{F_1} = \boldsymbol{\nabla}\times\mathbf{A_1})$. (But these are not unique.)

(b) $\boldsymbol{\nabla}\cdot\mathbf{F_3} = \frac{\partial}{\partial x}(yz) + \frac{\partial}{\partial y}(xz) + \frac{\partial}{\partial z}(xy) = 0;$ $\boldsymbol{\nabla}\times\mathbf{F_3} = \begin{vmatrix} \hat{\mathbf{x}} & \hat{\mathbf{y}} & \hat{\mathbf{z}} \\ \frac{\partial}{\partial x} & \frac{\partial}{\partial y} & \frac{\partial}{\partial z} \\ yz & xz & xy \end{vmatrix} = \hat{\mathbf{x}}\,(x - x) + \hat{\mathbf{y}}\,(y - y) + \hat{\mathbf{z}}\,(z - z) = 0$

So $\mathbf{F_3}$ can be written as the gradient of a scalar ($\mathbf{F_3} = \boldsymbol{\nabla} U_3$) and as the curl of a vector ($\mathbf{F_3} = \boldsymbol{\nabla}\times\mathbf{A_3}$). In fact, $\boxed{U_3 = xyz}$ does the job. For the vector potential, we have

$$\left\{\begin{array}{lll} \frac{\partial A_z}{\partial y} - \frac{\partial A_y}{\partial z} = yz, & \text{which suggests} & A_z = \frac{1}{4}y^2 z + f(x,z);\; A_y = -\frac{1}{4}yz^2 + g(x,y) \\ \frac{\partial A_x}{\partial z} - \frac{\partial A_z}{\partial x} = xz, & \text{suggesting} & A_x = \frac{1}{4}z^2 x + h(x,y);\; A_z = -\frac{1}{4}zx^2 + j(y,z) \\ \frac{\partial A_y}{\partial x} - \frac{\partial A_x}{\partial y} = xy, & \text{so} & A_y = \frac{1}{4}x^2 y + k(y,z);\; A_x = -\frac{1}{4}xy^2 + l(x,y) \end{array}\right\}$$

Putting this all together: $\boxed{\mathbf{A_3} = \frac{1}{4}\left\{x\left(z^2 - y^2\right)\hat{\mathbf{x}} + y\left(x^2 - z^2\right)\hat{\mathbf{y}} + z\left(y^2 - x^2\right)\hat{\mathbf{z}}\right\}}$ (again, not unique).

Problem 1.50

(d) $\Rightarrow$ (a): $\boldsymbol{\nabla}\times\mathbf{F} = \boldsymbol{\nabla}\times(-\boldsymbol{\nabla} U) = 0$ (Eq. 1.44 – curl of gradient is always zero).

(a) $\Rightarrow$ (c): $\oint \mathbf{F}\cdot d\mathbf{l} = \int(\boldsymbol{\nabla}\times\mathbf{F})\cdot d\mathbf{a} = 0$ (Eq. 1.57–Stokes' theorem).

(c) $\Rightarrow$ (b): $\int_{\mathbf{a}\,I}^{\mathbf{b}} \mathbf{F}\cdot d\mathbf{l} - \int_{\mathbf{a}\,II}^{\mathbf{b}} \mathbf{F}\cdot d\mathbf{l} = \int_{\mathbf{a}\,I}^{\mathbf{b}} \mathbf{F}\cdot d\mathbf{l} + \int_{\mathbf{b}\,II}^{\mathbf{a}} \mathbf{F}\cdot d\mathbf{l} = \oint \mathbf{F}\cdot d\mathbf{l} = 0$, so

$$\int_{\mathbf{a}\;I}^{\mathbf{b}} \mathbf{F}\cdot d\mathbf{l} = \int_{\mathbf{a}\;II}^{\mathbf{b}} \mathbf{F}\cdot d\mathbf{l}.$$

(b) $\Rightarrow$ (c): same as (c) $\Rightarrow$ (b), only in reverse; (c) $\Rightarrow$ (a): same as (a)$\Rightarrow$ (c).

Problem 1.51

(d) $\Rightarrow$ (a): $\boldsymbol{\nabla}\cdot\mathbf{F} = \boldsymbol{\nabla}\cdot(\boldsymbol{\nabla}\times\mathbf{W}) = 0$ (Eq 1.46—divergence of curl is always zero).

(a) $\Rightarrow$ (c): $\oint \mathbf{F}\cdot d\mathbf{a} = \int(\boldsymbol{\nabla}\cdot\mathbf{F})\,d\tau = 0$ (Eq. 1.56—divergence theorem).

(c) $\Rightarrow$ (b): $\int_I \mathbf{F}\cdot d\mathbf{a} - \int_{II} \mathbf{F}\cdot d\mathbf{a} = \oint \mathbf{F}\cdot d\mathbf{a} = 0$, so

$$\int_I \mathbf{F}\cdot d\mathbf{a} = \int_{II} \mathbf{F}\cdot d\mathbf{a}.$$

(*Note:* sign change because for $\oint \mathbf{F}\cdot d\mathbf{a}$, $d\mathbf{a}$ is *outward*, whereas for surface II it is *inward.*)

(b) $\Rightarrow$ (c): same as (c) $\Rightarrow$ (b), in reverse; (c)$\Rightarrow$ (a): same as (a)$\Rightarrow$ (c) .

Problem 1.52

In Prob. 1.15 we found that $\boldsymbol{\nabla}\cdot\mathbf{v}_a = 0$; in Prob. 1.18 we found that $\boldsymbol{\nabla}\times\mathbf{v}_c = 0$. So

$\boxed{\mathbf{v}_c \text{ can be written as the gradient of a scalar; } \mathbf{v}_a \text{ can be written as the curl of a vector.}}$

(a) To find t:

(1) $\frac{\partial t}{\partial x} = y^2 \Rightarrow t = y^2 x + f(y,z)$

(2) $\frac{\partial t}{\partial y} = \left(2xy + z^2\right)$

(3) $\frac{\partial t}{\partial z} = 2yz$

From (1) & (3) we get $\frac{\partial f}{\partial z} = 2yz \Rightarrow f = yz^2 + g(y) \Rightarrow t = y^2x + yz^2 + g(y)$, so $\frac{\partial t}{\partial y} = 2xy + z^2 + \frac{\partial g}{\partial y} = 2xy + z^2$ (from (2)) $\Rightarrow \frac{\partial g}{\partial y} = 0$. We may as well pick $g = 0$; then $\boxed{t = xy^2 + yz^2.}$

(b) To find **W**: $\frac{\partial W_z}{\partial y} - \frac{\partial W_y}{\partial z} = x^2$; $\frac{\partial W_x}{\partial z} - \frac{\partial W_z}{\partial x} = 3z^2x$; $\frac{\partial W_y}{\partial x} - \frac{\partial W_x}{\partial y} = -2xz$.
Pick $W_x = 0$; then

$$\begin{aligned}\frac{\partial W_z}{\partial x} &= -3xz^2 \Rightarrow W_z = -\frac{3}{2}x^2z^2 + f(y,z) \\ \frac{\partial W_y}{\partial x} &= -2xz \Rightarrow W_y = -x^2z + g(y,z).\end{aligned}$$

$\frac{\partial W_z}{\partial y} - \frac{\partial W_y}{\partial z} = \frac{\partial f}{\partial y} + x^2 - \frac{\partial g}{\partial z} = x^2 \Rightarrow \frac{\partial f}{\partial y} - \frac{\partial g}{\partial z} = 0$. May as well pick $f = g = 0$.

$\boxed{\mathbf{W} = -x^2z\,\hat{\mathbf{y}} - \frac{3}{2}x^2z^2\,\hat{\mathbf{z}}.}$

Check: $$\nabla\times\mathbf{W} = \begin{vmatrix} \hat{\mathbf{x}} & \hat{\mathbf{y}} & \hat{\mathbf{z}} \\ \frac{\partial}{\partial x} & \frac{\partial}{\partial y} & \frac{\partial}{\partial z} \\ 0 & -x^2z & -\frac{3}{2}x^2z^2 \end{vmatrix} = \hat{\mathbf{x}}\,(x^2) + \hat{\mathbf{y}}\,(3xz^2) + \hat{\mathbf{z}}\,(-2xz).\checkmark$$

You can add any gradient (∇t) to **W** without changing its curl, so this answer is far from unique. Some other solutions:

$\mathbf{W} = xz^3\,\hat{\mathbf{x}} - x^2z\,\hat{\mathbf{y}}$;

$\mathbf{W} = \left(2xyz + xz^3\right)\,\hat{\mathbf{x}} + x^2y\,\hat{\mathbf{z}}$;

$\mathbf{W} = xyz\,\hat{\mathbf{x}} - \frac{1}{2}x^2z\,\hat{\mathbf{y}} + \frac{1}{2}x^2\left(y - 3z^2\right)\,\hat{\mathbf{z}}$.

Probelm 1.53

$$\begin{aligned}\nabla\cdot\mathbf{v} &= \frac{1}{r^2}\frac{\partial}{\partial r}\left(r^2\,r^2\cos\theta\right) + \frac{1}{r\sin\theta}\frac{\partial}{\partial\theta}\left(\sin\theta\,r^2\cos\phi\right) = \frac{1}{r\sin\theta}\frac{\partial}{\partial\phi}\left(-r^2\cos\theta\sin\phi\right) \\ &= \frac{1}{r^2}4r^3\cos\theta + \frac{1}{r\sin\theta}\cos\theta\,r^2\cos\phi + \frac{1}{r\sin\theta}\left(-r^2\cos\theta\cos\phi\right) \\ &= \frac{r\cos\theta}{\sin\theta}\left[4\sin\theta + \cos\phi - \cos\phi\right] = 4r\cos\theta.\end{aligned}$$

$$\begin{aligned}\int(\nabla\cdot\mathbf{v})\,d\tau &= \int(4r\cos\theta)r^2\sin\theta\,dr\,d\theta\,d\phi = 4\int_0^R r^3\,dr\int_0^{\pi/2}\cos\theta\sin\theta\,d\theta\int_0^{\pi/2}d\phi \\ &= \left(R^4\right)\left(\frac{1}{2}\right)\left(\frac{\pi}{2}\right) = \boxed{\frac{\pi R^4}{4}.}\end{aligned}$$

Surface consists of four parts:
(1) *Curved:* $d\mathbf{a} = R^2\sin\theta\,d\theta\,d\phi\,\hat{\mathbf{r}}$; $r = R$. $\mathbf{v}\cdot d\mathbf{a} = \left(R^2\cos\theta\right)\left(R^2\sin\theta\,d\theta\,d\phi\right)$.

$$\int\mathbf{v}\cdot d\mathbf{a} = R^4\int_0^{\pi/2}\cos\theta\sin\theta\,d\theta\int_0^{\pi/2}d\phi = R^4\left(\frac{1}{2}\right)\left(\frac{\pi}{2}\right) = \frac{\pi R^4}{4}.$$

(2) *Left:* $\mathbf{da} = -r\,dr\,d\theta\,\hat{\boldsymbol{\phi}}$; $\phi = 0$. $\mathbf{v}\cdot\mathbf{da} = \left(r^2\cos\theta\sin\phi\right)(r\,dr\,d\theta) = 0$. $\int \mathbf{v}\cdot\mathbf{da} = 0$.
(3) *Back:* $\mathbf{da} = r\,dr\,d\theta\,\hat{\boldsymbol{\phi}}$; $\phi = \pi/2$. $\mathbf{v}\cdot\mathbf{da} = \left(-r^2\cos\theta\sin\phi\right)(r\,dr\,d\theta) = -r^3\cos\theta\,dr\,d\theta$.

$$\int \mathbf{v}\cdot\mathbf{da} = \int_0^R r^3\,dr \int_0^{\pi/2} \cos\theta\,d\theta = -\left(\frac{1}{4}R^4\right)(+1) = -\frac{1}{4}R^4.$$

(4) *Bottom:* $\mathbf{da} = r\sin\,dr\,d\phi\,\hat{\boldsymbol{\theta}}$; $\theta = \pi/2$. $\mathbf{v}\cdot\mathbf{da} = \left(r^2\cos\phi\right)(r\,dr\,d\phi)$.

$$\int \mathbf{v}\cdot\mathbf{da} = \int_0^R r^3\,dr \int_0^{\pi/2} \cos\phi\,d\phi = \frac{1}{4}R^4.$$

Total: $\oint \mathbf{v}\cdot\mathbf{da} = \pi R^4/4 + 0 - \frac{1}{4}R^4 + \frac{1}{4}R^4 = \frac{\pi R^4}{4}$. ✓

Problem 1.54

$$\boldsymbol{\nabla}\times\mathbf{v} = \begin{vmatrix} \hat{\mathbf{x}} & \hat{\mathbf{y}} & \hat{\mathbf{z}} \\ \frac{\partial}{\partial x} & \frac{\partial}{\partial y} & \frac{\partial}{\partial z} \\ ay & bx & 0 \end{vmatrix} = \hat{\mathbf{z}}\,(b-a). \quad \text{So} \quad \int(\boldsymbol{\nabla}\times\mathbf{v})\cdot\mathbf{da} = (b-a)\pi R^2.$$

$\mathbf{v}\cdot d\mathbf{l} = (ay\,\hat{\mathbf{x}} + bx\,\hat{\mathbf{y}})\cdot(dx\,\hat{\mathbf{x}} + dy\,\hat{\mathbf{y}} + dz\,\hat{\mathbf{z}}) = ay\,dx + bx\,dy$; $x^2 + y^2 = R^2 \Rightarrow 2x\,dx + 2y\,dy = 0$,
so $dy = -(x/y)\,dx$. So $\mathbf{v}\cdot d\mathbf{l} = ay\,dx + bx(-x/y)\,dx = \frac{1}{y}\left(ay^2 - bx^2\right)\,dx$.
For the "upper" semicircle, $y = \sqrt{R^2 - x^2}$, so $\mathbf{v}\cdot d\mathbf{l} = \frac{a\left(R^2-x^2\right)-bx^2}{\sqrt{R^2-x^2}}\,dx$.

$$\begin{aligned}\int \mathbf{v}\cdot d\mathbf{l} &= \int_R^{-R} \frac{aR^2 - (a+b)x^2}{\sqrt{R^2-x^2}}\,dx = \left\{aR^2\sin^{-1}\left(\frac{x}{R}\right) - (a+b)\left[-\frac{x}{2}\sqrt{R^2-x^2} + \frac{R^2}{2}\sin^{-1}\left(\frac{x}{R}\right)\right]\right\}\Bigg|_{+R}^{-R} \\ &= \frac{1}{2}R^2(a-b)\sin^{-1}(x/R)\Bigg|_{+R}^{-R} = \frac{1}{2}R^2(a-b)\left(\sin^{-1}(-1) - \sin^{-1}(+1)\right) = \frac{1}{2}R^2(a-b)\left(-\frac{\pi}{2} - \frac{\pi}{2}\right) \\ &= \frac{1}{2}\pi R^2(b-a).\end{aligned}$$

And the same for the lower semicircle (y changes sign, but the limits on the integral are reversed) so
$\oint \mathbf{v}\cdot d\mathbf{l} = \pi R^2(b-a)$. ✓

Problem 1.55

(1) $x = z = 0$; $dx = dz = 0$; $y: 0 \to 1$. $\mathbf{v}\cdot d\mathbf{l} = (y + 3x)\,dy = y\,dy$.

$$\int_0^1 \mathbf{v}\cdot d\mathbf{l} = \int_0^1 y\,dy = \frac{1}{2}.$$

(2) $x = 0$; $z = 2 - 2y$; $dz = -2\,dy$; $y: 1 \to 0$. $\mathbf{v}\cdot d\mathbf{l} = (y + 3x)\,dy + 6\,dz = y\,dy - 12\,dy = (y - 12)\,dy$.

$$\int \mathbf{v}\cdot d\mathbf{l} = \int_1^0 (y-12)\,dy = -\left(\frac{1}{2} - 12\right) = -\frac{1}{2} + 12.$$

(3) $x = y = 0$; $dx = dy = 0$; $z: 2 \to 0$. $\mathbf{v}\cdot d\mathbf{l} = 6\,dz$;

$$\int \mathbf{v}\cdot d\mathbf{l} = \int_2^0 6\,dz = -12.$$

Total: $\oint \mathbf{v}\cdot d\mathbf{l} = \frac{1}{2} - \frac{1}{2} + 12 - 12 = \boxed{0.}$
Meanwhile, Stokes' thereom says $\oint \mathbf{v}\cdot d\mathbf{l} = \int (\nabla\times\mathbf{v})\cdot d\mathbf{a}$. Here $d\mathbf{a} = dy\,dz\,\hat{\mathbf{x}}$, so all we need is $(\nabla\times\mathbf{v})_x = \frac{\partial}{\partial y}(6) - \frac{\partial}{\partial z}(y+3x) = 0$. Therefore $\int(\nabla\times\mathbf{v})\cdot d\mathbf{a} = 0$. ✓

Problem 1.56
Start at the origin.

(1) $\theta = \frac{\pi}{2}$, $\phi = 0$; $r: 0 \to 1$. $\mathbf{v}\cdot d\mathbf{l} = \left(r\cos^2\theta\right)(dr) = 0$. $\int \mathbf{v}\cdot d\mathbf{l} = 0$.

(2) $r = 1$, $\theta = \frac{\pi}{2}$; $\phi: 0 \to \pi/2$. $\mathbf{v}\cdot d\mathbf{l} = (3r)(r\sin\theta\,d\phi) = 3\,d\phi$. $\int \mathbf{v}\cdot d\mathbf{l} = 3\int_0^{\pi/2} d\phi = \frac{3\pi}{2}$.

(3) $\phi = \frac{\pi}{2}$; $r\sin\theta = y = 1$, so $r = \frac{1}{\sin\theta}$, $dr = \frac{-1}{\sin^2\theta}\cos\theta\,d\theta$, $\theta: \frac{\pi}{2} \to \frac{\pi}{4}$.

$$\begin{aligned}
\mathbf{v}\cdot d\mathbf{l} &= \left(r\cos^2\theta\right)(dr) - (r\cos\theta\sin\theta)(r\,d\theta) = \frac{\cos^2\theta}{\sin\theta}\left(-\frac{\cos\theta}{\sin^2\theta}\right) d\,theta - \frac{\cos\theta\sin\theta}{\sin^2\theta}\,d\theta \\
&= -\left(\frac{\cos^3\theta}{\sin^3\theta} + \frac{\cos\theta}{\sin\theta}\right) d\theta = -\frac{\cos\theta}{\sin\theta}\left(\frac{\cos^2\theta + \sin^2\theta}{\sin^2\theta}\right) d\theta = -\frac{\cos\theta}{\sin^3\theta}\,d\theta.
\end{aligned}$$

Therefore

$$\int \mathbf{v}\cdot d\mathbf{l} = -\int_{\pi/2}^{\pi/4} \frac{\cos\theta}{\sin^3\theta}\,d\theta = \left.\frac{1}{2\sin^2\theta}\right|_{\pi/2}^{\pi/4} = \frac{1}{2\cdot(1/2)} - \frac{1}{2\cdot(1)} = 1 - \frac{1}{2} = \frac{1}{2}.$$

(4) $\theta = \frac{\pi}{4}$, $\phi = \frac{\pi}{2}$; $r: \sqrt{2} \to 0$. $\mathbf{v}\cdot d\mathbf{l} = \left(r\cos^2\theta\right)(dr) = \frac{1}{2}r\,dr$.

$$\int \mathbf{v}\cdot d\mathbf{l} = \frac{1}{2}\int_{\sqrt{2}}^{0} r\,dr = \left.\frac{1}{2}\frac{r^2}{2}\right|_{\sqrt{2}}^{0} = -\frac{1}{4}\cdot 2 = -\frac{1}{2}.$$

Total:

$$\oint \mathbf{v}\cdot d\mathbf{l} = 0 + \frac{3\pi}{2} + \frac{1}{2} - \frac{1}{2} = \boxed{\tfrac{3\pi}{2}}.$$

Stokes' theorem says this should equal $\int(\nabla\times\mathbf{v})\cdot d\mathbf{a}$

$$\begin{aligned}
\nabla\times\mathbf{v} &= \frac{1}{r\sin\theta}\left[\frac{\partial}{\partial\theta}(\sin\theta\,3r) - \frac{\partial}{\partial\phi}(-r\sin\theta\cos\theta)\right]\hat{\mathbf{r}} + \frac{1}{r}\left[\frac{1}{\sin\theta}\frac{\partial}{\partial\phi}\left(r\cos^2\theta\right) - \frac{\partial}{\partial r}(r3r)\right]\hat{\boldsymbol{\theta}} \\
&\quad + \frac{1}{r}\left[\frac{\partial}{\partial r}(-rr\cos\theta\sin\theta) - \frac{\partial}{\partial\theta}\left(r\cos^2\theta\right)\right]\hat{\boldsymbol{\phi}} \\
&= \frac{1}{r\sin\theta}[3r\cos\theta]\,\hat{\mathbf{r}} + \frac{1}{r}[-6r]\,\hat{\boldsymbol{\theta}} + \frac{1}{r}[-2r\cos\theta\sin\theta + 2r\cos\theta\sin\theta]\,\hat{\boldsymbol{\phi}} \\
&= 3\cot\theta\,\hat{\mathbf{r}} - 6\,\hat{\boldsymbol{\theta}}.
\end{aligned}$$

(1) *Back face:* $d\mathbf{a} = -r\,dr\,d\theta\,\hat{\boldsymbol{\phi}}$; $(\nabla\times\mathbf{v})\cdot d\mathbf{a} = 0$. $\int(\nabla\times\mathbf{v})\cdot d\mathbf{a} = 0$.

(2) *Bottom:* $d\mathbf{a} = -r\sin\theta\,dr\,d\phi\,\hat{\boldsymbol{\theta}}$; $(\nabla\times\mathbf{v})\cdot d\mathbf{a} = 6r\sin\theta\,dr\,d\phi$. $\theta = \frac{\pi}{2}$, so $(\nabla\times\mathbf{v})\cdot d\mathbf{a} = 6r\,dr\,d\phi$

$$\int(\nabla\times\mathbf{v})\cdot d\mathbf{a} = \int_0^1 6r\,dr \int_0^{\pi/2} d\phi = 6\cdot\frac{1}{2}\cdot\frac{\pi}{2} = \frac{3\pi}{2}. \quad ✓$$

Problem 1.57
$\mathbf{v} \cdot d\mathbf{l} = y\,dz$.

(1) *Left side:* $z = a - x$; $dz = -dx$; $y = 0$. Therefore $\int \mathbf{v} \cdot d\mathbf{l} = 0$.

(2) *Bottom:* $dz = 0$. Therefore $\int \mathbf{v} \cdot d\mathbf{l} = 0$.

(3) *Back:* $z = a - \frac{1}{2}y$; $dz = -1/2\,dy$; $y : 2a \to 0$. $\int \mathbf{v} \cdot d\mathbf{l} = \int_{2a}^{0} y\left(-\frac{1}{2}\,dy\right) = -\frac{1}{2}\frac{y^2}{2}\Big|_{2a}^{0} = \frac{4a^2}{4} = \boxed{a^2.}$

Meanwhile, $\nabla\times\mathbf{v} = \hat{\mathbf{x}}$, so $\int(\nabla\times\mathbf{v}) \cdot d\mathbf{a}$ is the projection of this surface on the xy plane $= \frac{1}{2} \cdot a \cdot 2a = a^2$. ✓

Problem 1.58

$$\begin{aligned}
\nabla\cdot\mathbf{v} &= \frac{1}{r^2}\frac{\partial}{\partial r}\left(r^2 r^2 \sin\theta\right) + \frac{1}{r\sin\theta}\frac{\partial}{\partial\theta}\left(\sin\theta\, 4r^2\cos\theta\right) + \frac{1}{r\sin\theta}\frac{\partial}{\partial\phi}\left(r^2\tan\theta\right) \\
&= \frac{1}{r^2}4r^3\sin\theta + \frac{1}{r\sin\theta}4r^2\left(\cos^2\theta - \sin^2\theta\right) = \frac{4r}{\sin\theta}\left(\sin^2\theta + \cos^2\theta - \sin^2\theta\right) \\
&= 4r\frac{\cos^2\theta}{\sin\theta}.
\end{aligned}$$

$$\begin{aligned}
\int(\nabla\cdot\mathbf{v})\,d\tau &= \int\left(4r\frac{\cos^2\theta}{\sin\theta}\right)\left(r^2\sin\theta\,dr\,d\theta\,d\phi\right) = \int_0^R 4r^3\,dr\int_0^{\pi/6}\cos^2\theta\,d\theta\int_0^{2\pi}d\phi = \left(R^4\right)(2\pi)\left[\frac{\theta}{2} + \frac{\sin 2\theta}{4}\right]\Bigg|_0^{\pi/6} \\
&= 2\pi R^4\left(\frac{\pi}{12} + \frac{\sin 60^\circ}{4}\right) = \frac{\pi R^4}{6}\left(\pi + 3\frac{\sqrt{3}}{2}\right) = \boxed{\frac{\pi R^4}{12}\left(2\pi + 3\sqrt{3}\right).}
\end{aligned}$$

Surface consists of two parts:

(1) *The ice cream:* $r = R$; $\phi : 0 \to 2\pi$; $\theta : 0 \to \pi/6$; $d\mathbf{a} = R^2\sin\theta\,d\theta\,d\phi\,\hat{\mathbf{r}}$; $\mathbf{v}\cdot d\mathbf{a} = \left(R^2\sin\theta\right)\left(R^2\sin\theta\,d\theta\,d\phi\right) = R^4\sin^2\theta\,d\theta\,d\phi$.

$$\int\mathbf{v}\cdot d\mathbf{a} = R^4\int_0^{\pi/6}\sin^2\theta\,d\theta\int_0^{2\pi}d\phi = \left(R^4\right)(2\pi)\left[\frac{1}{2}\theta - \frac{1}{4}\sin 2\theta\right]_0^{\pi/6} = 2\pi R^4\left(\frac{\pi}{12} - \frac{1}{4}\sin 60^\circ\right) = \frac{\pi R^4}{6}\left(\pi - 3\frac{\sqrt{3}}{2}\right)$$

(2) *The cone:* $\theta = \frac{\pi}{6}$; $\phi : 0 \to 2\pi$; $r : 0 \to R$; $d\mathbf{a} = r\sin\theta\,d\phi\,dr\,\hat{\boldsymbol{\theta}} = \frac{\sqrt{3}}{2}r\,dr\,d\phi\,\hat{\boldsymbol{\theta}}$; $\mathbf{v}\cdot d\mathbf{a} = \sqrt{3}\,r^3\,dr\,d\phi$

$$\int\mathbf{v}\cdot d\mathbf{a} = \sqrt{3}\int_0^R r^3\,dr\int_0^{2\pi}d\phi = \sqrt{3}\cdot\frac{R^4}{4}\cdot 2\pi = \frac{\sqrt{3}}{2}\pi R^4.$$

Therefore $\int\mathbf{v}\cdot d\mathbf{a} = \frac{\pi R^4}{2}\left(\frac{\pi}{3} - \frac{\sqrt{3}}{2} + \sqrt{3}\right) = \frac{\pi R^4}{12}\left(2\pi + 3\sqrt{3}\right)$. ✓

Problem 1.59
(a) Corollary 2 says $\oint(\nabla T)\cdot d\mathbf{l} = 0$. Stokes' theorem says $\oint(\nabla T)\cdot d\mathbf{l} = \int[\nabla\times(\nabla T)]\cdot d\mathbf{a}$. So $\int[\nabla\times(\nabla T)]\cdot d\mathbf{a} = 0$, and since this is true for *any* surface, the integrand must vanish: $\nabla\times(\nabla T) = 0$, confirming Eq. 1.44.

(b) Corollary 2 says $\oint(\nabla\times\mathbf{v})\cdot d\mathbf{a} = 0$. Divergence theorem says $\oint(\nabla\times\mathbf{v})\cdot d\mathbf{a} = \int \nabla\cdot(\nabla\times\mathbf{v})\,d\tau$. So $\int \nabla\cdot(\nabla\times\mathbf{v})\,d\tau = 0$, and since this is true for *any* volume, the integrand must vanish: $\nabla(\nabla\times\mathbf{v}) = 0$, confirming Eq. 1.46.

Problem 1.60

(a) Divergence theorem: $\oint \mathbf{v}\cdot d\mathbf{a} = \int(\nabla\cdot\mathbf{v})\,d\tau$. Let $\mathbf{v} = \mathbf{c}T$, where $\mathbf{c}$ is a constant vector. Using product rule #5 in front cover: $\nabla\cdot\mathbf{v} = \nabla\cdot(\mathbf{c}T) = T(\nabla\cdot\mathbf{c}) + \mathbf{c}\cdot(\nabla T)$. But $\mathbf{c}$ is constant so $\nabla\cdot\mathbf{c} = 0$. Therefore we have: $\int \mathbf{c}\cdot(\nabla T)\,d\tau = \int T\mathbf{c}\cdot d\mathbf{a}$. Since $\mathbf{c}$ is constant, take it outside the integrals: $\mathbf{c}\cdot\int \nabla T\,d\tau = \mathbf{c}\cdot\int T\,d\mathbf{a}$. But $\mathbf{c}$ is *any* constant vector—in particular, it could be be $\hat{\mathbf{x}}$, or $\hat{\mathbf{y}}$, or $\hat{\mathbf{z}}$—so each ***component*** of the integral on left equals corresponding component on the right, and hence

$$\int \nabla T\,d\tau = \int T\,d\mathbf{a}. \qquad \text{qed}$$

(b) Let $\mathbf{v}\to(\mathbf{v}\times\mathbf{c})$ in divergence theorem. Then $\int \nabla\cdot(\mathbf{v}\times\mathbf{c})d\tau = \int(\mathbf{v}\times\mathbf{c})\cdot d\mathbf{a}$. Product rule #6 $\Rightarrow$ $\nabla\cdot(\mathbf{v}\times\mathbf{c}) = \mathbf{c}\cdot(\nabla\times\mathbf{v}) - \mathbf{v}\cdot(\nabla\times\mathbf{c}) = \mathbf{c}\cdot(\nabla\times\mathbf{v})$. (Note: $\nabla\times\mathbf{c} = 0$, since $\mathbf{c}$ is constant.) Meanwhile vector identity (1) says $d\mathbf{a}\cdot(\mathbf{v}\times\mathbf{c}) = \mathbf{c}\cdot(d\mathbf{a}\times\mathbf{v}) = -\mathbf{c}\cdot(\mathbf{v}\times d\mathbf{a})$. Thus $\int \mathbf{c}\cdot(\nabla\times\mathbf{v})\,d\tau = -\int \mathbf{c}\cdot(\mathbf{v}\times d\mathbf{a})$. Take $\mathbf{c}$ outside, and again let $\mathbf{c}$ be $\hat{\mathbf{x}}$, $\hat{\mathbf{y}}$, $\hat{\mathbf{z}}$ then:

$$\int(\nabla\times\mathbf{v})\,d\tau = -\int \mathbf{v}\times d\mathbf{a}. \quad \text{qed}$$

(c) Let $\mathbf{v} = T\nabla U$ in divergence theorem: $\int \nabla\cdot(T\nabla U)\,d\tau = \int T\nabla U\cdot d\mathbf{a}$. Product rule #(5) $\Rightarrow$ $\nabla\cdot(T\nabla U) = T\nabla\cdot(\nabla U) + (\nabla U)\cdot(\nabla T) = T\nabla^2 U + (\nabla U)\cdot(\nabla T)$. Therefore

$$\int \left(T\nabla^2 U + (\nabla U)\cdot(\nabla T)\right)\,d\tau = \int (T\nabla U)\cdot d\mathbf{a}. \quad \text{qed}$$

(d) Rewrite (c) with $T\leftrightarrow U$: $\int \left(U\nabla^2 T + (\nabla T)\cdot(\nabla U)\right)\,d\tau = \int (U\nabla T)\cdot d\mathbf{a}$. Subtract this from (c), noting that the $(\nabla U)\cdot(\nabla T)$ terms cancel:

$$\int \left(T\nabla^2 U - U\nabla^2 T\right)\,d\tau = \int (T\nabla U - U\nabla T)\cdot d\mathbf{a}. \quad \text{qed}$$

(e) Stoke's theorem: $\int(\nabla\times\mathbf{v})\cdot d\mathbf{a} = \oint \mathbf{v}\cdot d\mathbf{l}$. Let $\mathbf{v} = \mathbf{c}T$. By Product Rule #(7): $\nabla\times(\mathbf{c}T) = T(\nabla\times\mathbf{c}) - \mathbf{c}\times(\nabla T) = -\mathbf{c}\times(\nabla T)$ (since $\mathbf{c}$ is constant). Therefore, $-\int(\mathbf{c}\times(\nabla T))\cdot d\mathbf{a} = \oint T\mathbf{c}\cdot d\mathbf{l}$. Use vector indentity #1 to rewrite the first term $(\mathbf{c}\times(\nabla T))\cdot d\mathbf{a} = \mathbf{c}\cdot(\nabla T\times d\mathbf{a})$. So $-\int \mathbf{c}\cdot(\nabla T\times d\mathbf{a}) = \oint \mathbf{c}\cdot T\,d\mathbf{l}$. Pull $\mathbf{c}$ outside, and let $\mathbf{c}\to\hat{\mathbf{x}}$, $\hat{\mathbf{y}}$, and $\hat{\mathbf{z}}$ to prove:

$$\int \nabla T\times d\mathbf{a} = -\oint T\,d\mathbf{l}. \quad \text{qed}$$

Problem 1.61

(a) $d\mathbf{a} = R^2\sin\theta\,d\theta\,d\phi\,\hat{\mathbf{r}}$. Let the surface be the northern hemisphere. The $\hat{\mathbf{x}}$ and $\hat{\mathbf{y}}$ components clearly integrate to zero, and the $\hat{\mathbf{z}}$ component of $\hat{\mathbf{r}}$ is $\cos\theta$, so

$$\mathbf{a} = \int R^2\sin\theta\cos\theta\,d\theta\,d\phi\,\hat{\mathbf{z}} = 2\pi R^2\,\hat{\mathbf{z}}\int_0^{\pi/2}\sin\theta\cos\theta\,d\theta = 2\pi R^2\,\hat{\mathbf{z}}\frac{\sin^2\theta}{2}\Big|_0^{\pi/2} = \boxed{\pi R^2\,\hat{\mathbf{z}}.}$$

(b) Let $T = 1$ in Prob. 1.60(a). Then $\nabla T = 0$, so $\oint d\mathbf{a} = 0$. qed.

(c) This follows from (b). For suppose $\mathbf{a}_1 \neq \mathbf{a}_2$; then if you put them together to make a closed surface, $\oint d\mathbf{a} = \mathbf{a}_1 - \mathbf{a}_2 \neq 0$.

(d) For one such triangle, $d\mathbf{a} = \frac{1}{2}(\mathbf{r}\times d\mathbf{l})$ (since $\mathbf{r}\times d\mathbf{l}$ is the area of the parallelogram, and the direction is perpendicular to the surface), so for the entire conical surface, $\mathbf{a} = \frac{1}{2}\oint \mathbf{r}\times d\mathbf{l}$.

(e) Let $T = \mathbf{c}\cdot\mathbf{r}$, and use product rule #4: $\boldsymbol{\nabla}T = \boldsymbol{\nabla}(\mathbf{c}\cdot\mathbf{r}) = \mathbf{c}\times(\boldsymbol{\nabla}\times\mathbf{r}) + (\mathbf{c}\cdot\boldsymbol{\nabla})\mathbf{r}$. But $\boldsymbol{\nabla}\times\mathbf{r} = 0$, and $(\mathbf{c}\cdot\boldsymbol{\nabla})\mathbf{r} = (c_x\frac{\partial}{\partial x} + c_y\frac{\partial}{\partial y} + c_z\frac{\partial}{\partial z})(x\,\hat{\mathbf{x}} + y\,\hat{\mathbf{y}} = z\,\hat{\mathbf{z}}) = c_x\,\hat{\mathbf{x}} + c_y\,\hat{\mathbf{y}} + c_z\,\hat{\mathbf{z}} = \mathbf{c}$. So Prob. 1.60(e) says

$$\oint T\,d\mathbf{l} = \oint(\mathbf{c}\cdot\mathbf{r})\,d\mathbf{l} = -\int(\boldsymbol{\nabla}T)\times d\mathbf{a} = -\int \mathbf{c}\times d\mathbf{a} = -\mathbf{c}\times\int d\mathbf{a} = -\mathbf{c}\times\mathbf{a} = \mathbf{a}\times\mathbf{c}. \qquad \text{qed}$$

Problem 1.62

(1)

$$\boldsymbol{\nabla}\cdot\mathbf{v} = \frac{1}{r^2}\frac{\partial}{\partial r}\left(r^2\cdot\frac{1}{r}\right) = \frac{1}{r^2}\frac{\partial}{\partial r}(r) = \boxed{\frac{1}{r^2}.}$$

For a sphere of radius R:

$$\left.\begin{aligned} \int\mathbf{v}\cdot d\mathbf{a} &= \int\left(\tfrac{1}{R}\,\hat{\mathbf{r}}\right)\cdot\left(R^2\sin\theta\,d\theta\,d\phi\,\hat{\mathbf{r}}\right) = R\int\sin\theta\,d\theta\,d\phi = 4\pi R. \\ \int(\boldsymbol{\nabla}\cdot\mathbf{v})\,d\tau &= \int\left(\tfrac{1}{r^2}\right)\left(r^2\sin\theta\,dr\,d\theta\,d\phi\right) = \left(\int_0^R dr\right)\left(\int\sin\theta\,d\theta\,d\phi\right) = 4\pi R. \end{aligned}\right\}\ \text{So divergence theorem checks.}$$

Evidently there is ***no*** delta function at the origin.

$$\boldsymbol{\nabla}\times(r^n\,\hat{\mathbf{r}}) = \frac{1}{r^2}\frac{\partial}{\partial r}\left(r^2r^n\right) = \frac{1}{r^2}\frac{\partial}{\partial r}\left(r^{n+2}\right) = \frac{1}{r^2}(n+2)r^{n+1} = \boxed{(n+2)r^{n-1}}$$

(except for $n = -2$, for which we already know (Eq. 1.99) that the divergence is $4\pi\delta^3(\mathbf{r})$).

(2) *Geometrically,* it should be zero. Likewise, the curl in the spherical coordinates obviously gives $\boxed{\text{zero.}}$ To be certain there is no lurking delta function here, we integrate over a sphere of radius R, using Prob. 1.60(b): If $\boldsymbol{\nabla}\times(r^n\,\hat{\mathbf{r}}) = 0$, then $\int(\boldsymbol{\nabla}\times\mathbf{v})\,d\tau = 0 \stackrel{?}{=} -\oint\mathbf{v}\times d\mathbf{a}$. But $\mathbf{v} = r^n\,\hat{\mathbf{r}}$ and $d\mathbf{a} = R^2\sin\theta\,d\theta\,d\phi\,\hat{\mathbf{r}}$ are both in the $\hat{\mathbf{r}}$ directions, so $\mathbf{v}\times d\mathbf{a} = 0$. ✓

Chapter 2

Electrostatics

Problem 2.1

(a) $\boxed{\text{Zero.}}$

(b) $\boxed{F = \frac{1}{4\pi\epsilon_0}\frac{qQ}{r^2},}$ where r is the distance from center to each numeral. $\mathbf{F}$ points *toward* the missing q. *Explanation:* by superposition, this is equivalent to (a), with an extra $-q$ at 6 o'clock—since the force of all twelve is zero, the net force is that of $-q$ only.

(c) $\boxed{\text{Zero.}}$

(d) $\boxed{\frac{1}{4\pi\epsilon_0}\frac{qQ}{r^2},}$ pointing toward the missing q. Same reason as (b). Note, however, that if you explained (b) as a cancellation in pairs of opposite charges (1 o'clock against 7 o'clock; 2 against 8, etc.), with one unpaired q doing the job, then you'll need a *different* explanation for (d).

Problem 2.2

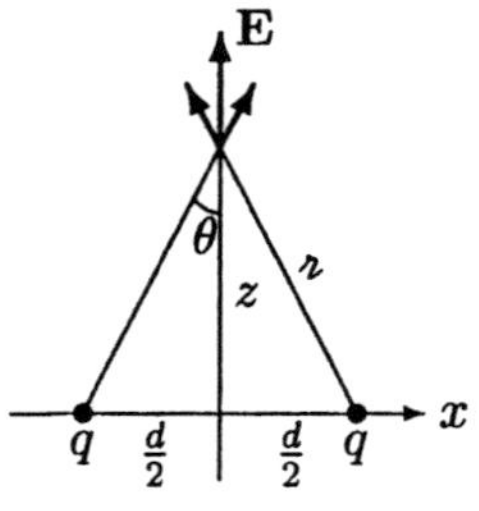

(a) "Horizontal" components cancel. Net vertical field is: $E_z = \frac{1}{4\pi\epsilon_0}2\frac{q}{\mathscr{r}^2}\cos\theta$.

Here $\mathscr{r}^2 = z^2 + \left(\frac{d}{2}\right)^2$; $\cos\theta = \frac{z}{\mathscr{r}}$, so $\boxed{\mathbf{E} = \frac{1}{4\pi\epsilon_0}\frac{2qz}{\left(z^2+\left(\frac{d}{2}\right)^2\right)^{3/2}}\,\hat{\mathbf{z}}.}$

When $z \gg d$ you're so far away it just looks like a single charge $2q$; the field should reduce to $\mathbf{E} = \frac{1}{4\pi\epsilon_0}\frac{2q}{z^2}\,\hat{\mathbf{z}}$. And it *does* (just set $d \to 0$ in the formula).

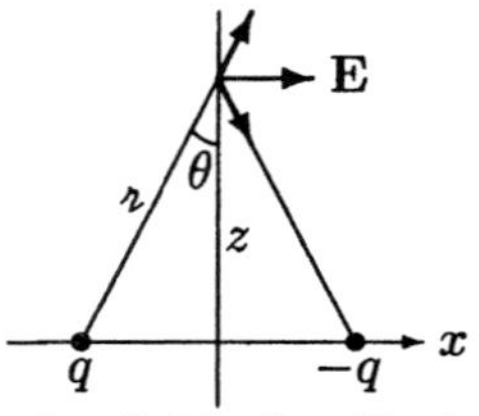

(b) This time the "vertical" components cancel, leaving

$\mathbf{E} = \frac{1}{4\pi\epsilon_0}2\frac{q}{\mathscr{r}^2}\sin\theta\,\hat{\mathbf{x}}$, or

$$\boxed{\mathbf{E} = \frac{1}{4\pi\epsilon_0}\frac{qd}{\left(z^2+\left(\frac{d}{2}\right)^2\right)^{3/2}}\,\hat{\mathbf{x}}.}$$

From far away, ($z \gg d$), the field goes like $\mathbf{E} \approx \frac{1}{4\pi\epsilon_0}\frac{qd}{z^3}\,\hat{\mathbf{z}}$, which, as we shall see, is the field of a *dipole*. (If we set $d \to 0$, we get $\mathbf{E} = 0$, as is appropriate; to the extent that this configuration looks like a single point charge from far away, the net charge is zero, so $\mathbf{E} \to 0$.)

Problem 2.3

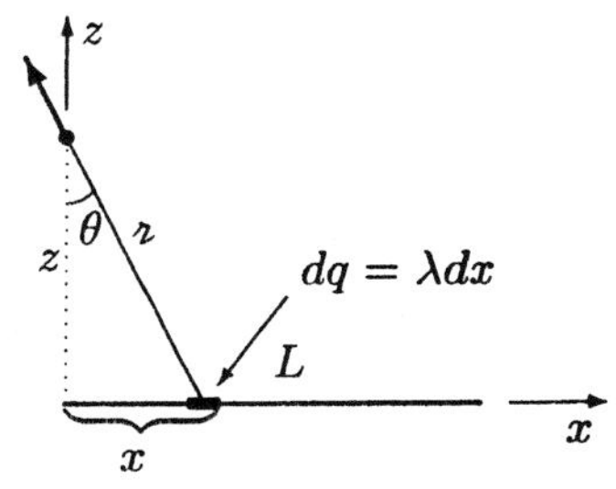

$$
\begin{aligned}
E_z &= \frac{1}{4\pi\epsilon_0}\int_0^L \frac{\lambda\,dx}{\mathscr{r}^2}\cos\theta;\ (\mathscr{r}^2 = z^2 + x^2;\ \cos\theta = \tfrac{z}{\mathscr{r}})\\
&= \frac{1}{4\pi\epsilon_0}\lambda z\int_0^L \frac{1}{(z^2+x^2)^{3/2}}dx\\
&= \frac{1}{4\pi\epsilon_0}\lambda z\left[\frac{1}{z^2}\frac{x}{\sqrt{z^2+x^2}}\right]\Bigg|_0^L = \frac{1}{4\pi\epsilon_0}\frac{\lambda}{z}\frac{L}{\sqrt{z^2+L^2}}.\\
E_x &= -\frac{1}{4\pi\epsilon_0}\int_0^L \frac{\lambda\,dx}{\mathscr{r}^2}\sin\theta = -\frac{1}{4\pi\epsilon_0}\lambda\int\frac{x\,dx}{(x^2+z^2)^{3/2}}\\
&= -\frac{1}{4\pi\epsilon_0}\lambda\left[-\frac{1}{\sqrt{x^2+z^2}}\right]\Bigg|_0^L = -\frac{1}{4\pi\epsilon_0}\lambda\left[\frac{1}{z}-\frac{1}{\sqrt{z^2+L^2}}\right].
\end{aligned}
$$

$$
\boxed{\mathbf{E} = \frac{1}{4\pi\epsilon_0}\frac{\lambda}{z}\left[\left(-1+\frac{z}{\sqrt{z^2+L^2}}\right)\hat{\mathbf{x}} + \left(\frac{L}{\sqrt{z^2+L^2}}\right)\hat{\mathbf{z}}\right].}
$$

For $z \gg L$ you expect it to look like a point charge $q = \lambda L$: $\mathbf{E} \to \frac{1}{4\pi\epsilon_0}\frac{\lambda L}{z^2}\hat{\mathbf{z}}$. It checks, for with $z \gg L$ the $\hat{\mathbf{x}}$ term $\to 0$, and the $\hat{\mathbf{z}}$ term $\to \frac{1}{4\pi\epsilon_0}\frac{\lambda}{z}\frac{L}{z}\hat{\mathbf{z}}$.

Problem 2.4

From Ex. 2.1, with $L \to \frac{a}{2}$ and $z \to \sqrt{z^2 + \left(\frac{a}{2}\right)^2}$ (distance from center of edge to P), field of *one* edge is:

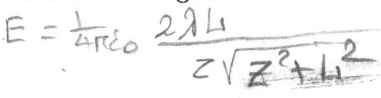

$$
E_1 = \frac{1}{4\pi\epsilon_0}\frac{\lambda a}{\sqrt{z^2+\frac{a^2}{4}}\sqrt{z^2+\frac{a^2}{4}+\frac{a^2}{4}}}.
$$

There are 4 sides, and we want vertical components only, so multiply by $4\cos\theta = 4\frac{z}{\sqrt{z^2+\frac{a^2}{4}}}$:

$$
\boxed{\mathbf{E} = \frac{1}{4\pi\epsilon_0}\frac{4\lambda a z}{\left(z^2+\frac{a^2}{4}\right)\sqrt{z^2+\frac{a^2}{2}}}\hat{\mathbf{z}}.}
$$

Problem 2.5

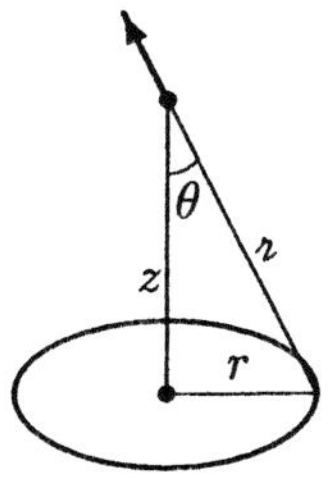

"Horizontal" components cancel, leaving: $\mathbf{E} = \frac{1}{4\pi\epsilon_0}\left\{\int\frac{\lambda dl}{\mathscr{r}^2}\cos\theta\right\}\hat{\mathbf{z}}$.
Here, $\mathscr{r}^2 = r^2 + z^2$, $\cos\theta = \frac{z}{\mathscr{r}}$ (both constants), while $\int dl = 2\pi r$. So

$$
\boxed{\mathbf{E} = \frac{1}{4\pi\epsilon_0}\frac{\lambda(2\pi r)z}{(r^2+z^2)^{3/2}}\hat{\mathbf{z}}.}
$$

Problem 2.6

Break it into rings of radius r, and thickness dr, and use Prob. 2.5 to express the field of each ring. Total charge of a ring is $\sigma\cdot 2\pi r\cdot dr = \lambda\cdot 2\pi r$, so $\lambda = \sigma dr$ is the "line charge" of each ring.

$$
E_{\text{ring}} = \frac{1}{4\pi\epsilon_0}\frac{(\sigma dr)2\pi r z}{(r^2+z^2)^{3/2}};\quad E_{\text{disk}} = \frac{1}{4\pi\epsilon_0}2\pi\sigma z\int_0^R \frac{r}{(r^2+z^2)^{3/2}}dr.
$$

$$
\boxed{\mathbf{E}_{\text{disk}} = \frac{1}{4\pi\epsilon_0}2\pi\sigma z\left[\frac{1}{z}-\frac{1}{\sqrt{R^2+z^2}}\right]\hat{\mathbf{z}}.}
$$

For $R \gg z$ the second term $\to 0$, so $\mathbf{E}_{\text{plane}} = \frac{1}{4\pi\epsilon_0} 2\pi\sigma\hat{\mathbf{z}} = \boxed{\frac{\sigma}{2\epsilon_0}\hat{\mathbf{z}}.}$

For $z \gg R$, $\frac{1}{\sqrt{R^2+z^2}} = \frac{1}{z}\left(1+\frac{R^2}{z^2}\right)^{-1/2} \approx \frac{1}{z}\left(1-\frac{1}{2}\frac{R^2}{z^2}\right)$, so $[\,] \approx \frac{1}{z} - \frac{1}{z} + \frac{1}{2}\frac{R^2}{z^3} = \frac{R^2}{2z^3}$,
and $E = \frac{1}{4\pi\epsilon_0}\frac{2\pi R^2\sigma}{2z^2} = \frac{1}{4\pi\epsilon_0}\frac{Q}{z^2}$, where $Q = \pi R^2\sigma$. ✓

Problem 2.7

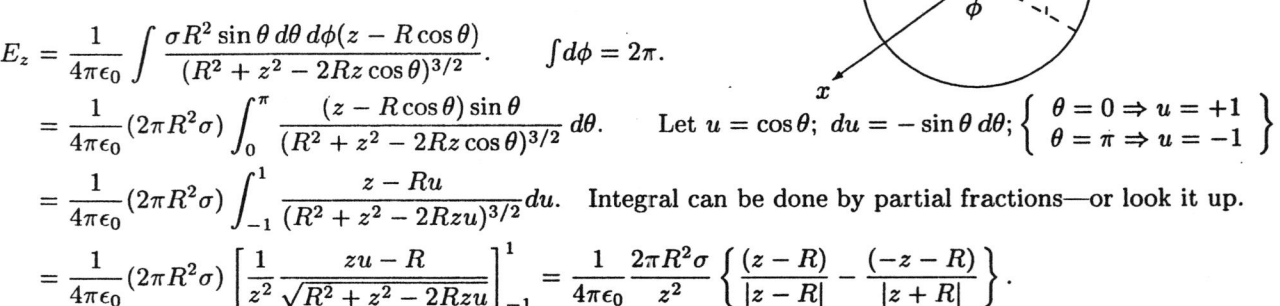

$\mathbf{E}$ is clearly in the z direction. From the diagram,
$dq = \sigma da = \sigma R^2 \sin\theta\, d\theta\, d\phi$,
$\mathscr{r}^2 = R^2 + z^2 - 2Rz\cos\theta$,
$\cos\psi = \frac{z - R\cos\theta}{\mathscr{r}}$.

So

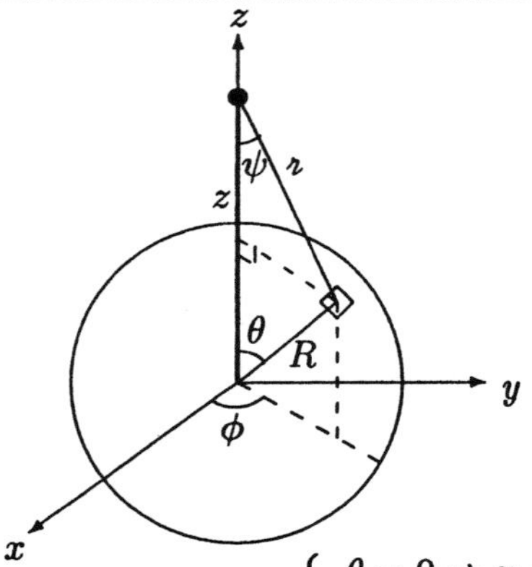

$$E_z = \frac{1}{4\pi\epsilon_0}\int \frac{\sigma R^2 \sin\theta\, d\theta\, d\phi (z - R\cos\theta)}{(R^2+z^2-2Rz\cos\theta)^{3/2}}. \qquad \int d\phi = 2\pi.$$

$$= \frac{1}{4\pi\epsilon_0}(2\pi R^2\sigma)\int_0^\pi \frac{(z - R\cos\theta)\sin\theta}{(R^2+z^2-2Rz\cos\theta)^{3/2}}\,d\theta. \qquad \text{Let } u = \cos\theta;\ du = -\sin\theta\, d\theta; \left\{\begin{array}{c}\theta = 0 \Rightarrow u = +1 \\ \theta = \pi \Rightarrow u = -1\end{array}\right\}.$$

$$= \frac{1}{4\pi\epsilon_0}(2\pi R^2\sigma)\int_{-1}^{1} \frac{z - Ru}{(R^2+z^2-2Rzu)^{3/2}}\,du. \quad \text{Integral can be done by partial fractions—or look it up.}$$

$$= \frac{1}{4\pi\epsilon_0}(2\pi R^2\sigma)\left[\frac{1}{z^2}\frac{zu - R}{\sqrt{R^2+z^2-2Rzu}}\right]_{-1}^{1} = \frac{1}{4\pi\epsilon_0}\frac{2\pi R^2\sigma}{z^2}\left\{\frac{(z-R)}{|z-R|} - \frac{(-z-R)}{|z+R|}\right\}.$$

For $z > R$ (outside the sphere), $E_z = \frac{1}{4\pi\epsilon_0}\frac{4\pi R^2\sigma}{z^2} = \frac{1}{4\pi\epsilon_0}\frac{q}{z^2}$, so $\boxed{\mathbf{E} = \frac{1}{4\pi\epsilon_0}\frac{q}{z^2}\hat{\mathbf{z}}.}$

For $z < R$ (inside), $E_z = 0$, so $\boxed{\mathbf{E} = 0.}$

Problem 2.8

According to Prob. 2.7, all shells *interior* to the point (i.e. at smaller r) contribute as though their charge were concentrated at the center, while all exterior shells contribute nothing. Therefore:

$$\mathbf{E}(r) = \frac{1}{4\pi\epsilon_0}\frac{Q_{\text{int}}}{r^2}\hat{\mathbf{r}},$$

where Q_{int} is the total charge interior to the point. *Outside* the sphere, ***all*** the charge is interior, so

$$\boxed{\mathbf{E} = \frac{1}{4\pi\epsilon_0}\frac{Q}{r^2}\hat{\mathbf{r}}.}$$

Inside the sphere, only that fraction of the total which is interior to the point counts:

$$Q_{\text{int}} = \frac{\frac{4}{3}\pi r^3}{\frac{4}{3}\pi R^3}Q = \frac{r^3}{R^3}Q, \quad \text{so } \mathbf{E} = \frac{1}{4\pi\epsilon_0}\frac{r^3}{R^3}Q\frac{1}{r^2}\hat{\mathbf{r}} = \boxed{\frac{1}{4\pi\epsilon_0}\frac{Q}{R^3}\mathbf{r}.}$$

Problem 2.9

(a) $\rho = \epsilon_0 \nabla\cdot\mathbf{E} = \epsilon_0\frac{1}{r^2}\frac{\partial}{\partial r}\left(r^2\cdot kr^3\right) = \epsilon_0\frac{1}{r^2}k(5r^4) = \boxed{5\epsilon_0 kr^2.}$

(b) *By Gauss's law:* $Q_{\text{enc}} = \epsilon_0 \oint \mathbf{E} \cdot d\mathbf{a} = \epsilon_0 (kR^3)(4\pi R^2) = \boxed{4\pi\epsilon_0 k R^5.}$

By direct integration: $Q_{\text{enc}} = \int \rho\, d\tau = \int_0^R (5\epsilon_0 k r^2)(4\pi r^2 dr) = 20\pi\epsilon_0 k \int_0^R r^4 dr = 4\pi\epsilon_0 k R^5.$✓

Problem 2.10

Think of this cube as one of 8 surrounding the charge. Each of the 24 squares which make up the surface of this larger cube gets the same flux as every other one, so:

$$\int_{\substack{\text{one}\\\text{face}}} \mathbf{E} \cdot d\mathbf{a} = \frac{1}{24} \int_{\substack{\text{whole}\\\text{large}\\\text{cube}}} \mathbf{E} \cdot d\mathbf{a}.$$

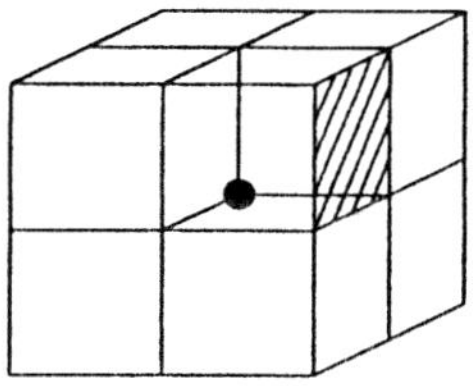

The latter is $\frac{1}{\epsilon_0} q$, by Gauss's law. Therefore $\boxed{\int_{\substack{\text{one}\\\text{face}}} \mathbf{E} \cdot d\mathbf{a} = \frac{q}{24\epsilon_0}.}$

Problem 2.11

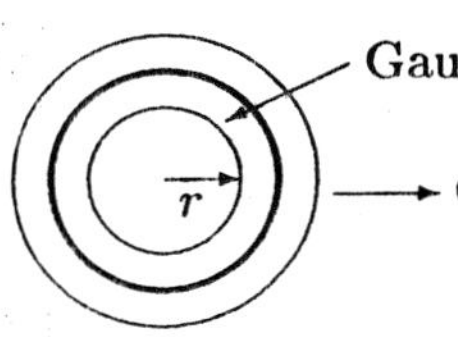

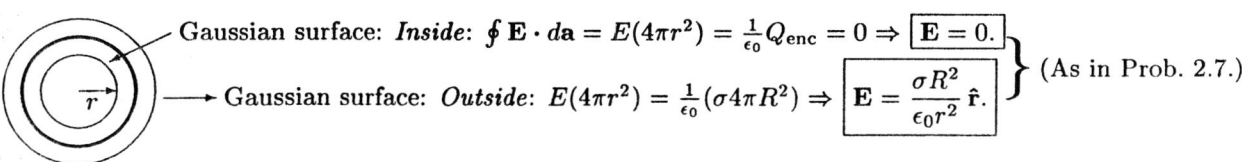

Gaussian surface: ***Inside***: $\oint \mathbf{E} \cdot d\mathbf{a} = E(4\pi r^2) = \frac{1}{\epsilon_0} Q_{\text{enc}} = 0 \Rightarrow \boxed{\mathbf{E} = 0.}$

Gaussian surface: ***Outside***: $E(4\pi r^2) = \frac{1}{\epsilon_0}(\sigma 4\pi R^2) \Rightarrow \boxed{\mathbf{E} = \frac{\sigma R^2}{\epsilon_0 r^2}\hat{\mathbf{r}}.}$

} (As in Prob. 2.7.)

Problem 2.12

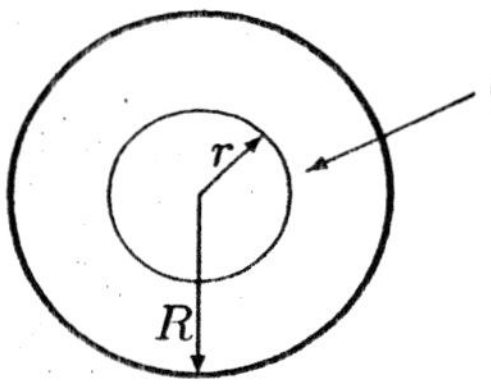

Gaussian surface

$\oint \mathbf{E} \cdot d\mathbf{a} = E \cdot 4\pi r^2 = \frac{1}{\epsilon_0} Q_{\text{enc}} = \frac{1}{\epsilon_0} \frac{4}{3} \pi r^3 \rho.$ So

$$\boxed{\mathbf{E} = \frac{1}{3\epsilon_0} \rho r \hat{\mathbf{r}}.}$$

Since $Q_{\text{tot}} = \frac{4}{3}\pi R^2 \rho$, $\mathbf{E} = \frac{1}{4\pi\epsilon_0} \frac{Q}{R^3} \hat{\mathbf{r}}$ (as in Prob. 2.8).

Problem 2.13

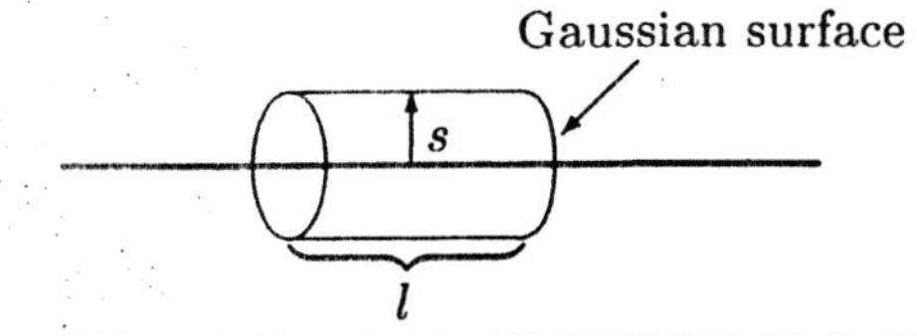

$\oint \mathbf{E} \cdot d\mathbf{a} = E \cdot 2\pi s \cdot l = \frac{1}{\epsilon_0} Q_{\text{enc}} = \frac{1}{\epsilon_0} \lambda l.$ So

$\boxed{\mathbf{E} = \frac{\lambda}{2\pi\epsilon_0 s} \hat{\mathbf{s}}}$ (same as Ex. 2.1).

Problem 2.14

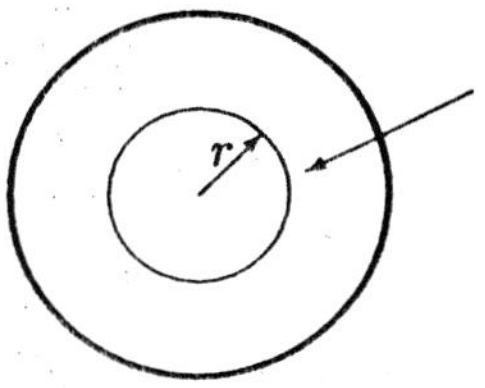

Gaussian surface $\oint \mathbf{E} \cdot d\mathbf{a} = E \cdot 4\pi r^2 = \frac{1}{\epsilon_0} Q_{\text{enc}} = \frac{1}{\epsilon_0} \int \rho\, d\tau = \frac{1}{\epsilon_0} \int (k\bar{r})(\bar{r}^2 \sin\theta\, d\bar{r}\, d\theta\, d\phi)$

$= \frac{1}{\epsilon_0} k\, 4\pi \int_0^r \bar{r}^3 d\bar{r} = \frac{4\pi k}{\epsilon_0} \frac{r^4}{4} = \frac{\pi k}{\epsilon_0} r^4.$

$$\therefore \boxed{\mathbf{E} = \frac{1}{4\pi\epsilon_0} \pi k r^2 \hat{\mathbf{r}}.}$$

Problem 2.15

(i) $Q_{\text{enc}} = 0$, so $\boxed{\mathbf{E} = 0.}$

(ii) $\oint \mathbf{E} \cdot d\mathbf{a} = E(4\pi r^2) = \frac{1}{\epsilon_0} Q_{\text{enc}} = \frac{1}{\epsilon_0} \int \rho \, d\tau = \frac{1}{\epsilon_0} \int \frac{k}{\bar{r}^2} \bar{r}^2 \sin\theta \, d\bar{r} \, d\theta \, d \; phi$

$= \frac{4\pi k}{\epsilon_0} \int_a^r d\bar{r} = \frac{4\pi k}{\epsilon_0}(r-a) \; \therefore \; \boxed{\mathbf{E} = \frac{k}{\epsilon_0}\left(\frac{r-a}{r^2}\right)\hat{\mathbf{r}}.}$

(iii) $E(4\pi r^2) = \frac{4\pi k}{\epsilon_0} \int_a^b d\bar{r} = \frac{4\pi k}{\epsilon_0}(b-a)$, so

$$\boxed{\mathbf{E} = \frac{k}{\epsilon_0}\left(\frac{b-a}{r^2}\right)\hat{\mathbf{r}}.}$$

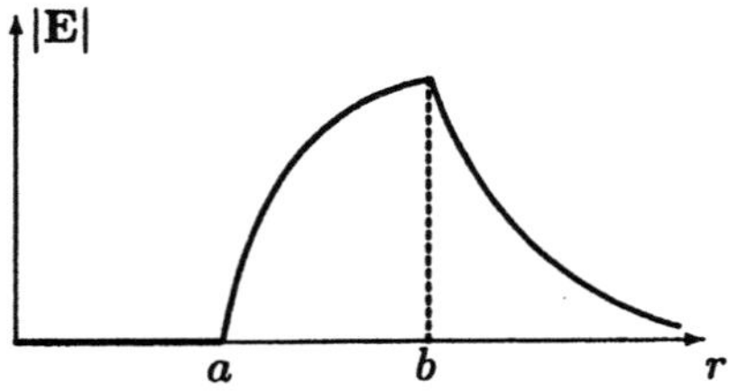

Problem 2.16

(i)

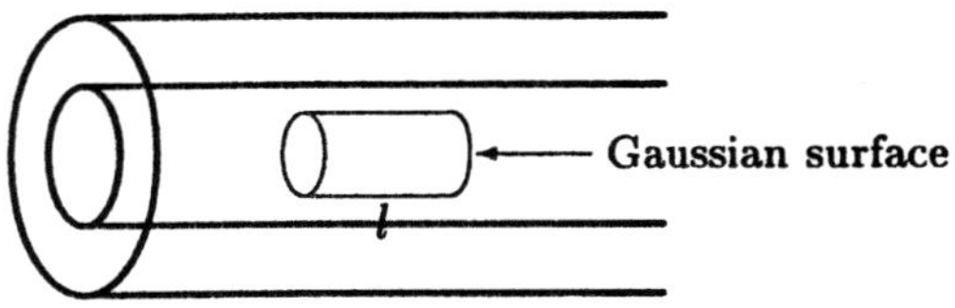

$\oint \mathbf{E} \cdot d\mathbf{a} = E \cdot 2\pi s \cdot l = \frac{1}{\epsilon_0} Q_{\text{enc}} = \frac{1}{\epsilon_0} \rho \pi s^2 l;$

$$\boxed{\mathbf{E} = \frac{\rho s}{2\epsilon_0}\hat{\mathbf{s}}.}$$

(ii)

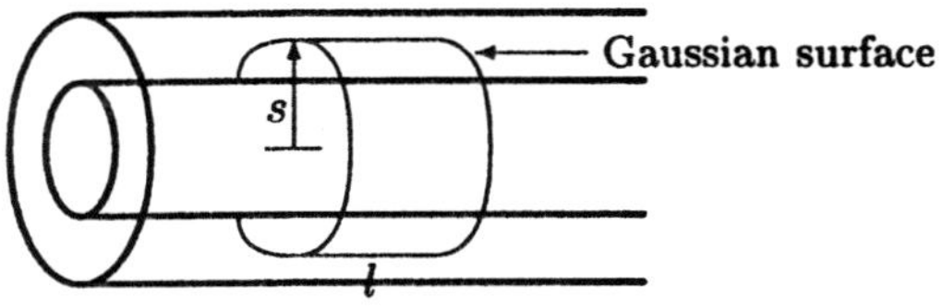

$\oint \mathbf{E} \cdot d\mathbf{a} = E \cdot 2\pi s \cdot l = \frac{1}{\epsilon_0} Q_{\text{enc}} = \frac{1}{\epsilon_0} \rho \pi a^2 l;$

$$\boxed{\mathbf{E} = \frac{\rho a^2}{2\epsilon_0 s}\hat{\mathbf{s}}.}$$

(iii)

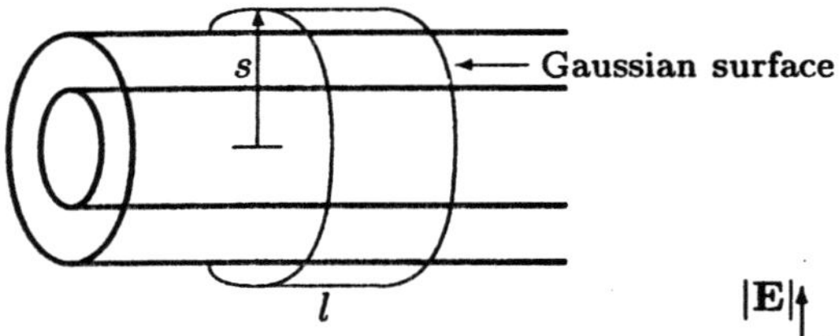

$\oint \mathbf{E} \cdot d\mathbf{a} = E \cdot 2\pi s \cdot l = \frac{1}{\epsilon_0} Q_{\text{enc}} = 0;$

$$\boxed{\mathbf{E} = 0.}$$

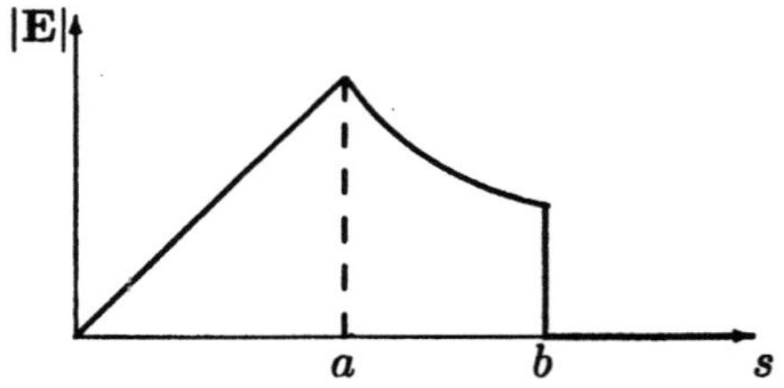

Problem 2.17

On the $x\,z$ plane $E = 0$ by symmetry. Set up a Gaussian "pillbox" with one face in this plane and the other at y.

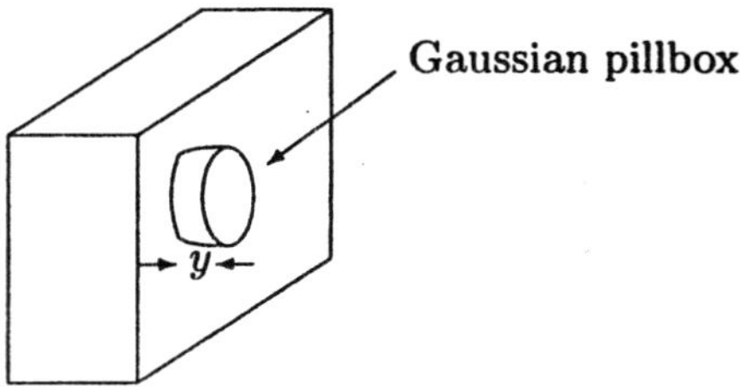

$\int \mathbf{E} \cdot d\mathbf{a} = E \cdot A = \frac{1}{\epsilon_0} Q_{\text{enc}} = \frac{1}{\epsilon_0} A y \rho;$

$$\boxed{\mathbf{E} = \frac{\rho}{\epsilon_0} y \, \hat{\mathbf{y}}} \text{ (for } |y| < d).$$

$Q_{enc} = \frac{1}{\epsilon_0}Ad\rho \Rightarrow \boxed{\mathbf{E} = \frac{\rho}{\epsilon_0}d\,\hat{\mathbf{y}}}$ (for $y > d$).

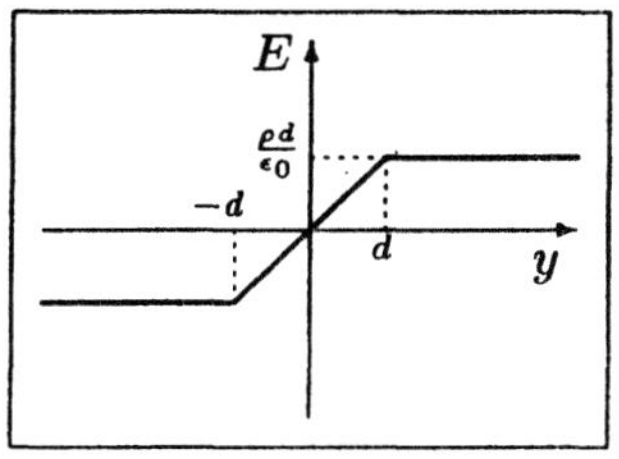

Problem 2.18

From Prob. 2.12, the field inside the positive sphere is $\mathbf{E}_+ = \frac{\rho}{3\epsilon_0}\mathbf{r}_+$, where $\mathbf{r}_+$ is the vector from the positive center to the point in question. Likewise, the field of the negative sphere is $-\frac{\rho}{3\epsilon_0}\mathbf{r}_-$. So the *total* field is

$$\mathbf{E} = \frac{\rho}{3\epsilon_0}(\mathbf{r}_+ - \mathbf{r}_-)$$

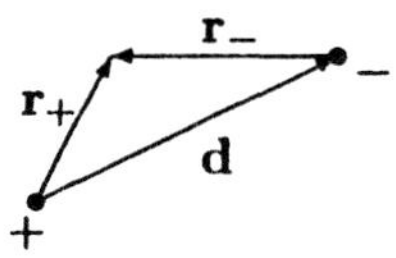

But (see diagram) $\mathbf{r}_+ - \mathbf{r}_- = \mathbf{d}$. So $\boxed{\mathbf{E} = \frac{\rho}{3\epsilon_0}\mathbf{d}.}$

Problem 2.19

$$\nabla\times\mathbf{E} = \frac{1}{4\pi\epsilon_0}\nabla\times\int\frac{\hat{\boldsymbol{\imath}}}{\imath^2}\rho\,d\tau = \frac{1}{4\pi\epsilon_0}\int\left[\nabla\times\left(\frac{\hat{\boldsymbol{\imath}}}{\imath^2}\right)\right]\rho\,d\tau \quad \text{(since } \rho \text{ depends on } \mathbf{r}', \text{ not } \mathbf{r})$$

$$= 0 \quad \left(\text{since } \nabla\times\left(\frac{\hat{\boldsymbol{\imath}}}{\imath^2}\right) = 0, \text{ from Prob. 1.62}\right).$$

Problem 2.20

$$(1)\ \nabla\times\mathbf{E}_1 = k\begin{vmatrix}\hat{\mathbf{x}} & \hat{\mathbf{y}} & \hat{\mathbf{z}}\\ \frac{\partial}{\partial x} & \frac{\partial}{\partial y} & \frac{\partial}{\partial z}\\ xy & 2yz & 3zx\end{vmatrix} = k\left[\hat{\mathbf{x}}(0-2y) + \hat{\mathbf{y}}(0-3z) + \hat{\mathbf{z}}(0-x)\right] \neq 0,$$

so $\mathbf{E}_1$ is an *impossible* electrostatic field.

$$(2)\ \nabla\times\mathbf{E}_2 = k\begin{vmatrix}\hat{\mathbf{x}} & \hat{\mathbf{y}} & \hat{\mathbf{z}}\\ \frac{\partial}{\partial x} & \frac{\partial}{\partial y} & \frac{\partial}{\partial z}\\ y^2 & 2xy+z^2 & 2yz\end{vmatrix} = k\left[\hat{\mathbf{x}}(2z-2z) + \hat{\mathbf{y}}(0-0) + \hat{\mathbf{z}}(2y-2y)\right] = 0,$$

so $\mathbf{E}_2$ is a *possible* electrostatic field.

Let's go by the indicated path:

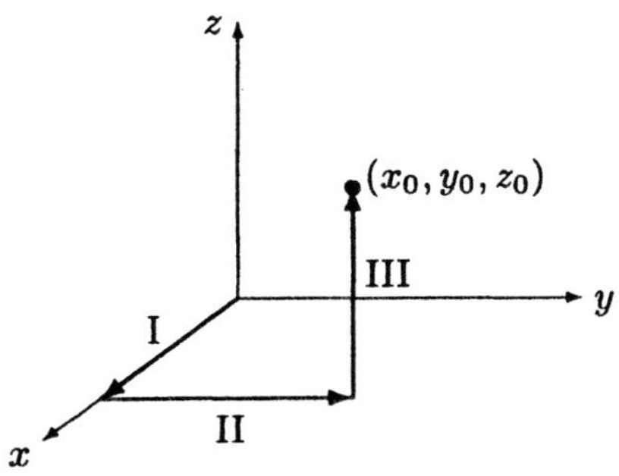

$$\mathbf{E}\cdot d\mathbf{l} = (y^2\,dx + (2xy+z^2)dy + 2yz\,dz)k$$

Step I: $y = z = 0$; $dy = dz = 0$. $\mathbf{E}\cdot d\mathbf{l} = ky^2\,dx = 0$.

Step II: $x = x_0$, $y: 0 \to y_0$, $z = 0$. $dx = dz = 0$.
$\mathbf{E}\cdot d\mathbf{l} = k(2xy+z^2)dy = 2kx_0y\,dy$.
$\int_{II}\mathbf{E}\cdot d\mathbf{l} = 2kx_0\int_0^{y_0}y\,dy = kx_0y_0^2$.

Step III: $x = x_0$, $y = y_0$, $z: 0 \to z_0$; $dx = dy = 0$.

$\mathbf{E}\cdot d\mathbf{l} = 2kyz\,dz = 2ky_0z\,dz.$
$\int_{III}\mathbf{E}\cdot d\mathbf{l} = 2y_0k\int_0^{z_0} z\,dz = ky_0z_0^2.$

$$V(x_0,y_0,z_0) = -\int_0^{(x_0,y_0,z_0)}\mathbf{E}\cdot d\mathbf{l} = -k(x_0y_0^2+y_0z_0^2),\text{ or } \boxed{V(x,y,z) = -k(xy^2+yz^2).}$$

Check: $-\nabla V = k[\frac{\partial}{\partial x}(xy^2+yz^2)\,\hat{\mathbf{x}} + \frac{\partial}{\partial y}(xy^2+yz^2)\,\hat{\mathbf{y}} + \frac{\partial}{\partial z}(xy^2+yz^2)\,\hat{\mathbf{z}}] = k[y^2\,\hat{\mathbf{x}} + (2xy+z^2)\,\hat{\mathbf{y}} + 2yz\,\hat{\mathbf{z}}] = \mathbf{E}.$ ✓

Problem 2.21

$$V(r) = -\int_\infty^r \mathbf{E}\cdot d\mathbf{l}. \quad \begin{cases}\text{Outside the sphere } (r>R): & \mathbf{E} = \frac{1}{4\pi\epsilon_0}\frac{q}{r^2}\hat{\mathbf{r}}.\\ \text{Inside the sphere } (r<R): & \mathbf{E} = \frac{1}{4\pi\epsilon_0}\frac{q}{R^3}r\hat{\mathbf{r}}.\end{cases}$$

So for $r>R$: $V(r) = -\int_\infty^r\left(\frac{1}{4\pi\epsilon_0}\frac{q}{\bar{r}^2}\right)d\bar{r} = \frac{1}{4\pi\epsilon_0}q\left(\frac{1}{\bar{r}}\right)\Big|_\infty^r = \boxed{\frac{q}{4\pi\epsilon_0}\frac{1}{r},}$

and for $r<R$: $V(r) = -\int_\infty^R\left(\frac{1}{4\pi\epsilon_0}\frac{q}{\bar{r}^2}\right)d\bar{r} - \int_R^r\left(\frac{1}{4\pi\epsilon_0}\frac{q}{R^3}\bar{r}\right)d\bar{r} = \frac{q}{4\pi\epsilon_0}\left[\frac{1}{R} - \frac{1}{R^3}\left(\frac{r^2-R^2}{2}\right)\right]$

$$= \boxed{\frac{q}{4\pi\epsilon_0}\frac{1}{2R}\left(3-\frac{r^2}{R^2}\right).}$$

When $r>R$, $\nabla V = \frac{q}{4\pi\epsilon_0}\frac{\partial}{\partial r}\left(\frac{1}{r}\right)\hat{\mathbf{r}} = -\frac{q}{4\pi\epsilon_0}\frac{1}{r^2}\hat{\mathbf{r}}$, so $\mathbf{E} = -\nabla V = \frac{q}{4\pi\epsilon_0}\frac{1}{r^2}\hat{\mathbf{r}}.$ ✓

When $r<R$, $\nabla V = \frac{q}{4\pi\epsilon_0}\frac{1}{2R}\frac{\partial}{\partial r}\left(3-\frac{r^2}{R^2}\right)\hat{\mathbf{r}} = \frac{q}{4\pi\epsilon_0}\frac{1}{2R}\left(-\frac{2r}{R^2}\right)\hat{\mathbf{r}} = -\frac{q}{4\pi\epsilon_0}\frac{r}{R^3}\hat{\mathbf{r}}$; so $\mathbf{E} = -\nabla V = \frac{1}{4\pi\epsilon_0}\frac{q}{R^3}r\hat{\mathbf{r}}.$ ✓

Problem 2.22

$\mathbf{E} = \frac{1}{4\pi\epsilon_0}\frac{2\lambda}{s}\hat{\mathbf{s}}$ (Prob. 2.13). In this case we cannot set the reference point at ∞, since the charge itself extends to ∞. Let's set it at $s=a$. Then

$$V(s) = -\int_a^s\left(\frac{1}{4\pi\epsilon_0}\frac{2\lambda}{\bar{s}}\right)d\bar{s} = \boxed{-\frac{1}{4\pi\epsilon_0}2\lambda\ln\left(\frac{s}{a}\right).}$$

(In this form it is clear why $a=\infty$ would be no good—likewise the other "natural" point, $a=0$.)

$\nabla V = -\frac{1}{4\pi\epsilon_0}2\lambda\frac{\partial}{\partial s}\left(\ln\left(\frac{s}{a}\right)\right)\hat{\mathbf{s}} = -\frac{1}{4\pi\epsilon_0}2\lambda\frac{1}{s}\hat{\mathbf{s}} = -\mathbf{E}.$ ✓

Problem 2.23

$V(0) = -\int_\infty^0\mathbf{E}\cdot d\mathbf{l} = -\int_\infty^b\left(\frac{k}{\epsilon_0}\frac{(b-a)}{r^2}\right)dr - \int_b^a\left(\frac{k}{\epsilon_0}\frac{(r-a)}{r^2}\right)dr - \int_a^0(0)dr = \frac{k}{\epsilon_0}\frac{(b-a)}{b} - \frac{k}{\epsilon_0}\left(\ln\left(\frac{a}{b}\right) + a\left(\frac{1}{a}-\frac{1}{b}\right)\right)$

$$= \frac{k}{\epsilon_0}\left\{1-\frac{a}{b}-\ln\left(\frac{a}{b}\right)-1+\frac{a}{b}\right\} = \boxed{\frac{k}{\epsilon_0}\ln\left(\frac{b}{a}\right).}$$

Problem 2.24

Using Eq. 2.22 and the fields from Prob. 2.16:

$V(b)-V(0) = -\int_0^b\mathbf{E}\cdot d\mathbf{l} = -\int_0^a\mathbf{E}\cdot d\mathbf{l} - \int_a^b\mathbf{E}\cdot d\mathbf{l} = -\frac{\rho}{2\epsilon_0}\int_0^a s\,ds - \frac{\rho a^2}{2\epsilon_0}\int_a^b\frac{1}{s}ds$

$$= -\left(\frac{\rho}{2\epsilon_0}\right)\frac{s^2}{2}\Big|_0^a + \frac{\rho a^2}{2\epsilon_0}\ln s\Big|_a^b = \boxed{-\frac{\rho a^2}{4\epsilon_0}\left(1+2\ln\left(\frac{b}{a}\right)\right).}$$

Problem 2.25

(a) $\boxed{V = \frac{1}{4\pi\epsilon_0}\frac{2q}{\sqrt{z^2+\left(\frac{d}{2}\right)^2}}.}$

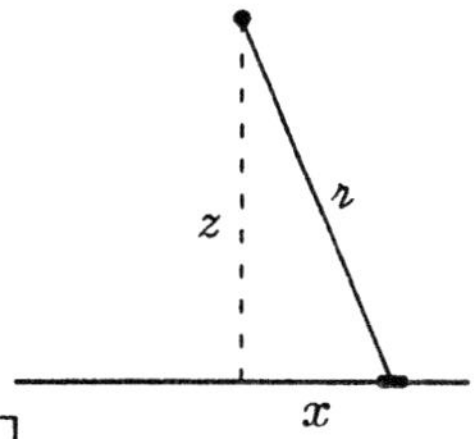

(b) $V = \frac{1}{4\pi\epsilon_0}\int_{-L}^{L}\frac{\lambda\,dx}{\sqrt{z^2+x^2}} = \frac{\lambda}{4\pi\epsilon_0}\ln(x+\sqrt{z^2+x^2})\Big|_{-L}^{L}$

$$= \boxed{\frac{\lambda}{4\pi\epsilon_0}\ln\left[\frac{L+\sqrt{z^2+L^2}}{-L+\sqrt{z^2+L^2}}\right]} = \frac{\lambda}{2\pi\epsilon_0}\ln\left(\frac{L+\sqrt{z^2+L^2}}{z}\right).$$

(c) $V = \frac{1}{4\pi\epsilon_0}\int_0^R\frac{\sigma\,2\pi r\,dr}{\sqrt{r^2+z^2}} = \frac{1}{4\pi\epsilon_0}2\pi\sigma\left(\sqrt{r^2+z^2}\right)\Big|_0^R = \boxed{\frac{\sigma}{2\epsilon_0}\left(\sqrt{R^2+z^2}-z\right).}$

In each case, by symmetry $\frac{\partial V}{\partial y} = \frac{\partial V}{\partial x} = 0$. $\therefore\ \mathbf{E} = -\frac{\partial V}{\partial z}\hat{\mathbf{z}}$.

(a) $\mathbf{E} = -\frac{1}{4\pi\epsilon_0}2q\left(-\frac{1}{2}\right)\frac{2z}{\left(z^2+\left(\frac{d}{2}\right)^2\right)^{3/2}}\hat{\mathbf{z}} = \boxed{\frac{1}{4\pi\epsilon_0}\frac{2qz}{\left(z^2+\left(\frac{d}{2}\right)^2\right)^{3/2}}\hat{\mathbf{z}}}$ (agrees with Prob. 2.2a).

(b) $\mathbf{E} = -\frac{\lambda}{4\pi\epsilon_0}\left\{\frac{1}{(L+\sqrt{z^2+L^2})}\frac{1}{2}\frac{1}{\sqrt{z^2+L^2}}2z - \frac{1}{(-L+\sqrt{z^2+L^2})}\frac{1}{2}\frac{1}{\sqrt{z^2+L^2}}2z\right\}\hat{\mathbf{z}}$

$= -\frac{\lambda}{4\pi\epsilon_0}\frac{z}{\sqrt{z^2+L^2}}\left\{\frac{-L+\sqrt{z^2+L^2}-L-\sqrt{z^2+L^2}}{(z^2+L^2)-L^2}\right\}\hat{\mathbf{z}} = \boxed{\frac{2L\lambda}{4\pi\epsilon_0}\frac{1}{z\sqrt{z^2+L^2}}\hat{\mathbf{z}}}$ (agrees with Ex. 2.1).

(c) $\mathbf{E} = -\frac{\sigma}{2\epsilon_0}\left\{\frac{1}{2}\frac{1}{\sqrt{R^2+z^2}}2z-1\right\}\hat{\mathbf{z}} = \boxed{\frac{\sigma}{2\epsilon_0}\left[1-\frac{z}{\sqrt{R^2+z^2}}\right]\hat{\mathbf{z}}}$ (agrees with Prob. 2.6).

If the right-hand charge in (a) is $-q$, then $\boxed{V=0}$, which, naively, suggests $\mathbf{E} = -\nabla V = 0$, in contradiction with the answer to Prob. 2.2b. The point is that we only know V *on the z axis*, and from this we cannot hope to compute $E_x = -\frac{\partial V}{\partial x}$ or $E_y = -\frac{\partial V}{\partial y}$. That was OK in part (a), because we knew from symmetry that $E_x = E_y = 0$. But now $\mathbf{E}$ points in the x direction, so knowing V on the z axis is insufficient to determine $\mathbf{E}$.

Problem 2.26

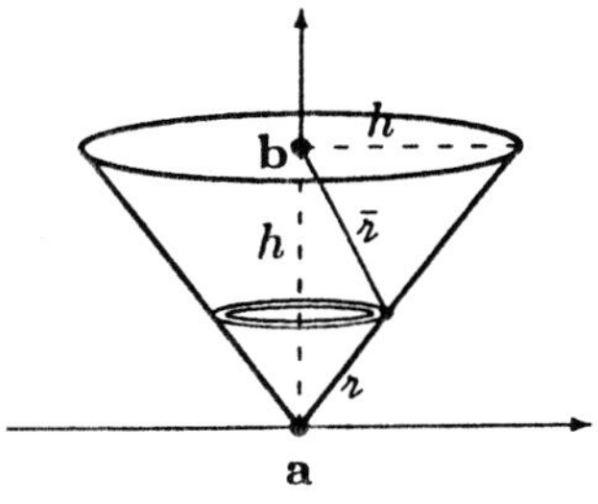

$$V(\mathbf{a}) = \frac{1}{4\pi\epsilon_0}\int_0^{\sqrt{2}h}\left(\frac{\sigma 2\pi r}{\mathscr{r}}\right)d\mathscr{r} = \frac{2\pi\sigma}{4\pi\epsilon_0}\frac{1}{\sqrt{2}}(\sqrt{2}h) = \frac{\sigma h}{2\epsilon_0}.$$

(where $r = \mathscr{r}/\sqrt{2}$)

$$V(\mathbf{b}) = \frac{1}{4\pi\epsilon_0}\int_0^{\sqrt{2}h}\left(\frac{\sigma 2\pi r}{\bar{\mathscr{r}}}\right)d\mathscr{r},\quad \text{where } \bar{\mathscr{r}} = \sqrt{h^2+\mathscr{r}^2-\sqrt{2}h\mathscr{r}}.$$

$$= \frac{2\pi\sigma}{4\pi\epsilon_0}\frac{1}{\sqrt{2}}\int_0^{\sqrt{2}h}\frac{\mathscr{r}}{\sqrt{h^2+\mathscr{r}^2-\sqrt{2}h\mathscr{r}}}d\mathscr{r}$$

$$= \frac{\sigma}{2\sqrt{2}\epsilon_0}\left[\sqrt{h^2+\mathscr{r}^2-\sqrt{2}h\mathscr{r}}+\frac{h}{\sqrt{2}}\ln(2\sqrt{h^2+\mathscr{r}^2-\sqrt{2}h\mathscr{r}}+2\mathscr{r}-\sqrt{2}h)\right]_0^{\sqrt{2}h}$$

$$= \frac{\sigma}{2\sqrt{2}\epsilon_0}\left[h+\frac{h}{\sqrt{2}}\ln(2h+2\sqrt{2}h-\sqrt{2}h)-h-\frac{h}{\sqrt{2}}\ln(2h-\sqrt{2}h)\right] = \frac{\sigma}{2\sqrt{2}\epsilon_0}\frac{h}{\sqrt{2}}\left[\ln(2h+\sqrt{2}h)-\ln(2h-\sqrt{2}h)\right]$$

$$= \frac{\sigma h}{4\epsilon_0}\ln\left(\frac{2+\sqrt{2}}{2-\sqrt{2}}\right) = \frac{\sigma h}{4\epsilon_0}\ln\left(\frac{(2+\sqrt{2})^2}{2}\right) = \frac{\sigma h}{2\epsilon_0}\ln(1+\sqrt{2}).$$

$$\therefore \boxed{V(\mathbf{a})-V(\mathbf{b}) = \frac{\sigma h}{2\epsilon_0}\left[1-\ln(1+\sqrt{2})\right].}$$

Problem 2.27

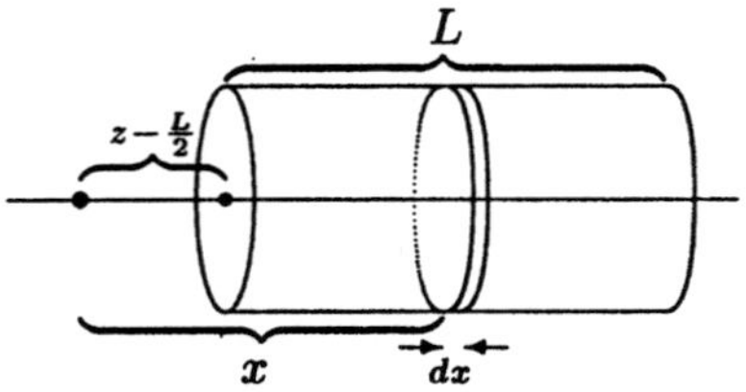

Cut the cylinder into slabs, as shown in the figure, and use result of Prob. 2.25c, with $z \to x$ and $\sigma \to \rho\, dx$:

$$V = \frac{\rho}{2\epsilon_0} \int_{z-L/2}^{z+L/2} \left(\sqrt{R^2+x^2} - x\right) dx$$

$$= \frac{\rho}{2\epsilon_0}\frac{1}{2}\left[x\sqrt{R^2+x^2} + R^2 \ln(x+\sqrt{R^2+x^2}) - x^2\right]\Big|_{z-L/2}^{z+L/2}$$

$$= \boxed{\frac{\rho}{4\epsilon_0}\left\{\left(z+\tfrac{L}{2}\right)\sqrt{R^2+\left(z+\tfrac{L}{2}\right)^2} - \left(z-\tfrac{L}{2}\right)\sqrt{R^2+\left(z-\tfrac{L}{2}\right)^2} + R^2 \ln\left[\frac{z+\frac{L}{2}+\sqrt{R^2+\left(z+\frac{L}{2}\right)^2}}{z-\frac{L}{2}+\sqrt{R^2+\left(z-\frac{L}{2}\right)^2}}\right] - 2zL\right\}.}$$

(*Note:* $-\left(z+\frac{L}{2}\right)^2 + \left(z-\frac{L}{2}\right)^2 = -z^2 - zL - \frac{L^2}{4} + z^2 - zL + \frac{L^2}{4} = -2zL$.)

$$\mathbf{E} = -\nabla V = -\hat{\mathbf{z}}\frac{\partial V}{\partial z} = -\frac{\hat{\mathbf{z}}\rho}{4\epsilon_0}\left\{\sqrt{R^2+\left(z+\frac{L}{2}\right)^2} + \frac{\left(z+\frac{L}{2}\right)^2}{\sqrt{R^2+\left(z+\frac{L}{2}\right)^2}} - \sqrt{R^2+\left(z-\frac{L}{2}\right)^2} - \frac{\left(z-\frac{L}{2}\right)^2}{\sqrt{R^2+\left(z-\frac{L}{2}\right)^2}}\right.$$

$$\left. + R^2 \underbrace{\left[\frac{1+\frac{z+\frac{L}{2}}{\sqrt{R^2+\left(z+\frac{L}{2}\right)^2}}}{z+\frac{L}{2}+\sqrt{R^2+\left(z+\frac{L}{2}\right)^2}} - \frac{1+\frac{z-\frac{L}{2}}{\sqrt{R^2+\left(z-\frac{L}{2}\right)^2}}}{z-\frac{L}{2}+\sqrt{R^2+\left(z-\frac{L}{2}\right)^2}}\right]}_{\frac{1}{\sqrt{R^2+\left(z+\frac{L}{2}\right)^2}} - \frac{1}{\sqrt{R^2+\left(z-\frac{L}{2}\right)^2}}} - 2L\right\}$$

$$\mathbf{E} = -\frac{\hat{\mathbf{z}}\rho}{4\epsilon_0}\left\{2\sqrt{R^2+\left(z+\frac{L}{2}\right)^2} - 2\sqrt{R^2+\left(z-\frac{L}{2}\right)^2} - 2L\right\}$$

$$= \boxed{\frac{\rho}{2\epsilon_0}\left[L - \sqrt{R^2+\left(z+\frac{L}{2}\right)^2} + \sqrt{R^2+\left(z-\frac{L}{2}\right)^2}\right]\hat{\mathbf{z}}.}$$

Problem 2.28

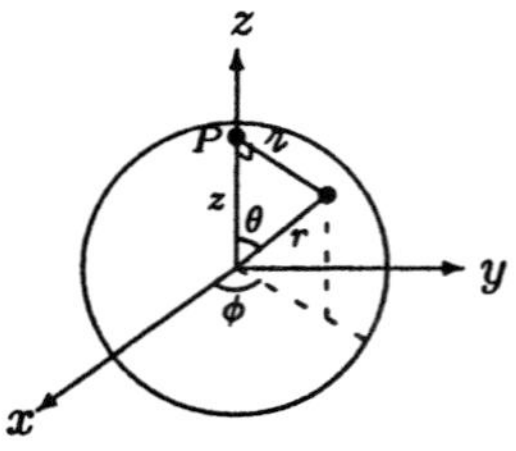

Orient axes so P is on z axis.

$$V = \frac{1}{4\pi\epsilon_0}\int \frac{\rho}{\mathscr{r}} d\tau. \qquad \left\{ \begin{array}{l} \text{Here } \rho \text{ is constant, } d\tau = r^2 \sin\theta\, dr\, d\theta\, d\phi, \\ \mathscr{r} = \sqrt{z^2+r^2-2rz\cos\theta}. \end{array} \right.$$

$$V = \frac{\rho}{4\pi\epsilon_0}\int \frac{r^2 \sin\theta\, dr\, d\theta\, d\phi}{\sqrt{z^2+r^2-2rz\cos\theta}}\;; \quad \int_0^{2\pi} d\phi = 2\pi.$$

$$\int_0^{\pi} \frac{\sin\theta}{\sqrt{z^2+r^2-2rz\cos\theta}}\, d\theta = \frac{1}{rz}\left(\sqrt{r^2+z^2-2rz\cos\theta}\right)\Big|_0^{\pi} = \frac{1}{rz}\left(\sqrt{r^2+z^2+2rz} - \sqrt{r^2+z^2-2rz}\right)$$

$$= \frac{1}{rz}\left(r+z-|r-z|\right) = \left\{ \begin{array}{l} 2/z\ , \text{ if } r < z, \\ 2/r\ , \text{ if } r > z. \end{array} \right\}$$

$$\therefore V = \frac{\rho}{4\pi\epsilon_0} \cdot 2\pi \cdot 2\left\{\int_0^z \frac{1}{z} r^2 dr + \int_z^R \frac{1}{r} r^2 dr\right\} = \frac{\rho}{\epsilon_0}\left\{\frac{1}{z}\frac{z^3}{3} + \frac{R^2 - z^2}{2}\right\} = \frac{\rho}{2\epsilon_0}\left(R^2 - \frac{z^2}{3}\right).$$

But $\rho = \frac{q}{\frac{4}{3}\pi R^3}$, so $V(z) = \frac{1}{2\epsilon_0}\frac{3q}{4\pi R^3}\left(R^2 - \frac{z^2}{3}\right) = \frac{q}{8\pi\epsilon_0 R}\left(3 - \frac{z^2}{R^2}\right)$; $\boxed{V(r) = \frac{q}{8\pi\epsilon_0 R}\left(3 - \frac{r^2}{R^2}\right).}$ ✓

Problem 2.29

$\nabla^2 V = \frac{1}{4\pi\epsilon_0}\nabla^2 \int\left(\frac{\rho}{\imath}\right) d\tau = \frac{1}{4\pi\epsilon_0}\int \rho(\mathbf{r}')\left(\nabla^2 \frac{1}{\imath}\right) d\tau$ (since ρ is a function of $\mathbf{r}'$, not $\mathbf{r}$)

$= \frac{1}{4\pi\epsilon_0}\int \rho(\mathbf{r}')[-4\pi\delta^3(\mathbf{r} - \mathbf{r}')]\, d\tau = -\frac{1}{\epsilon_0}\rho(\mathbf{r})$. ✓

Problem 2.30.

(a) Ex. 2.4: $\mathbf{E}_{\text{above}} = \frac{\sigma}{2\epsilon_0}\hat{\mathbf{n}}$; $\mathbf{E}_{\text{below}} = -\frac{\sigma}{2\epsilon_0}\hat{\mathbf{n}}$ ($\hat{\mathbf{n}}$ always pointing *up*); $\mathbf{E}_{\text{above}} - \mathbf{E}_{\text{below}} = \frac{\sigma}{\epsilon_0}\hat{\mathbf{n}}$. ✓

Ex. 2.5: At each surface, $E = 0$ one side and $E = \frac{\sigma}{\epsilon_0}$ other side, so $\Delta E = \frac{\sigma}{\epsilon_0}$. ✓

Prob. 2.11: $\mathbf{E}_{\text{out}} = \frac{\sigma R^2}{\epsilon_0 r^2}\hat{\mathbf{r}} = \frac{\sigma}{\epsilon_0}\hat{\mathbf{r}}$; $\mathbf{E}_{\text{in}} = 0$; so $\Delta\mathbf{E} = \frac{\sigma}{\epsilon_0}\hat{\mathbf{r}}$. ✓

(b) 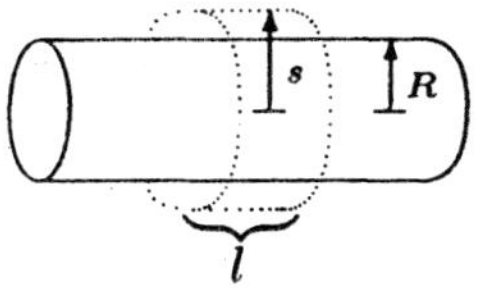

Outside: $\oint \mathbf{E}\cdot d\mathbf{a} = E(2\pi s)l = \frac{1}{\epsilon_0}Q_{\text{enc}} = \frac{1}{\epsilon_0}(2\pi R)l \Rightarrow \mathbf{E} = \frac{\sigma}{\epsilon_0}\frac{R}{s}\hat{\mathbf{s}} = \frac{\sigma}{\epsilon_0}\hat{\mathbf{s}}$ (at surface).

Inside: $Q_{\text{enc}} = 0$, so $\mathbf{E} = 0$. $\therefore \Delta\mathbf{E} = \frac{\sigma}{\epsilon_0}\hat{\mathbf{s}}$. ✓

(c) $V_{\text{out}} = \frac{R^2\sigma}{\epsilon_0 r} = \frac{R\sigma}{\epsilon_0}$ (at surface); $V_{\text{in}} = \frac{R\sigma}{\epsilon_0}$; so $V_{\text{out}} = V_{\text{in}}$. ✓

$\frac{\partial V_{\text{out}}}{\partial r} = -\frac{R^2\sigma}{\epsilon_0 r^2} = -\frac{\sigma}{\epsilon_0}$ (at surface); $\frac{\partial V_{\text{in}}}{\partial r} = 0$; so $\frac{\partial V_{\text{out}}}{\partial r} - \frac{\partial V_{\text{in}}}{\partial r} = -\frac{\sigma}{\epsilon_0}$. ✓

Problem 2.31

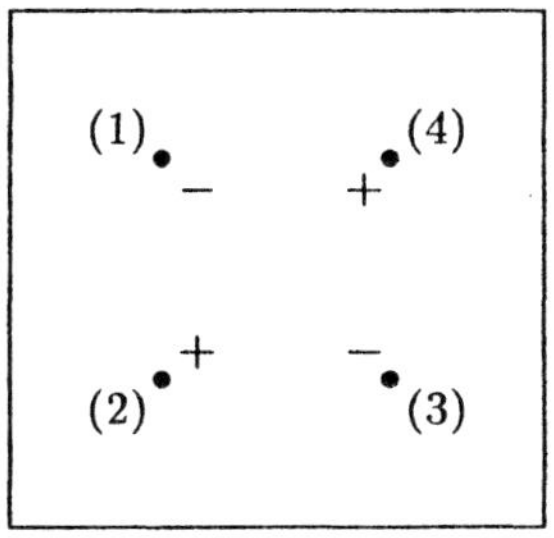

(a) $V = \frac{1}{4\pi\epsilon_0}\sum \frac{q_i}{r_{ij}} = \frac{1}{4\pi\epsilon_0}\left\{\frac{-q}{a} + \frac{q}{\sqrt{2}a} + \frac{-q}{a}\right\} = \frac{q}{4\pi\epsilon_0 a}\left(-2 + \frac{1}{\sqrt{2}}\right)$.

$\therefore W_4 = qV = \boxed{\frac{q^2}{4\pi\epsilon_0 a}\left(-2 + \frac{1}{\sqrt{2}}\right).}$

(b) $W_1 = 0$, $W_2 = \frac{1}{4\pi\epsilon_0}\left(\frac{-q^2}{a}\right)$; $W_3 = \frac{1}{4\pi\epsilon_0}\left(\frac{q^2}{\sqrt{2}\,a} - \frac{q^2}{a}\right)$; $W_4 =$ (see (a)).

$W_{\text{tot}} = \frac{1}{4\pi\epsilon_0}\frac{q^2}{a}\left\{-1 + \frac{1}{\sqrt{2}} - 1 - 2 + \frac{1}{\sqrt{2}}\right\} = \boxed{\frac{1}{4\pi\epsilon_0}\frac{2q^2}{a}\left(-2 + \frac{1}{\sqrt{2}}\right).}$

Problem 2.32

(a) $W = \frac{1}{2}\int \rho V d\tau$. From Prob. 2.21 (or Prob. 2.28): $V = \frac{\rho}{2\epsilon_0}\left(R^2 - \frac{r^2}{3}\right) = \frac{1}{4\pi\epsilon_0}\frac{q}{2R}\left(3 - \frac{r^2}{R^2}\right)$

$$W = \frac{1}{2}\rho\frac{1}{4\pi\epsilon_0}\frac{q}{2R}\int_0^R\left(3 - \frac{r^2}{R^2}\right)4\pi r^2 dr = \frac{q\rho}{4\epsilon_0 R}\left[3\frac{r^3}{3} - \frac{1}{R^2}\frac{r^5}{5}\right]\Bigg|_0^R = \frac{q\rho}{4\epsilon_0 R}\left(R^3 - \frac{R^3}{5}\right)$$

$$= \frac{q\rho}{5\epsilon_0}R^2 = \frac{qR^2}{5\epsilon_0}\frac{q}{\frac{4}{3}\pi R^3} = \boxed{\frac{1}{4\pi\epsilon_0}\left(\frac{3}{5}\frac{q^2}{R}\right).}$$

(b) $W = \frac{\epsilon_0}{2}\int E^2 d\tau$. *Outside* $(r > R)$ $\mathbf{E} = \frac{1}{4\pi\epsilon_0}\frac{q}{r^2}\hat{\mathbf{r}}$; *Inside* $(r < R)$ $\mathbf{E} = \frac{1}{4\pi\epsilon_0}\frac{q}{R^3}r\hat{\mathbf{r}}$.

$$\therefore W = \frac{\epsilon_0}{2}\frac{1}{(4\pi\epsilon_0)^2}q^2\left\{\int_R^\infty \frac{1}{r^4}(r^2 4\pi\, dr) + \int_0^R\left(\frac{r}{R^3}\right)^2(4\pi r^2 dr)\right\}$$

$$= \frac{1}{4\pi\epsilon_0}\frac{q^2}{2}\left\{\left(-\frac{1}{r}\right)\Bigg|_R^\infty + \frac{1}{R^6}\left(\frac{r^5}{5}\right)\Bigg|_0^R\right\} = \frac{1}{4\pi\epsilon_0}\frac{q^2}{2}\left(\frac{1}{R} + \frac{1}{5R}\right) = \frac{1}{4\pi\epsilon_0}\frac{3}{5}\frac{q^2}{R}.\checkmark$$

(c) $W = \frac{\epsilon_0}{2}\left\{\oint_S V\mathbf{E}\cdot d\mathbf{a} + \int_\mathcal{V} E^2 d\tau\right\}$, where $\mathcal{V}$ is large enough to enclose all the charge, but otherwise arbitrary. Let's use a sphere of radius $a > R$. Here $V = \frac{1}{4\pi\epsilon_0}\frac{q}{r}$.

$$W = \frac{\epsilon_0}{2}\left\{\int_{r=a}\left(\frac{1}{4\pi\epsilon_0}\frac{q}{r}\right)\left(\frac{1}{4\pi\epsilon_0}\frac{q}{r^2}\right)r^2\sin\theta\, d\theta\, d\phi + \int_0^R E^2 d\tau + \int_R^a\left(\frac{1}{4\pi\epsilon_0}\frac{q}{r^2}\right)^2(4\pi r^2 dr)\right\}$$

$$= \frac{\epsilon_0}{2}\left\{\frac{q^2}{(4\pi\epsilon_0)^2}\frac{1}{a}4\pi + \frac{q^2}{(4\pi\epsilon_0)^2}\frac{4\pi}{5R} + \frac{1}{(4\pi\epsilon_0)^2}4\pi q^2\left(-\frac{1}{r}\right)\Bigg|_R^a\right\}$$

$$= \frac{1}{4\pi\epsilon_0}\frac{q^2}{2}\left\{\frac{1}{a} + \frac{1}{5R} - \frac{1}{a} + \frac{1}{R}\right\} = \frac{1}{4\pi\epsilon_0}\frac{3}{5}\frac{q^2}{R}.\checkmark$$

As $a \to \infty$, the contribution from the surface integral $\left(\frac{1}{4\pi\epsilon_0}\frac{q^2}{2a}\right)$ goes to zero, while the volume integral $\left(\frac{1}{4\pi\epsilon_0}\frac{q^2}{2a}\left(\frac{6a}{5R} - 1\right)\right)$ picks up the slack.

Problem 2.33

$$dW = d\bar{q}\, V = d\bar{q}\left(\frac{1}{4\pi\epsilon_0}\right)\frac{\bar{q}}{r}, \quad (\bar{q} = \text{charge on sphere of radius } r).$$

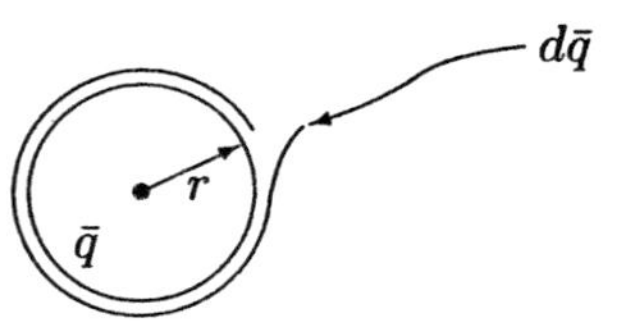

$$\bar{q} = \frac{4}{3}\pi r^3\rho = q\frac{r^3}{R^3} \quad (q = \text{total charge on sphere}).$$

$$d\bar{q} = 4\pi r^2 dr\, \rho = \frac{4\pi r^2}{\frac{4}{3}\pi R^3}q\, dr = \frac{3q}{R^3}r^2 dr.$$

$$dW = \frac{1}{4\pi\epsilon_0}\left(\frac{qr^3}{R^3}\right)\frac{1}{r}\left(\frac{3q}{R^3}r^2 dr\right) = \frac{1}{4\pi\epsilon_0}\frac{3q^2}{R^6}r^4 dr$$

$$W = \frac{1}{4\pi\epsilon_0}\frac{3q^2}{R^6}\int_0^R r^4 dr = \frac{1}{4\pi\epsilon_0}\frac{3q^2}{R^6}\frac{R^5}{5} = \frac{1}{4\pi\epsilon_0}\left(\frac{3}{5}\frac{q^2}{R}\right).\checkmark$$

Problem 2.34

(a) $W = \frac{\epsilon_0}{2}\int E^2\,d\tau$. $\quad \mathbf{E} = \frac{1}{4\pi\epsilon_0}\frac{q}{r^2}$ $(a < r < b)$, zero elsewhere.

$$W = \frac{\epsilon_0}{2}\left(\frac{q}{4\pi\epsilon_0}\right)^2\int_a^b\left(\frac{1}{r^2}\right)^2 4\pi r^2 dr = \frac{q^2}{8\pi\epsilon_0}\int_a^b\frac{1}{r^2} = \boxed{\frac{q^2}{8\pi\epsilon_0}\left(\frac{1}{a}-\frac{1}{b}\right)}.$$

(b) $W_1 = \frac{1}{8\pi\epsilon_0}\frac{q^2}{a}$, $W_2 = \frac{1}{8\pi\epsilon_0}\frac{q^2}{b}$, $\quad \mathbf{E}_1 = \frac{1}{4\pi\epsilon_0}\frac{q}{r^2}\hat{\mathbf{r}}$ $(r > a)$, $\mathbf{E}_2 = \frac{1}{4\pi\epsilon_0}\frac{-q}{r^2}\hat{\mathbf{r}}$ $(r > b)$. So
$\mathbf{E}_1\cdot\mathbf{E}_2 = \left(\frac{1}{4\pi\epsilon_0}\right)^2\frac{-q^2}{r^4}$, $(r > b)$, and hence $\int \mathbf{E}_1\cdot\mathbf{E}_2\,d\tau = -\left(\frac{1}{4\pi\epsilon_0}\right)^2 q^2\int_b^\infty\frac{1}{r^4}4\pi r^2 dr = -\frac{q^2}{4\pi\epsilon_0 b}$.
$W_{\text{tot}} = W_1 + W_2 + \epsilon_0\int \mathbf{E}_1\cdot\mathbf{E}_2\,d\tau = \frac{1}{8\pi\epsilon_0}q^2\left(\frac{1}{a}+\frac{1}{b}-\frac{2}{b}\right) = \frac{q^2}{8\pi\epsilon_0}\left(\frac{1}{a}-\frac{1}{b}\right)$.✓

Problem 2.35

(a) $\boxed{\sigma_R = \frac{q}{4\pi R^2};\ \sigma_a = \frac{-q}{4\pi a^2};\ \sigma_b = \frac{q}{4\pi b^2}.}$

(b) $V(0) = -\int_\infty^0 \mathbf{E}\cdot d\mathbf{l} = -\int_\infty^b\left(\frac{1}{4\pi\epsilon_0}\frac{q}{r^2}\right)dr - \int_b^a(0)dr - \int_a^R\left(\frac{1}{4\pi\epsilon_0}\frac{q}{r^2}\right)dr - \int_R^0(0)dr = \boxed{\frac{1}{4\pi\epsilon_0}\left(\frac{q}{b}+\frac{q}{R}-\frac{q}{a}\right).}$

(c) $\boxed{\sigma_b \to 0}$ (the charge "drains off"); $V(0) = -\int_\infty^a(0)dr - \int_a^R\left(\frac{1}{4\pi\epsilon_0}\frac{q}{r^2}\right)dr - \int_R^0(0)dr = \boxed{\frac{1}{4\pi\epsilon_0}\left(\frac{q}{R}-\frac{q}{a}\right).}$

Problem 2.36

(a) $\boxed{\sigma_a = -\frac{q_a}{4\pi a^2};}\ \boxed{\sigma_b = -\frac{q_b}{4\pi b^2};}\ \boxed{\sigma_R = \frac{q_a+q_b}{4\pi R^2}.}$

(b) $\boxed{\mathbf{E}_{\text{out}} = \frac{1}{4\pi\epsilon_0}\frac{q_a+q_b}{r^2}\hat{\mathbf{r}},}$ where $\mathbf{r}$ = vector from center of large sphere.

(c) $\boxed{\mathbf{E}_a = \frac{1}{4\pi\epsilon_0}\frac{q_a}{r_a^2}\hat{\mathbf{r}}_a,\quad \mathbf{E}_b = \frac{1}{4\pi\epsilon_0}\frac{q_b}{r_b^2}\hat{\mathbf{r}}_b,}$ where $\mathbf{r}_a$ ($\mathbf{r}_b$) is the vector from center of cavity a (b).

(d) $\boxed{\text{Zero.}}$

(e) σ_R changes (but not σ_a or σ_b); $\mathbf{E}_{\text{outside}}$ changes (but not $\mathbf{E}_a$ or $\mathbf{E}_b$); force on q_a and q_b still zero.

Problem 2.37

Between the plates, $E = 0$; outside the plates $E = \sigma/\epsilon_0 = Q/\epsilon_0 A$. So

$$P = \frac{\epsilon_0}{2}E^2 = \frac{\epsilon_0}{2}\frac{Q^2}{\epsilon_0^2 A^2} = \boxed{\frac{Q^2}{2\epsilon_0 A^2}}.$$

Problem 2.38

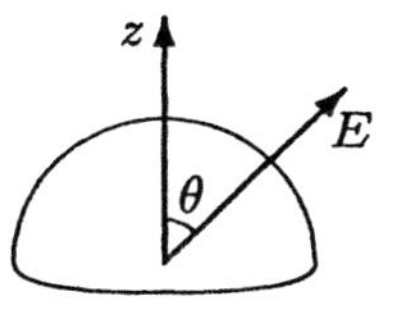

Inside, $\mathbf{E} = 0$; outside, $\mathbf{E} = \frac{1}{4\pi\epsilon_0}\frac{Q}{r^2}\hat{\mathbf{r}}$; so
$\mathbf{E}_{\text{ave}} = \frac{1}{2}\frac{1}{4\pi\epsilon_0}\frac{Q}{R^2}\hat{\mathbf{r}}$; $f_z = \sigma(E_{\text{ave}})_z$; $\sigma = \frac{Q}{4\pi R^2}$.

$F_z = \int f_z da = \int\left(\frac{Q}{4\pi R^2}\right)\frac{1}{2}\left(\frac{1}{4\pi\epsilon_0}\frac{Q}{R^2}\right)\cos\theta\, R^2\sin\theta\,d\theta\,d\phi$

$= \frac{1}{2\epsilon_0}\left(\frac{Q}{4\pi R}\right)^2 2\pi\int_0^{\pi/2}\sin\theta\cos\theta\,d\theta = \frac{1}{\pi\epsilon_0}\left(\frac{Q}{4R}\right)^2\left(\frac{1}{2}\sin^2\theta\right)\Big|_0^{\pi/2} = \frac{1}{2\pi\epsilon_0}\left(\frac{Q}{4R}\right)^2 = \boxed{\frac{Q^2}{32\pi R^2\epsilon_0}}.$

Problem 2.39

Say the charge on the inner cylinder is Q, for a length L. The field is given by Gauss's law: $\oint \mathbf{E}\cdot d\mathbf{a} = E\cdot 2\pi s\cdot L = \frac{1}{\epsilon_0}Q_{\text{enc}} = \frac{1}{\epsilon_0}Q \Rightarrow \mathbf{E} = \frac{Q}{2\pi\epsilon_0 L}\frac{1}{s}\,\hat{\mathbf{s}}$. Potential difference between the cylinders is

$$V(b)-V(a) = -\int_a^b \mathbf{E}\cdot d\mathbf{l} = -\frac{Q}{2\pi\epsilon_0 L}\int_a^b \frac{1}{s}ds = -\frac{Q}{2\pi\epsilon_0 L}\ln\left(\frac{b}{a}\right).$$

As set up here, a is at the higher potential, so $V = V(a)-V(b) = \frac{Q}{2\pi\epsilon_0 L}\ln\left(\frac{b}{a}\right)$.

$C = \frac{Q}{V} = \frac{2\pi\epsilon_0 L}{\ln\left(\frac{b}{a}\right)}$, so capacitance *per unit length* is $\boxed{\dfrac{2\pi\epsilon_0}{\ln\left(\frac{b}{a}\right)}}$.

Problem 2.40

(a) $W =$ (force)×(distance) = (pressure)×(area)×(distance) = $\boxed{\dfrac{\epsilon_0}{2}E^2 A\epsilon.}$

(b) $W =$ (energy per unit volume)×(decrease in volume) = $\left(\epsilon_0\frac{E^2}{2}\right)(A\epsilon)$. Same as (a), confirming that the energy lost is equal to the work done.

Problem 2.41

From Prob. 2.4, the field at height z above the center of a square loop (side a) is

$$\mathbf{E} = \frac{1}{4\pi\epsilon_0}\frac{4\lambda a z}{\left(z^2+\frac{a^2}{4}\right)\sqrt{z^2+\frac{a^2}{2}}}\,\hat{\mathbf{z}}.$$

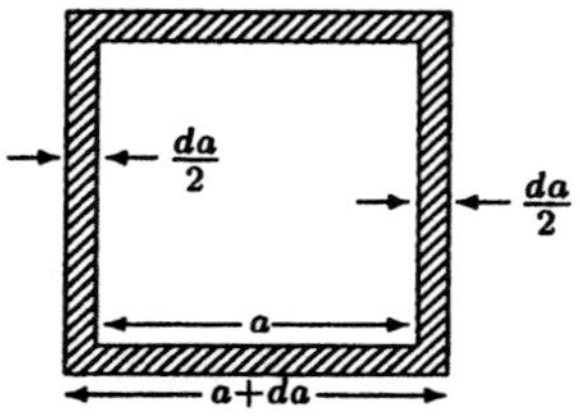

Here $\lambda \to \sigma\frac{da}{2}$ (see figure), and we integrate over a from 0 to $\bar{a}$:

$$E = \frac{1}{4\pi\epsilon_0}2\sigma z\int_0^{\bar{a}}\frac{a\,da}{\left(z^2+\frac{a^2}{4}\right)\sqrt{z^2+\frac{a^2}{2}}}.\quad \text{Let } u=\frac{a^2}{4}, \text{ so } a\,da = 2\,du.$$

$$= \frac{1}{4\pi\epsilon_0}4\sigma z\int_0^{\bar{a}^2/4}\frac{du}{(u+z^2)\sqrt{2u+z^2}} = \frac{\sigma z}{\pi\epsilon_0}\left[\frac{2}{z}\tan^{-1}\left(\frac{\sqrt{2u+z^2}}{z}\right)\right]_0^{\bar{a}^2/4}$$

$$= \frac{2\sigma}{\pi\epsilon_0}\left\{\tan^{-1}\left(\frac{\sqrt{\frac{\bar{a}^2}{2}+z^2}}{z}\right) - \tan^{-1}(1)\right\};$$

$$\boxed{\mathbf{E} = \frac{2\sigma}{\pi\epsilon_0}\left[\tan^{-1}\sqrt{1+\frac{a^2}{2z^2}} - \frac{\pi}{4}\right]\hat{\mathbf{z}}.}$$

$a\to\infty$ (infinite plane): $E = \frac{2\sigma}{\pi\epsilon_0}\left[\tan^{-1}(\infty)-\frac{\pi}{4}\right] = \frac{2\sigma}{\pi\epsilon_0}\left(\frac{\pi}{2}-\frac{\pi}{4}\right) = \frac{\sigma}{2\epsilon_0}$. ✓

$z \gg a$ (point charge): Let $f(x) = \tan^{-1}\sqrt{1+x} - \frac{\pi}{4}$, and expand as a Taylor series:

$$f(x) = f(0) + xf'(0) + \frac{1}{2}x^2 f''(0) + \cdots$$

Here $f(0) = \tan^{-1}(1) - \frac{\pi}{4} = \frac{\pi}{4} - \frac{\pi}{4} = 0$; $f'(x) = \frac{1}{1+(1+x)}\frac{1}{2}\frac{1}{\sqrt{1+x}} = \frac{1}{2(2+x)\sqrt{1+x}}$, so $f'(0) = \frac{1}{4}$, so

$$f(x) = \frac{1}{4}x + (\ )x^2 + (\ )x^3 + \cdots$$

Thus (since $\frac{a^2}{2z^2} = x \ll 1$), $E \approx \frac{2\sigma}{\pi\epsilon_0}\left(\frac{1}{4}\frac{a^2}{2z^2}\right) = \frac{1}{4\pi\epsilon_0}\frac{\sigma a^2}{z^2} = \frac{1}{4\pi\epsilon_0}\frac{q}{z^2}$. ✓

Problem 2.42

$$\rho = \epsilon_0\,\nabla\cdot\mathbf{E} = \epsilon_0\left\{\frac{1}{r^2}\frac{\partial}{\partial r}\left(r^2\frac{A}{r}\right) + \frac{1}{r\sin\theta}\frac{\partial}{\partial\phi}\left(\frac{B\sin\theta\cos\phi}{r}\right)\right\}$$
$$= \epsilon_0\left[\frac{1}{r^2}A + \frac{1}{r\sin\theta}\frac{B\sin\theta}{r}(-\sin\phi)\right] = \boxed{\frac{\epsilon_0}{r^2}(A - B\sin\phi).}$$

Problem 2.43

From Prob. 2.12, the field inside a uniformly charged sphere is: $\mathbf{E} = \frac{1}{4\pi\epsilon_0}\frac{Q}{R^3}\mathbf{r}$. So the force per unit volume is $\mathbf{f} = \rho\mathbf{E} = \left(\frac{Q}{\frac{4}{3}\pi R^3}\right)\left(\frac{Q}{4\pi\epsilon_0 R^3}\right)\mathbf{r} = \frac{3}{\epsilon_0}\left(\frac{Q}{4\pi R^3}\right)^2\mathbf{r}$, and the force in the z direction on $d\tau$ is:

$$dF_z = f_z\,d\tau = \frac{3}{\epsilon_0}\left(\frac{Q}{4\pi R^3}\right)^2 r\cos\theta(r^2\sin\theta\,dr\,d\theta\,d\phi).$$

The total force on the "northern" hemisphere is:

$$F_z = \int f_z\,d\tau = \frac{3}{\epsilon_0}\left(\frac{Q}{4\pi R^3}\right)^2\int_0^R r^3 dr\int_0^{\pi/2}\cos\theta\sin\theta\,d\theta\int_0^{2\pi}d\phi$$
$$= \frac{3}{\epsilon_0}\left(\frac{Q}{4\pi R^3}\right)^2\left(\frac{R^4}{4}\right)\left(\left.\frac{\sin^2\theta}{2}\right|_0^{\pi/2}\right)(2\pi) = \boxed{\frac{3Q^2}{64\pi\epsilon_0 R^2}.}$$

Problem 2.44

$$V_{\text{center}} = \frac{1}{4\pi\epsilon_0}\int\frac{\sigma}{\mathscr{r}}da = \frac{1}{4\pi\epsilon_0}\frac{\sigma}{R}\int da = \frac{1}{4\pi\epsilon_0}\frac{\sigma}{R}(2\pi R^2) = \frac{\sigma R}{2\epsilon_0}$$

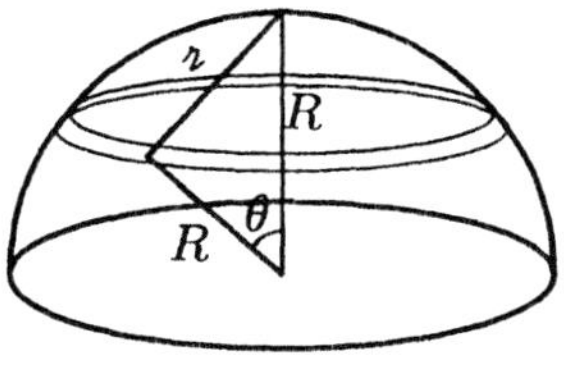

$$V_{\text{pole}} = \frac{1}{4\pi\epsilon_0}\int\frac{\sigma}{\mathscr{r}}da\ ,\ \text{with}\begin{cases}da = 2\pi R^2\sin\theta\,d\theta,\\ \mathscr{r}^2 = R^2 + R^2 - 2R^2\cos\theta = 2R^2(1-\cos\theta).\end{cases}$$
$$= \frac{1}{4\pi\epsilon_0}\frac{\sigma(2\pi R^2)}{R\sqrt{2}}\int_0^{\pi/2}\frac{\sin\theta\,d\theta}{\sqrt{1-\cos\theta}} = \frac{\sigma R}{2\sqrt{2}\epsilon_0}\left.(2\sqrt{1-\cos\theta})\right|_0^{\pi/2}$$
$$= \frac{\sigma R}{\sqrt{2}\epsilon_0}(1-0) = \frac{\sigma R}{\sqrt{2}\epsilon_0}. \qquad \therefore V_{\text{pole}} - V_{\text{center}} = \boxed{\frac{\sigma R}{2\epsilon_0}(\sqrt{2}-1).}$$

Problem 2.45

First let's determine the electric field inside and outside the sphere, using Gauss's law:

$$\epsilon_0\oint\mathbf{E}\cdot d\mathbf{a} = \epsilon_0 4\pi r^2 E = Q_{\text{enc}} = \int\rho\,d\tau = \int(k\bar{r})\bar{r}^2\sin\theta\,d\bar{r}\,d\theta\,d\phi = 4\pi k\int_0^r\bar{r}^3 d\bar{r} = \begin{cases}\pi k r^4 & (r<R),\\ \pi k R^4 & (r>R).\end{cases}$$

So $\mathbf{E} = \frac{k}{4\epsilon_0} r^2 \,\hat{\mathbf{r}}$ $(r < R)$; $\quad \mathbf{E} = \frac{kR^4}{4\epsilon_0 r^2}\,\hat{\mathbf{r}}$ $(r > R)$.

Method I:

$$W = \frac{\epsilon_0}{2}\int E^2 d\tau \text{ (Eq. 2.45)} = \frac{\epsilon_0}{2}\int_0^R \left(\frac{kr^2}{4\epsilon_0}\right)^2 4\pi r^2 dr + \frac{\epsilon_0}{2}\int_R^\infty \left(\frac{kR^4}{4\epsilon_0 r^2}\right)^2 4\pi r^2 dr$$
$$= 4\pi\frac{\epsilon_0}{2}\left(\frac{k}{4\epsilon_0}\right)^2 \left\{\int_0^R r^6 dr + R^8\int_R^\infty \frac{1}{r^2}dr\right\} = \frac{\pi k^2}{8\epsilon_0}\left\{\frac{R^7}{7} + R^8\left(-\frac{1}{r}\right)\Big|_R^\infty\right\} = \frac{\pi k^2}{8\epsilon_0}\left(\frac{R^7}{7}+R^7\right)$$
$$= \boxed{\frac{\pi k^2 R^7}{7\epsilon_0}.}$$

Method II:

$$W = \frac{1}{2}\int \rho V\, d\tau \quad \text{(Eq. 2.43)}.$$

For $r < R$, $V(r) = -\int_\infty^r \mathbf{E}\cdot d\mathbf{l} = -\int_\infty^R \left(\frac{kR^4}{4\epsilon_0 r^2}\right) dr - \int_R^r \left(\frac{kr^2}{4\epsilon_0}\right) dr = -\frac{k}{4\epsilon_0}\left\{R^4\left(-\frac{1}{r}\right)\Big|_\infty^R + \frac{r^3}{3}\Big|_R^r\right\}$

$$= -\frac{k}{4\epsilon_0}\left(-R^3 + \frac{r^3}{3} - \frac{R^3}{3}\right) = \frac{k}{3\epsilon_0}\left(R^3 - \frac{r^3}{4}\right).$$
$$\therefore W = \frac{1}{2}\int_0^R (kr)\left[\frac{k}{3\epsilon_0}\left(R^3 - \frac{r^3}{4}\right)\right] 4\pi r^2 dr = \frac{2\pi k^2}{3\epsilon_0}\int_0^R \left(R^3 r^3 - \frac{1}{4}r^6\right) dr$$
$$= \frac{2\pi k^2}{3\epsilon_0}\left\{R^3\frac{R^4}{4} - \frac{1}{4}\frac{R^7}{7}\right\} = \frac{\pi k^2 R^7}{2\cdot 3\epsilon_0}\left(\frac{6}{7}\right) = \frac{\pi k^2 R^7}{7\epsilon_0}. \checkmark$$

Problem 2.46

$$\mathbf{E} = -\nabla V = -A\frac{\partial}{\partial r}\left(\frac{e^{-\lambda r}}{r}\right)\hat{\mathbf{r}} = -A\left\{\frac{r(-\lambda)e^{-\lambda r} - e^{-\lambda r}}{r^2}\right\}\hat{\mathbf{r}} = \boxed{Ae^{-\lambda r}(1+\lambda r)\frac{\hat{\mathbf{r}}}{r^2}.}$$

$\rho = \epsilon_0\, \nabla\cdot\mathbf{E} = \epsilon_0 A\left\{e^{-\lambda r}(1+\lambda r)\,\nabla\cdot\left(\frac{\hat{\mathbf{r}}}{r^2}\right) + \frac{\hat{\mathbf{r}}}{r^2}\cdot\nabla\left(e^{-\lambda r}(1+\lambda r)\right)\right\}$. But $\nabla\cdot\left(\frac{\hat{\mathbf{r}}}{r^2}\right) = 4\pi\delta^3(\mathbf{r})$ (Eq. 1.99), and $e^{-\lambda r}(1+\lambda r)\delta^3(\mathbf{r}) = \delta^3(\mathbf{r})$ (Eq. 1.88). Meanwhile,

$\nabla\left(e^{-\lambda r}(1+\lambda r)\right) = \hat{\mathbf{r}}\frac{\partial}{\partial r}\left(e^{-\lambda r}(1+\lambda r)\right) = \hat{\mathbf{r}}\left\{-\lambda e^{-\lambda r}(1+\lambda r) + e^{-\lambda r}\lambda\right\} = \hat{\mathbf{r}}(-\lambda^2 r e^{-\lambda r})$.

So $\frac{\hat{\mathbf{r}}}{r^2}\cdot\nabla\left(e^{-\lambda r}(1+\lambda r)\right) = -\frac{\lambda^2}{r}e^{-\lambda r}$, and $\boxed{\rho = \epsilon_0 A\left[4\pi\delta^3(\mathbf{r}) - \frac{\lambda^2}{r}e^{-\lambda r}\right].}$

$$Q = \int \rho\, d\tau = \epsilon_0 A\left\{4\pi\int\delta^3(\mathbf{r})\,d\tau - \lambda^2\int\frac{e^{-\lambda r}}{r}4\pi r^2 dr\right\} = \epsilon_0 A\left(4\pi - \lambda^2 4\pi\int_0^\infty r e^{-\lambda r} dr\right).$$

But $\int_0^\infty r e^{-\lambda r} dr = \frac{1}{\lambda^2}$, so $Q = 4\pi\epsilon_0 A\left(1 - \frac{\lambda^2}{\lambda^2}\right) = \boxed{\text{zero.}}$

Problem 2.47

(a) Potential of $+\lambda$ is $V_+ = -\frac{\lambda}{2\pi\epsilon_0}\ln\left(\frac{s_+}{a}\right)$, where s_+ is distance from λ_+ (Prob. 2.22).
Potential of $-\lambda$ is $V_- = +\frac{\lambda}{2\pi\epsilon_0}\ln\left(\frac{s_-}{a}\right)$, where s_- is distance from λ_-.

$\therefore$ Total $\boxed{V = \frac{\lambda}{2\pi\epsilon_0}\ln\left(\frac{s_-}{s_+}\right).}$

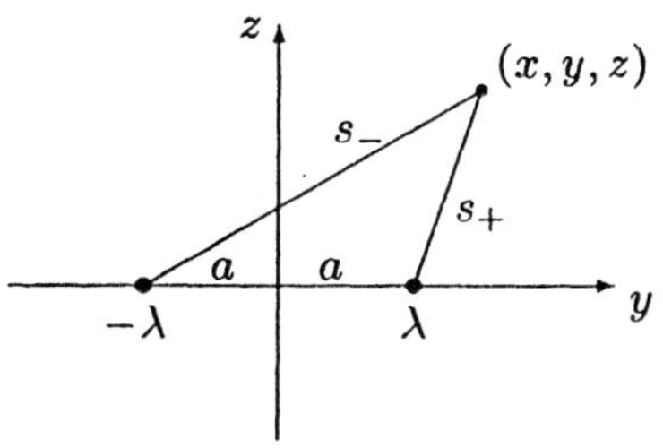

Now $s_+ = \sqrt{(y-a)^2 + z^2}$, and $s_- = \sqrt{(y+a)^2 + z^2}$, so

$V(x,y,z) = \frac{\lambda}{2\pi\epsilon_0}\ln\left(\frac{\sqrt{(y+a)^2+z^2}}{\sqrt{(y-a)^2+z^2}}\right) = \boxed{\frac{\lambda}{4\pi\epsilon_0}\ln\left[\frac{(y+a)^2+z^2}{(y-a)^2+z^2}\right].}$

(b) Equipotentials are given by $\frac{(y+a)^2+z^2}{(y-a)^2+z^2} = e^{(4\pi\epsilon_0 V_0/\lambda)} = k =$ constant. That is:
$y^2 + 2ay + a^2 + z^2 = k(y^2 - 2ay + a^2 + z^2) \Rightarrow y^2(k-1) + z^2(k-1) + a^2(k-1) - 2ay(k+1) = 0$, or
$y^2 + z^2 + a^2 - 2ay\left(\frac{k+1}{k-1}\right) = 0$. The equation for a *circle*, with center at $(y_0, 0)$ and radius R, is
$(y-y_0)^2 + z^2 = R^2$, or $y^2 + z^2 + (y_0^2 - R^2) - 2yy_0 = 0$.
Evidently the equipotentials *are* circles, with $y_0 = a\left(\frac{k+1}{k-1}\right)$ and
$a^2 = y_0^2 - R^2 \Rightarrow R^2 = y_0^2 - a^2 = a^2\left(\frac{k+1}{k-1}\right)^2 - a^2 = a^2\frac{(k^2+2k+1-k^2+2k-1)}{(k-1)^2} = a^2\frac{4k}{(k-1)^2}$, or
$R = \frac{2a\sqrt{k}}{|k-1|}$; or, in terms of V_0:

$$y_0 = a\frac{e^{4\pi\epsilon_0 V_0/\lambda}+1}{e^{4\pi\epsilon_0 V_0/\lambda}-1} = a\frac{e^{2\pi\epsilon_0 V_0/\lambda}+e^{-2\pi\epsilon_0 V_0/\lambda}}{e^{2\pi\epsilon_0 V_0/\lambda}-e^{-2\pi\epsilon_0 V_0/\lambda}} = \boxed{a\coth\left(\frac{2\pi\epsilon_0 V_0}{\lambda}\right).}$$

$$R = 2a\frac{e^{2\pi\epsilon_0 V_0/\lambda}}{e^{4\pi\epsilon_0 V_0/\lambda}-1} = a\frac{2}{(e^{2\pi\epsilon_0 V_0/\lambda}-e^{-2\pi\epsilon_0 V_0/\lambda})} = \frac{a}{\sinh\left(\frac{2\pi\epsilon_0 V_0}{\lambda}\right)} = \boxed{a\,\text{csch}\left(\frac{2\pi\epsilon_0 V_0}{\lambda}\right).}$$

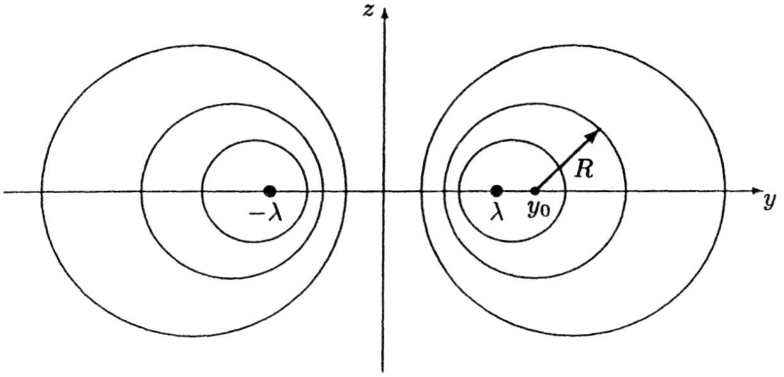

Problem 2.48

(a) $\nabla^2 V = -\frac{\rho}{\epsilon_0}$ (Eq. 2.24), so $\boxed{\frac{d^2V}{dx^2} = -\frac{1}{\epsilon_0}\rho.}$

(b) $qV = \frac{1}{2}mv^2 \to \boxed{v = \sqrt{\frac{2qV}{m}}.}$

(c) $dq = A\rho\,dx$; $\frac{dq}{dt} = a\rho\frac{dx}{dt} = \boxed{A\rho v = I}$ (constant). (*Note*: ρ, hence also I, is *negative*.)

(d) $\frac{d^2V}{dx^2} = -\frac{1}{\epsilon_0}\rho = -\frac{1}{\epsilon_0}\frac{I}{Av} = -\frac{I}{\epsilon_0 A}\sqrt{\frac{m}{2qV}} \Rightarrow \boxed{\frac{d^2V}{dx^2} = \beta V^{-1/2}}$, where $\beta = -\frac{I}{\epsilon_0 A}\sqrt{\frac{m}{2q}}$.

(*Note:* I is *negative*, so β is positive; q is *positive*.)

(e) Multiply by $V' = \frac{dV}{dx}$:

$$V'\frac{dV'}{dx} = \beta V^{-1/2}\frac{dV}{dx} \Rightarrow \int V'\,dV' = \beta\int V^{-1/2}\,dV \Rightarrow \frac{1}{2}V'^2 = 2\beta V^{1/2} + \text{constant}.$$

But $V(0) = V'(0) = 0$ (cathode is at potential zero, and field at cathode is zero), so the constant is zero, and

$$V'^2 = 4\beta V^{1/2} \Rightarrow \frac{dV}{dx} = 2\sqrt{\beta}\,V^{1/4} \Rightarrow V^{-1/4}dV = 2\sqrt{\beta}\,dx;$$

$$\int V^{-1/4}\,dV = 2\sqrt{\beta}\int dx \Rightarrow \frac{4}{3}V^{3/4} = 2\sqrt{\beta}\,x + \text{constant}.$$

But $V(0) = 0$, so this constant is also zero.

$$V^{3/4} = \frac{3}{2}\sqrt{\beta}\,x, \text{ so } V(x) = \left(\frac{3}{2}\sqrt{\beta}\right)^{4/3}x^{4/3}, \text{ or } V(x) = \left(\frac{9}{4}\beta\right)^{2/3}x^{4/3} = \left(\frac{81I^2m}{32\epsilon_0^2A^2q}\right)^{1/3}x^{4/3}.$$

Interms of V_0 (instead of I): $\boxed{V(x) = V_0\left(\frac{x}{d}\right)^{4/3}}$ (see graph).

Without space-charge, V would increase linearly: $V(x) = V_0\left(\frac{x}{d}\right)$.

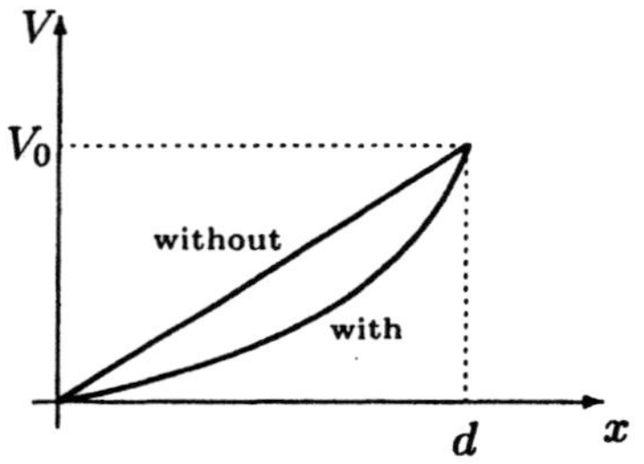

$$\rho = -\epsilon_0\frac{d^2V}{dx^2} = -\epsilon_0 V_0\frac{1}{d^{4/3}}\frac{4}{3}\cdot\frac{1}{3}x^{-2/3} = \boxed{-\frac{4\epsilon_0 V_0}{9(d^2x)^{2/3}}.}$$

$$v = \sqrt{\frac{2q}{m}}\sqrt{V} = \boxed{\sqrt{2qV_0/m}\left(\frac{x}{d}\right)^{2/3}.}$$

(f) $V(d) = V_0 = \left(\frac{81I^2m}{32\epsilon_0^2A^2q}\right)^{1/3}d^{4/3} \Rightarrow V_0^3 = \frac{81md^4}{32\epsilon_0^2A^2q}I^2$; $I^2 = \frac{32\epsilon_0^2A^2q}{81md^4}V_0^3$;

$I = \frac{4\sqrt{2}\,\epsilon_0 A\sqrt{q}}{9\sqrt{m}\,d^2}V_0^{3/2} = KV_0^{3/2}$, where $\boxed{K = \frac{4\epsilon_0 A}{9d^2}\sqrt{\frac{2q}{m}}.}$

Problem 2.49

(a) $\boxed{\mathbf{E} = \frac{1}{4\pi\epsilon_0}\int\frac{\rho\hat{\mathscr{r}}}{\mathscr{r}^2}\left(1+\frac{\mathscr{r}}{\lambda}\right)e^{-\mathscr{r}/\lambda}d\tau.}$

(b) $\boxed{\text{Yes.}}$ The field of a point charge at the origin is radial and symmetric, so $\nabla\times\mathbf{E} = 0$, and hence this is also true (by superposition) for any *collection* of charges.

(c)
$$V = -\int_\infty^r \mathbf{E}\cdot d\mathbf{l} = -\frac{1}{4\pi\epsilon_0}q\int_\infty^r\frac{1}{r^2}\left(1+\frac{r}{\lambda}\right)e^{-r/\lambda}dr$$
$$= \frac{1}{4\pi\epsilon_0}q\int_r^\infty\frac{1}{r^2}\left(1+\frac{r}{\lambda}\right)e^{-r/\lambda}dr = \frac{q}{4\pi\epsilon_0}\left\{\int_r^\infty\frac{1}{r^2}e^{-r/\lambda}dr + \frac{1}{\lambda}\int_r^\infty\frac{1}{r}e^{-r/\lambda}dr\right\}.$$

Now $\int \frac{1}{r^2} e^{-r/\lambda} dr = -\frac{e^{-r/\lambda}}{r} - \frac{1}{\lambda}\int \frac{e^{-r/\lambda}}{r} dr \longleftarrow$ exactly right to kill the last term. Therefore

$$V(r) = \frac{q}{4\pi\epsilon_0}\left\{ -\frac{e^{-r/\lambda}}{r}\Big|_r^\infty \right\} = \boxed{\frac{q}{4\pi\epsilon_0}\frac{e^{-r/\lambda}}{r}.}$$

(d)

$$\oint_{\mathcal{S}} \mathbf{E}\cdot d\mathbf{a} = \frac{1}{\cancel{4\pi}\epsilon_0} q \frac{1}{\cancel{R^2}}\left(1+\frac{R}{\lambda}\right)e^{-R/\lambda}\cancel{4\pi}\cancel{R^2} = \frac{q}{\epsilon_0}\left(1+\frac{R}{\lambda}\right)e^{-R/\lambda}.$$

$$\int_{\mathcal{V}} V\,d\tau = \frac{q}{\cancel{4\pi}\epsilon_0}\int_0^R \frac{e^{-r/\lambda}}{r} r^2 \cancel{4\pi}\,dr = \frac{q}{\epsilon_0}\int_0^R r e^{-r/\lambda} dr = \frac{q}{\epsilon_0}\left[\frac{e^{-r/\lambda}}{(1/\lambda)^2}\left(-\frac{r}{\lambda}-1\right)\right]_0^R$$

$$= \lambda^2 \frac{q}{\epsilon_0}\left\{-e^{-R/\lambda}\left(1+\frac{R}{\lambda}\right)+1\right\}.$$

$$\therefore \oint_{\mathcal{S}} \mathbf{E}\cdot d\mathbf{a} + \frac{1}{\lambda^2}\int_{\mathcal{V}} V\,d\tau = \frac{q}{\epsilon_0}\left\{\left(1+\frac{R}{\lambda}\right)e^{-R/\lambda} - \left(1+\frac{R}{\lambda}\right)e^{-R/\lambda}+1\right\} = \frac{q}{\epsilon_0}. \quad \text{qed}$$

(e) Does the result in (d) hold for a *non*spherical surface? Suppose we make a "dent" in the sphere—pushing a patch (area $R^2 \sin\theta\, d\theta\, d\phi$) from radius R out to radius S (area $S^2 \sin\theta\, d\theta\, d\phi$).

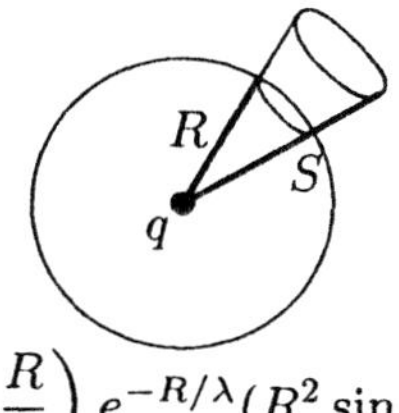

$$\Delta \oint \mathbf{E}\cdot d\mathbf{a} = \frac{q}{4\pi\epsilon_0}\left\{\frac{1}{S^2}\left(1+\frac{S}{\lambda}\right)e^{-S/\lambda}(S^2\sin\theta\,d\theta\,d\phi) - \frac{1}{R^2}\left(1+\frac{R}{\lambda}\right)e^{-R/\lambda}(R^2\sin\theta\,d\theta\,d\phi)\right\}$$

$$= \frac{q}{4\pi\epsilon_0}\left[\left(1+\frac{S}{\lambda}\right)e^{-S/\lambda} - \left(1+\frac{R}{\lambda}\right)e^{-R/\lambda}\right]\sin\theta\,d\theta\,d\phi.$$

$$\Delta\frac{1}{\lambda^2}\int V\,d\tau = \frac{1}{\lambda^2}\frac{q}{4\pi\epsilon_0}\int \frac{e^{-r/\lambda}}{r} r^2 \sin\theta\,, dr\,d\theta\,d\phi = \frac{1}{\lambda^2}\frac{q}{4\pi\epsilon_0}\sin\theta\,d\theta\,d\phi\int_R^S r e^{-r/\lambda} dr$$

$$= -\frac{q}{4\pi\epsilon_0}\sin\theta\,d\theta\,d\phi\left(e^{-r/\lambda}\left(1+\frac{r}{\lambda}\right)\right)\Big|_R^S$$

$$= -\frac{q}{4\pi\epsilon_0}\left[\left(1+\frac{S}{\lambda}\right)e^{-S/\lambda} - \left(1+\frac{R}{\lambda}\right)e^{-R/\lambda}\right]\sin\theta\,d\theta\,d\phi.$$

So the change in $\frac{1}{\lambda^2}\int V\,d\tau$ exactly compensates for the change in $\oint \mathbf{E}\cdot d\mathbf{a}$, and we get $\frac{1}{\epsilon_0}q$ for the total using the dented sphere, just as we did with the perfect sphere. Any closed surface can be built up by successive distortions of the sphere, so the result holds for all shapes. By superposition, if there are many charges inside, the total is $\frac{1}{\epsilon_0}Q_{\text{enc}}$. Charges *outside* do not contribute (in the argument above we found that  for this volume $\oint \mathbf{E}\cdot d\mathbf{a} + \frac{1}{\lambda^2}\int V\,d\tau = 0$—and, again, the sum is not changed by distortions of the surface, as long as q remains outside). So the new "Gauss's Law" holds for *any* charge configuration.

(f) In differential form, "Gauss's law" reads: $\boxed{\nabla\cdot\mathbf{E} + \frac{1}{\lambda^2}V = \frac{1}{\epsilon_0}\rho,}$ or, putting it all in terms of $\mathbf{E}$:

$\nabla\cdot\mathbf{E} - \frac{1}{\lambda^2}\int \mathbf{E}\cdot d\mathbf{l} = \frac{1}{\epsilon_0}\rho$. Since $\mathbf{E} = -\nabla V$, this also yields "Poisson's equation": $-\nabla^2 V + \frac{1}{\lambda^2}V = \frac{1}{\epsilon_0}\rho$.

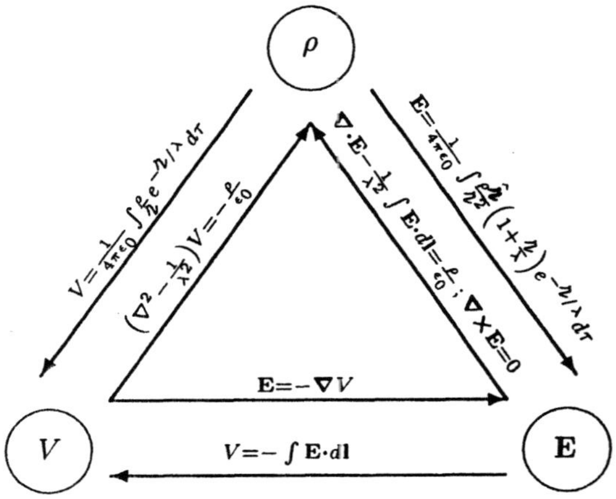

Problem 2.50

$\rho = \epsilon_0 \,\nabla\cdot\mathbf{E} = \epsilon_0 \frac{\partial}{\partial x}(ax) = \boxed{\epsilon_0 a}$ (constant everywhere).

The same charge density would be compatible (as far as Gauss's law is concerned) with $\mathbf{E} = ay\hat{\mathbf{y}}$, for instance, or $\mathbf{E} = (\frac{a}{3})\mathbf{r}$, etc. The point is that Gauss's law (and $\nabla\times\mathbf{E} = 0$) by themselves ***do not determine the field***—like any differential equations, they must be supplemented by appropriate ***boundary conditions.*** Ordinarily, these are so "obvious" that we impose them almost subconsciously ("E must go to zero far from the source charges")—or we appeal to symmetry to resolve the ambiguity ("the field must be the same—in magnitude—on both sides of an infinite plane of surface charge"). But in this case there are ***no*** natural boundary conditions, and no persuasive symmetry conditions, to fix the answer. The question "What is the electric field produced by a uniform charge density filling all of space?" is simply ***ill-posed***: it does not give us sufficient information to determine the answer. (Incidentally, it won't help to appeal to Coulomb's law $\left(\mathbf{E} = \frac{1}{4\pi\epsilon_0}\int \rho \frac{\hat{\boldsymbol{\imath}}}{\imath^2} d\tau\right)$—the integral is hopelessly indefinite, in this case.)

Problem 2.51

Compare Newton's law of universal gravitation to Coulomb's law:

$$\mathbf{F} = -G\frac{m_1 m_2}{r^2}\,\hat{\mathbf{r}}; \qquad \mathbf{F} = \frac{1}{4\pi\epsilon_0}\frac{q_1 q_2}{r^2}\,\hat{\mathbf{r}}.$$

Evidently $\frac{1}{4\pi\epsilon_0} \to G$ and $q \to m$. The gravitational energy of a sphere (translating Prob. 2.32) is therefore

$$\boxed{W_{\text{grav}} = \frac{3}{5}G\frac{M^2}{R}.}$$

Now, $G = 6.67\times10^{-11}$ N m^2/kg^2, and for the sun $M = 1.99\times10^{30}$ kg, $R = 6.96\times10^{8}$ m, so the sun's gravitational energy is $W = 2.28\times10^{41}$ J. At the current rate, this energy would be dissipated in a time

$$t = \frac{W}{P} = \frac{2.28\times10^{41}}{3.86\times10^{26}} = 5.90\times10^{14}\,\text{s} = \boxed{1.87\times10^{7}\text{ years.}}$$

Problem 2.52

First eliminate z, using the formula for the ellipsoid:

$$\sigma(x,y)=\frac{Q}{4\pi ab}\frac{1}{\sqrt{c^2(x^2/a^4)+c^2(y^2/b^4)+1-(x^2/a^2)-(y^2/b^2)}}.$$

Now (for parts (a) and (b)) set $c\to 0$, "squashing" the ellipsoid down to an ellipse in the xy plane:

$$\sigma(x,y)=\frac{Q}{2\pi ab}\frac{1}{\sqrt{1-(x/a)^2-(y/b)^2}}.$$

(I multiplied by 2 to count both surfaces.)

(a) For the circular disk, set $a=b=R$ and let $r\equiv\sqrt{x^2+y^2}$. $\boxed{\sigma(r)=\frac{Q}{2\pi R}\frac{1}{\sqrt{R^2-r^2}}.}$

(b) For the ribbon, let $Q/b\equiv\Lambda$, and then take the limit $b\to\infty$: $\boxed{\sigma(x)=\frac{\Lambda}{2\pi}\frac{1}{\sqrt{a^2-x^2}}.}$

(c) Let $b=c$, $r\equiv\sqrt{y^2+z^2}$, making an ellipsoid of revolution:

$$\frac{x^2}{a^2}+\frac{r^2}{c^2}=1,\quad\text{with } \sigma=\frac{Q}{4\pi ac^2}\frac{1}{\sqrt{x^2/a^4+r^2/c^4}}.$$

The charge on a ring of width dx is

$$dq=\sigma 2\pi r\,ds,\quad\text{where } ds=\sqrt{dx^2+dr^2}=dx\sqrt{1+(dr/dx)^2}.$$

Now $\frac{2x\,dx}{a^2}+\frac{2r\,dr}{c^2}=0\Rightarrow\frac{dr}{dx}=-\frac{c^2x}{a^2r}$, so $ds=dx\sqrt{1+\frac{c^4x^2}{a^4r^2}}=dx\frac{c^2}{r}\sqrt{x^2/a^4+r^2/c^4}$. Thus

$$\lambda(x)=\frac{dq}{dx}=2\pi r\frac{Q}{4\pi ac^2}\frac{1}{\sqrt{x^2/a^4+r^2/c^4}}\frac{c^2}{r}\sqrt{x^2/a^4+r^2/c^4}=\boxed{\frac{Q}{2a}.\quad\textit{(Constant!)}}$$

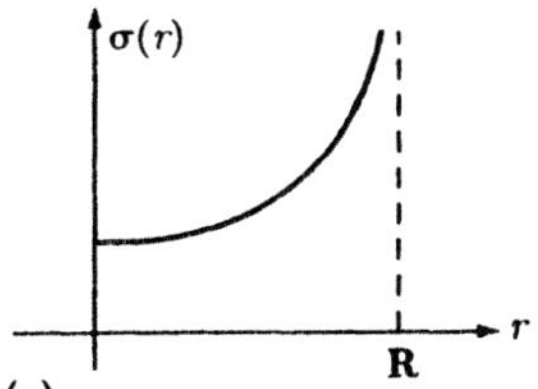

(a)

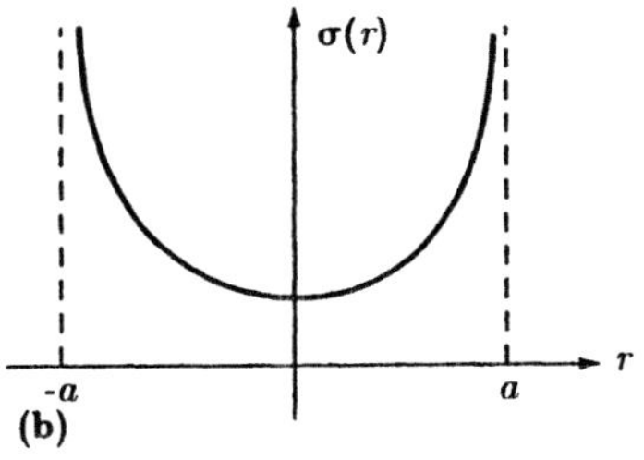

(b)

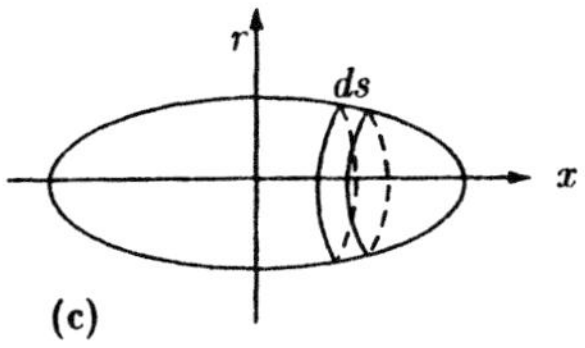

(c)

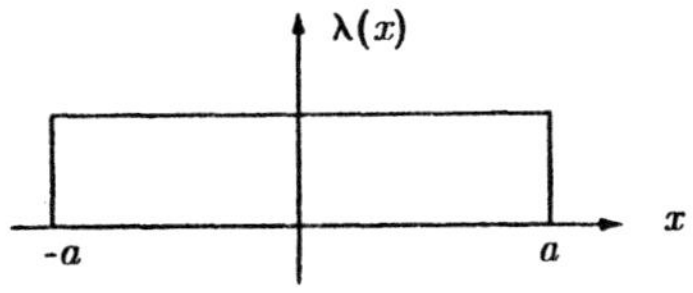

(d)

Chapter 3

Special Techniques

Problem 3.1

The argument is exactly the same as in Sect. 3.1.4, except that since $z < R$, $\sqrt{z^2 + R^2 - 2zR} = (R - z)$, instead of $(z - R)$. Hence $V_{\text{ave}} = \frac{q}{4\pi\epsilon_0}\frac{1}{2zR}[(z+R)-(R-z)] = \boxed{\frac{1}{4\pi\epsilon_0}\frac{q}{R}}.$ If there is more than one charge inside the sphere, the average potential due to interior charges is $\frac{1}{4\pi\epsilon_0}\frac{Q_{\text{enc}}}{R}$, and the average due to exterior charges is V_{center}, so $V_{\text{ave}} = V_{\text{center}} + \frac{Q_{\text{enc}}}{4\pi\epsilon_0 R}$. ✓

Problem 3.2

A stable equilibrium is a point of local minimum in the potential energy. Here the potential energy is qV. But we know that Laplace's equation allows no local minima for V. What *looks* like a minimum, in the figure, must in fact be a saddle point, and the box "leaks" through the center of each face.

Problem 3.3

Laplace's equation in *spherical* coordinates, for V dependent only on r, reads:

$$\nabla^2 V = \frac{1}{r^2}\frac{d}{dr}\left(r^2\frac{dV}{dr}\right) = 0 \Rightarrow r^2\frac{dV}{dr} = c \text{ (constant)} \Rightarrow \frac{dV}{dr} = \frac{c}{r^2} \Rightarrow \boxed{V = -\frac{c}{r} + k.}$$

Example: potential of a uniformly charged sphere.

In *cylindrical* coordinates: $\nabla^2 V = \frac{1}{s}\frac{d}{ds}\left(s\frac{dV}{ds}\right) = 0 \Rightarrow s\frac{dV}{ds} = c \Rightarrow \frac{dV}{ds} = \frac{c}{s} \Rightarrow \boxed{V = c\ln s + k.}$

Example: potential of a long wire.

Problem 3.4

Same as proof of second uniqueness theorem, up to the equation $\oint_S V_3\mathbf{E}_3 \cdot d\mathbf{a} = -\int_{\mathcal{V}}(E_3)^2\,d\tau$. But on each surface, either $V_3 = 0$ (if V is specified on the surface), or else $E_{3\perp} = 0$ (if $\frac{\partial V}{\partial n} = -E_\perp$ is specified). So $\int_{\mathcal{V}}(E_3)^2 = 0$, and hence $\mathbf{E}_2 = \mathbf{E}_1$. qed

Problem 3.5

Putting $U = T = V_3$ into Green's identity:

$\int_{\mathcal{V}}\left[V_3\nabla^2 V_3 + \boldsymbol{\nabla} V_3 \cdot \boldsymbol{\nabla} V_3\right]d\tau = \oint_S V_3\boldsymbol{\nabla} V_3 \cdot d\mathbf{a}$. But $\nabla^2 V_3 = \nabla^2 V_1 - \nabla^2 V_2 = -\frac{\rho}{\epsilon_0} + \frac{\rho}{\epsilon_0} = 0$, and $\boldsymbol{\nabla} V_3 = -\mathbf{E}_3$.

So $\int_{\mathcal{V}} E_3^2\,d\tau = -\oint_S V_2\mathbf{E}_3 \cdot d\mathbf{a}$, and the rest is the same as before.

Problem 3.6

Place image charges $+2q$ at $z = -d$ and $-q$ at $z = -3d$. Total force on $+q$ is

$$\mathbf{F} = \frac{q}{4\pi\epsilon_0}\left[\frac{-2q}{(2d)^2} + \frac{2q}{(4d)^2} + \frac{-q}{(6d)^2}\right]\hat{\mathbf{z}} = \frac{q^2}{4\pi\epsilon_0 d^2}\left(-\frac{1}{2} + \frac{1}{8} - \frac{1}{36}\right)\hat{\mathbf{z}} = \boxed{-\frac{1}{4\pi\epsilon_0}\left(\frac{29q^2}{72d^2}\right)\hat{\mathbf{z}}.}$$

Problem 3.7

(a) From Fig. 3.13: $\mathscr{r} = \sqrt{r^2 + a^2 - 2ra\cos\theta}$; $\mathscr{r}' = \sqrt{r^2 + b^2 - 2rb\cos\theta}$. Therefore:

$$\begin{aligned}
\frac{q'}{\mathscr{r}'} &= -\frac{R}{a}\frac{q}{\sqrt{r^2 + b^2 - 2rb\cos\theta}} \quad \text{(Eq. 3.15), while } b = \frac{R^2}{a} \text{ (Eq. 3.16).} \\
&= -\frac{q}{\left(\frac{a}{R}\right)\sqrt{r^2 + \frac{R^4}{a^2} - 2r\frac{R^2}{a}\cos\theta}} = -\frac{q}{\sqrt{\left(\frac{ar}{R}\right)^2 + R^2 - 2ra\cos\theta}}.
\end{aligned}$$

Therefore:

$$V(r,\theta) = \frac{1}{4\pi\epsilon_0}\left(\frac{q}{\mathscr{r}} + \frac{q'}{\mathscr{r}'}\right) = \boxed{\frac{q}{4\pi\epsilon_0}\left\{\frac{1}{\sqrt{r^2 + a^2 - 2ra\cos\theta}} - \frac{1}{\sqrt{R^2 + (ra/R)^2 - 2ra\cos\theta}}\right\}.}$$

Clearly, when $r = R$, $V \to 0$.

(b) $\sigma = -\epsilon_0\frac{\partial V}{\partial n}$ (Eq. 2.49). In this case, $\frac{\partial V}{\partial n} = \frac{\partial V}{\partial r}$ at the point $r = R$. Therefore,

$$\begin{aligned}
\sigma(\theta) &= -\epsilon_0\left(\frac{q}{4\pi\epsilon_0}\right)\left\{-\frac{1}{2}(r^2 + a^2 - 2ra\cos\theta)^{-3/2}(2r - 2a\cos\theta)\right. \\
&\quad \left.\left. + \frac{1}{2}\left(R^2 + (ra/R)^2 - 2ra\cos\theta\right)^{-3/2}\left(\frac{a^2}{R^2}2r - 2a\cos\theta\right)\right\}\right|_{r=R} \\
&= -\frac{q}{4\pi}\left\{-(R^2 + a^2 - 2Ra\cos\theta)^{-3/2}(R - a\cos\theta) + \left(R^2 + a^2 - 2Ra\cos\theta\right)^{-3/2}\left(\frac{a^2}{R} - a\cos\theta\right)\right\} \\
&= \frac{q}{4\pi}(R^2 + a^2 - 2Ra\cos\theta)^{-3/2}\left[R - a\cos\theta - \frac{a^2}{R} + a\cos\theta\right] \\
&= \boxed{\frac{q}{4\pi R}(R^2 - a^2)(R^2 + a^2 - 2Ra\cos\theta)^{-3/2}.}
\end{aligned}$$

$$\begin{aligned}
q_{\text{induced}} &= \int \sigma\, da = \frac{q}{4\pi R}(R^2 - a^2)\int(R^2 + a^2 - 2Ra\cos\theta)^{-3/2}R^2\sin\theta\, d\theta\, d\phi \\
&= \frac{q}{4\pi R}(R^2 - a^2)2\pi R^2\left[-\frac{1}{Ra}(R^2 + a^2 - 2Ra\cos\theta)^{-1/2}\right]\Big|_0^\pi \\
&= \frac{q}{2a}(a^2 - R^2)\left[\frac{1}{\sqrt{R^2 + a^2 + 2Ra}} - \frac{1}{\sqrt{R^2 + a^2 - 2Ra}}\right].
\end{aligned}$$

But $a > R$ (else q would be *inside*), so $\sqrt{R^2 + a^2 - 2Ra} = a - R$.

$$\begin{aligned}
&= \frac{q}{2a}(a^2 - R^2)\left[\frac{1}{(a+R)} - \frac{1}{(a-R)}\right] = \frac{q}{2a}\left[(a - R) - (a + R)\right] = \frac{q}{2a}(-2R) \\
&= \boxed{-\frac{qR}{a} = q'.}
\end{aligned}$$

(c) The force on q, due to the sphere, is the same as the force of the image charge q', to wit:

$$F = \frac{1}{4\pi\epsilon_0}\frac{qq'}{(a-b)^2} = \frac{1}{4\pi\epsilon_0}\left(-\frac{R}{a}q^2\right)\frac{1}{(a-R^2/a)^2} = -\frac{1}{4\pi\epsilon_0}\frac{q^2Ra}{(a^2-R^2)^2}.$$

To bring q in from infinity to a, then, we do work

$$W = \frac{q^2R}{4\pi\epsilon_0}\int_\infty^a \frac{\bar{a}}{(\bar{a}^2-R^2)^2}\,d\bar{a} = \frac{q^2R}{4\pi\epsilon_0}\left[-\frac{1}{2}\frac{1}{(\bar{a}^2-R^2)}\right]\Bigg|_\infty^a = \boxed{-\frac{1}{4\pi\epsilon_0}\frac{q^2R}{2(a^2-R^2)}.}$$

Problem 3.8

Place a second image charge, q'', at the *center* of the sphere; this will not alter the fact that the sphere is an *equipotential*, but merely *increase* that potential from zero to $V_0 = \dfrac{1}{4\pi\epsilon_0}\dfrac{q''}{R}$;

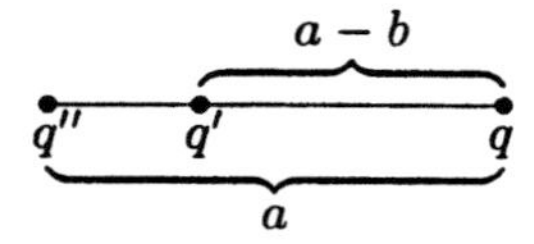

$\boxed{q'' = 4\pi\epsilon_0 V_0 R \text{ at center of sphere.}}$

For a *neutral* sphere, $q' + q'' = 0$.

$$\begin{aligned}
F &= \frac{1}{4\pi\epsilon_0}q\left(\frac{q''}{a^2}+\frac{q'}{(a-b)^2}\right) = \frac{qq'}{4\pi\epsilon_0}\left(-\frac{1}{a^2}+\frac{1}{(a-b)^2}\right)\\
&= \frac{qq'}{4\pi\epsilon_0}\frac{b(2a-b)}{a^2(a-b)^2} = \frac{q(-Rq/a)}{4\pi\epsilon_0}\frac{(R^2/a)(2a-R^2/a)}{a^2(a-R^2/a)^2}\\
&= -\boxed{\frac{q^2}{4\pi\epsilon_0}\left(\frac{R}{a}\right)^3\frac{(2a^2-R^2)}{(a^2-R^2)^2}.}
\end{aligned}$$

(Drop the minus sign, because the problem asks for the force of *attraction*.)

Problem 3.9

(a) Image problem: λ above, $-\lambda$ below. Potential was found in Prob. 2.47:

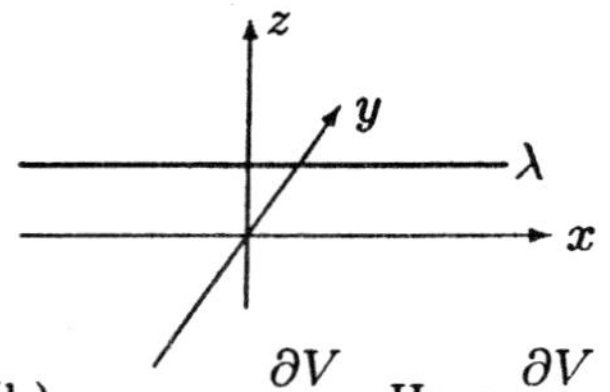

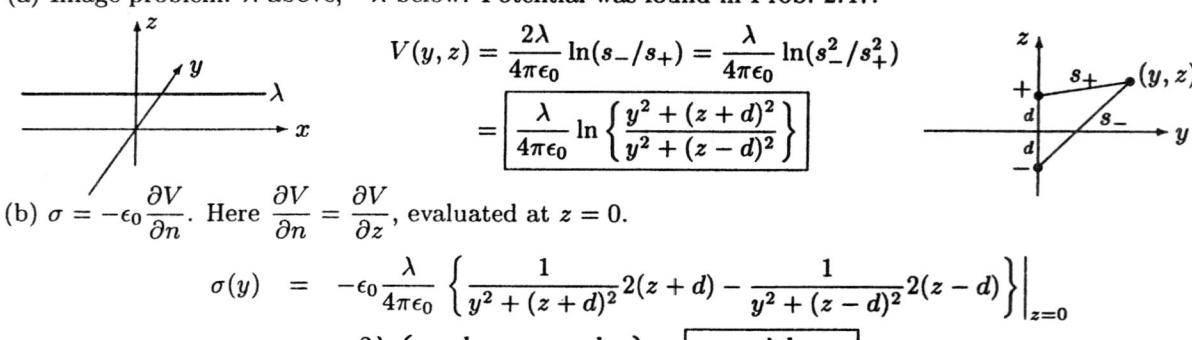

$$V(y,z) = \frac{2\lambda}{4\pi\epsilon_0}\ln(s_-/s_+) = \frac{\lambda}{4\pi\epsilon_0}\ln(s_-^2/s_+^2)$$
$$= \boxed{\frac{\lambda}{4\pi\epsilon_0}\ln\left\{\frac{y^2+(z+d)^2}{y^2+(z-d)^2}\right\}}$$

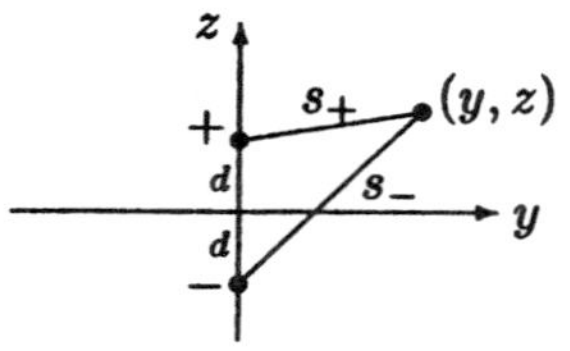

(b) $\sigma = -\epsilon_0\dfrac{\partial V}{\partial n}$. Here $\dfrac{\partial V}{\partial n} = \dfrac{\partial V}{\partial z}$, evaluated at $z=0$.

$$\begin{aligned}
\sigma(y) &= -\epsilon_0\frac{\lambda}{4\pi\epsilon_0}\left\{\frac{1}{y^2+(z+d)^2}2(z+d) - \frac{1}{y^2+(z-d)^2}2(z-d)\right\}\Bigg|_{z=0}\\
&= -\frac{2\lambda}{4\pi}\left\{\frac{d}{y^2+d^2} - \frac{-d}{y^2+d^2}\right\} = \boxed{-\frac{\lambda d}{\pi(y^2+d^2)}.}
\end{aligned}$$

Check: Total charge induced on a strip of width l parallel to the y axis:

$$\begin{aligned}
q_{\text{ind}} &= -\frac{l\lambda d}{\pi}\int_{-\infty}^{\infty}\frac{1}{y^2+d^2}\,dy = -\frac{l\lambda d}{\pi}\left[\frac{1}{d}\tan^{-1}\left(\frac{y}{d}\right)\right]\Bigg|_{-\infty}^{\infty} = -\frac{l\lambda d}{\pi}\left[\frac{\pi}{2}-\left(-\frac{\pi}{2}\right)\right]\\
&= -\lambda l. \quad \text{Therefore } \lambda_{\text{ind}} = -\lambda, \text{ as it should be.}
\end{aligned}$$

Problem 3.10

The image configuration is as shown.

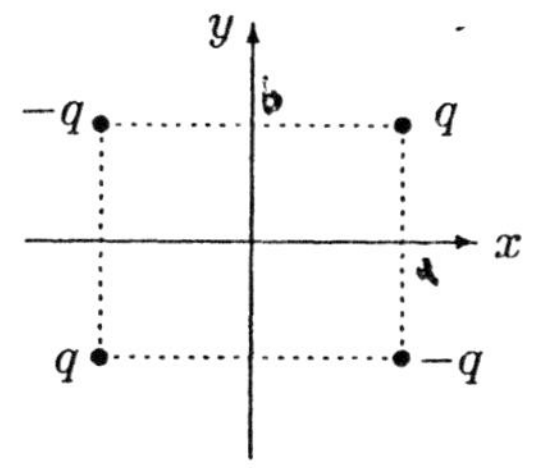

$$\boxed{\begin{aligned} V(x,y) &= \frac{q}{4\pi\epsilon_0}\left\{\frac{1}{\sqrt{(x-a)^2+(y-b)^2+z^2}}+\frac{1}{\sqrt{(x+a)^2+(y+b)^2+z^2}}\right. \\ &\left.-\frac{1}{\sqrt{(x+a)^2+(y-b)^2+z^2}}-\frac{1}{\sqrt{(x-a)^2+(y+b)^2+z^2}}\right\}. \end{aligned}}$$

For this to work, $\boxed{\theta \text{ must be and integer divisor of } 180^\circ.}$ Thus 180°, 90°, 60°, 45°, etc., are OK, but no others. It works for 45°, say, with the charges as shown.

(Note the strategy: to make the x axis an equipotential ($V = 0$), you place the image charge (1) in the reflection point. To make the 45° line an equipotential, you place charge (2) at the image point. But that screws up the x axis, so you must now insert image (3) to balance (2). Moreover, to make the 45° line $V = 0$ you also need (4), to balance (1). But now, to restore the x axis to $V = 0$ you need (5) to balance (4), and so on.

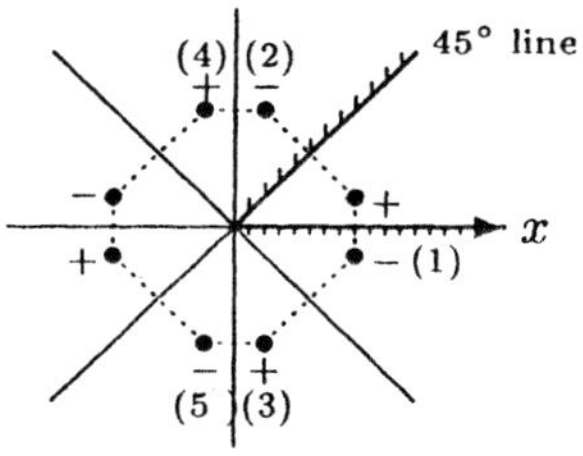

why it works for $\theta = 45°$

The reason this doesn't work for ***arbitrary*** angles is that you are eventually forced to place an image charge ***within the original region of interest***, and that's not allowed—***all*** images must go ***outside*** the region, or you're no longer dealing with the same problem at all.)

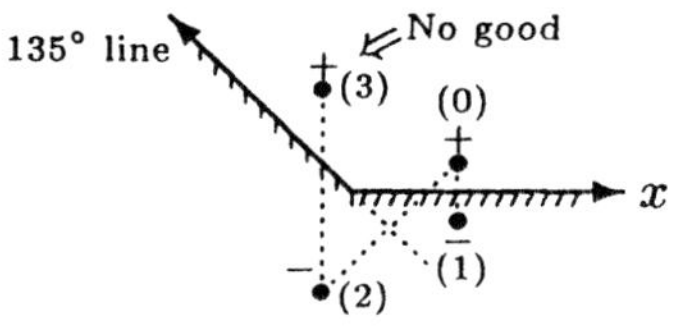

why it *doesn't* work for $\theta = 135°$

Problem 3.11

From Prob. 2.47 (with $y_0 \to d$): $\boxed{V = \frac{\lambda}{4\pi\epsilon_0}\ln\left[\frac{(x+a)^2+y^2}{(x-a)^2+y^2}\right],}$ where $a^2 = {y_0}^2 - R^2 \Rightarrow$ $\boxed{a = \sqrt{d^2-R^2},}$

and

$$\left\{\begin{array}{l} a\coth(2\pi\epsilon_0 V_0/\lambda) = d \\ a\,\mathrm{csch}(2\pi\epsilon_0 V_0/\lambda) = R \end{array}\right\} \Rightarrow \text{(dividing)}\ \frac{d}{R} = \cosh\left(\frac{2\pi\epsilon_0 V_0}{\lambda}\right), \text{ or } \boxed{\lambda = \frac{2\pi\epsilon_0 V_0}{\cosh^{-1}(d/R)}.}$$

Problem 3.12

$$V(x,y) = \sum_{n=1}^{\infty} C_n e^{-n\pi x/a}\sin(n\pi y/a) \quad \text{(Eq. 3.30)}, \quad \text{where } C_n = \frac{2}{a}\int_0^a V_0(y)\sin(n\pi y/a)\,dy \quad \text{(Eq. 3.34)}.$$

In this case $V_0(y) = \left\{\begin{array}{ll} +V_0, & \text{for } 0 < y < a/2 \\ -V_0, & \text{for } a/2 < y < a \end{array}\right\}$. Therefore,

$$\begin{aligned} C_n &= \frac{2}{a}V_0\left\{\int_0^{a/2}\sin(n\pi y/a)\,dy - \int_{a/2}^{a}\sin(n\pi y/a)\,dy\right\} = \frac{2V_0}{a}\left\{-\left.\frac{\cos(n\pi y/a)}{(n\pi/a)}\right|_0^{a/2} + \left.\frac{\cos(n\pi y/a)}{(n\pi/a)}\right|_{a/2}^{a}\right\} \\ &= \frac{2V_0}{n\pi}\left\{-\cos\left(\frac{n\pi}{2}\right) + \cos(0) + \cos(n\pi) - \cos\left(\frac{n\pi}{2}\right)\right\} = \frac{2V_0}{n\pi}\left\{1 + (-1)^n - 2\cos\left(\frac{n\pi}{2}\right)\right\}. \end{aligned}$$

The term in curly brackets is:

$$\left\{\begin{array}{lcl} n=1 & : & 1-1-2\cos(\pi/2)=0, \\ n=2 & : & 1+1-2\cos(\pi)=4, \\ n=3 & : & 1-1-2\cos(3\pi/2)=0, \\ n=4 & : & 1+1-2\cos(2\pi)=0, \end{array}\right\} \text{etc. (Zero if } n \text{ is odd or divisible by 4, otherwise 4.)}$$

Therefore

$$C_n = \begin{cases} 8V_0/n\pi, & n = 2, 6, 10, 14, \text{etc. (in general, } 4j+2, \text{ for } j = 0, 1, 2, \ldots), \\ 0, & \text{otherwise.} \end{cases}$$

So

$$\boxed{V(x,y) = \frac{8V_0}{\pi} \sum_{n=2,6,10,\ldots} \frac{e^{-n\pi x/a}\sin(n\pi y/a)}{n}} = \frac{8V_0}{\pi} \sum_{j=0}^{\infty} \frac{e^{-(4j+2)\pi x/a}\sin[(4j+2)\pi y/a]}{(4j+2)}.$$

Problem 3.13

$$V(x,y) = \frac{4V_0}{\pi} \sum_{n=1,3,5,\ldots} \frac{1}{n} e^{-n\pi x/a}\sin(n\pi y/a) \quad \text{(Eq. 3.36)}; \quad \sigma = -\epsilon_0 \frac{\partial V}{\partial n} \quad \text{(Eq. 2.49)}.$$

So

$$\begin{aligned} \sigma(y) &= -\epsilon_0 \frac{\partial}{\partial x}\left\{\frac{4V_0}{\pi}\sum \frac{1}{n} e^{-n\pi x/a}\sin(n\pi y/a)\right\}\Bigg|_{x=0} = -\epsilon_0 \frac{4V_0}{\pi} \sum \frac{1}{n}\left(-\frac{n\pi}{a}\right) e^{-n\pi x/a}\sin(n\pi y/a)\Bigg|_{x=0} \\ &= \boxed{\frac{4\epsilon_0 V_0}{a} \sum_{n=1,3,5,\ldots} \sin(n\pi y/a).} \end{aligned}$$

Or, using the closed form 3.37:

$$\begin{aligned} V(x,y) &= \frac{2V_0}{\pi}\tan^{-1}\left(\frac{\sin(\pi y/a)}{\sinh(\pi x/a)}\right) \Rightarrow \sigma = -\epsilon_0 \frac{2V_0}{\pi} \frac{1}{1+\frac{\sin^2(\pi y/a)}{\sinh^2(\pi x/a)}} \left(\frac{-\sin(\pi y/a)}{\sinh^2(\pi x/a)}\right) \frac{\pi}{a}\cosh(\pi x/a)\Bigg|_{x=0} \\ &= \frac{2\epsilon_0 V_0}{a} \frac{\sin(\pi y/a)\cosh(\pi x/a)}{\sin^2(\pi y/a) + \sinh^2(\pi x/a)}\Bigg|_{x=0} = \boxed{\frac{2\epsilon_0 V_0}{a}\frac{1}{\sin(\pi y/a)}.} \end{aligned}$$

Summation of series Eq. 3.36

$$V(x,y) = \frac{4V_0}{\pi} I, \text{ where } I \equiv \sum_{n=1,3,5,\ldots} \frac{1}{n} e^{-n\pi x/a}\sin(n\pi y/a).$$

Now $\sin w = \mathcal{I}m\left(e^{iw}\right)$, so

$$I = \mathcal{I}m \sum \frac{1}{n} e^{-n\pi x/a} e^{in\pi y/a} = \mathcal{I}m \sum \frac{1}{n} \mathcal{Z}^n,$$

where $\mathcal{Z} \equiv e^{-\pi(x-iy)/a}$. Now

$$\begin{aligned} \sum_{1,3,5,\ldots} \frac{1}{n}\mathcal{Z}^n &= \sum_{j=0}^{\infty} \frac{1}{(2j+1)} \mathcal{Z}^{(2j+1)} = \int_0^{\mathcal{Z}} \left\{\sum_{j=0}^{\infty} u^{2j}\right\} du \\ &= \int_0^{\mathcal{Z}} \frac{1}{1-u^2}\, du = \frac{1}{2}\ln\left(\frac{1+\mathcal{Z}}{1-\mathcal{Z}}\right) = \frac{1}{2}\ln\left(Re^{i\theta}\right) = \frac{1}{2}\left(\ln R + i\theta\right), \end{aligned}$$

where $Re^{i\theta} = \frac{1+\mathcal{Z}}{1-\mathcal{Z}}$. Therefore

$$\begin{aligned} I &= \mathcal{I}m\left\{\frac{1}{2}(\ln R + i\,\theta)\right\} = \frac{1}{2}\theta. \quad \text{But } \frac{1+\mathcal{Z}}{1-\mathcal{Z}} = \frac{1+e^{-\pi(x-iy)/a}}{1-e^{-\pi(x-iy)/a}} = \frac{\left(1+e^{-\pi(x-iy)/a}\right)\left(1-e^{-\pi(x+iy)/a}\right)}{\left(1-e^{-\pi(x-iy)/a}\right)\left(1-e^{-\pi(x+iy)/a}\right)} \\ &= \frac{1+e^{-\pi x/a}\left(e^{i\pi y/a}-e^{-i\pi y/a}\right)-e^{-2\pi x/a}}{\left|1-e^{-\pi(x-iy)/a}\right|^2} = \frac{1+2ie^{-\pi x/a}\sin(\pi y/a)-e^{-2\pi x/a}}{\left|1-e^{-\pi(x-iy)/a}\right|^2}, \end{aligned}$$

so

$$\tan\theta = \frac{2e^{-\pi x/a}\sin(\pi y/a)}{1-e^{-2\pi x/a}} = \frac{2\sin(\pi y/a)}{e^{\pi x/a}-e^{-\pi x/a}} = \frac{\sin(\pi y/a)}{\sinh(\pi x/a)}.$$

Therefore

$$I = \frac{1}{2}\tan^{-1}\left(\frac{\sin(\pi y/a)}{\sinh(\pi x/a)}\right), \text{ and } \boxed{V(x,y) = \frac{2V_0}{\pi}\tan^{-1}\left(\frac{\sin(\pi y/a)}{\sinh(\pi x/a)}\right).}$$

Problem 3.14

(a) $\frac{\partial^2 V}{\partial x^2} + \frac{\partial^2 V}{\partial y^2} = 0$, with boundary conditions

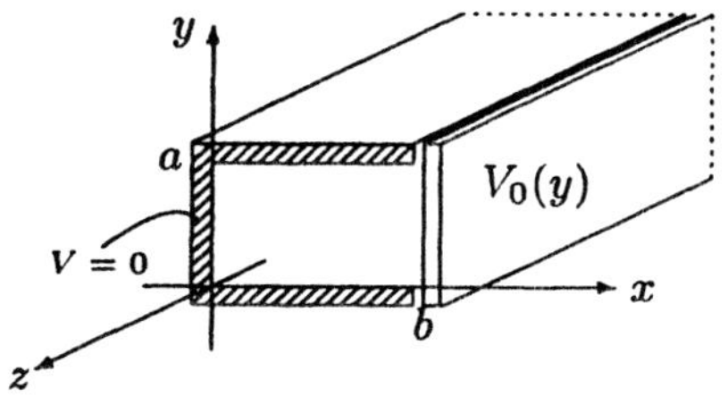

$$\left\{\begin{array}{ll} \text{(i)} & V(x,0) = 0, \\ \text{(ii)} & V(x,a) = 0, \\ \text{(iii)} & V(0,y) = 0, \\ \text{(iv)} & V(b,y) = V_0(y). \end{array}\right\}$$

As in Ex. 3.4, separation of variables yields

$$V(x,y) = \left(Ae^{kx} + Be^{-kx}\right)(C\sin ky + D\cos ky).$$

Here (i)$\Rightarrow$ $D = 0$, (iii)$\Rightarrow$ $B = -A$, (ii)$\Rightarrow$ ka is an integer multiple of π:

$$V(x,y) = AC\left(e^{n\pi x/a} - e^{-n\pi x/a}\right)\sin(n\pi y/a) = (2AC)\sinh(n\pi x/a)\sin(n\pi y/a).$$

But $(2AC)$ is a constant, and the most general linear combination of separable solutions consistent with (i), (ii), (iii) is

$$\boxed{V(x,y) = \sum_{n=1}^{\infty} C_n \sinh(n\pi x/a)\sin(n\pi y/a).}$$

It remains to determine the coefficients C_n so as to fit boundary condition (iv):

$$\sum C_n \sinh(n\pi b/a)\sin(n\pi y/a) = V_0(y). \text{ Fourier's trick} \Rightarrow C_n \sinh(n\pi b/a) = \frac{2}{a}\int_0^a V_0(y)\sin(n\pi y/a)\,dy.$$

Therefore

$$\boxed{C_n = \frac{2}{a\sinh(n\pi b/a)}\int_0^a V_0(y)\sin(n\pi y/a)\,dy.}$$

(b) $$C_n = \frac{2}{a\sinh(n\pi b/a)} V_0 \int_0^a \sin(n\pi y/a)\,dy = \frac{2V_0}{a\sinh(n\pi b/a)} \times \left\{ \begin{array}{ll} 0, & \text{if } n \text{ is even,} \\ \frac{2a}{n\pi}, & \text{if } n \text{ is odd.} \end{array} \right\}$$

$$\boxed{V(x,y) = \frac{4V_0}{\pi} \sum_{n=1,3,5,\ldots} \frac{\sinh(n\pi x/a)\sin(n\pi y/a)}{n\sinh(n\pi b/a)}.}$$

Problem 3.15

Same format as Ex. 3.5, only the boundary conditions are:

$$\left\{ \begin{array}{llll} \text{(i)} & V=0 & \text{when} & x=0, \\ \text{(ii)} & V=0 & \text{when} & x=a, \\ \text{(iii)} & V=0 & \text{when} & y=0, \\ \text{(iv)} & V=0 & \text{when} & y=a, \\ \text{(v)} & V=0 & \text{when} & z=0, \\ \text{(vi)} & V=V_0 & \text{when} & z=a. \end{array} \right\}$$

This time we want sinusoidal functions in x and y, exponential in z:

$$X(x) = A\sin(kx) + B\cos(kx), \quad Y(y) = C\sin(ly) + D\cos(ly), \quad Z(z) = Ee^{\sqrt{k^2+l^2}z} + Ge^{-\sqrt{k^2+l^2}z}.$$

(i)$\Rightarrow B=0$; (ii)$\Rightarrow k=n\pi/a$; (iii)$\Rightarrow D=0$; (iv)$\Rightarrow l=m\pi/a$; (v)$\Rightarrow E+G=0$. Therefore

$$Z(z) = 2E\sinh(\pi\sqrt{n^2+m^2}z/a).$$

Putting this all together, and combining the constants, we have:

$$V(x,y,z) = \sum_{n=1}^{\infty}\sum_{m=1}^{\infty} C_{n,m}\sin(n\pi x/a)\sin(m\pi y/a)\sinh(\pi\sqrt{n^2+m^2}z/a).$$

It remains to evaluate the constants $C_{n,m}$, by imposing boundary condition (vi):

$$V_0 = \sum\sum\left[C_{n,m}\sinh(\pi\sqrt{n^2+m^2})\right]\sin(n\pi x/a)\sin(m\pi y/a).$$

According to Eqs. 3.50 and 3.51:

$$C_{n,m}\sinh\left(\pi\sqrt{n^2+m^2}\right) = \left(\frac{2}{a}\right)^2 V_0 \int_0^a\int_0^a \sin(n\pi x/a)\sin(m\pi y/a)\,dx\,dy = \left\{ \begin{array}{ll} 0, & \text{if } n \text{ or } m \text{ is even,} \\ \frac{16V_0}{\pi^2 nm}, & \text{if both are odd.} \end{array} \right\}$$

Therefore

$$\boxed{V(x,y,z) = \frac{16V_0}{\pi^2}\sum_{n=1,3,5,\ldots}\sum_{m=1,3,5,\ldots}\frac{1}{nm}\sin(n\pi x/a)\sin(m\pi y/a)\frac{\sinh\left(\pi\sqrt{n^2+m^2}z/a\right)}{\sinh\left(\pi\sqrt{n^2+m^2}\right)}.}$$

Problem 3.16

$$\begin{aligned}
P_3(x) &= \frac{1}{8\cdot 6}\frac{d^3}{dx^3}\left(x^2-1\right)^3 = \frac{1}{48}\frac{d^2}{dx^2}3\left(x^2-1\right)^2 2x = \frac{1}{8}\frac{d^2}{dx^2}x\left(x^2-1\right)^2\\
&= \frac{1}{8}\frac{d}{dx}\left[\left(x^2-1\right)^2 + 2x\left(x^2-1\right)2x\right] = \frac{1}{8}\frac{d}{dx}\left[\left(x^2-1\right)\left(x^2-1+4x^2\right)\right]\\
&= \frac{1}{8}\frac{d}{dx}\left[\left(x^2-1\right)\left(5x^2-1\right)\right] = \frac{1}{8}\left[2x\left(5x^2-1\right)+\left(x^2-1\right)10x\right]\\
&= \frac{1}{4}\left(5x^3-x+5x^3-5x\right) = \frac{1}{4}\left(10x^3-6x\right) = \boxed{\frac{5}{2}x^3-\frac{3}{2}x.}
\end{aligned}$$

We need to show that $P_3(\cos\theta)$ satisfies

$$\frac{1}{\sin\theta}\frac{d}{d\theta}\left(\sin\theta\frac{dP}{d\theta}\right) = -l(l+1)P, \text{ with } l=3,$$

where $P_3(\cos\theta) = \frac{1}{2}\cos\theta\left(5\cos^2\theta-3\right)$.

$$\begin{aligned}
\frac{dP_3}{d\theta} &= \frac{1}{2}\left[-\sin\theta\left(5\cos^2\theta-3\right)+\cos\theta(10\cos\theta(-\sin\theta)\right] = -\frac{1}{2}\sin\theta\left(5\cos^2\theta-3+10\cos^2\theta\right)\\
&= -\frac{3}{2}\sin\theta\left(5\cos^2\theta-1\right).
\end{aligned}$$

$$\begin{aligned}
\frac{\partial}{\partial\theta}\left(\sin\theta\frac{dP_3}{d\theta}\right) &= -\frac{3}{2}\frac{d}{d\theta}\left[\sin^2\theta\left(5\cos^2\theta-1\right)\right] = -\frac{3}{2}\left[2\sin\theta\cos\theta\left(5\cos^2\theta-1\right)+\sin^2\theta\left(-10\cos\theta\sin\theta\right)\right]\\
&= -3\sin\theta\cos\theta\left[5\cos^2\theta-1-5\sin^2\theta\right].
\end{aligned}$$

$$\begin{aligned}
\frac{1}{\sin\theta}\frac{d}{d\theta}\left(\sin\theta\frac{dP}{d\theta}\right) &= -3\cos\theta\left[5\cos^2-1-5\left(1-\cos^2\theta\right)\right] = -3\cos\theta\left(10\cos^2\theta-6\right)\\
&= -3\cdot 4\cdot\frac{1}{2}\cos\theta\left(5\cos^2\theta-3\right) = -l(l+1)P_3. \quad \text{qed}
\end{aligned}$$

$$\int_{-1}^{1} P_1(x)P_3(x)\,dx = \int_{-1}^{1}(x)\frac{1}{2}\left(5x^3-3x\right)\,dx = \frac{1}{2}\left(x^5-x^3\right)\Big|_{-1}^{1} = \frac{1}{2}(1-1+1-1) = 0. \checkmark$$

Problem 3.17

(a) *Inside:* $V(r,\theta) = \sum_{l=0}^{\infty} A_l r^l P_l(\cos\theta)$ (Eq. 3.66) where

$$A_l = \frac{(2l+1)}{2R^l}\int_0^\pi V_0(\theta)P_l(\cos\theta)\sin\theta\,d\theta \quad \text{(Eq. 3.69)}.$$

In this case $V_0(\theta) = V_0$ comes outside the integral, so

$$A_l = \frac{(2l+1)V_0}{2R^l}\int_0^\pi P_l(\cos\theta)\sin\theta\,d\theta.$$

But $P_0(\cos\theta) = 1$, so the integral can be written

$$\int_0^\pi P_0(\cos\theta)P_l(\cos\theta)\sin\theta\,d\theta = \left\{\begin{array}{ll} 0, & \text{if } l \neq 0 \\ 2, & \text{if } l = 0 \end{array}\right\} \quad \text{(Eq. 3.68)}.$$

Therefore

$$A_l = \left\{\begin{array}{ll} 0, & \text{if } l \neq 0 \\ V_0, & \text{if } l = 0 \end{array}\right\}.$$

Plugging this into the general form:

$$V(r,\theta) = A_0\, r^0 P_0(\cos\theta) = \boxed{V_0.}$$

The potential is *constant throughout the sphere.*

Outside: $V(r,\theta) = \sum_{l=0}^{\infty} \frac{B_l}{r^{l+1}} P_l(\cos\theta)$ (Eq. 3.72), where

$$\begin{aligned} B_l &= \frac{(2l+1)}{2}R^{l+1}\int_0^\pi V_0(\theta)P_l(\cos\theta)\sin\theta\,d\theta \quad \text{(Eq. 3.73)}. \\ &= \frac{(2l+1)}{2}R^{l+1}V_0\int_0^\pi P_l(\cos\theta)\sin\theta\,d\theta = \left\{\begin{array}{ll} 0, & \text{if } l \neq 0 \\ RV_0, & \text{if } l = 0 \end{array}\right\}. \end{aligned}$$

Therefore $\boxed{V(r,\theta) = V_0\frac{R}{r}}$ (i.e. equals V_0 at $r = R$, then falls off like $\frac{1}{r}$).

(b)

$$V(r,\theta) = \left\{\begin{array}{lll} \sum_{l=0}^{\infty} A_l r^l P_l(\cos\theta), & \text{for } r \leq R & \text{(Eq. 3.78)} \\ \sum_{l=0}^{\infty} \frac{B_l}{r^{l+1}} P_l(\cos\theta); & \text{for } r \geq R & \text{(Eq. 3.79)} \end{array}\right\},$$

where

$$B_l = R^{2l+1}A_l \quad \text{(Eq. 3.81)}$$

and

$$\begin{aligned} A_l &= \frac{1}{2\epsilon_0 R^{l-1}}\int_0^\pi \sigma_0(\theta)P_l(\cos\theta)\sin\theta\,d\theta \quad \text{(Eq. 3.84)} \\ &= \frac{1}{2\epsilon_0 R^{l-1}}\sigma_0\int_0^\pi P_l(\cos\theta)\sin\theta\,d\theta = \left\{\begin{array}{ll} 0, & \text{if } l \neq 0 \\ R\sigma_0/\epsilon_0, & \text{if } l = 0 \end{array}\right\}. \end{aligned}$$

Therefore

$$\boxed{V(r,\theta) = \left\{\begin{array}{ll} \dfrac{R\sigma_0}{\epsilon_0}, & \text{for } r \leq R \\ \dfrac{R^2\sigma_0}{\epsilon_0}\dfrac{1}{r}; & \text{for } r \geq R \end{array}\right\}.}$$

Note: in terms of the total charge $Q = 4\pi R^2\sigma_0$,

$$V(r,\theta) = \left\{ \begin{array}{ll} \frac{1}{4\pi\epsilon_0}\frac{Q}{R}, & \text{for } r \le R \\ \frac{1}{4\pi\epsilon_0}\frac{Q}{r}, & \text{for } r \ge R \end{array} \right\}$$

Problem 3.18

$$V_0(\theta) = k\cos(3\theta) = k\left[4\cos^3\theta - 3\cos\theta\right] = k\left[\alpha P_3(\cos\theta) + \beta P_1(\cos\theta)\right].$$

(I know that any 3rd order polynomial can be expressed as a linear combination of the first four Legendre polynomials; in this case, since the polynomial is *odd,* I only need P_1 and P_3.)

$$4\cos^3\theta - 3\cos\theta = \alpha\left[\frac{1}{2}\left(5\cos^3\theta - 3\cos\theta\right)\right] + \beta\cos\theta = \frac{5\alpha}{2}\cos^3\theta + \left(\beta - \frac{3}{2}\alpha\right)\cos\theta,$$

so

$$4 = \frac{5\alpha}{2} \Rightarrow \alpha = \frac{8}{5}; \quad -3 = \beta - \frac{3}{2}\alpha = \beta - \frac{3}{2}\cdot\frac{8}{5} = \beta - \frac{12}{5} \Rightarrow \beta = \frac{12}{5} - 3 = -\frac{3}{5}.$$

Therefore

$$V_0(\theta) = \frac{k}{5}\left[8P_3(\cos\theta) - 3P_1(\cos\theta)\right].$$

Now

$$V(r,\theta) = \left\{ \begin{array}{lll} \sum_{l=0}^{\infty} A_l r^l P_l(\cos\theta), & \text{for } r \le R & \text{(Eq. 3.66)} \\ \sum_{l=0}^{\infty} \frac{B_l}{r^{l+1}} P_l(\cos\theta), & \text{for } r \ge R & \text{(Eq. 3.71)} \end{array} \right\},$$

where

$$\begin{aligned} A_l &= \frac{(2l+1)}{2R^l}\int_0^\pi V_0(\theta)P_l(\cos\theta)\sin\theta\,d\theta \quad \text{(Eq. 3.69)} \\ &= \frac{(2l+1)}{2R^l}\frac{k}{5}\left\{8\int_0^\pi P_3(\cos\theta)P_l(\cos\theta)\sin\theta\,d\theta - 3\int_0^\pi P_1(\cos\theta)P_l(\cos\theta)\sin\theta\,d\theta\right\} \\ &= \frac{k}{5}\frac{(2l+1)}{2R^l}\left\{8\frac{2}{(2l+1)}\delta_{l3} - 3\frac{2}{(2l+1)}\delta_{l1}\right\} = \frac{k}{5}\frac{1}{R^l}\left[8\,\delta_{l3} - 3\,\delta_{l1}\right] \\ &= \left\{ \begin{array}{ll} 8k/5R^3, & \text{if } l = 3 \\ -3k/5R, & \text{if } l = 1 \end{array} \right\} \text{(zero otherwise)}. \end{aligned}$$

Therefore

$$V(r,\theta) = -\frac{3k}{5R}rP_1(\cos\theta) + \frac{8k}{5R^3}r^3P_3(\cos\theta) = \boxed{\frac{k}{5}\left[8\left(\frac{r}{R}\right)^3 P_3(\cos\theta) - 3\left(\frac{r}{R}\right)P_1(\cos\theta)\right]},$$

or

$$\frac{k}{5}\left\{8\left(\frac{r}{R}\right)^3\frac{1}{2}\left[5\cos^3\theta - 3\cos\theta\right] - 3\left(\frac{r}{R}\right)\cos\theta\right\} \Rightarrow \boxed{V(r,\theta) = \frac{k}{5}\frac{r}{R}\cos\theta\left\{4\left(\frac{r}{R}\right)^2\left[5\cos^2\theta - 3\right] - 3\right\}}$$

(for $r \leq R$). Meanwhile, $B_l = A_l R^{2l+1}$ (Eq. 3.81—this follows from the continuity of V at R). Therefore

$$B_l = \left\{ \begin{array}{ll} 8kR^4/5, & \text{if } l = 3 \\ -3kR^2/5, & \text{if } l = 1 \end{array} \right\} \quad \text{(zero otherwise).}$$

So

$$V(r,\theta) = \frac{-3kR^2}{5}\frac{1}{r^2}P_1(\cos\theta) + \frac{8kR^4}{5}\frac{1}{r^4}P_3(\cos\theta) = \boxed{\frac{k}{5}\left[8\left(\frac{R}{r}\right)^4 P_3(\cos\theta) - 3\left(\frac{R}{r}\right)^2 P_1(\cos\theta)\right]},$$

or

$$\boxed{V(r,\theta) = \frac{k}{5}\left(\frac{R}{r}\right)^2 \cos\theta \left\{4\left(\frac{R}{r}\right)^2 \left[5\cos^2\theta - 3\right] - 3\right\}}$$

(for $r \geq R$). Finally, using Eq. 3.83:

$$\begin{aligned}
\sigma(\theta) &= \epsilon_0 \sum_{l=0}^{\infty}(2l+1)A_l R^{l-1} P_l(\cos\theta) = \epsilon_0\left[3A_1P_1 + 7A_3R^2P_3\right] \\
&= \epsilon_0\left[3\left(-\frac{3k}{5R}\right)P_1 + 7\left(\frac{8k}{5R^3}\right)R^2P_3\right] = \boxed{\frac{\epsilon_0 k}{5R}\left[-9P_1(\cos\theta) + 56P_3(\cos\theta)\right]} \\
&= \frac{\epsilon_0 k}{5R}\left[-9\cos\theta + \frac{56}{2}\left(5\cos^3\theta - 3\cos\theta\right)\right] = \frac{\epsilon_0 k}{5R}\cos\theta[-9 + 28\cdot 5\cos^2\theta - 28\cdot 3] \\
&= \boxed{\frac{\epsilon_0 k}{5R}\cos\theta\left[140\cos^2\theta - 93\right].}
\end{aligned}$$

Problem 3.19

Use Eq. 3.83: $\sigma(\theta) = \epsilon_0 \sum_{l=0}^{\infty}(2l+1)A_l R^{l-1}P_l(\cos\theta)$. But Eq. 3.69 says: $A_l = \frac{2l+1}{2R^l}\int_0^\pi V_0(\theta)P_l(\cos\theta)\sin\theta\, d\theta$.

Putting them together:

$$\sigma(\theta) = \frac{\epsilon_0}{2R}\sum_{l=0}^{\infty}(2l+1)^2 C_l P_l(\cos\theta), \quad \text{with } C_l = \int_0^\pi V_0(\theta)P_l(\cos\theta)\sin\theta\, d\theta. \quad \text{qed}$$

Problem 3.20

Set $V = 0$ on the equatorial plane, far from the sphere. Then the potential is the same as Ex. 3.8 ***plus*** the potential of a uniformly charged spherical shell:

$$\boxed{V(r,\theta) = -E_0\left(r - \frac{R^3}{r^2}\right)\cos\theta + \frac{1}{4\pi\epsilon_0}\frac{Q}{r}.}$$

Problem 3.21

(a) $V(r,\theta) = \sum_{l=0}^{\infty} \frac{B_l}{r^{l+1}} P_l(\cos\theta)$ $(r > R)$, so $V(r,0) = \sum_{l=0}^{\infty} \frac{B_l}{r^{l+1}} P_l(1) = \sum_{l=0}^{\infty} \frac{B_l}{r^{l+1}} = \frac{\sigma}{2\epsilon_0}\left[\sqrt{r^2+R^2} - r\right]$.

Since $r > R$ in this region, $\sqrt{r^2+R^2} = r\sqrt{1+(R/r)^2} = r\left[1 + \frac{1}{2}(R/r)^2 - \frac{1}{8}(R/r)^4 + \dots\right]$, so

$$\sum_{l=0}^{\infty} \frac{B_l}{r^{l+1}} = \frac{\sigma}{2\epsilon_0} r\left[1 + \frac{1}{2}\frac{R^2}{r^2} - \frac{1}{8}\frac{R^4}{r^4} + \dots - 1\right] = \frac{\sigma}{2\epsilon_0}\left(\frac{R^2}{2r} - \frac{R^4}{8r^3} + \dots\right).$$

Comparing like powers of r, I see that $B_0 = \frac{\sigma R^2}{4\epsilon_0}$, $B_1 = 0$, $B_2 = -\frac{\sigma R^4}{16\epsilon_0}, \dots$. Therefore

$$\boxed{\begin{aligned} V(r,\theta) &= \frac{\sigma R^2}{4\epsilon_0}\left[\frac{1}{r} - \frac{R^2}{4r^3}P_2(\cos\theta) + \dots\right], \\ &= \frac{\sigma R^2}{4\epsilon_0 r}\left[1 - \frac{1}{8}\left(\frac{R}{r}\right)^2 (3\cos^2\theta - 1) + \dots\right], \end{aligned}} \quad \text{(for } r > R\text{)}.$$

(b) $V(r,\theta) = \sum_{l=0}^{\infty} A_l r^l P_l(\cos\theta)$ $(r < R)$. In the northern hemispere, $0 \le \theta \le \pi/2$,

$$V(r,0) = \sum_{l=0}^{\infty} A_l r^l = \frac{\sigma}{2\epsilon_0}\left[\sqrt{r^2+R^2} - r\right].$$

Since $r < R$ in this region, $\sqrt{r^2+R^2} = R\sqrt{1+(r/R)^2} = R\left[1 + \frac{1}{2}(r/R)^2 - \frac{1}{8}(r/R)^4 + \dots\right]$. Therefore

$$\sum_{l=0}^{\infty} A_l r^l = \frac{\sigma}{2\epsilon_0}\left[R + \frac{1}{2}\frac{r^2}{R} - \frac{1}{8}\frac{r^4}{R^3} + \dots - r\right].$$

Comparing like powers: $A_0 = \frac{\sigma}{2\epsilon_0}R$, $A_1 = -\frac{\sigma}{2\epsilon_0}$, $A_2 = \frac{\sigma}{2\epsilon_0 R}, \dots$, so

$$\boxed{\begin{aligned} V(r,\theta) &= \frac{\sigma}{2\epsilon_0}\left[R - rP_1(\cos\theta) + \frac{1}{2R}P_2(\cos\theta) + \dots\right], \\ &= \frac{\sigma R}{2\epsilon_0}\left[1 - \left(\frac{r}{R}\right)\cos\theta + \frac{1}{4}\left(\frac{r}{R}\right)^2 (3\cos^2\theta - 1) + \dots\right], \end{aligned}} \quad \text{(for } r < R\text{, northern hemisphere)}.$$

In the southern hemisphere we'll have to go for $\theta = \pi$, using $P_l(-1) = (-1)^l$.

$$V(r,\pi) = \sum_{l=0}^{\infty} (-1)^l \overline{A}_l r^l = \frac{\sigma}{2\epsilon_0}\left[\sqrt{r^2+R^2} - r\right].$$

(I put an overbar on $\overline{A}_l$ to distinguish it from the northern A_l). The only difference is the sign of $\overline{A}_1$: $\overline{A}_1 = +(\sigma/2\epsilon_0)$, $\overline{A}_0 = A_0$, $\overline{A}_2 = A_2$. So:

$$\boxed{\begin{aligned} V(r,\theta) &= \frac{\sigma}{2\epsilon_0}\left[R + rP_1(\cos\theta) + \frac{1}{2R}r^2 P_2(\cos\theta) + \dots\right], \\ &= \frac{\sigma R}{2\epsilon_0}\left[1 + \left(\frac{r}{R}\right)\cos\theta + \frac{1}{4}\left(\frac{r}{R}\right)^2(3\cos^2\theta - 1) + \dots\right], \end{aligned}}$$ (for $r < R$, southern hemisphere).

Problem 3.22

$$V(r,\theta) = \begin{cases} \displaystyle\sum_{l=0}^{\infty} A_l r^l P_l(\cos\theta), & (r \le R) \text{ (Eq. 3.78)}, \\ \displaystyle\sum_{l=0}^{\infty} \frac{B_l}{r^{l+1}} P_l(\cos\theta), & (r \ge R) \text{ (Eq. 3.79)}, \end{cases}$$

where $B_l = A_l R^{2l+1}$ (Eq. 3.81) and

$$\begin{aligned} A_l &= \frac{1}{2\epsilon_0 R^{l-1}} \int_0^{\pi} \sigma_0(\theta) P_l(\cos\theta)\sin\theta\, d\theta \quad \text{(Eq. 3.84)} \\ &= \frac{1}{2\epsilon_0 R^{l-1}}\sigma_0 \left\{ \int_0^{\pi/2} P_l(\cos\theta)\sin\theta\, d\theta - \int_{\pi/2}^{\pi} P_l(\cos\theta)\sin\theta\, d\theta \right\} \quad (\text{let } x = \cos\theta) \\ &= \frac{\sigma_0}{2\epsilon_0 R^{l-1}} \left\{ \int_0^1 P_l(x)\,dx - \int_{-1}^0 P_l(x)\,dx \right\}. \end{aligned}$$

Now $P_l(-x) = (-1)^l P_l(x)$, since $P_l(x)$ is even, for even l, and odd, for odd l. Therefore

$$\int_{-1}^{0} P_l(x)\,dx = \int_{1}^{0} P_l(-x)\,d(-x) = (-1)^l \int_0^1 P_l(x)\,dx,$$

and hence

$$A_l = \frac{\sigma_0}{2\epsilon_0 R^{l-1}}\left[1 - (-1)^l\right]\int_0^1 P_l(x)\,dx = \begin{cases} 0, & \text{if } l \text{ is even} \\ \displaystyle\frac{\sigma_0}{\epsilon_0 R^{l-1}}\int_0^1 P_l(x)\,dx, & \text{if } l \text{ is odd} \end{cases}.$$

So $A_0 = A_2 = A_4 = A_6 = 0$, and all we need are A_1, A_3, and A_5.

$$\begin{aligned}
\int_0^1 P_1(x)\,dx &= \int_0^1 x\,dx = \left.\frac{x^2}{2}\right|_0^1 = \frac{1}{2}.\\
\int_0^1 P_3(x)\,dx &= \frac{1}{2}\int_0^1 (5x^3 - 3x)\,dx = \frac{1}{2}\left.\left(5\frac{x^4}{4} - 3\frac{x^2}{2}\right)\right|_0^1 = \frac{1}{2}\left(\frac{5}{4} - \frac{3}{2}\right) = -\frac{1}{8}.\\
\int_0^1 P_5(x)\,dx &= \frac{1}{8}\int_0^1 (63x^5 - 70x^3 + 15x)\,dx = \frac{1}{8}\left.\left(63\frac{x^6}{6} - 70\frac{x^4}{4} + 15\frac{x^2}{2}\right)\right|_0^1\\
&= \frac{1}{8}\left(\frac{21}{2} - \frac{35}{2} + \frac{15}{2}\right) = \frac{1}{16}(36 - 35) = \frac{1}{16}.
\end{aligned}$$

Therefore

$$\boxed{A_1 = \frac{\sigma_0}{\epsilon_0}\left(\frac{1}{2}\right);\ A_3 = \frac{\sigma_0}{\epsilon_0 R^2}\left(-\frac{1}{8}\right);\ A_5 = \frac{\sigma_0}{\epsilon_0 R^4}\left(\frac{1}{16}\right);\ \text{etc.}}$$

and

$$\boxed{B_1 = \frac{\sigma_0}{\epsilon_0}R^3\left(\frac{1}{2}\right);\ B_3 = \frac{\sigma_0}{\epsilon_0}R^5\left(-\frac{1}{8}\right);\ B_5 = \frac{\sigma_0}{\epsilon_0}R^7\left(\frac{1}{16}\right);\ \text{etc.}}$$

Thus

$$\boxed{V(r,\theta) = \left\{\begin{array}{ll}
\dfrac{\sigma_0 r}{2\epsilon_0}\left[P_1(\cos\theta) - \dfrac{1}{4}\left(\dfrac{r}{R}\right)^2 P_3(\cos\theta) + \dfrac{1}{8}\left(\dfrac{r}{R}\right)^4 P_5(\cos\theta) + \ldots\right], & (r \le R),\\
\dfrac{\sigma_0 R^3}{2\epsilon_0 r^2}\left[P_1(\cos\theta) - \dfrac{1}{4}\left(\dfrac{R}{r}\right)^2 P_3(\cos\theta) + \dfrac{1}{8}\left(\dfrac{R}{r}\right)^4 P_5(\cos\theta) + \ldots\right], & (r \ge R).
\end{array}\right\}}$$

Problem 3.23

$$\frac{1}{s}\frac{\partial}{\partial s}\left(s\frac{\partial V}{\partial s}\right) + \frac{1}{s^2}\frac{\partial^2 V}{\partial \phi^2} = 0.$$

Look for solutions of the form $V(s,\phi) = S(s)\Phi(\phi)$:

$$\frac{1}{s}\Phi\frac{d}{ds}\left(s\frac{dS}{ds}\right) + \frac{1}{s^2}S\frac{d^2\Phi}{d\phi^2} = 0.$$

Multiply by s^2 and divide by $V = S\Phi$:

$$\frac{s}{S}\Phi\frac{d}{ds}\left(s\frac{dS}{ds}\right) + \frac{1}{\Phi}\frac{d^2\Phi}{d\phi^2} = 0.$$

Since the first term involves s only, and the second ϕ only, each is a constant:

$$\frac{s}{S}\frac{d}{ds}\left(s\frac{dS}{ds}\right) = C_1, \quad \frac{1}{\Phi}\frac{d^2\Phi}{d\phi^2} = C_2, \quad \text{with } C_1 + C_2 = 0.$$

Now C_2 must be negative (else we get exponentials for Φ, which do not return to their original value—as geometrically they *must*— when ϕ is increased by 2π).

$$C_2 = -k^2. \quad \text{Then } \frac{d^2\Phi}{d\phi^2} = -k^2\Phi \Rightarrow \Phi = A\cos k\phi + B\sin k\phi.$$

Moreover, since $\Phi(\phi + 2\pi) = \Phi(\phi)$, *k must be an integer:* $k = 0, 1, 2, 3, \ldots$ (negative integers are just repeats, but $k = 0$ must be included, since $\Phi = A$ (a constant) is OK).

$s\frac{d}{ds}\left(s\frac{dS}{ds}\right) = k^2 S$ can be solved by $S = s^n$, provided n is chosen right:

$$s\frac{d}{ds}\left(sns^{n-1}\right) = ns\frac{d}{ds}\left(s^n\right) = n^2 ss^{n-1} = n^2 s^n = k^2 S \Rightarrow n = \pm k.$$

Evidently the general solution is $S(s) = Cs^k + Ds^{-k}$, *unless* $k = 0$, in which case we have only *one* solution to a *second-order* equation—namely, $S =$ constant. So we must treat $k = 0$ separately. *One* solution is a constant—but what's the other? Go back to the differential equation for S, and put in $k = 0$:

$$s\frac{d}{ds}\left(s\frac{dS}{ds}\right) = 0 \Rightarrow s\frac{dS}{ds} = \text{constant} = C \Rightarrow \frac{dS}{ds} = \frac{C}{s} \Rightarrow dS = C\frac{ds}{s} \Rightarrow S = C\ln s + D \text{ (another constant).}$$

So the second solution in this case is $\ln s$. [How about Φ? That *too* reduces to a single solution, $\Phi = A$, in the case $k = 0$. What's the second solution here? Well, putting $k = 0$ into the Φ equation:

$$\frac{d^2\Phi}{d\phi^2} = 0 \Rightarrow \frac{d\Phi}{d\phi} = \text{constant} = B \Rightarrow \Phi = B\phi + A.$$

But a term of the form $B\phi$ is unacceptable, since it does not return to its initial value when ϕ is augmented by 2π.] *Conclusion:* The general solution with cylindrical symmetry is

$$\boxed{V(s,\phi) = a_0 + b_0\ln s + \sum_{k=1}^{\infty}\left[s^k(a_k\cos k\phi + b_k\sin k\phi) + s^{-k}(c_k\cos k\phi + d_k\sin k\phi)\right].}$$

Yes: the potential of a line charge goes like $\ln s$, which *is* included.

Problem 3.24

Picking $V = 0$ on the yz plane, with $\mathbf{E}_0$ in the x direction, we have (Eq. 3.74):

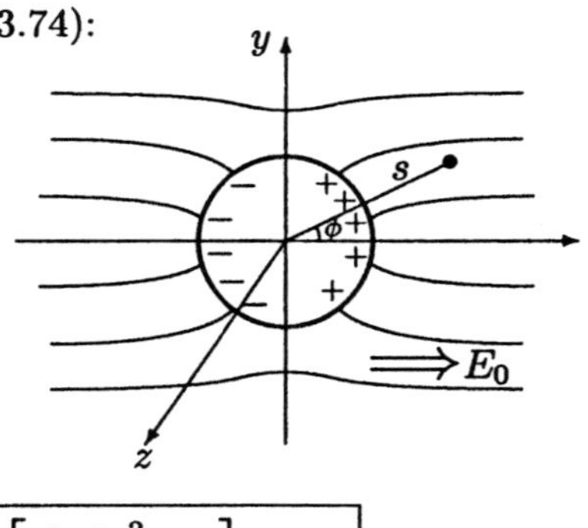

$$\begin{cases} \text{(i)} & V = 0, & \text{when } s = R, \\ \text{(ii)} & V \to -E_0 x = -E_0 s\cos\phi, & \text{for } s \gg R. \end{cases}$$

Evidently $a_0 = b_0 = b_k = d_k = 0$, and $a_k = c_k = 0$ except for $k = 1$:

$$V(s,\phi) = \left(a_1 s + \frac{c_1}{s}\right)\cos\phi.$$

(i)$\Rightarrow c_1 = -a_1 R^2$; (ii)$\to a_1 = -E_0$. Therefore

$$V(s,\phi) = \left(-E_0 s + \frac{E_0 R^2}{s}\right)\cos\phi, \quad \text{or} \quad \boxed{V(s,\phi) = -E_0 s\left[\left(\frac{R}{s}\right)^2 - 1\right]\cos\phi.}$$

$$\sigma = -\epsilon_0 \left.\frac{\partial V}{\partial s}\right|_{s=R} = -\epsilon_0 E_0 \left(-\frac{R^2}{s^2} - 1\right) \cos\phi \Big|_{s=R} = \boxed{2\epsilon_0 E_0 \cos\phi.}$$

Problem 3.25

Inside: $V(s,\phi) = a_0 + \sum_{k=1}^{\infty} s^k (a_k \cos k\phi + b_k \sin k\phi)$. (In this region $\ln s$ and s^{-k} are no good—they blow up at $s = 0$.)

Outside: $V(s,\phi) = \bar{a}_0 + \sum_{k=1}^{\infty} \frac{1}{s^k} (c_k \cos k\phi + d_k \sin k\phi)$. (Here $\ln s$ and s^k are no good at $s \to \infty$).

$$\sigma = -\epsilon_0 \left(\frac{\partial V_{\text{out}}}{\partial s} - \frac{\partial V_{\text{in}}}{\partial s}\right)\Bigg|_{s=R} \qquad \text{(Eq. 2.36).}$$

Thus

$$a \sin 5\phi = -\epsilon_0 \sum_{k=1}^{\infty} \left\{-\frac{k}{R^{k+1}} (c_k \cos k\phi + d_k \sin k\phi) - kR^{k-1} (a_k \cos k\phi + b_k \sin k\phi)\right\}.$$

Evidently $a_k = c_k = 0$; $b_k = d_k = 0$ except $k = 5$; $a = 5\epsilon_0 \left(\frac{1}{R^6} d_5 + R^4 b_5\right)$. Also, V is continuous at $s = R$: $a_0 + R^5 b_5 \sin 5\phi = \bar{a}_0 + \frac{1}{R^5} d_5 \sin 5\phi$. So $a_0 = \bar{a}_0$ (might as well choose both zero); $R^5 b_5 = R^{-5} d_5$, or $d_5 = R^{10} b_5$. Combining these results: $a = 5\epsilon_0 (R^4 b_5 + R^4 b_5) = 10\epsilon_0 R^4 b_5$; $b_5 = \frac{a}{10\epsilon_0 R^4}$; $d_5 = \frac{aR^6}{10\epsilon_0}$. Therefore

$$\boxed{V(s,\phi) = \frac{a \sin 5\phi}{10\epsilon_0} \left\{ \begin{array}{ll} s^5/R^4, & \text{for } s < R, \\ R^6/s^5, & \text{for } s > R. \end{array} \right\}}$$

Problem 3.26

Monopole term:

$$Q = \int \rho \, d\tau = kR \int \left[\frac{1}{r^2}(R - 2r) \sin\theta\right] r^2 \sin\theta \, dr \, d\theta \, d\phi.$$

But the r integral is

$$\int_0^R (R - 2r) \, dr = \left(Rr - r^2\right)\Big|_0^R = R^2 - R^2 = 0. \quad \text{So } Q = 0.$$

Dipole term:

$$\int r \cos\theta \rho \, d\tau = kR \int (r\cos\theta) \left[\frac{1}{r^2}(R - 2r) \sin\theta\right] r^2 \sin\theta \, dr \, d\theta \, d\phi.$$

But the θ integral is

$$\int_0^\pi \sin^2\theta \cos\theta \, d\theta = \left.\frac{\sin^3\theta}{3}\right|_0^\pi = \frac{1}{3}(0 - 0) = 0.$$

So the dipole contribution is likewise zero.

Quadrupole term:

$$\int r^2 \left(\frac{3}{2}\cos^2\theta - \frac{1}{2}\right) \rho \, d\tau = \frac{1}{2} kR \int\int r^2 \left(3\cos^2\theta - 1\right) \left[\frac{1}{r^2}(R - 2r) \sin\theta\right] r^2 \sin\theta \, dr \, d\theta.$$

r integral:

$$\int_0^R r^2(R-2r)\,dr = \left(\frac{r^3}{3}R - \frac{r^4}{2}\right)\Big|_0^R = \frac{R^4}{3} - \frac{R^4}{2} = -\frac{R^4}{6}.$$

θ integral:

$$\int_0^\pi \underbrace{(3\cos^2\theta - 1)}_{3(1-\sin^2\theta)-1=2-3\sin^2\theta} \sin^2\theta\,d\theta = 2\int_0^\pi \sin^2\theta\,d\theta - 3\int_0^\pi \sin^4\theta\,d\theta$$

$$= 2\left(\frac{\pi}{2}\right) - 3\left(\frac{3\pi}{8}\right) = \pi\left(1-\frac{9}{8}\right) = -\frac{\pi}{8}.$$

φ integral:

$$\int_0^{2\pi} d\phi = 2\pi.$$

The whole integral is:

$$\frac{1}{2}kR\left(-\frac{R^4}{6}\right)\left(-\frac{\pi}{8}\right)(2\pi) = \frac{k\pi^2R^5}{48}.$$

For point P on the z axis ($r \to z$ in Eq. 3.95) the approximate potential is

$$\boxed{V(z) \cong \frac{1}{4\pi\epsilon_0}\frac{k\pi^2R^5}{48z^3}.} \quad \text{(Quadrupole.)}$$

Problem 3.27

$\mathbf{p} = (3qa - qa)\,\hat{\mathbf{z}} + (-2qa - 2q(-a))\,\hat{\mathbf{y}} = 2qa\,\hat{\mathbf{z}}$. Therefore

$$V \cong \frac{1}{4\pi\epsilon_0}\frac{\mathbf{p}\cdot\hat{\mathbf{r}}}{r^2},$$

and $\mathbf{p}\cdot\hat{\mathbf{r}} = 2qa\,\hat{\mathbf{z}}\cdot\hat{\mathbf{r}} = 2qa\cos\theta$, so

$$V \cong \boxed{\frac{1}{4\pi\epsilon_0}\frac{2qa\cos\theta}{r^2}.} \quad \text{(Dipole.)}$$

Problem 3.28

(a) By symmetry, $\mathbf{p}$ is clearly in the z direction: $\mathbf{p} = p\,\hat{\mathbf{z}}$; $p = \int z\rho\,d\tau \Rightarrow \int z\sigma\,da$.

$$p = \int (R\cos\theta)(k\cos\theta)R^3\sin\theta\,d\theta\,d\phi = 2\pi R^3k\int_0^\pi \cos^2\theta\sin\theta\,d\theta = 2\pi R^3k\left(-\frac{\cos^3\theta}{3}\right)\Big|_0^\pi$$

$$= \frac{2}{3}\pi R^3k[1-(-1)] = \frac{4\pi R^3k}{3}; \quad \boxed{\mathbf{p} = \frac{4\pi R^3k}{3}\,\hat{\mathbf{z}}.}$$

(b)

$$V \cong \frac{1}{4\pi\epsilon_0}\frac{4\pi R^3k}{3}\frac{\cos\theta}{r^2} = \boxed{\frac{kR^3}{3\epsilon_0}\frac{\cos\theta}{r^2}.} \quad \text{(Dipole.)}$$

This is *also* the *exact* potential. *Conclusion:* all multiple moments of this distribution (except the dipole) are exactly zero.

Problem 3.29

Using Eq. 3.94 with $r' = d/2$:

$$\frac{1}{\imath_+} = \frac{1}{r}\sum_{n=0}^{\infty}\left(\frac{d}{2r}\right)^n P_n(\cos\theta);$$

for $\imath_-$, we let $\theta \to 180° + \theta$, so $\cos\theta \to -\cos\theta$:

$$\frac{1}{\imath_-} = \frac{1}{r}\sum_{n=0}^{\infty}\left(\frac{d}{2r}\right)^n P_n(-\cos\theta).$$

But $P_n(-x) = (-1)^n P_n(x)$, so

$$V = \frac{1}{4\pi\epsilon_0}q\left(\frac{1}{\imath_+} - \frac{1}{\imath_-}\right) = \frac{1}{4\pi\epsilon_0}q\frac{1}{r}\sum_{n=0}^{\infty}\left(\frac{d}{2r}\right)^n [P_n(\cos\theta) - P_n(-\cos\theta)] = \frac{2q}{4\pi\epsilon_0 r}\sum_{n=1,3,5,\ldots}\left(\frac{d}{2r}\right)^n P_n(\cos\theta).$$

Therefore

$$V_{\text{dip}} = \frac{2q}{4\pi\epsilon_0}\frac{1}{r}\frac{d}{2r}P_1(\cos\theta) = \frac{qd\cos\theta}{4\pi\epsilon_0 r^2}, \quad \text{while} \quad \boxed{V_{\text{quad}} = 0.}$$

$$V_{\text{oct}} = \frac{2q}{4\pi\epsilon_0 r}\left(\frac{d}{2r}\right)^3 P_3(\cos\theta) = \frac{2q}{4\pi\epsilon_0}\frac{d^3}{8r^4}\frac{1}{2}\left(5\cos^3\theta - 3\cos\theta\right) = \boxed{\frac{qd^3}{4\pi\epsilon_0}\frac{1}{8r^4}\left(5\cos^3\theta - 3\cos\theta\right).}$$

Problem 3.30

(a) (i) $Q = \boxed{2q,}$ (ii) $\mathbf{p} = \boxed{3qa\,\hat{\mathbf{z}},}$ (iii) $V \cong \frac{1}{4\pi\epsilon_0}\left[\frac{Q}{r} + \frac{\mathbf{p}\cdot\hat{\mathbf{r}}}{r^2}\right] = \boxed{\frac{1}{4\pi\epsilon_0}\left[\frac{2q}{r} + \frac{3qa\cos\theta}{r^2}\right].}$

(b) (i) $Q = \boxed{2q,}$ (ii) $\mathbf{p} = \boxed{qa\,\hat{\mathbf{z}},}$ (iii) $V \cong \boxed{\frac{1}{4\pi\epsilon_0}\left[\frac{2q}{r} + \frac{qa\cos\theta}{r^2}\right].}$

(c) (i) $Q = \boxed{2q,}$ (ii) $\mathbf{p} = \boxed{3qa\,\hat{\mathbf{y}},}$ (iii) $V \cong \boxed{\frac{1}{4\pi\epsilon_0}\left[\frac{2q}{r} + \frac{3qa\sin\theta\sin\phi}{r^2}\right]}$ (from Eq. 1.64, $\hat{\mathbf{y}}\cdot\hat{\mathbf{r}} = \sin\theta\sin\phi$).

Problem 3.31

(a) This point is at $r = a$, $\theta = \frac{\pi}{2}$, $\phi = 0$, so $\mathbf{E} = \frac{p}{4\pi\epsilon_0 a^3}\hat{\boldsymbol{\theta}} = \frac{p}{4\pi\epsilon_0 a^3}(-\hat{\mathbf{z}})$; $\mathbf{F} = q\mathbf{E} = \boxed{-\frac{pq}{4\pi\epsilon_0 a^3}\hat{\mathbf{z}}.}$

(b) Here $r = a$, $\theta = 0$, so $\mathbf{E} = \frac{p}{4\pi\epsilon_0 a^3}(2\hat{\mathbf{r}}) = \frac{2p}{4\pi\epsilon_0 a^3}\hat{\mathbf{z}}$. $\boxed{\mathbf{F} = \frac{2pq}{4\pi\epsilon_0 a^3}\hat{\mathbf{z}}.}$

(c) $V = q\left[V(0,0,a) - V(a,0,0)\right] = \frac{qp}{4\pi\epsilon_0 a^2}\left[\cos(0) - \cos\left(\frac{\pi}{2}\right)\right] = \boxed{\frac{pq}{4\pi\epsilon_0 a^2}.}$

Problem 3.32

$Q = -q$, so $V_{\text{mono}} = \frac{1}{4\pi\epsilon_0}\frac{-q}{r}$; $\mathbf{p} = qa\,\hat{\mathbf{z}}$, so $V_{\text{dip}} = \frac{1}{4\pi\epsilon_0}\frac{qa\cos\theta}{r^2}$. Therefore

$$\boxed{V(r,\theta) \cong \frac{q}{4\pi\epsilon_0}\left(-\frac{1}{r} + \frac{a\cos\theta}{r^2}\right).} \quad \boxed{\mathbf{E}(r,\theta) \cong \frac{q}{4\pi\epsilon_0}\left[-\frac{1}{r^2}\hat{\mathbf{r}} + \frac{a}{r^3}\left(2\cos\theta\,\hat{\mathbf{r}} + \sin\theta\,\hat{\boldsymbol{\theta}}\right)\right].}$$

Problem 3.33

$\mathbf{p} = (\mathbf{p}\cdot\hat{\mathbf{r}})\,\hat{\mathbf{r}} + (\mathbf{p}\cdot\hat{\boldsymbol{\theta}})\,\hat{\boldsymbol{\theta}} = p\cos\theta\,\hat{\mathbf{r}} - p\sin\theta\,\hat{\boldsymbol{\theta}}$ (Fig. 3.36). So $3(\mathbf{p}\cdot\hat{\mathbf{r}})\,\hat{\mathbf{r}} - \mathbf{p} = 3p\cos\theta\,\hat{\mathbf{r}} - p\cos\theta\,\hat{\mathbf{r}} + p\sin\theta\,\hat{\boldsymbol{\theta}} = 2p\cos\theta\,\hat{\mathbf{r}} + p\sin\theta\,\hat{\boldsymbol{\theta}}$. So Eq. 3.104 ≡ Eq. 3.103. ✓

Problem 3.34

At height x above the plane, the force on q is given by Eq. 3.12: $F = -\dfrac{1}{4\pi\epsilon_0}\dfrac{q^2}{4z^2} = m\dfrac{d^2x}{dt^2}$; $\dfrac{d^2x}{dt^2} = -A/x^2$, where $A \equiv \dfrac{q^2}{16\pi\epsilon_0 m}$. Multiply by $v = \dfrac{dx}{dt}$: $v\dfrac{dv}{dt} = -\dfrac{A}{x^2}\dfrac{dx}{dt} \Rightarrow \dfrac{d}{dt}\left(\dfrac{1}{2}v^2\right) = \dfrac{d}{dt}\left(\dfrac{A}{x}\right) \Rightarrow \dfrac{1}{2}v^2 = \dfrac{A}{x} +$ constant. But $v = 0$ when $x = d$, so constant $= -A/d$, and hence $v^2 = 2A\left(\dfrac{1}{x} - \dfrac{1}{d}\right)$; $-\dfrac{dx}{dt} = \sqrt{2A}\sqrt{\dfrac{1}{x} - \dfrac{1}{d}} = \sqrt{\dfrac{2A}{d}}\sqrt{\dfrac{d-x}{x}}$.

$$\int_d^0 \frac{\sqrt{x}}{\sqrt{d-x}}\,dx = -\sqrt{\frac{2A}{d}}\int_0^t dt = -\sqrt{\frac{2A}{d}}\,t.$$

This integral can also be integrated directly. Let $x = u^2$; $dx = 2u\,du$.

$$\int_d^0 \frac{\sqrt{x}}{\sqrt{d-x}}\,dx = 2\int_{\sqrt{d}}^0 \frac{u^2}{\sqrt{d-u^2}}\,du = 2\left\{-\frac{u}{2}\sqrt{d-u^2} + \frac{d}{2}\sin^{-1}\left(\frac{u}{\sqrt{d}}\right)\right\}\Bigg|_{\sqrt{d}}^{0} = -d\sin^{-1}(1) = -d\frac{\pi}{2}.$$

Therefore

$$t = \sqrt{\frac{d}{2A}}\,\frac{\pi d}{2} = \sqrt{\frac{\pi^2 d^2}{4}\,\frac{d}{2q^2}\,16\pi\epsilon_0 m} = \boxed{\sqrt{\frac{2\pi^3 d^3\epsilon_0 m}{q^2}}.}$$

Problem 3.35

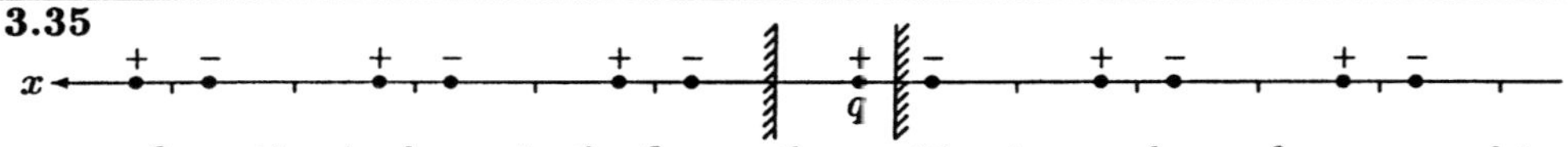

The image configuration is shown in the figure; the positive image charge forces cancel in pairs. The net force of the negative image charges is:

$$\begin{aligned} F &= \frac{1}{4\pi\epsilon_0}q^2\left\{\frac{1}{[2(a-x)]^2} + \frac{1}{[2a+2(a-x)]^2} + \frac{1}{[4a+2(a-x)]^2} + \dots \right.\\ &\quad \left. - \frac{1}{(2x)^2} - \frac{1}{(2a+2x)^2} - \frac{1}{(4a+2x)^2} - \dots\right\} \\ &= \boxed{\frac{1}{4\pi\epsilon_0}\frac{q^2}{4}\left\{\left[\frac{1}{(a-x)^2} + \frac{1}{(2a-x)^2} + \frac{1}{(3a-x)^2} + \dots\right] - \left[\frac{1}{x^2} + \frac{1}{(a+x)^2} + \frac{1}{(2a+x)^2} + \dots\right]\right\}.} \end{aligned}$$

When $a \to \infty$ (i.e. $a \gg x$) only the $\dfrac{1}{x^2}$ term survives: $F = -\dfrac{1}{4\pi\epsilon_0}\dfrac{q^2}{(2x)^2}$ ✓ (same as for only *one* plane—Eq. 3.12). When $x = a/2$,

$$F = \frac{1}{4\pi\epsilon_0}\frac{q^2}{4}\left\{\left[\frac{1}{(a/2)^2} + \frac{1}{(3a/2)^2} + \frac{1}{(5a/2)^2} + \dots\right] - \left[\frac{1}{(a/2)^2} + \frac{1}{(3a/2)^2} + \frac{1}{(5a/2)^2} + \dots\right]\right\} = 0. ✓$$

Problem 3.36

Following Prob. 2.47, we place image line charges $-\lambda$ at $y = b$ and $+\lambda$ at $y = -b$ (here y is the horizontal axis, z vertical).

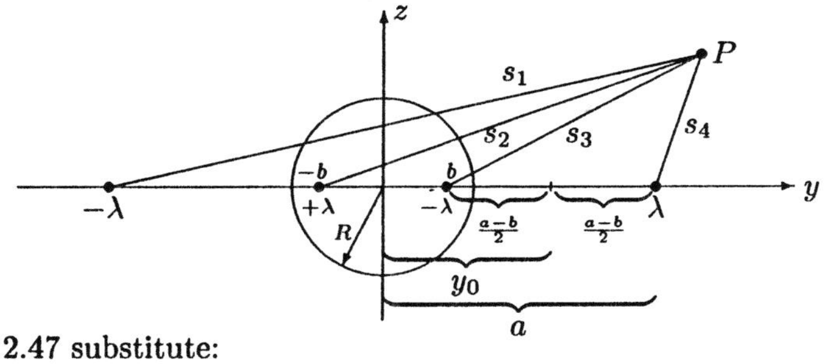

In the solution to Prob. 2.47 substitute:

$$a \to \frac{a-b}{2},\ y_0 \to \frac{a+b}{2} \text{ so } \left(\frac{a-b}{2}\right)^2 = \left(\frac{a+b}{2}\right)^2 - R^2 \Rightarrow b = \frac{R^2}{a}.$$

$$\begin{aligned} V &= \frac{\lambda}{4\pi\epsilon_0}\left[\ln\left(\frac{s_3^2}{s_4^2}\right) + \ln\left(\frac{s_1^2}{s_2^2}\right)\right] = \frac{\lambda}{4\pi\epsilon_0}\ln\left(\frac{s_1^2 s_3^2}{s_4^2 s_2^2}\right) \\ &= \frac{\lambda}{4\pi\epsilon_0}\ln\left\{\frac{[(y+a)^2+z^2][(y-b)^2+z^2]}{[(y-a)^2+z^2][(y+b)^2+z^2]}\right\}, \quad \text{or, using } y = s\cos\phi,\ z = s\sin\phi, \\ &= \boxed{\frac{\lambda}{4\pi\epsilon_0}\ln\left\{\frac{(s^2+a^2+2as\cos\phi)[(as/R)^2+R^2-2as\cos\phi]}{(a^2+a^2-2as\cos\phi)[(as/R)^2+R^2+2as\cos\phi]}\right\}}. \end{aligned}$$

Problem 3.37

Since the configuration is azimuthally symmetric, $V(r,\theta) = \sum\left(A_l r^l + \frac{B_l}{r^{l+1}}\right)P_l(\cos\theta)$.

(a) $r > b$: $A_l = 0$ for all l, since $V \to 0$ at ∞. Therefore $V(r,\theta) = \sum \frac{B_l}{r^{l+1}}P_l(\cos\theta)$.

$a < r < b$: $V(r,\theta) = \sum\left(C_l r^l + \frac{D_l}{r^{l+1}}\right)P_l(\cos\theta)$. $r < a$: $V(r,\theta) = V_0$.

We need to determine B_l, C_l, D_l, and V_0. To do this, invoke boundary conditions as follows: (i) V is continuous at a, (ii) V is continuous at b, (iii) $\Delta\left(\frac{\partial V}{\partial r}\right) = -\frac{1}{\epsilon_0}\sigma(\theta)$ at b.

(ii) $\Rightarrow \sum \frac{B_l}{b^{l+1}}P_l(\cos\theta) = \sum\left(C_l b^l + \frac{D_l}{b^{l+1}}\right)P_l(\cos\theta)$; $\frac{B_l}{b^{l+1}} = C_l b^l + \frac{D_l}{b^{l+1}} \Rightarrow \boxed{B_l = b^{2l+1}C_l + D_l.}$ (1)

$$\text{(i)} \Rightarrow \sum\left(C_l a^l + \frac{D_l}{a^{l+1}}\right)P_l(\cos\theta) = V_0; \left\{\begin{array}{ll} C_l a^l + \frac{D_l}{a^{l+1}} = 0, & \text{if } l \neq 0, \\ C_0 a^0 + \frac{D_0}{a^1} = V_0, & \text{if } l = 0; \end{array}\right\} \boxed{\begin{array}{l} D_l = -a^{2l+1}C_l,\ l \neq 0, \\ D_0 = aV_0 - aC_0. \end{array}} \quad (2)$$

Putting (2) into (1) gives $B_l = b^{2l+1}C_l - a^{2l+1}C_l,\ l \neq 0,\quad B_0 = bC_0 + aV_0 - aC_0$. Therefore

$$\boxed{\begin{array}{l} B_l = \left(b^{2l+1} - a^{2l+1}\right)C_l,\ l \neq 0, \\ B_0 = (b-a)C_0 + aV_0. \end{array}} \quad (1')$$

(iii) $\Rightarrow \sum B_l[-(l+1)]\frac{1}{b^{l+2}}P_l(\cos\theta) - \sum\left(C_l l b^{l-1} + D_l\frac{-(l+1)}{b^{l+2}}\right)P_l(\cos\theta) = \frac{-k}{\epsilon_0}P_1(\cos\theta)$. So

$$-\frac{(l+1)}{b^{l+2}}B_l - \left(C_l l b^{l-1} + D_l\frac{-(l+1)}{b^{l+2}}\right) = 0, \text{ if } l \neq 1;$$

or

$$-(l+1)B_l - lC_l b^{2l+1} + (l+1)D_l = 0; \quad (l+1)(B_l - D_l) = -lb^{2l+1}C_l.$$

$$B_1(+2)\frac{1}{b^2} + \left(C_1 + D_1\frac{-2}{b^2}\right) = \frac{k}{\epsilon_0}, \text{ for } l = 1; \quad C_1 + \frac{2}{b^3}(B_1 - D_1) = k.$$

Therefore

$$\boxed{\begin{array}{l}(l+1)(B_l - D_l) + lb^{2l+1}C_l = 0, \text{for } l \neq 1,\\ C_1 + \frac{2}{b^3}(B_1 - D_1) = \frac{k}{\epsilon_0}.\end{array}} \quad (3)$$

Plug (2) and (1′) into (3):

For $l \neq 0$ or 1:

$(l+1)\left[\left(b^{2l+1} - a^{2l+1}\right)C_l + a^{2l+1}C_l\right] + lb^{2l+1}C_l = 0; \quad (l+1)b^{2l+1}C_l + lb^{2l+1}C_l = 0; \ (2l+1)C_l = 0 \Rightarrow C_l = 0.$

Therefore (1′) and (2) $\Rightarrow \boxed{B_l = C_l = D_l = 0 \text{ for } l > 1.}$

For $l = 1$: $C_1 + \frac{2}{b^3}\left[\left(b^3 - a^3\right)C_1 + a^3C_1\right] = k; \quad C_1 + 2C_1 = k \Rightarrow \boxed{C_1 = k/3\epsilon_0;} \ D_1 = -a^3C_1 \Rightarrow \boxed{D_1 = -a^3k/3\epsilon_0;} \quad B_1 = \left(b^3 - a^3\right)C_1 \Rightarrow \boxed{B_1 = \left(b^3 - a^3\right)k/3\epsilon_0.}$

For $l = 0$: $B_0 - D_0 = 0 \Rightarrow B_0 = D_0 \Rightarrow (b-a)C_0 + aV_0 = aV_0 - aC_0$, so $bC_0 = 0 \Rightarrow \boxed{C_0 = 0; \ D_0 = aV_0 = B_0.}$

Conclusion: $\boxed{V(r,\theta) = \frac{aV_0}{r} + \frac{\left(b^3 - a^3\right)k}{3r^2\epsilon_0}\cos\theta, \ r \geq b.} \quad \boxed{V(r,\theta) = \frac{aV_0}{r} + \frac{k}{3\epsilon_0}\left(r - \frac{a^3}{r^2}\right)\cos\theta,} \quad a \leq r \leq b.$

(b) $\sigma_i(\theta) = -\epsilon_0 \left.\frac{\partial V}{\partial r}\right|_a = -\epsilon_0\left[-\frac{aV_0}{a^2} + \frac{k}{3\epsilon_0}\left(1 + 2\frac{a^3}{a^3}\right)\cos\theta\right] = -\epsilon_0\left(-\frac{V_0}{a} + \frac{k}{\epsilon_0}\cos\theta\right) = \boxed{-k\cos\theta + V_0\frac{\epsilon_0}{a}.}$

(c) $q_i = \int \sigma_i \, da = \frac{V_0\epsilon_0}{a}4\pi a^2 = \boxed{4\pi a\epsilon_0 V_0 = Q_{\text{tot}}.}$ At large r: $V \approx \frac{aV_0}{r} \stackrel{?}{=} \frac{1}{4\pi\epsilon_0}\frac{Q}{r} = \frac{1}{4\pi\epsilon_0}\frac{4\pi a\epsilon_0 V_0}{r} = \frac{aV_0}{r}.$ ✓

Problem 3.38

Use multipole expansion (Eq. 3.95): $\rho\, d\tau \to \lambda\, dz = \frac{Q}{2a}dz$, and $r' \to z$:

$$V(\mathbf{r}) = \frac{1}{4\pi\epsilon_0}\sum_{n=0}^{\infty}\frac{1}{r^{n+1}}\int_{-a}^{a} z^n P_n(\cos\theta)\frac{Q}{2a}\, dz.$$

The integral is

$$\frac{Q}{2a}P_n(\cos\theta)\int_{-a}^{a} z^n\, dz = \frac{Q}{2a}P_n(\cos\theta)\left.\frac{z^{n+1}}{n+1}\right|_{-a}^{a} = \frac{Q}{2a}P_n(\cos\theta)\frac{2a^{n+1}}{n+1} \quad \text{for } n \text{ even, zero for } n \text{ odd.}$$

Therefore

$$V = \frac{Q}{4\pi\epsilon_0}\frac{1}{r}\sum_{n=0,2,4,\ldots}\left[\frac{1}{n+1}\left(\frac{a}{r}\right)^n P_n(\cos\theta)\right]. \quad \text{qed}$$

Problem 3.39

Use separation of variables in cylindrical coordinates (Prob. 3.23):

$$V(s,\phi) = a_0 + b_0\ln s + \sum_{k=1}^{\infty}\left[s^k(a_k\cos k\phi + b_k\sin k\phi) + s^{-k}(c_k\cos k\phi + d_k\sin k\phi)\right].$$

$$s < R: \quad V(s,\phi) = \sum_{k=1}^{\infty} s^k(a_k \cos k\phi + b_k \sin k\phi) \quad (\ln s \text{ and } s^{-k} \text{ blow up at } s = 0);$$
$$s > R: \quad V(s,\phi) = \sum_{k=1}^{\infty} s^{-k}(c_k \cos k\phi + d_k \sin k\phi) \quad (\ln s \text{ and } s^k \text{ blow up as } s \to \infty).$$

(We may as well pick constants so $V \to 0$ as $s \to \infty$, and hence $a_0 = 0$.) Continuity at $s = R \Rightarrow \sum R^k(a_k \cos k\phi + b_k \sin k\phi) = \sum R^{-k}(c_k \cos k\phi + d_k \sin k\phi)$, so $c_k = R^{2k}a_k$, $d_k = R^{2k}b_k$. Eq. 2.36 says: $\left.\frac{\partial V}{\partial s}\right|_{R^+} - \left.\frac{\partial V}{\partial s}\right|_{R^-} = -\frac{1}{\epsilon_0}\sigma$. Therefore

$$\sum \frac{-k}{R^{k+1}}(c_k \cos k\phi + d_k \sin k\phi) - \sum kR^{k-1}(a_k \cos k\phi + b_k \sin k\phi) = -\frac{1}{\epsilon_0}\sigma,$$

or:

$$\sum 2kR^{k-1}(a_k \cos k\phi + b_k \sin k\phi) = \left\{ \begin{array}{ll} \sigma_0/\epsilon_0 & (0 < \phi < \pi) \\ -\sigma_0/\epsilon_0 & (\pi < \phi < 2\pi) \end{array} \right\}.$$

Fourier's trick: multiply by $(\cos l\phi)\, d\phi$ and integrate from 0 to 2π, using

$$\int_0^{2\pi} \sin k\phi \cos l\phi \, d\phi = 0; \quad \int_0^{2\pi} \cos k\phi \cos l\phi \, d\phi = \left\{ \begin{array}{ll} 0, & k \neq l \\ \pi, & k = l \end{array} \right\}.$$

Then

$$2lR^{l-1}\pi a_l = \frac{\sigma_0}{\epsilon_0}\left[\int_0^{\pi} \cos l\phi \, d\phi - \int_{\pi}^{2\pi} \cos l\phi \, d\phi\right] = \frac{\sigma_0}{\epsilon_0}\left\{ \left.\frac{\sin l\phi}{l}\right|_0^{\pi} - \left.\frac{\sin l\phi}{l}\right|_{\pi}^{2\pi} \right\} = 0; \quad a_l = 0.$$

Multiply by $(\sin l\phi)\, d\phi$ and integrate, using $\int_0^{2\pi} \sin k\phi \sin l\phi \, d\phi = \left\{ \begin{array}{ll} 0, & k \neq l \\ \pi, & k = l \end{array} \right\}$:

$$\begin{aligned} 2lR^{l-1}\pi b_l &= \frac{\sigma_0}{\epsilon_0}\left[\int_0^{\pi} \sin l\phi \, d\phi - \int_{\pi}^{2\pi} \sin l\phi \, d\phi\right] = \frac{\sigma_0}{\epsilon_0}\left\{ -\left.\frac{\cos l\phi}{l}\right|_0^{\pi} + \left.\frac{\cos l\phi}{l}\right|_{\pi}^{2\pi} \right\} = \frac{\sigma_0}{l\epsilon_0}(2 - 2\cos l\pi) \\ &= \left\{ \begin{array}{ll} 0, & \text{if } l \text{ is even} \\ 4\sigma_0/l\epsilon_0, & \text{if } l \text{ is odd} \end{array} \right\} \Rightarrow b_l = \left\{ \begin{array}{ll} 0, & \text{if } l \text{ is even} \\ 2\sigma_0/\pi\epsilon_0 l^2 R^{l-1}, & \text{if } l \text{ is odd} \end{array} \right\}. \end{aligned}$$

Conclusion:

$$\boxed{V(s,\phi) = \frac{2\sigma_0 R}{\pi\epsilon_0} \sum_{k=1,3,5,\ldots} \frac{1}{k^2} \sin k\phi \left\{ \begin{array}{ll} (s/R)^k & (s < R) \\ (R/s)^k & (s > R) \end{array} \right\}.}$$

Problem 3.40

Use Eq. 3.95, in the form $V(\mathbf{r}) = \frac{1}{4\pi\epsilon_0} \sum_{n=0}^{\infty} \frac{P_n(\cos\theta)}{r^{n+1}} I_n; \quad I_n = \int_{-a}^{a} z^n \lambda(z)\, dz.$

(a) $I_0 = k \int_{-a}^{a} \cos\left(\frac{\pi z}{2a}\right) dz = k\left[\frac{2a}{\pi} \sin\left(\frac{\pi z}{2a}\right)\right]\Bigg|_{-a}^{a} = \frac{2ak}{\pi}\left[\sin\left(\frac{\pi}{2}\right) - \sin\left(-\frac{\pi}{2}\right)\right] = \frac{4ak}{\pi}$. Therefore:

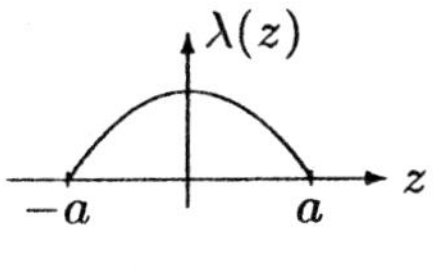

$$\boxed{V(r,\theta) \cong \frac{1}{4\pi\epsilon_0}\left(\frac{4ak}{\pi}\right)\frac{1}{r}.} \quad \text{(Monopole.)}$$

(b)

$$I_0 = 0.$$

$$I_1 = k\int_{-a}^{a} z\sin(\pi z/a)\,dz = k\left\{\left(\frac{a}{\pi}\right)^2\sin\left(\frac{\pi z}{a}\right) - \frac{az}{\pi}\cos\left(\frac{\pi z}{a}\right)\right\}\Bigg|_{-a}^{a}$$

$$= k\left\{\left(\frac{a}{\pi}\right)^2[\sin(\pi) - \sin(-\pi)] - \frac{a^2}{\pi}\cos(\pi) - \frac{a^2}{\pi}\cos(-\pi)\right\} = k\frac{2a^2}{\pi};$$

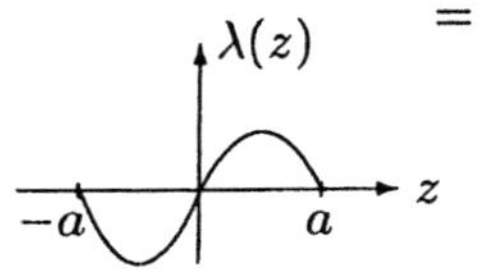

$$\boxed{V(r,\theta) \cong \frac{1}{4\pi\epsilon_0}\left(\frac{2a^2k}{\pi}\right)\frac{1}{r^2}\cos\theta.}\quad \text{(Dipole.)}$$

(c)

$$I_0 = I_1 = 0.$$

$$I_2 = k\int_{-a}^{a} z^2\cos\left(\frac{\pi z}{a}\right)dz = k\left\{\frac{2z\cos(\pi z/a)}{(\pi/a)^2} + \frac{(\pi z/a)^2 - 2}{(\pi/a)^3}\sin\left(\frac{\pi z}{a}\right)\right\}\Bigg|_{-a}^{a}$$

$$= 2k\left(\frac{a}{\pi}\right)^2[a\cos(\pi) + a\cos(-\pi)] = -\frac{4a^3k}{\pi^2}.$$

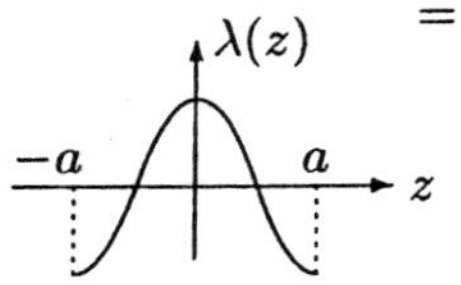

$$\boxed{V(r,\theta) \cong \frac{1}{4\pi\epsilon_0}\left(-\frac{4a^3k}{\pi^2}\right)\frac{1}{2r^3}(3\cos^2\theta - 1).}\quad \text{(Quadrupole.)}$$

Problem 3.41

(a) The average field due to a point charge q at $\mathbf{r}$ is

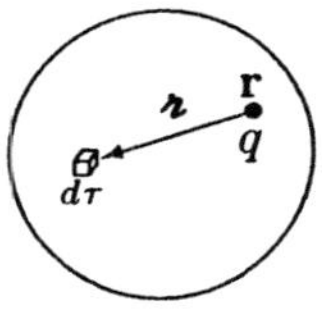

$$\mathbf{E}_{\text{ave}} = \frac{1}{\left(\frac{4}{3}\pi\epsilon_0 R^3\right)}\int \mathbf{E}\,d\tau, \quad \text{where } \mathbf{E} = \frac{1}{4\pi\epsilon_0}\frac{q}{\mathscr{r}^2}\hat{\boldsymbol{\mathscr{r}}},$$

$$\text{so}\quad \mathbf{E}_{\text{ave}} = \frac{1}{\left(\frac{4}{3}\pi\epsilon_0 R^3\right)}\frac{1}{4\pi\epsilon_0}\int \rho\frac{\hat{\boldsymbol{\mathscr{r}}}}{\mathscr{r}^2}\,d\tau.$$

(Here $\mathbf{r}$ is the source point, $d\tau$ is the field point, so $\boldsymbol{\mathscr{r}}$ goes from $\mathbf{r}$ to $d\tau$.) The field at $\mathbf{r}$ due to uniform charge ρ over the sphere is $\mathbf{E}_\rho = \frac{1}{4\pi\epsilon_0}\int \rho\frac{\hat{\boldsymbol{\mathscr{r}}}}{\mathscr{r}^2}\,d\tau$. This time $d\tau$ is the source point and $\mathbf{r}$ is the field point, so $\boldsymbol{\mathscr{r}}$ goes from $d\tau$ to $\mathbf{r}$, and hence carries the opposite sign. So with $\rho = -q/\left(\frac{4}{3}\pi R^3\right)$, the two expressions agree: $\mathbf{E}_{\text{ave}} = \mathbf{E}_\rho$.

(b) From Prob. 2.12:

$$\mathbf{E}_\rho = \frac{1}{3\epsilon_0}\rho\,\hat{\mathbf{r}} = -\frac{q}{4\pi\epsilon_0}\frac{\hat{\mathbf{r}}}{R^3} = -\frac{\mathbf{p}}{4\pi\epsilon_0 R^3}.$$

(c) If there are many charges inside the sphere, $\mathbf{E}_{\text{ave}}$ is the sum of the individual averages, and $\mathbf{p}_{\text{tot}}$ is the sum of the individual dipole moments. So $\mathbf{E}_{\text{ave}} = -\frac{\mathbf{p}}{4\pi\epsilon_0 R^3}$. qed

(d) The same argument, only with q placed at $\mathbf{r}$ *outside* the sphere, gives

$$\mathbf{E}_{\text{ave}} = \mathbf{E}_\rho = \frac{1}{4\pi\epsilon_0}\frac{\left(\frac{4}{3}\pi R^3\rho\right)}{r^2}\hat{\mathbf{r}}\ \text{(field at } \mathbf{r} \text{ due to uniformly charged sphere)} = \frac{1}{4\pi\epsilon_0}\frac{-q}{r^2}\hat{\mathbf{r}}.$$

But this is precisely the field produced by q (at $\mathbf{r}$) at the *center* of the sphere. So the average field (over the sphere) due to a point charge *outside* the sphere is the same as the field that same charge produces at the center. And by superposition, this holds for any *collection* of exterior charges.

Problem 3.42

(a)

$$\begin{aligned}\mathbf{E}_{\text{dip}} &= \frac{p}{4\pi\epsilon_0 r^3}(2\cos\theta\,\hat{\mathbf{r}} + \sin\theta\,\hat{\boldsymbol{\theta}}) \\ &= \frac{p}{4\pi\epsilon_0 r^3}[2\cos\theta(\sin\theta\cos\phi\,\hat{\mathbf{x}} + \sin\theta\sin\phi\,\hat{\mathbf{y}} + \cos\theta\,\hat{\mathbf{z}}) \\ &\quad + \sin\theta(\cos\theta\cos\phi\,\hat{\mathbf{x}} + \cos\theta\sin\phi\,\hat{\mathbf{y}} - \sin\theta\,\hat{\mathbf{z}})] \\ &= \frac{p}{4\pi\epsilon_0 r^3}\left[3\sin\theta\cos\theta\cos\phi\,\hat{\mathbf{x}} + 3\sin\theta\cos\theta\sin\phi\,\hat{\mathbf{y}} + \underbrace{\left(2\cos^2\theta - \sin^2\theta\right)}_{=3\cos^2\theta-1}\hat{\mathbf{z}}\right].\end{aligned}$$

$$\begin{aligned}\mathbf{E}_{\text{ave}} &= \frac{1}{\left(\frac{4}{3}\pi R^3\right)}\int \mathbf{E}_{\text{dip}}\,d\tau \\ &= \frac{1}{\left(\frac{4}{3}\pi R^3\right)}\left(\frac{p}{4\pi\epsilon_0}\right)\int\frac{1}{r^3}\left[3\sin\theta\cos\theta(\cos\phi\,\hat{\mathbf{x}} + \sin\phi\,\hat{\mathbf{y}}) + \left(3\cos^2\theta - 1\right)\hat{\mathbf{z}}\right] r^2\sin\theta\,dr\,d\theta\,d\phi.\end{aligned}$$

But $\int_0^{2\pi}\cos\phi\,d\phi = \int_0^{2\pi}\sin\phi\,d\phi = 0$, so the $\hat{\mathbf{x}}$ and $\hat{\mathbf{y}}$ terms drop out, and $\int_0^{2\pi} d\phi = 2\pi$, so

$$\mathbf{E}_{\text{ave}} = \frac{1}{\left(\frac{4}{3}\pi R^3\right)}\left(\frac{p}{4\pi\epsilon_0}\right)2\pi\int_0^R\frac{1}{r}\,dr\;\underbrace{\int_0^\pi\left(3\cos^2\theta - 1\right)\sin\theta\,d\theta}_{(-\cos^3\theta+\cos\theta)|_0^\pi=1-1+1-1=0}.$$

Evidently $\boxed{\mathbf{E}_{\text{ave}} = 0,}$ which contradicts the result of Prob. 3.41. [Note, however, that the r integral, $\int_0^R \frac{1}{r}\,dr$, blows up, since $\ln r \to -\infty$ as $r \to 0$. If, as suggested, we truncate the r integral at $r = \epsilon$, then it is finite, and the θ integral gives $\mathbf{E}_{\text{ave}} = 0$.]

(b) We want $\mathbf{E}$ within the ϵ-sphere to be a delta function: $\mathbf{E} = \mathbf{A}\delta^3(\mathbf{r})$, with $\mathbf{A}$ selected so that the *average* field is consistent with the general theorem in Prob. 3.41:

$$\mathbf{E}_{\text{ave}} = \frac{1}{\left(\frac{4}{3}\pi R^3\right)}\int \mathbf{A}\delta^3(\mathbf{r})\,d\tau = \frac{\mathbf{A}}{\left(\frac{4}{3}\pi R^3\right)} = -\frac{\mathbf{p}}{4\pi\epsilon_0 R^3} \Rightarrow \mathbf{A} = -\frac{\mathbf{p}}{3\epsilon_0}, \text{ and hence } \boxed{\mathbf{E} = -\frac{\mathbf{p}}{3\epsilon_0}\delta^3(\mathbf{r}).}$$

Problem 3.43

(a) $I = \int(\nabla V_1)\cdot(\nabla V_2)\,d\tau$. But $\nabla\cdot(V_1\nabla V_2) = (\nabla V_1)\cdot(\nabla V_2) + V_1(\nabla^2 V_2)$, so

$$I = \int\nabla\cdot(V_1\nabla V_2)\,d\tau - \int V_1(\nabla^2 V_2) = \oint_S V_1(\nabla V_2)\cdot d\mathbf{a} + \frac{1}{\epsilon_0}\int V_1\rho_2\,d\tau.$$

But the surface integral is over a huge sphere "at infinity", where V_1 and $V_2 \to 0$. So $I = \frac{1}{\epsilon_0}\int V_1\rho_2\,d\tau$. By the same argument, with 1 and 2 reversed, $I = \frac{1}{\epsilon_0}\int V_2\rho_1\,d\tau$. So $\int V_1\rho_2\,d\tau = \int V_2\rho_1\,d\tau$. qed

(b) $\left\{ \begin{array}{l} \textit{Situation (1)}: Q_a = \int_a \rho_1\, d\tau = Q;\ Q_b = \int_b \rho_1\, d\tau = 0;\ V_{1b} \equiv V_{ab}. \\ \textit{Situation (2)}: Q_a = \int_a \rho_2\, d\tau = 0;\ Q_b = \int_b \rho_2\, d\tau = Q;\ V_{2a} \equiv V_{ba}. \end{array} \right\}$

$$\left\{ \begin{array}{l} \int V_1\rho_2\, d\tau = V_{1a}\int_a \rho_2\, d\tau + V_{1b}\int_b \rho_2\, d\tau = V_{ab}Q. \\ \int V_2\rho_1\, d\tau = V_{2a}\int_a \rho_1\, d\tau + V_{2b}\int_b \rho_1\, d\tau = V_{ba}Q. \end{array} \right\}$$

Green's reciprocity theorem says $QV_{ab} = QV_{ba}$, so $V_{ab} = V_{ba}$. qed

Problem 3.44

(a) *Situation (1):* actual. *Situation (2):* right plate at V_0, left plate at $V = 0$, no charge at x.

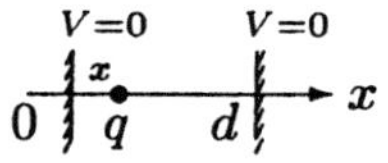

$$\int V_1\rho_2\, d\tau = V_{l_1}Q_{l_2} + V_{x_1}Q_{x_2} + V_{r_1}Q_{r_2}.$$

But $V_{l_1} = V_{r_1} = 0$ and $Q_{x_2} = 0$, so $\int V_1\rho_2\, d\tau = 0$.

$$\int V_2\rho_1\, d\tau = V_{l_2}Q_{l_1} + V_{x_2}Q_{x_1} + V_{r_2}Q_{r_1}.$$

But $V_{l_2} = 0$ $Q_{x_1} = q$, $V_{r_2} = V_0$, $Q_{r_1} = Q_2$, and $V_{x_2} = V_0(x/d)$. So $0 = V_0(x/d)q + V_0Q_2$, and hence

$$\boxed{Q_2 = -qx/d.}$$

Situation (1): actual. *Situation (2):* left plate at V_0, right plate at $V = 0$, no charge at x.

$$\int V_1\rho_2\, d\tau = 0 = \int V_2\rho_1\, d\tau = V_{l_2}Q_{l_1} + V_{x_2}Q_{x_1} + V_{r_2}Q_{r_1} = V_0Q_1 + qV_{x_2} + 0.$$

But $V_{x_2} = V_0\left(1 - \frac{x}{d}\right)$, so

$$\boxed{Q_1 = -q(1 - x/d).}$$

(b) *Situation (1):* actual. *Situation (2):* inner sphere at V_0, outer sphere at zero, no charge at r.

$$\int V_1\rho_2\, d\tau = V_{a_1}Q_{a_2} + V_{r_1}Q_{r_2} + V_{b_1}Q_{b_2}.$$

But $V_{a_1} = V_{b_1} = 0$, $Q_{r_2} = 0$. So $\int V_1\rho_2\, d\tau = 0$.

$$\int V_2\rho_1\, d\tau = V_{a_2}Q_{a_1} + V_{r_2}Q_{r_1} + V_{b_2}Q_{b_1} = Q_aV_0 + qV_{r_2} + 0.$$

But V_{r_2} is the potential at r in configuration 2: $V(r) = A + B/r$, with $V(a) = V_0 \Rightarrow A + B/a = V_0$, or $aA + B = aV_0$, and $V(b) = 0 \Rightarrow A + B/b = 0$, or $bA + B = 0$. Subtract: $(b-a)A = -aV_0 \Rightarrow A = -aV_0/(b-a)$; $B\left(\frac{1}{a} - \frac{1}{b}\right) = V_0 = B\frac{(b-a)}{ab} \Rightarrow B = abV_0/(b-a)$. So $V(r) = \frac{aV_0}{(b-a)}\left(\frac{b}{r} - 1\right)$. Therefore

$$Q_aV_0 + q\frac{aV_0}{(b-a)}\left(\frac{b}{r} - 1\right) = 0; \quad \boxed{Q_a = -\frac{qa}{(b-a)}\left(\frac{b}{r} - 1\right).}$$

Now let *Situation (2)* be: inner sphere at zero, outer at V_0, no charge at r.

$$\int V_1\rho_2\,d\tau = 0 = \int V_2\rho_1\,d\tau = V_{a_2}Q_{a_1} + V_{r_2}Q_{r_1} + V_{b_2}Q_{b_1} = 0 + qV_{r_2} + Q_bV_0.$$

This time įsplaystyle$V(r) = A + \frac{B}{r}$ with $V(a) = 0 \Rightarrow A + B/a = 0$; $V(b) = V_0 \Rightarrow A + B/b = V_0$, so $V(r) = \frac{bV_0}{(b-a)}\left(1 - \frac{a}{r}\right)$. Therefore, $q\frac{bV_0}{(b-a)}\left(1-\frac{a}{r}\right) + Q_bV_0 = 0$; $\boxed{Q_b = -\frac{qb}{(b-a)}\left(1-\frac{a}{r}\right).}$

Problem 3.45

(a) $$\frac{1}{2}\sum_{i,j=1}^{3}\hat{\mathbf{r}}_i\hat{\mathbf{r}}_jQ_{ij} = \frac{1}{2}\int\left\{3\sum_{i=1}^{3}\hat{\mathbf{r}}_ir_i'\sum_{j=1}^{3}\hat{\mathbf{r}}_jr_j' - (r')^2\sum_{i,j}\hat{\mathbf{r}}_i\hat{\mathbf{r}}_j\delta_{ij}\right\}\rho\,d\tau'$$

But $\sum_{i=1}^{3}\hat{\mathbf{r}}_ir_i' = \hat{\mathbf{r}}\cdot\mathbf{r}' = r'\cos\theta' = \sum_{j=1}^{3}\hat{\mathbf{r}}_jr_j'$; $\sum_{i,j}\hat{\mathbf{r}}_i\hat{\mathbf{r}}_j\delta_{ij} = \sum\hat{\mathbf{r}}_j\hat{\mathbf{r}}_j = \hat{\mathbf{r}}\cdot\hat{\mathbf{r}} = 1$. So

$V_{\text{quad}} = \frac{1}{4\pi\epsilon_0}\frac{1}{r^3}\int\frac{1}{2}\left(r'^2\cos^2\theta' - r'^2\right)\rho\,d\tau' = \frac{1}{4\pi\epsilon_0}\frac{1}{r^3}\int r'^2P_2(\cos\theta')\rho\,d\tau'$ (the $n = 2$ term in Eq. 3.95).

(b) Because $x^2 = y^2 = (a/2)^2$ for all four charges, $Q_{xx} = Q_{yy} = \left[3(a/2)^2 - (\sqrt{2}a/2)^2\right](q - q - q + q) = 0$. Because $z = 0$ for all four charges, $Q_{zz} = -(\sqrt{2}a/2)^2(q-q-q+q) = 0$ and $Q_{xz} = Q_{yz} = Q_{zx} = Q_{zy} = 0$. This leaves only

$$Q_{xy} = Q_{yx} = 3\left[\left(\frac{a}{2}\right)\left(\frac{a}{2}\right)q + \left(\frac{a}{2}\right)\left(-\frac{a}{2}\right)(-q) + \left(-\frac{a}{2}\right)\left(\frac{a}{2}\right)(-q) + \left(-\frac{a}{2}\right)\left(-\frac{a}{2}\right)q\right] = \boxed{3a^2q.}$$

(c)

$$\begin{aligned}\overline{Q}_{ij} &= \int\left[3(r_i - d_i)(r_j - d_j) - (\mathbf{r} - \mathbf{d})^2\delta_{ij}\right]\rho\,d\tau \quad \text{(I'll drop the primes, for simplicity.)}\\ &= \int\left[3r_ir_j - r^2\delta_{ij}\right]\rho\,d\tau - 3d_i\int r_j\rho\,d\tau - 3d_j\int r_i\rho\,d\tau + 3d_id_j\int\rho\,d\tau + 2\mathbf{d}\cdot\int\mathbf{r}\rho\,d\tau\,\delta_{ij}\\ &\quad - d^2\delta_{ij}\int\rho\,d\tau = Q_{ij} - 3(d_ip_j + d_jp_i) + 3d_id_jQ + 2\delta_{ij}\mathbf{d}\cdot\mathbf{p} - d^2\delta_{ij}Q.\end{aligned}$$

So if $\mathbf{p} = 0$ and $Q = 0$ then $\overline{Q}_{ij} = Q_{ij}$. qed

(d) Eq. 3.95 with $n = 3$:

$$V_{\text{oct}} = \frac{1}{4\pi\epsilon_0}\frac{1}{r^4}\int(r')^3P_3(\cos\theta')\rho\,d\tau'; \quad P_3(\cos\theta) = \frac{1}{2}\left(5\cos^3\theta - 3\cos\theta\right).$$

$$\boxed{V_{\text{oct}} = \frac{1}{4\pi\epsilon_0}\frac{\left(\frac{1}{2}\sum_{i,j,k}\hat{\mathbf{r}}_i\hat{\mathbf{r}}_j\hat{\mathbf{r}}_kQ_{ijk}\right)}{r^4},}$$

Define the "octopole moment" as

$$\boxed{Q_{ijk} \equiv \int\left(5r_i'r_j'r_k' - (r')^2(r_i'\delta_{jk} + r_j'\delta_{ik} + r_k'\delta_{ij})\right)\rho(\mathbf{r}')\,d\tau'.}$$

Problem 3.46

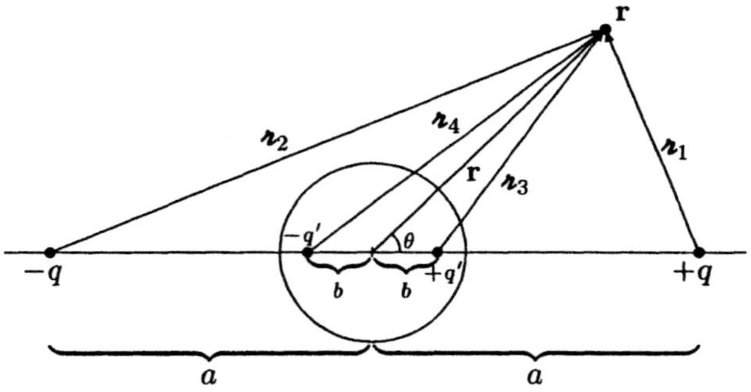

$$V = \frac{1}{4\pi\epsilon_0}\left\{q\left(\frac{1}{\imath_1}-\frac{1}{\imath_2}\right)+q'\left(\frac{1}{\imath_3}-\frac{1}{\imath_4}\right)\right\}$$

$$\begin{aligned}\imath_1 &= \sqrt{r^2+a^2-2ra\cos\theta},\\ \imath_2 &= \sqrt{r^2+a^2+2ra\cos\theta},\\ \imath_3 &= \sqrt{r^2+b^2-2rb\cos\theta},\\ \imath_4 &= \sqrt{r^2+b^2+2rb\cos\theta}.\end{aligned}$$

Expanding as in Ex. 3.10: $\left(\frac{1}{\imath_1}-\frac{1}{\imath_2}\right) \cong \frac{2r}{a^2}\cos\theta$ (we want $a \gg r$, not $r \gg a$, this time).

$$\begin{aligned}\left(\frac{1}{\imath_3}-\frac{1}{\imath_4}\right) &\cong \frac{2b}{r^2}\cos\theta \text{ (here we want } b \ll r, \text{ because } b = R^2/a, \text{ Eq. 3.16)}\\ &= \frac{2}{a}\frac{R^2}{r^2}\cos\theta.\end{aligned}$$

But $q' = -\frac{R}{a}q$ (Eq. 3.15), so

$$V(r,\theta) \cong \frac{1}{4\pi\epsilon_0}\left[q\frac{2r}{a^2}\cos\theta - \frac{R}{a}q\frac{2}{a}\frac{R^2}{r^2}\cos\theta\right] = \frac{1}{4\pi\epsilon_0}\left(\frac{2q}{a^2}\right)\left(r-\frac{R^3}{r^2}\right)\cos\theta.$$

Set $E_0 = -\frac{1}{4\pi\epsilon_0}\frac{2q}{a^2}$ (field in the vicinity of the sphere produced by $\pm q$):

$$\boxed{V(r,\theta) = -E_0\left(r-\frac{R^3}{r^2}\right)\cos\theta} \quad \text{(agrees with Eq. 3.76).}$$

Problem 3.47

The boundary conditions are

$$\left.\begin{array}{ll}\text{(i)} & V=0 \text{ when } y=0,\\ \text{(ii)} & V=V_0 \text{ when } y=a,\\ \text{(iii)} & V=0 \text{ when } x=b,\\ \text{(iv)} & V=0 \text{ when } x=-b.\end{array}\right\}$$

Go back to Eq. 3.26 and examine the case $k=0$: $d^2X/dx^2 = d^2Y/dy^2 = 0$, so $X(x) = Ax+B$, $Y(y) = Cy+D$. But this configuration is symmetric in x, so $A=0$, and hence the $k=0$ solution is $V(x,y) = Cy+D$. Pick $D=0$, $C=V_0/a$, and subtract off this part:

$$V(x,y) = V_0\frac{y}{a} + \bar{V}(x,y).$$

The *remainder* ($\bar{V}(x,y)$) satisfies boundary conditions similar to Ex. 3.4:

$$\left.\begin{array}{ll}\text{(i)} & \bar{V}=0 \text{ when } y=0,\\ \text{(ii)} & \bar{V}=0 \text{ when } y=a,\\ \text{(iii)} & \bar{V}=-V_0(y/a) \text{ when } x=b,\\ \text{(iv)} & \bar{V}=-V_0(y/a) \text{ when } x=-b.\end{array}\right\}$$

(The point of peeling off $V_0(y/a)$ was to recover (ii), on which the constraint $k = n\pi/a$ depends.)
The solution (following Ex. 3.4) is

$$\bar{V}(x,y) = \sum_{n=1}^{\infty} C_n \cosh(n\pi x/a)\sin(n\pi y/a),$$

and it remains to fit condition (iii):

$$\bar{V}(b,y) = \sum C_n \cosh(n\pi b/a)\sin(n\pi y/a) = -V_0(y/a).$$

Invoke Fourier's trick:

$$\sum C_n \cosh(n\pi b/a)\int_0^a \sin(n\pi y/a)\sin(n'\pi y/a)\,dy = -\frac{V_0}{a}\int_0^a y\sin(n'\pi y/a)\,dy,$$

$$\frac{a}{2}C_n \cosh(n\pi b/a) = -\frac{V_0}{a}\int_0^a y\sin(n\pi y/a)\,dy.$$

$$\begin{aligned} C_n &= -\frac{2V_0}{a^2\cosh(n\pi b/a)}\left[\left(\frac{a}{n\pi}\right)^2\sin(n\pi y/a) - \left(\frac{ay}{n\pi}\right)\cos(n\pi y/a)\right]\Bigg|_0^a \\ &= \frac{2V_0}{a^2\cosh(n\pi b/a)}\left(\frac{a^2}{n\pi}\right)\cos(n\pi) = \frac{2V_0}{n\pi}\frac{(-1)^n}{\cosh(n\pi b/a)}. \end{aligned}$$

$$V(x,y) = \boxed{V_0\left[\frac{y}{a} + \frac{2}{\pi}\sum_{n=1}^{\infty}\frac{(-1)^n}{n}\frac{\cosh(n\pi x/a)}{\cosh(n\pi b/a)}\sin(n\pi y/a)\right]}.$$

Problem 3.48

(a) Using Prob. 3.14b (with $b = a$):

$$V(x,y) = \frac{4V_0}{\pi}\sum_{n\ \mathrm{odd}}\frac{\sinh(n\pi x/a)\sin(n\pi y/a)}{n\sinh(n\pi)}.$$

$$\begin{aligned} \sigma(y) &= -\epsilon_0\frac{\partial V}{\partial x}\bigg|_{x=0} = -\epsilon_0\frac{4V_0}{\pi}\sum_{n\ \mathrm{odd}}\left(\frac{n\pi}{a}\right)\frac{\cosh(n\pi x/a)\sin(n\pi y/a)}{n\sinh(n\pi)}\bigg|_{x=0} \\ &= -\frac{4\epsilon_0 V_0}{a}\sum_{n\ \mathrm{odd}}\frac{\sin(n\pi y/a)}{\sinh(n\pi)}. \\ \lambda &= \int_0^a \sigma(y)\,dy = -\frac{4\epsilon_0 V_0}{a}\sum_{n\ \mathrm{odd}}\frac{1}{\sinh(n\pi)}\int_0^a \sin(n\pi y/a)\,dy. \end{aligned}$$

But $\displaystyle\int_0^a \sin(n\pi y/a)\,dy = -\frac{a}{n\pi}\cos(n\pi y/a)\Big|_0^a = \frac{a}{n\pi}[1-\cos(n\pi)] = \frac{2a}{n\pi}$ (since n is odd).

$$= -\frac{8\epsilon_0 V_0}{\pi}\sum_{n\ \mathrm{odd}}\frac{1}{n\sinh(n\pi)} = \boxed{-\frac{\epsilon_0 V_0}{\pi}\ln 2.}$$

[I have not found a way to sum this series analytically. Mathematica gives the numerical value 0.0866434, which agrees precisely with $\ln 2/8$.]

Using Prob. 3.47 (with $b = a/2$):

$$V(x,y) = V_0 \left[\frac{y}{a} + \frac{2}{\pi}\sum_n \frac{(-1)^n \cosh(n\pi x/a)\sin(n\pi y/a)}{n\cosh(n\pi/2)}\right].$$

$$\begin{aligned}
\sigma(x) &= -\epsilon_0 \frac{\partial V}{\partial y}\bigg|_{y=0} = -\epsilon_0 V_0 \left[\frac{1}{a} + \frac{2}{\pi}\sum_n \left(\frac{n\pi}{a}\right)\frac{(-1)^n \cosh(n\pi x/a)\cos(n\pi y/a)}{n\cosh(n\pi/2)}\right]\bigg|_{y=0} \\
&= -\epsilon_0 V_0 \left[\frac{1}{a} + \frac{2}{a}\sum_n \frac{(-1)^n \cosh(n\pi x/a)}{\cosh(n\pi/2)}\right] = -\frac{\epsilon_0 V_0}{a}\left[1 + 2\sum_n \frac{(-1)^n \cosh(n\pi x/a)}{\cosh(n\pi/2)}\right]. \\
\lambda &= \int_{-a/2}^{a/2} \sigma(x)\,dx = -\frac{\epsilon_0 V_0}{a}\left[a + 2\sum_n \frac{(-1)^n}{\cosh(n\pi/2)}\int_{-a/2}^{a/2}\cosh(n\pi x/a)\,dx\right].
\end{aligned}$$

But $\displaystyle\int_{-a/2}^{a/2}\cosh(n\pi x/a)\,dx = \frac{a}{n\pi}\sinh(n\pi x/a)\bigg|_{-a/2}^{a/2} = \frac{2a}{n\pi}\sinh(n\pi/2).$

$$\begin{aligned}
&= -\frac{\epsilon_0 V_0}{a}\left[a + \frac{4a}{\pi}\sum_n \frac{(-1)^n \tanh(n\pi/2)}{n}\right] = -\epsilon_0 V_0\left[1 + \frac{4}{\pi}\sum_n \frac{(-1)^n \tanh(n\pi/2)}{n}\right] \\
&= \boxed{-\frac{\epsilon_0 V_0}{\pi}\ln 2.}
\end{aligned}$$

[Again, I have not found a way to sum this series analytically. The numerical value is -0.612111, which agrees with the expected value $(\ln 2 - \pi)/4$.]

(b) From Prob. 3.23:

$$V(s,\phi) = a_0 + b_0 \ln s + \sum_{k=1}^{\infty}\left(a_k s^k + b_k \frac{1}{s^k}\right)[c_k\cos(k\phi) + d_k\sin(k\phi)].$$

In the interior ($s < R$) b_0 and b_k must be zero ($\ln s$ and $1/s$ blow up at the origin). Symmetry $\Rightarrow d_k = 0$. So

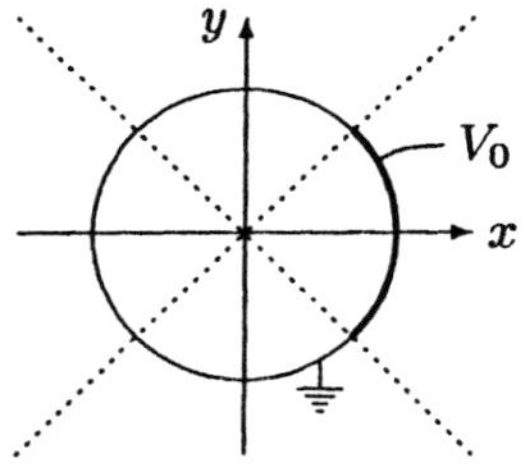

$$V(s,\phi) = a_0 + \sum_{k=1}^{\infty} a_k s^k \cos(k\phi).$$

At the surface:

$$V(R,\phi) = \sum_{k=0} a_k R^k \cos(k\phi) = \begin{cases} V_0, & \text{if } -\pi/4 < \phi < \pi/4, \\ 0, & \text{otherwise.}\end{cases}$$

Fourier's trick: multiply by $\cos(k'\phi)$ and integrate from $-\pi$ to π:

$$\sum_{k=0}^{\infty} a_k R^k \int_{-\pi}^{\pi}\cos(k\phi)\cos(k'\phi)\,d\phi = V_0\int_{-\pi/4}^{\pi/4}\cos(k'\phi)\,d\phi = \begin{cases} V_0 \sin(k'\phi)/k'\Big|_{-\pi/4}^{\pi/4} = (V_0/k')\sin(k'\pi/4), & \text{if } k' \neq 0, \\ V_0\pi/2, & \text{if } k' = 0.\end{cases}$$

But

$$\int_{-\pi}^{\pi}\cos(k\phi)\cos(k'\phi)\,d\phi = \begin{cases} 0, & \text{if } k \neq k' \\ 2\pi, & \text{if } k = k' = 0, \\ \pi, & \text{if } k = k' \neq 0.\end{cases}$$

So $2\pi a_0 = V_0\pi/2 \Rightarrow a_0 = V_0/4$; $\pi a_k R^k = (2V_0/k)\sin(k\pi/4) \Rightarrow a_k = (2V_0/\pi k R^k)\sin(k\pi/4)$ $(k \neq 0)$; hence

$$V(s,\phi) = V_0\left[\frac{1}{4} + \frac{2}{\pi}\sum_{k=1}^{\infty}\frac{\sin(k\pi/4)}{k}\left(\frac{s}{R}\right)^k\cos(k\phi)\right].$$

Using Eq. 2.49, and noting that in this case $\hat{\mathbf{n}} = -\hat{\mathbf{s}}$:

$$\sigma(\phi) = \epsilon_0\left.\frac{\partial V}{\partial s}\right|_{s=R} = \epsilon_0 V_0\frac{2}{\pi}\sum_{k=1}^{\infty}\left.\frac{\sin(k\pi/4)}{kR^k}ks^{k-1}\cos(k\phi)\right|_{s=R} = \frac{2\epsilon_0 V_0}{\pi R}\sum_{k=1}^{\infty}\sin(k\pi/4)\cos(k\phi).$$

We want the net (line) charge on the segment opposite to V_0 $(-\pi < \phi < -3\pi/4$ and $3\pi/4 < \phi < \pi)$:

$$\begin{aligned}\lambda &= \int\sigma(\phi)R\,d\phi = 2R\int_{3\pi/4}^{\pi}\sigma(\phi)\,d\phi = \frac{4\epsilon_0 V_0}{\pi}\sum_{k=1}^{\infty}\sin(k\pi/4)\int_{3\pi/4}^{\pi}\cos(k\phi)\,d\phi \\ &= \frac{4\epsilon_0 V_0}{\pi}\sum_{k=1}^{\infty}\sin(k\pi/4)\left[\left.\frac{\sin(k\phi)}{k}\right|_{3\pi/4}^{\pi}\right] = -\frac{4\epsilon_0 V_0}{\pi}\sum_{k=1}^{\infty}\frac{\sin(k\pi/4)\sin(3k\pi/4)}{k}.\end{aligned}$$

k	$\sin(k\pi/4)$	$\sin(3k\pi/4)$	product
1	$1/\sqrt{2}$	$1/\sqrt{2}$	1/2
2	1	-1	-1
3	$1/\sqrt{2}$	$1/\sqrt{2}$	1/2
4	0	0	0
5	$-1/\sqrt{2}$	$-1/\sqrt{2}$	1/2
6	-1	1	-1
7	$-1/\sqrt{2}$	$-1/\sqrt{2}$	1/2
8	0	0	0

$$\lambda = -\frac{4\epsilon_0 V_0}{\pi}\left[\frac{1}{2}\sum_{1,3,5\ldots}\frac{1}{k} - \sum_{2,6,10,\ldots}\frac{1}{k}\right] = -\frac{4\epsilon_0 V_0}{\pi}\left[\frac{1}{2}\sum_{1,3,5\ldots}\frac{1}{k} - \frac{1}{2}\sum_{1,3,5,\ldots}\frac{1}{k}\right] = 0.$$

Ouch! What went wrong? The problem is that the series $\sum(1/k)$ is divergent, so the "subtraction" $\infty - \infty$ is suspect. One way to avoid this is to go back to $V(s,\phi)$, calculate $\epsilon_0(\partial V/\partial s)$ at $s \neq R$, and save the limit

$s \to R$ until the end:

$$\begin{aligned}
\sigma(\phi, s) &\equiv \epsilon_0 \frac{\partial V}{\partial s} = \frac{2\epsilon_0 V_0}{\pi} \sum_{k=1}^{\infty} \frac{\sin(k\pi/4)}{k} \frac{k s^{k-1}}{R^k} \cos(k\phi) \\
&= \frac{2\epsilon_0 V_0}{\pi R} \sum_{k=1}^{\infty} x^{k-1} \sin(k\pi/4)\cos(k\phi) \quad \text{(where } x \equiv s/R \to 1 \text{ at the end).} \\
\lambda(x) &\equiv \sigma(\phi, s) R\, d\phi = -\frac{4\epsilon_0 V_0}{\pi} \sum_{k=1}^{\infty} \frac{1}{k} x^{k-1} \sin(k\pi/4)\sin(3k\pi/4) \\
&= -\frac{4\epsilon_0 V_0}{\pi}\left[\frac{1}{2x}\left(x + \frac{x^3}{3} + \frac{x^5}{5} + \cdots\right) - \frac{1}{x}\left(\frac{x^2}{2} + \frac{x^6}{6} + \frac{x^{10}}{10} + \cdots\right)\right] \\
&= -\frac{2\epsilon_0 V_0}{\pi x}\left[\left(x + \frac{x^3}{3} + \frac{x^5}{5} + \cdots\right) - \left(x^2 + \frac{x^6}{3} + \frac{x^{10}}{5} + \cdots\right)\right].
\end{aligned}$$

$$\text{But (see math tables)} : \ln\left(\frac{1+x}{1-x}\right) = 2\left(x + \frac{x^3}{3} + \frac{x^5}{5} + \cdots\right).$$

$$\begin{aligned}
&= -\frac{2\epsilon_0 V_0}{\pi x}\left[\frac{1}{2}\ln\left(\frac{1+x}{1-x}\right) - \frac{1}{2}\ln\left(\frac{1+x^2}{1-x^2}\right)\right] = -\frac{\epsilon_0 V_0}{\pi x}\ln\left[\left(\frac{1+x}{1-x}\right)\left(\frac{1+x^2}{1-x^2}\right)\right] \\
&= -\frac{\epsilon_0 V_0}{\pi x}\ln\left[\frac{(1+x)^2}{1+x^2}\right]; \quad \lambda = \lim_{x\to 1}\lambda(x) = \boxed{\frac{-\epsilon_0 V_0}{\pi}\ln 2.}
\end{aligned}$$

Problem 3.49

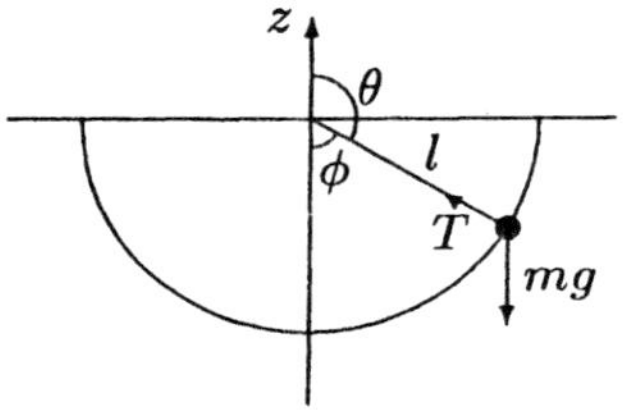

$$\mathbf{F} = q\mathbf{E} = \frac{qp}{4\pi\epsilon_0 r^3}(2\cos\theta\,\hat{\mathbf{r}} + \sin\theta\,\hat{\boldsymbol{\theta}}).$$

Now consider the pendulum: $\mathbf{F} = -mg\,\hat{\mathbf{z}} - T\,\hat{\mathbf{r}}$, where $T - mg\cos\phi = mv^2/l$ and (by conservation of energy) $mgl\cos\phi = (1/2)mv^2 \Rightarrow v^2 = 2gl\cos\phi$ (assuming it started from rest at $\phi = 90°$, as stipulated). But $\cos\phi = -\cos\theta$, so $T = mg(-\cos\theta) + (m/l)(-2gl\cos\theta) = -3mg\cos\theta$, and hence

$$\mathbf{F} = -mg(\cos\theta\,\hat{\mathbf{r}} - \sin\theta\,\hat{\boldsymbol{\theta}}) + 3mg\cos\theta\,\hat{\mathbf{r}} = mg(2\cos\theta\,\hat{\mathbf{r}} + \sin\theta\,\hat{\boldsymbol{\theta}}).$$

This total force is such as to keep the pendulum on a circular arc, and it is identical to the force on q in the field of a dipole, with $mg \leftrightarrow qp/4\pi\epsilon_0 l^3$. Evidently q also executes semicircular motion, as though it were on a tether of fixed length l.

Chapter 4

Electrostatic Fields in Matter

Problem 4.1

$E = V/x = 500/10^{-3} = 5\times10^5$. Table 4.1: $\alpha/4\pi\epsilon_0 = 0.66\times10^{-30}$, so $\alpha = 4\pi(8.85\times10^{-12})(0.66\times10^{-30}) = 7.34\times10^{-41}$. $p = \alpha E = ed \Rightarrow d = \alpha E/e = (7.34\times10^{-41})(5\times10^5)/(1.6\times10^{-19}) = 2.29\times10^{-16}\,\text{m}$. $d/R = (2.29\times10^{-16})/(0.5\times10^{-10}) = \boxed{4.6\times10^{-6}.}$ To ionize, say $d = R$. Then $R = \alpha E/e = \alpha V/ex \Rightarrow V = Rex/\alpha = (0.5\times10^{-10})(1.6\times10^{-19})(10^{-3})/(7.34\times10^{-41}) = \boxed{10^8\,\text{V}.}$

Problem 4.2

First find the field, at radius r, using Gauss' law: $\int \mathbf{E}\cdot d\mathbf{a} = \frac{1}{\epsilon_0}Q_{\text{enc}}$, or $E = \frac{1}{4\pi\epsilon_0}\frac{1}{r^2}Q_{\text{enc}}$.

$$\begin{aligned} Q_{\text{enc}} &= \int_0^r \rho\,d\tau = \frac{4\pi q}{\pi a^3}\int_0^r e^{-2\bar{r}/a}\bar{r}^2\,d\bar{r} = \frac{4q}{a^3}\left[-\frac{a}{2}e^{-2\bar{r}/a}\left(\bar{r}^2 + a\bar{r} + \frac{a^2}{2}\right)\right]\Bigg|_0^r \\ &= -\frac{2q}{a^2}\left[e^{-2r/a}\left(r^2 + ar + \frac{a^2}{2}\right) - \frac{a^2}{2}\right] = q\left[1 - e^{-2r/a}\left(1 + 2\frac{r}{a} + 2\frac{r^2}{a^2}\right)\right]. \end{aligned}$$

[*Note:* $Q_{\text{enc}}(r\to\infty) = q$.] So the field of the electron cloud is $E_e = \frac{1}{4\pi\epsilon_0}\frac{q}{r^2}\left[1 - e^{-2r/a}\left(1 + 2\frac{r}{a} + 2\frac{r^2}{a^2}\right)\right]$. The proton will be shifted from $r = 0$ to the point d where $E_e = E$ (the external field):

$$E = \frac{1}{4\pi\epsilon_0}\frac{q}{d^2}\left[1 - e^{-2d/a}\left(1 + 2\frac{d}{a} + 2\frac{d^2}{a^2}\right)\right].$$

Expanding in powers of (d/a):

$$\begin{aligned} e^{-2d/a} &= 1 - \left(\frac{2d}{a}\right) + \frac{1}{2}\left(\frac{2d}{a}\right)^2 - \frac{1}{3!}\left(\frac{2d}{a}\right)^3 + \cdots = 1 - 2\frac{d}{a} + 2\left(\frac{d}{a}\right)^2 - \frac{4}{3}\left(\frac{d}{a}\right)^3 + \cdots \\ 1 - e^{-2d/a}\left(1 + 2\frac{d}{a} + 2\frac{d^2}{a^2}\right) &= 1 - \left(1 - 2\frac{d}{a} + 2\left(\frac{d}{a}\right)^2 - \frac{4}{3}\left(\frac{d}{a}\right)^3 + \cdots\right)\left(1 + 2\frac{d}{a} + 2\frac{d^2}{a^2}\right) \\ &= \cancel{1} - \cancel{1} - \cancel{2\frac{d}{a}} - \cancel{2\frac{d^2}{a^2}} + \cancel{2\frac{d}{a}} + \cancel{4\frac{d^2}{a^2}} + \cancel{4\frac{d^3}{a^3}} - \cancel{2\frac{d^2}{a^2}} - \cancel{4\frac{d^3}{a^3}} + \frac{4}{3}\frac{d^3}{a^3} + \cdots \\ &= \frac{4}{3}\left(\frac{d}{a}\right)^3 + \text{higher order terms.} \end{aligned}$$

$$E = \frac{1}{4\pi\epsilon_0}\frac{q}{d^2}\left(\frac{4}{3}\frac{d^3}{a^3}\right) = \frac{1}{4\pi\epsilon_0}\frac{4}{3a^3}(qd) = \frac{1}{3\pi\epsilon_0 a^3}p. \quad \boxed{\alpha = 3\pi\epsilon_0 a^3.}$$

[Not so different from the *uniform* sphere model of Ex. 4.1 (see Eq. 4.2). Note that this result predicts $\frac{1}{4\pi\epsilon_0}\alpha = \frac{3}{4}a^3 = \frac{3}{4}\left(0.5\times 10^{-10}\right)^3 = 0.09\times 10^{-30}\,\mathrm{m}^3$, compared with an experimental value (Table 4.1) of $0.66\times 10^{-30}\,\mathrm{m}^3$. Ironically the "classical" formula (Eq. 4.2) is slightly *closer* to the empirical value.]

Problem 4.3

$\rho(r) = Ar$. Electric field (by Gauss's Law): $\oint \mathbf{E}\cdot d\mathbf{a} = E\left(4\pi r^2\right) = \frac{1}{\epsilon_0}Q_{\text{enc}} = \frac{1}{\epsilon_0}\int_0^r A\bar{r}\,4\pi\bar{r}^2\,d\bar{r}$, or $E = \frac{1}{4\pi r^2}\frac{4\pi A}{\epsilon_0}\frac{r^4}{4} = \frac{Ar^2}{4\epsilon_0}$. This "internal" field balances the external field $\mathbf{E}$ when nucleus is "off-center" an amount d: $ad^2/4\epsilon_0 = E \Rightarrow d = \sqrt{4\epsilon_0 E/A}$. So the induced dipole moment is $p = ed = 2e\sqrt{\epsilon_0/A}\sqrt{E}$. Evidently $\boxed{p \text{ is proportional to } E^{1/2}.}$

For Eq. 4.1 to hold in the weak-field limit, *E must be proportional to r*, for small r, which means that ρ must go to a constant (not zero) at the origin: $\boxed{\rho(0)\neq 0}$ (nor infinite).

Problem 4.4

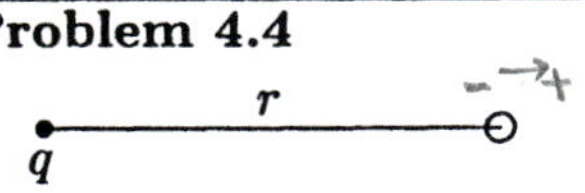

Field of q: $\frac{1}{4\pi\epsilon_0}\frac{q}{r^2}\,\hat{\mathbf{r}}$. Induced dipole moment of atom: $\mathbf{p} = \alpha\mathbf{E} = \frac{\alpha q}{4\pi\epsilon_0 r^2}\,\hat{\mathbf{r}}$.

Field of this dipole, at location of q ($\theta = \pi$, in Eq. 3.103): $E = \frac{1}{4\pi\epsilon_0}\frac{1}{r^3}\left(\frac{2\alpha q}{4\pi\epsilon_0 r^2}\right)$ (to the right).

Force on q due to this field: $\boxed{F = 2\alpha\left(\frac{q}{4\pi\epsilon_0}\right)^2\frac{1}{r^3}}$ (attractive).

Problem 4.5

Field of $\mathbf{p}_1$ at $\mathbf{p}_2$ ($\theta = \pi/2$ in Eq. 3.103): $\mathbf{E}_1 = \frac{p_1}{4\pi\epsilon_0 r^3}\,\hat{\boldsymbol{\theta}}$ (points *down*).

Torque on $\mathbf{p}_2$: $\mathbf{N}_2 = \mathbf{p}_2\times\mathbf{E}_1 = p_2E_1\sin 90^\circ = p_2E_1 = \boxed{\frac{p_1p_2}{4\pi\epsilon_0 r^3}}$ (points *into* the page).

Field of $\mathbf{p}_2$ at $\mathbf{p}_1$ ($\theta = \pi$ in Eq. 3.103): $\mathbf{E}_2 = \frac{p_2}{4\pi\epsilon_0 r^3}(-2\,\hat{\mathbf{r}})$ (points to the *right*).

Torque on $\mathbf{p}_1$: $\mathbf{N}_1 = \mathbf{p}_1\times\mathbf{E}_2 = \boxed{\frac{2p_1p_2}{4\pi\epsilon_0 r^3}}$ (points *into* the page).

Problem 4.6

(a)

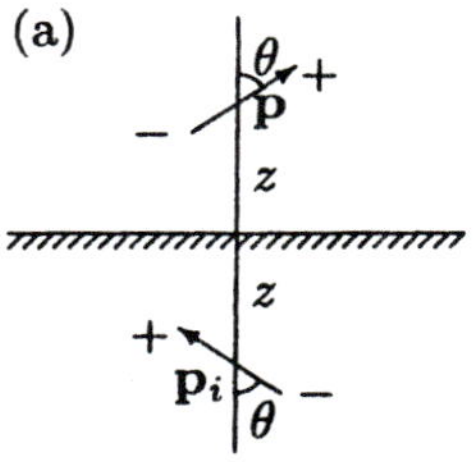

Use image dipole as shown in Fig. (a). Redraw, placing $\mathbf{p}_i$ at the origin, Fig. (b).

$$\mathbf{E}_i = \frac{p}{4\pi\epsilon_0(2z)^3}(2\cos\theta\,\hat{\mathbf{r}} + \sin\theta\,\hat{\boldsymbol{\theta}}); \quad \mathbf{p} = p\cos\theta\,\hat{\mathbf{r}} + p\sin\theta\,\hat{\boldsymbol{\theta}}.$$

(b)

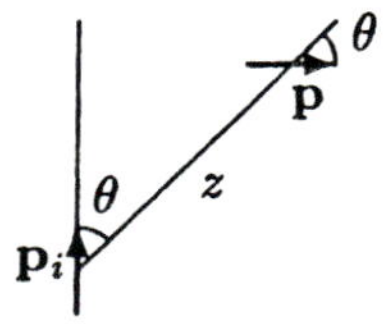

$$\begin{aligned}\mathbf{N} &= \mathbf{p}\times\mathbf{E}_i = \frac{p^2}{4\pi\epsilon_0(2z)^3}\left[(\cos\theta\,\hat{\mathbf{r}} + \sin\theta\,\hat{\boldsymbol{\theta}})\times(2\cos\theta\,\hat{\mathbf{r}} + \sin\theta\,\hat{\boldsymbol{\theta}})\right]\\ &= \frac{p^2}{4\pi\epsilon_0(2z)^3}\left[\cos\theta\sin\theta\,\hat{\boldsymbol{\phi}} + 2\sin\theta\cos\theta(-\hat{\boldsymbol{\phi}})\right]\\ &= \frac{p^2\sin\theta\cos\theta}{4\pi\epsilon_0(2z)^3}(-\hat{\boldsymbol{\phi}}) \quad \text{(out of the page).}\end{aligned}$$

But $\sin\theta\cos\theta = (1/2)\sin 2\theta$, so $\boxed{N = \dfrac{p^2 \sin 2\theta}{4\pi\epsilon_0(16z^3)}}$ (out of the page).

For $0 < \theta < \pi/2$, $\mathbf{N}$ tends to rotate $\mathbf{p}$ counterclockwise; for $\pi/2 < \theta < \pi$, $\mathbf{N}$ rotates $\mathbf{p}$ clockwise. Thus the stable orientation is perpendicular to the surface—either $\uparrow$ or $\downarrow$.

Problem 4.7

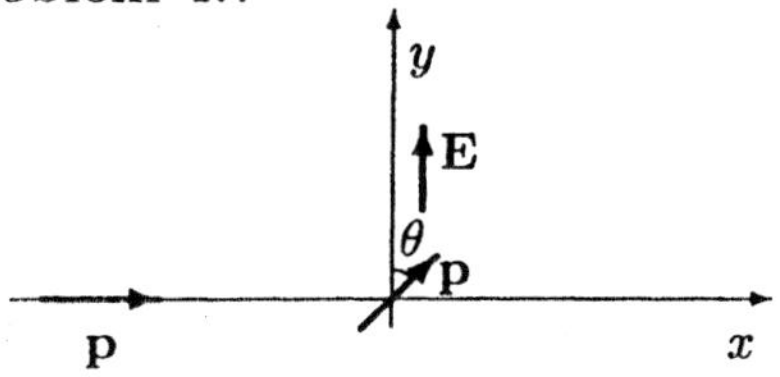

Say the field is uniform and points in the y direction. First slide $\mathbf{p}$ in from infinity along the x axis—this takes no work, since $\mathbf{F}$ is $\perp$ $d\mathbf{l}$. (If $\mathbf{E}$ is *not* uniform, slide $\mathbf{p}$ in along a trajectory $\perp$ the field.) Now rotate (counterclockwise) into final position. The torque exerted by $\mathbf{E}$ is $\mathbf{N} = \mathbf{p}\times\mathbf{E} = pE\sin\theta\,\hat{\mathbf{z}}$. The torque *we* exert is $N = pE\sin\theta$ *clockwise*, and $d\theta$ is *counterclockwise*, so the net work done by us is *negative*:

$U = \int_{\pi/2}^{\theta} pE\sin\bar\theta\, d\bar\theta = pE\left(-\cos\bar\theta\right)\Big|_{\pi/2}^{\theta} = -pE\left(\cos\theta - \cos\frac{\pi}{2}\right) = -pE\cos\theta = -\mathbf{p}\cdot\mathbf{E}$. qed

Problem 4.8

$U = -\mathbf{p}_1\cdot\mathbf{E}_2$, but $\mathbf{E}_2 = \frac{1}{4\pi\epsilon_0}\frac{1}{r^3}\left[3(\mathbf{p}_2\cdot\hat{\mathbf{r}})\hat{\mathbf{r}} - \mathbf{p}_2\right]$. So $U = \frac{1}{4\pi\epsilon_0}\frac{1}{r^3}\left[\mathbf{p}_1\cdot\mathbf{p}_2 - 3(\mathbf{p}_1\cdot\hat{\mathbf{r}})(\mathbf{p}_2\cdot\hat{\mathbf{r}})\right]$. qed

Problem 4.9

(a) $\mathbf{F} = (\mathbf{p}\cdot\boldsymbol{\nabla})\mathbf{E}$ (Eq. 4.5); $\mathbf{E} = \dfrac{1}{4\pi\epsilon_0}\dfrac{q}{r^2}\hat{\mathbf{r}} = \dfrac{q}{4\pi\epsilon_0}\dfrac{x\,\hat{\mathbf{x}} + y\,\hat{\mathbf{y}} + z\,\hat{\mathbf{z}}}{(x^2+y^2+z^2)^{3/2}}$.

$$\begin{aligned}
F_x &= \left(p_x\frac{\partial}{\partial x} + p_y\frac{\partial}{\partial y} + p_z\frac{\partial}{\partial z}\right)\frac{q}{4\pi\epsilon_0}\frac{x}{(x^2+y^2+z^2)^{3/2}}\\
&= \frac{q}{4\pi\epsilon_0}\left\{p_x\left[\frac{1}{(x^2+y^2+z^2)^{3/2}} - \frac{3}{2}x\frac{2x}{(x^2+y^2+z^2)^{5/2}}\right] + p_y\left[-\frac{3}{2}x\frac{2y}{(x^2+y^2+z^2)^{5/2}}\right]\right.\\
&\quad\left. + p_z\left[-\frac{3}{2}x\frac{2z}{(x^2+y^2+z^2)^{5/2}}\right]\right\} = \frac{q}{4\pi\epsilon_0}\left[\frac{p_x}{r^3} - \frac{3x}{r^5}(p_x x + p_y y + p_z z)\right] = \frac{q}{4\pi\epsilon_0}\left[\frac{\mathbf{p}}{r^3} - \frac{3\mathbf{r}(\mathbf{p}\cdot\mathbf{r})}{r^5}\right]_x.\\
\mathbf{F} &= \boxed{\frac{1}{4\pi\epsilon_0}\frac{q}{r^3}\left[\mathbf{p} - 3(\mathbf{p}\cdot\hat{\mathbf{r}})\hat{\mathbf{r}}\right].}
\end{aligned}$$

(b) $\mathbf{E} = \dfrac{1}{4\pi\epsilon_0}\dfrac{1}{r^3}\left\{3\left[\mathbf{p}\cdot(-\hat{\mathbf{r}})\right](-\hat{\mathbf{r}}) - \mathbf{p}\right\} = \dfrac{1}{4\pi\epsilon_0}\dfrac{1}{r^3}\left[3(\mathbf{p}\cdot\hat{\mathbf{r}})\hat{\mathbf{r}} - \mathbf{p}\right]$. (This is from Eq. 3.104; the minus signs are because $\mathbf{r}$ points *toward* $\mathbf{p}$, in this problem.)

$$\mathbf{F} = q\mathbf{E} = \boxed{\frac{1}{4\pi\epsilon_0}\frac{q}{r^3}\left[3(\mathbf{p}\cdot\hat{\mathbf{r}})\hat{\mathbf{r}} - \mathbf{p}\right].}$$

[Note that the forces are equal and opposite, as you would expect from Newton's third law.]

Problem 4.10

(a) $\sigma_b = \mathbf{P}\cdot\hat{\mathbf{n}} = \boxed{kR;}\ \rho_b = -\boldsymbol{\nabla}\cdot\mathbf{P} = -\dfrac{1}{r^2}\dfrac{\partial}{\partial r}(r^2 kr) = -\dfrac{1}{r^2}3kr^2 = \boxed{-3k.}$

(b) For $r < R$, $\mathbf{E} = \frac{1}{3\epsilon_0}\rho r\,\hat{\mathbf{r}}$ (Prob. 2.12), so $\mathbf{E} = \boxed{-(k/\epsilon_0)\,\mathbf{r}.}$

For $r > R$, same as if all charge at center; but $Q_{\text{tot}} = (kR)(4\pi R^2) + (-3k)(\frac{4}{3}\pi R^3) = 0$, so $\boxed{\mathbf{E} = 0.}$

Problem 4.11

$\rho_b = 0$; $\sigma_b = \mathbf{P}\cdot\hat{\mathbf{n}} = \pm P$ (plus sign at one end—the one $\mathbf{P}$ points *toward*; minus sign at the other—the one $\mathbf{P}$ points *away* from).

(i) $L \gg a$. Then the ends look like point charges, and the whole thing is like a physical dipole, of length L and charge $P\pi a^2$. See Fig. (a).

(ii) $L \ll a$. Then it's like a circular parallel-plate capacitor. Field is nearly uniform inside; nonuniform "fringing field" at the edges. See Fig. (b).

(iii) $L \approx a$. See Fig. (c).

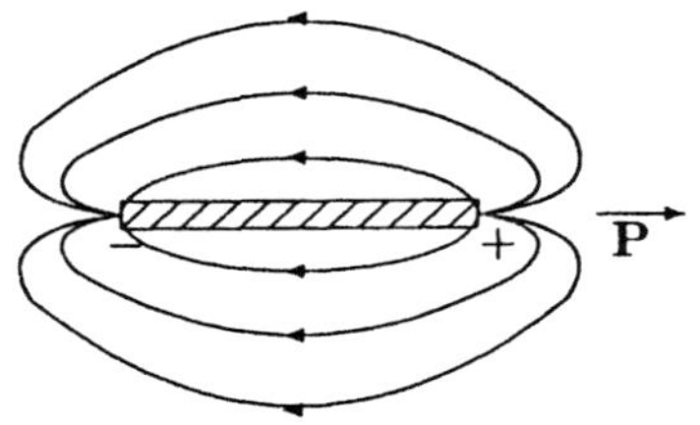

(a) Like a dipole

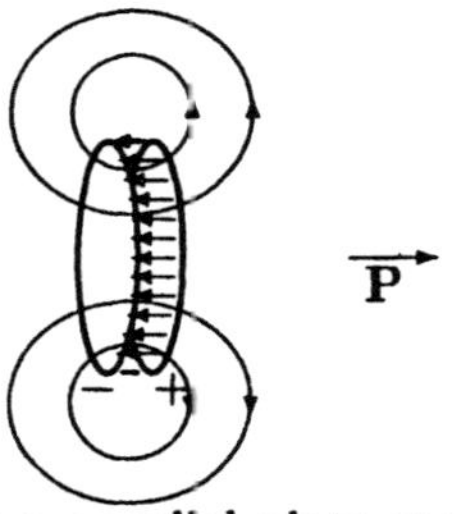

(b) Like a parallel-plate capacitor

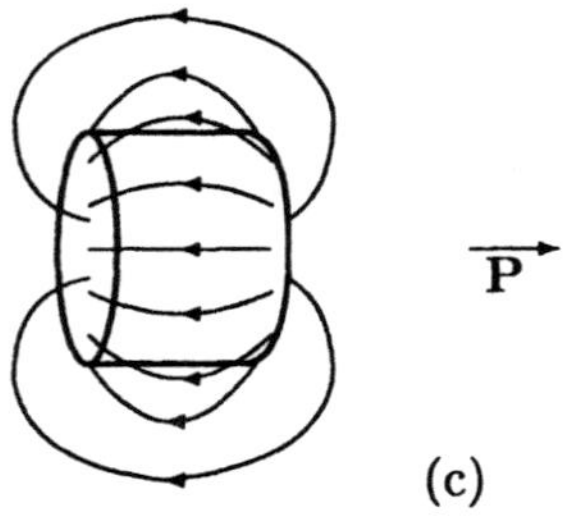

(c)

Problem 4.12

$V = \frac{1}{4\pi\epsilon_0}\int \frac{\mathbf{P}\cdot\hat{\boldsymbol{\mathscr{r}}}}{\mathscr{r}^2}d\tau = \mathbf{P}\cdot\left\{\frac{1}{4\pi\epsilon_0}\int \frac{\hat{\boldsymbol{\mathscr{r}}}}{\mathscr{r}^2}d\tau\right\}$. But the term in curly brackets is precisely the *field* of a uniformly *charged* sphere, divided by ρ. The integral was done explicitly in Prob. 2.7 and 2.8:

$$\frac{1}{4\pi\epsilon_0}\int \frac{\hat{\boldsymbol{\mathscr{r}}}}{\mathscr{r}^2}d\tau = \frac{1}{\rho}\left\{\begin{array}{ll} \frac{1}{4\pi\epsilon_0}\frac{(4/3)\pi R^3\rho}{r^2}\hat{\mathbf{r}}, & (r > R), \\ \frac{1}{4\pi\epsilon_0}\frac{(4/3)\pi R^3\rho}{R^3}\mathbf{r}, & (r < R). \end{array}\right\} \quad \text{So } V(r,\theta) = \left\{\begin{array}{ll} \frac{R^3}{3\epsilon_0 r^2}\mathbf{P}\cdot\hat{\mathbf{r}} = \boxed{\frac{R^3 P\cos\theta}{3\epsilon_0 r^2},} & (r > R), \\ \frac{1}{3\epsilon_0}\mathbf{P}\cdot\mathbf{r} = \boxed{\frac{Pr\cos\theta}{3\epsilon_0},} & (r < R). \end{array}\right\}$$

Problem 4.13

Think of it as two cylinders of opposite uniform charge density $\pm\rho$. *Inside*, the field at a distance s from the axis of a uniformly charge cylinder is given by Gauss's law: $E2\pi s\ell = \frac{1}{\epsilon_0}\rho\pi s^2\ell \Rightarrow \mathbf{E} = (\rho/2\epsilon_0)\mathbf{s}$. For *two* such cylinders, one plus and one minus, the net field (inside) is $\mathbf{E} = \mathbf{E}_+ + \mathbf{E}_- = (\rho/2\epsilon_0)(\mathbf{s}_+ - \mathbf{s}_-)$. But $\mathbf{s}_+ - \mathbf{s}_- = -\mathbf{d}$, so $\mathbf{E} = \boxed{-\rho\mathbf{d}/(2\epsilon_0),}$ where $\mathbf{d}$ is the vector from the negative axis to positive axis. In this case the total dipole moment of a chunk of length ℓ is $\mathbf{P}\left(\pi a^2\ell\right) = \left(\rho\pi a^2\ell\right)\mathbf{d}$. So $\rho\mathbf{d} = \mathbf{P}$, and $\boxed{\mathbf{E} = -\mathbf{P}/(2\epsilon_0),}$ for $s < a$.

Outside, Gauss's law gives $E 2\pi s \ell = \frac{1}{\epsilon_0}\rho\pi a^2 \ell \Rightarrow \mathbf{E} = \frac{\rho a^2}{2\epsilon_0}\frac{\hat{\mathbf{s}}}{s}$, for *one* cylinder. For the combination, $\mathbf{E} = \mathbf{E}_+ + \mathbf{E}_- = \frac{\rho a^2}{2\epsilon_0}\left(\frac{\hat{\mathbf{s}}_+}{s_+} - \frac{\hat{\mathbf{s}}_-}{s_-}\right)$, where

$$\begin{aligned}
\mathbf{s}_\pm &= \mathbf{s} \mp \frac{\mathbf{d}}{2};\\
\frac{\mathbf{s}_\pm}{s_\pm^2} &= \left(\mathbf{s} \mp \frac{\mathbf{d}}{2}\right)\left(s^2 + \frac{d^2}{4} \mp \mathbf{s}\cdot\mathbf{d}\right)^{-1} \cong \frac{1}{s^2}\left(\mathbf{s} \mp \frac{\mathbf{d}}{2}\right)\left(1 \mp \frac{\mathbf{s}\cdot\mathbf{d}}{s^2}\right)^{-1} \cong \frac{1}{s^2}\left(\mathbf{s} \mp \frac{\mathbf{d}}{2}\right)\left(1 \pm \frac{\mathbf{s}\cdot\mathbf{d}}{s^2}\right)\\
&= \frac{1}{s^2}\left(\mathbf{s} \pm \mathbf{s}\frac{(\mathbf{s}\cdot\mathbf{d})}{s^2} \mp \frac{\mathbf{d}}{2}\right) \quad \text{(keeping only 1st order terms in } \mathbf{d}).\\
\left(\frac{\hat{\mathbf{s}}_+}{s_+} - \frac{\hat{\mathbf{s}}_-}{s_-}\right) &= \frac{1}{s^2}\left[\left(\mathbf{s} + \mathbf{s}\frac{(\mathbf{s}\cdot\mathbf{d})}{s^2} - \frac{\mathbf{d}}{2}\right) - \left(\mathbf{s} - \mathbf{s}\frac{(\mathbf{s}\cdot\mathbf{d})}{s^2} + \frac{\mathbf{s}}{2}\right)\right] = \frac{1}{s^2}\left(2\frac{\mathbf{s}(\mathbf{s}\cdot\mathbf{d})}{s^2} - \mathbf{d}\right).
\end{aligned}$$

$$\boxed{\mathbf{E}(\mathbf{s}) = \frac{a^2}{2\epsilon_0}\frac{1}{s^2}\left[2(\mathbf{P}\cdot\hat{\mathbf{s}})\,\hat{\mathbf{s}} - \mathbf{P}\right],} \quad \text{for } s > a.$$

Problem 4.14

Total charge on the dielectric is $Q_{\text{tot}} = \oint_{\mathcal{S}} \sigma_b\, da + \int_{\mathcal{V}} \rho_b\, d\tau = \oint_{\mathcal{S}} \mathbf{P}\cdot d\mathbf{a} - \int_{\mathcal{V}} \nabla\cdot\mathbf{P}\, d\tau$. But the divergence theorem says $\oint_{\mathcal{S}} \mathbf{P}\cdot d\mathbf{a} = \int_{\mathcal{V}} \nabla\cdot\mathbf{P}\, d\tau$, so $Q_{\text{enc}} = 0$. qed

Problem 4.15

(a) $\rho_b = -\nabla\cdot\mathbf{P} = -\frac{1}{r^2}\frac{\partial}{\partial r}\left(r^2\frac{k}{r}\right) = -\frac{k}{r^2}; \quad \sigma_b = \mathbf{P}\cdot\hat{\mathbf{n}} = \left\{\begin{array}{ll} +\mathbf{P}\cdot\hat{\mathbf{r}} = k/b & (\text{at } r = b),\\ -\mathbf{P}\cdot\hat{\mathbf{r}} = -k/a & (\text{at } r = a).\end{array}\right\}$

Gauss's law $\Rightarrow \mathbf{E} = \frac{1}{4\pi\epsilon_0}\frac{Q_{\text{enc}}}{r^2}\hat{\mathbf{r}}$. For $r < a$, $Q_{\text{enc}} = 0$, so $\boxed{\mathbf{E} = 0.}$ For $r > b$, $Q_{\text{enc}} = 0$ (Prob. 4.14), so $\boxed{\mathbf{E} = 0.}$ For $a < r < b$, $Q_{\text{enc}} = \left(\frac{-k}{a}\right)(4\pi a^2) + \int_a^r \left(\frac{-k}{\bar{r}^2}\right) 4\pi\bar{r}^2 d\bar{r} = -4\pi k a - 4\pi k(r-a) = -4\pi k r$; so $\boxed{\mathbf{E} = -(k/\epsilon_0 r)\,\hat{\mathbf{r}}.}$

(b) $\oint \mathbf{D}\cdot d\mathbf{a} = Q_{f_{\text{enc}}} = 0 \Rightarrow \mathbf{D} = 0$ everywhere. $\mathbf{D} = \epsilon_0\mathbf{E} + \mathbf{P} = 0 \Rightarrow \mathbf{E} = (-1/\epsilon_0)\mathbf{P}$, so $\boxed{\mathbf{E} = 0 \text{ (for } r < a \text{ and } r > b);}$ $\boxed{\mathbf{E} = -(k/\epsilon_0 r)\,\hat{\mathbf{r}} \text{ (for } a < r < b).}$

Problem 4.16

(a) Same as $\mathbf{E}_0$ minus the field at the center of a sphere with uniform polarization $\mathbf{P}$. The latter (Eq. 4.14) is $-\mathbf{P}/3\epsilon_0$. So $\boxed{\mathbf{E} = \mathbf{E}_0 + \frac{1}{3\epsilon_0}\mathbf{P}.}$ $\mathbf{D} = \epsilon_0\mathbf{E} = \epsilon_0\mathbf{E}_0 + \frac{1}{3}\mathbf{P} = \mathbf{D}_0 - \mathbf{P} + \frac{1}{3}\mathbf{P}$, so $\boxed{\mathbf{D} = \mathbf{D}_0 - \frac{2}{3}\mathbf{P}.}$

(b) Same as $\mathbf{E}_0$ minus the field of $\pm$ charges at the two ends of the "needle"—but these are small, and far away, so $\boxed{\mathbf{E} = \mathbf{E}_0.}$ $\mathbf{D} = \epsilon_0\mathbf{E} = \epsilon_0\mathbf{E}_0 = \mathbf{D}_0 - \mathbf{P}$, so $\boxed{\mathbf{D} = \mathbf{D}_0 - \mathbf{P}.}$

(c) Same as $\mathbf{E}_0$ minus the field of a parallel-plate capacitor with upper plate at $\sigma = P$. The latter is $-(1/\epsilon_0)P$, so $\boxed{\mathbf{E} = \mathbf{E}_0 + \frac{1}{\epsilon_0}\mathbf{P}.}$ $\mathbf{D} = \epsilon_0\mathbf{E} = \epsilon_0\mathbf{E}_0 + \mathbf{P}$, so $\boxed{\mathbf{D} = \mathbf{D}_0.}$

Problem 4.17

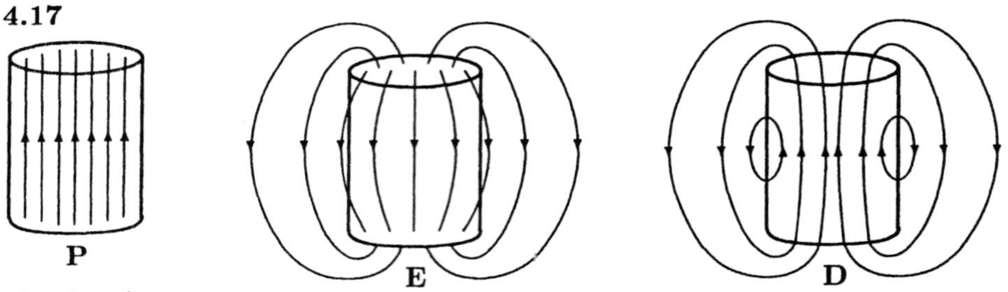

P (uniform)

E (field of two circular plates)

D (same as **E** *outside*, but lines *continuous*, since $\nabla\cdot\mathbf{D} = 0$)

Problem 4.18

(a) Apply $\int \mathbf{D}\cdot d\mathbf{a} = Q_{f_{enc}}$ to the gaussian surface shown. $DA = \sigma A \Rightarrow \boxed{D = \sigma.}$ (*Note:* $\mathbf{D} = 0$ inside the metal plate.) This is true in both slabs; $\mathbf{D}$ points *down*.

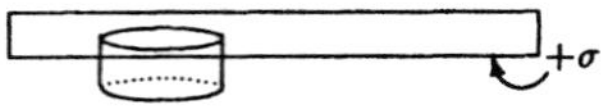

(b) $\mathbf{D} = \epsilon\mathbf{E} \Rightarrow E = \sigma/\epsilon_1$ in slab 1, $E = \sigma/\epsilon_2$ in slab 2. But $\epsilon = \epsilon_0\epsilon_r$, so $\epsilon_1 = 2\epsilon_0$; $\epsilon_2 = \frac{3}{2}\epsilon_0$. $\boxed{E_1 = \sigma/2\epsilon_0,}$ $\boxed{E_2 = 2\sigma/3\epsilon_0.}$

(c) $\mathbf{P} = \epsilon_0\chi_e\mathbf{E}$, so $P = \epsilon_0\chi_e d/(\epsilon_0\epsilon_r) = (\chi_e/\epsilon_r)\sigma$; $\chi_e = \epsilon_r - 1 \Rightarrow P = (1-\epsilon_r^{-1})\sigma$. $\boxed{P_1 = \sigma/2,}$ $\boxed{P_2 = \sigma/3.}$

(d) $V = E_1 a + E_2 a = (\sigma a/6\epsilon_0)(3+4) = \boxed{7\sigma a/6\epsilon_0.}$

(e) $\rho_b = 0$; $\boxed{\begin{array}{l}\sigma_b = +P_1 \text{ at bottom of slab (1)} = \sigma/2,\\ \sigma_b = -P_1 \text{ at top of slab (1)} = -\sigma/2;\end{array}}$ $\boxed{\begin{array}{l}\sigma_b = +P_2 \text{ at bottom of slab (2)} = \sigma/3,\\ \sigma_b = -P_2 \text{ at top of slab (2)} = -\sigma/3.\end{array}}$

(f) In slab 1: $\left\{\begin{array}{l}\text{total surface charge } \textit{above}\text{: } \sigma - (\sigma/2) = \sigma/2,\\ \text{total surface charge } \textit{below}\text{: } (\sigma/2) - (\sigma/3) + (\sigma/3) - \sigma = -\sigma/2,\end{array}\right\} \Longrightarrow E_1 = \dfrac{\sigma}{2\epsilon_0}.\ \checkmark$

In slab 2: $\left\{\begin{array}{l}\text{total surface charge } \textit{above}\text{: } \sigma - (\sigma/2) + (\sigma/2) - (\sigma/3) = 2\sigma/3,\\ \text{total surface charge } \textit{below}\text{: } (\sigma/3) - \sigma = -2\sigma/3,\end{array}\right\} \Longrightarrow E_2 = \dfrac{2\sigma}{3\epsilon_0}.\ \checkmark$

Problem 4.19

With no dielectric, $C_0 = A\epsilon_0/d$ (Eq. 2.54).

In configuration (a), with $+\sigma$ on upper plate, $-\sigma$ on lower, $D = \sigma$ between the plates. $E = \sigma/\epsilon_0$ (in air) and $E = \sigma/\epsilon$ (in dielectric). So $V = \frac{\sigma}{\epsilon_0}\frac{d}{2} + \frac{\sigma}{\epsilon}\frac{d}{2} = \frac{Qd}{2\epsilon_0 A}\left(1 + \frac{\epsilon_0}{\epsilon}\right)$.

$C_a = \frac{Q}{V} = \frac{\epsilon_0 A}{d}\left(\frac{2}{1+1/\epsilon_r}\right) \Longrightarrow \boxed{\dfrac{C_a}{C_0} = \dfrac{2\epsilon_r}{1+\epsilon_r}.}$

In configuration (b), with potential difference V: $E = V/d$, so $\sigma = \epsilon_0 E = \epsilon_0 V/d$ (in air).

$P = \epsilon_0\chi_e E = \epsilon_0\chi_e V/d$ (in dielectric), so $\sigma_b = -\epsilon_0\chi_e V/d$ (at top surface of dielectric).
$\sigma_{\text{tot}} = \epsilon_0 V/d = \sigma_f + \sigma_b = \sigma_f - \epsilon_0\chi_e V/d$, so $\sigma_f = \epsilon_0 V(1+\chi_e)/d = \epsilon_0\epsilon_r V/d$ (on top plate above dielectric).

$$\Longrightarrow C_b = \frac{Q}{V} = \frac{1}{V}\left(\sigma\frac{A}{2} + \sigma_f\frac{A}{2}\right) = \frac{A}{2V}\left(\epsilon_0\frac{V}{d} + \epsilon_0\frac{V}{d}\epsilon_r\right) = \frac{A\epsilon_0}{d}\left(\frac{1+\epsilon_r}{2}\right). \quad \boxed{\frac{C_b}{C_0} = \frac{1+\epsilon_r}{2}.}$$

[Which is greater? $\frac{C_b}{C_0} - \frac{C_a}{C_0} = \frac{1+\epsilon_r}{2} - \frac{2\epsilon_r}{1+\epsilon_r} = \frac{(1+\epsilon_r)^2 - 4\epsilon_r}{2(1+\epsilon_r)} = \frac{1+2\epsilon_r+4\epsilon_r^2-4\epsilon_r}{2(1+\epsilon_r)} = \frac{(1-\epsilon_r)^2}{2(1+\epsilon_r)} > 0$. So $C_b > C_a$.]
If the x axis points *down*:

	$\mathbf{E}$	$\mathbf{D}$	$\mathbf{P}$	σ_b (top surface)	σ_f (top plate)
(a) air	$\frac{2\epsilon_r}{(\epsilon_r+1)}\frac{V}{d}\,\hat{\mathbf{x}}$	$\frac{2\epsilon_r}{(\epsilon_r+1)}\frac{\epsilon_0 V}{d}\,\hat{\mathbf{x}}$	0	0	$\frac{2\epsilon_r}{(\epsilon_r+1)}\frac{V}{d}$
(a) dielectric	$\frac{2}{(\epsilon_r+1)}\frac{V}{d}\,\hat{\mathbf{x}}$	$\frac{2\epsilon_r}{(\epsilon_r+1)}\frac{\epsilon_0 V}{d}\,\hat{\mathbf{x}}$	$\frac{2(\epsilon_r-1)}{(\epsilon_r+1)}\frac{\epsilon_0 V}{d}\,\hat{\mathbf{x}}$	$-\frac{2(\epsilon_r-1)}{(\epsilon_r+1)}\frac{\epsilon_0 V}{d}$	—
(b) air	$\frac{V}{d}\,\hat{\mathbf{x}}$	$\frac{\epsilon_0 V}{d}\,\hat{\mathbf{x}}$	0	0	$\frac{\epsilon_0 V}{d}$ (left)
(b) dielectric	$\frac{V}{d}\,\hat{\mathbf{x}}$	$\epsilon_r\frac{\epsilon_0 V}{d}\,\hat{\mathbf{x}}$	$(\epsilon_r-1)\frac{\epsilon_0 V}{d}\,\hat{\mathbf{x}}$	$-(\epsilon_r-1)\frac{\epsilon_0 V}{d}$	$\epsilon_r\frac{\epsilon_0 V}{d}$ (right)

Problem 4.20

$\int \mathbf{D}\cdot d\mathbf{a} = Q_{f_{\text{enc}}} \Rightarrow D4\pi r^2 = \rho\frac{4}{3}\pi r^3 \Rightarrow D = \frac{1}{3}\rho r \Rightarrow \mathbf{E} = (\rho r/3\epsilon)\,\hat{\mathbf{r}}$, for $r < R$; $D4\pi r^2 = \rho\frac{4}{3}\pi R^3 \Rightarrow D = \rho R^3/3r^2 \Rightarrow \mathbf{E} = (\rho R^3/3\epsilon_0 r^2)\,\hat{\mathbf{r}}$, for $r > R$.

$$V = -\int_\infty^0 \mathbf{E}\cdot d\mathbf{l} = \frac{\rho R^3}{3\epsilon_0}\left.\frac{1}{r}\right|_\infty^R - \frac{\rho}{3\epsilon}\int_R^0 r\,dr = \frac{\rho R^2}{3\epsilon_0} + \frac{\rho}{3\epsilon}\frac{R^2}{2} = \boxed{\frac{\rho R^2}{3\epsilon_0}\left(1 + \frac{1}{2\epsilon_r}\right).}$$

Problem 4.21

Let Q be the charge on a length ℓ of the inner conductor.

$$\oint \mathbf{D}\cdot d\mathbf{a} = D2\pi s\ell = Q \Rightarrow D = \frac{Q}{2\pi s\ell};\quad E = \frac{Q}{2\pi\epsilon_0 s\ell}\ (a < s < b),\quad E = \frac{Q}{2\pi\epsilon s\ell}\ (b < r < c).$$

$$V = -\int_c^a \mathbf{E}\cdot d\mathbf{l} = \int_a^b\left(\frac{Q}{2\pi\epsilon_0\ell}\right)\frac{ds}{s} + \int_b^c\left(\frac{Q}{2\pi\epsilon\ell}\right)\frac{ds}{s} = \frac{Q}{2\pi\epsilon_0\ell}\left[\ln\left(\frac{b}{a}\right) + \frac{\epsilon_0}{\epsilon}\ln\left(\frac{c}{b}\right)\right].$$

$$\frac{C}{\ell} = \frac{Q}{V\ell} = \boxed{\frac{2\pi\epsilon_0}{\ln(b/a) + (1/\epsilon_r)\ln(c/b)}.}$$

Problem 4.22

Same method as Ex. 4.7: solve Laplace's equation for $V_{\text{in}}(s,\phi)$ $(s < a)$ and $V_{\text{out}}(s,\phi)$ $(s > a)$, subject to the boundary conditions

$$\begin{cases} \text{(i)}\ V_{\text{in}} = V_{\text{out}} & \text{at } s = a, \\ \text{(ii)}\ \epsilon\frac{\partial V_{\text{in}}}{\partial s} = \epsilon_0\frac{\partial V_{\text{out}}}{\partial s} & \text{at } s = a, \\ \text{(iii)}\ V_{\text{out}} \to -E_0 s\cos\phi & \text{for } s \gg a. \end{cases}$$

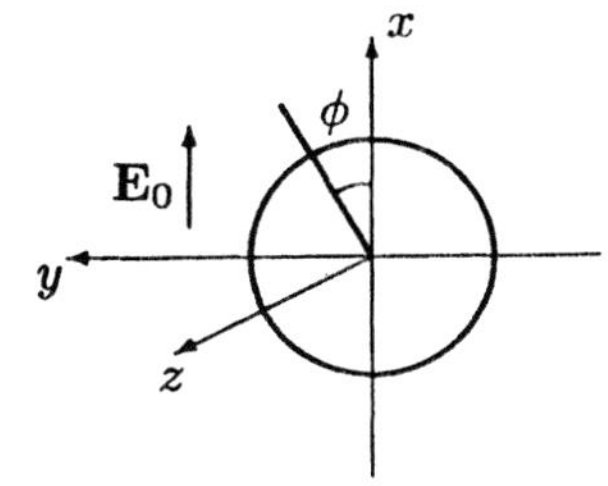

From Prob. 3.23 (invoking boundary condition (iii)):

$$V_{\text{in}}(s,\phi) = \sum_{k=1}^{\infty} s^k(a_k\cos k\phi + b_k\sin k\phi),\qquad V_{\text{out}}(s,\phi) = -E_0 s\cos\phi + \sum_{k=1}^{\infty} s^{-k}(c_k\cos k\phi + d_k\sin k\phi).$$

(I eliminated the constant terms by setting $V = 0$ on the $y\,z$ plane.) Condition (i) says

$$\sum a^k(a_k \cos k\phi + b_k \sin k\phi) = -E_0 s \cos\phi + \sum a^{-k}(c_k \cos k\phi + d_k \sin k\phi),$$

while (ii) says

$$\epsilon_r \sum k a^{k-1}(a_k \cos k\phi + b_k \sin k\phi) = -E_0 \cos\phi - \sum k a^{-k-1}(c_k \cos k\phi + d_k \sin k\phi).$$

Evidently $b_k = d_k = 0$ for all k, $a_k = c_k = 0$ unless $k = 1$, whereas for $k = 1$,

$$a a_1 = -E_0 a + a^{-1} c_1, \quad \epsilon_r a_1 = -E_0 - a^{-2} c_1.$$

Solving for a_1,

$$a_1 = -\frac{E_0}{(1+\chi_e/2)}, \quad \text{so } V_{\text{in}}(s,\phi) = -\frac{E_0}{(1+\chi_e/2)} s\cos\phi = -\frac{E_0}{(1+\chi_e/2)} x,$$

and hence $\mathbf{E}_{\text{in}}(s,\phi) = -\dfrac{\partial V_{\text{in}}}{\partial x}\hat{\mathbf{x}} = \boxed{\dfrac{\mathbf{E}_0}{(1+\chi_e/2)}}.$ As in the spherical case (Ex. 4.7), the field inside is *uniform*.

Problem 4.23

$$\mathbf{P}_0 = \epsilon_0\chi_e\mathbf{E}_0;\ \mathbf{E}_1 = -\frac{1}{3\epsilon_0}\mathbf{P}_0 = -\frac{\chi_e}{3}\mathbf{E}_0;\ \mathbf{P}_1 = \epsilon_0\chi_e\mathbf{E}_1 = -\frac{\epsilon_0\chi_e^2}{3}\mathbf{E}_0;\ \mathbf{E}_2 = -\frac{1}{3\epsilon_0}\mathbf{P}_1 = \frac{\chi_e^2}{9}\mathbf{E}_0;\ \dots.$$

Evidently $\mathbf{E}_n = \left(-\frac{\chi_e}{3}\right)^n \mathbf{E}_0$, so

$$\mathbf{E} = \mathbf{E}_0 + \mathbf{E}_1 + \mathbf{E}_2 + \cdots = \left[\sum_{n=0}^{\infty}\left(-\frac{\chi_e}{3}\right)^n\right]\mathbf{E}_0.$$

The geometric series can be summed explicitly:

$$\sum_{n=0}^{\infty} x^n = \frac{1}{1-x}, \quad \text{so} \quad \boxed{\mathbf{E} = \frac{1}{(1+\chi_e/3)}\mathbf{E}_0,}$$

which agrees with Eq. 4.49. [Curiously, this method formally requires that $\chi_e < 3$ (else the infinite series diverges), yet the *result* is subject to no such restriction, since we can also get it by the method of Ex. 4.7.]

Problem 4.24

Potentials:

$$\begin{cases} V_{\text{out}}(r,\theta) &= -E_0 r\cos\theta + \sum \frac{B_l}{r^{l+1}} P_l(\cos\theta), & (r > b); \\ V_{\text{med}}(r,\theta) &= \sum \left(A_l r^l + \frac{B_l}{r^{l+1}}\right) P_l(\cos\theta), & (a < r < b); \\ V_{\text{in}}(r,\theta) &= 0, & (r < a). \end{cases}$$

Boundary Conditions:

$$\begin{cases} \text{(i)}\ V_{\text{out}} &= V_{\text{med}}, & (r = b); \\ \text{(ii)}\ \epsilon\frac{\partial V_{\text{med}}}{\partial r} &= \epsilon_0\frac{\partial V_{\text{out}}}{\partial r}, & (r = b); \\ \text{(iii)}\ V_{\text{med}} &= 0, & (r = a). \end{cases}$$

$$\text{(i)} \;\Rightarrow\; -E_0 b\cos\theta + \sum \frac{B_l}{b^{l+1}} P_l(\cos\theta) = \sum \left(A_l b^l + \frac{\bar{B}_l}{b^{l+1}}\right) P_l(\cos\theta);$$

$$\text{(ii)} \;\Rightarrow\; \epsilon_r \sum \left[l A_l b^{l-1} - (l+1)\frac{\bar{B}_l}{b^{l+2}}\right] P_l(\cos\theta) = -E_0\cos\theta - \sum (l+1)\frac{B_l}{b^{l+2}} P_l(\cos\theta);$$

$$\text{(iii)} \;\Rightarrow\; A_l a^l + \frac{\bar{B}_l}{a^{l+1}} = 0 \;\Rightarrow\; \bar{B}_l = -a^{2l+1} A_l.$$

For $l \neq 1$:

$$\text{(i)} \quad \frac{B_l}{b^{l+1}} = \left(A_l b^l - \frac{a^{2l+1}A_l}{b^{l+1}}\right) \;\Rightarrow\; B_l = A_l\left(b^{2l+1} - a^{2l+1}\right);$$

$$\text{(ii)} \quad \epsilon_r \left[l A_l b^{l-1} + (l+1)\frac{a^{2l+1}A_l}{b^{l+2}}\right] = -(l+1)\frac{B_l}{b^{l+2}} \;\Rightarrow\; B_l = -\epsilon_r A_l \left[\left(\frac{l}{l+1}\right) b^{2l+1} + a^{2l+1}\right] \;\Rightarrow\; A_l = B_l = 0.$$

For $l = 1$:

$$\text{(i)} \quad -E_0 b + \frac{B_1}{b^2} = A_1 b - \frac{a^3 A_1}{b^2} \;\Rightarrow\; B_1 - E_0 b^3 = A_1 2\left(b^3 - a^3\right);$$

$$\text{(ii)} \quad \epsilon_r\left(A_1 + 2\frac{a^3 A_1}{b^3}\right) = -E_0 - 2\frac{B_1}{b^3} \;\Rightarrow\; -2B_1 - E_0 b^3 = \epsilon_r A_1\left(b^3 + 2a^3\right).$$

So $-3E_0 b^3 = A_1\left[2\left(b^3 - a^3\right) + \epsilon_r\left(b^3 + 2a^3\right)\right]; \quad A_1 = \dfrac{-3E_0}{2[1-(a/b)^3] + \epsilon_r[1 + 2(a/b)^3]}.$

$$V_{\text{med}}(r,\theta) = \frac{-3E_0}{2[1-(a/b)^3] + \epsilon_r[1+2(a/b)^3]}\left(r - \frac{a^3}{r^2}\right)\cos\theta,$$

$$\mathbf{E}(r,\theta) = -\nabla V_{\text{med}} = \boxed{\frac{3E_0}{2[1-(a/b)^3] + \epsilon_r[1+2(a/b)^3]}\left\{\left(1 + \frac{2a^3}{r^3}\right)\cos\theta\,\hat{\mathbf{r}} - \left(1 - \frac{a^3}{r^3}\right)\sin\theta\,\hat{\boldsymbol{\theta}}\right\}.}$$

Problem 4.25

There are four charges involved: (i) q, (ii) polarization charge surrounding q, (iii) surface charge (σ_b) on the top surface of the lower dielectric, (iv) surface charge (σ_b') on the lower surface of the upper dielectric. In view of Eq. 4.39, the bound charge (ii) is $q_p = -q(\chi_e'/(1+\chi_e'))$, so the *total* (point) charge at $(0,0,d)$ is $q_t = q + q_p = q/(1+\chi_e') = q/\epsilon_r'$. As in Ex. 4.8,

$$\text{(a) } \sigma_b = \epsilon_0\chi_e\left[\frac{-1}{4\pi\epsilon_0}\frac{qd/\epsilon_r'}{(r^2+d^2)^{\frac{3}{2}}} - \frac{\sigma_b}{2\epsilon_0} - \frac{\sigma_b'}{2\epsilon_0}\right] \quad (\text{here } \sigma_b = \mathbf{P}\cdot\hat{\mathbf{n}} = +P_z = \epsilon_0\chi_e E_z);$$

$$\text{(b) } \sigma_b' = \epsilon_0\chi_e'\left[\frac{1}{4\pi\epsilon_0}\frac{qd/\epsilon_r'}{(r^2+d^2)^{\frac{3}{2}}} - \frac{\sigma_b}{2\epsilon_0} - \frac{\sigma_b'}{2\epsilon_0}\right] \quad (\text{here } \sigma_b = -P_z = -\epsilon_0\chi_e' E_z).$$

Solve for σ_b, σ_b': first divide by χ_e and χ_e' (respectively) and subtract:

$$\frac{\sigma_b'}{\chi_e'} - \frac{\sigma_b}{\chi_e} = \frac{1}{2\pi}\frac{qd/\epsilon_r'}{(r^2+d^2)^{\frac{3}{2}}} \;\Rightarrow\; \sigma_b' = \chi_e'\left[\frac{\sigma_b}{\chi_e} + \frac{1}{2\pi}\frac{qd/\epsilon_r'}{(r^2+d^2)^{\frac{3}{2}}}\right].$$

Plug this into (a) and solve for σ_b, using $\epsilon_r' = 1 + \chi_e'$:

$$\sigma_b = \frac{-1}{4\pi}\frac{qd/\epsilon_r'}{(r^2+d^2)^{\frac{3}{2}}}\chi_e(1+\chi_e') - \frac{\sigma_b}{2}(\chi_e+\chi_e'),\ \text{so}\ \boxed{\sigma_b = \frac{-1}{4\pi}\frac{qd}{(r^2+d^2)^{\frac{3}{2}}}\frac{\chi_e}{[1+(\chi_e+\chi_e')/2]};}$$

$$\sigma_b' = \chi_e'\left\{\frac{-1}{4\pi}\frac{qd}{(r^2+d^2)^{\frac{3}{2}}}\frac{1}{[1+(\chi_e+\chi_e')/2]} + \frac{1}{2\pi}\frac{qd/\epsilon_r'}{(r^2+d^2)^{\frac{3}{2}}}\right\},\ \text{so}\ \boxed{\sigma_b' = \frac{1}{4\pi}\frac{qd}{(r^2+d^2)^{\frac{3}{2}}}\frac{\epsilon_r\chi_e'/\epsilon_r'}{[1+(\chi_e+\chi_e')/2]}.}$$

The *total* bound surface charge is $\sigma_t = \sigma_b + \sigma_b' = \frac{1}{4\pi}\frac{qd}{(r^2+d^2)^{\frac{3}{2}}}\frac{(\chi_e'-\chi_e)}{\epsilon_r'[1+(\chi_e+\chi_e')/2]}$ (which vanishes, as it should, when $\chi_e' = \chi_e$). The total bound charge is (compare Eq. 4.51):

$$q_t = \frac{(\chi_e'-\chi_e)q}{2\epsilon_r'[1+(\chi_e+\chi_e')/2]} = \boxed{\left(\frac{\epsilon_r'-\epsilon_r}{\epsilon_r'+\epsilon_r}\right)\frac{q}{\epsilon_r'},}\ \text{and hence}$$

$$\boxed{V(\mathbf{r}) = \frac{1}{4\pi\epsilon_0}\left\{\frac{q/\epsilon_r'}{\sqrt{x^2+y^2+(z-d)^2}} + \frac{q_t}{\sqrt{x^2+y^2+(z+d)^2}}\right\}}\ (\text{for } z>0).$$

Meanwhile, since $\frac{q}{\epsilon_r'} + q_t = \frac{q}{\epsilon_r'}\left[1 + \frac{\epsilon_r'-\epsilon_r}{\epsilon_r'+\epsilon_r}\right] = \frac{2q}{\epsilon_r'+\epsilon_r}$, $\boxed{V(\mathbf{r}) = \frac{1}{4\pi\epsilon_0}\frac{[2q/(\epsilon_r'+\epsilon_r)]}{\sqrt{x^2+y^2+(z-d)^2}}}$ (for $z<0$).

Problem 4.26

From Ex. 4.5:

$$\mathbf{D} = \left\{\begin{array}{ll} 0, & (r<a) \\ \frac{Q}{4\pi r^2}\hat{\mathbf{r}}, & (r>a)\end{array}\right\},\quad \mathbf{E} = \left\{\begin{array}{ll} 0, & (r<a) \\ \frac{Q}{4\pi\epsilon r^2}\hat{\mathbf{r}}, & (a<r<b) \\ \frac{Q}{4\pi\epsilon_0 r^2}\hat{\mathbf{r}}, & (r>b)\end{array}\right\}.$$

$$W = \frac{1}{2}\int \mathbf{D}\cdot\mathbf{E}\,d\tau = \frac{1}{2}\frac{Q^2}{(4\pi)^2}4\pi\left\{\frac{1}{\epsilon}\int_a^b \frac{1}{r^2}\frac{1}{r^2}r^2 dr + \frac{1}{\epsilon_0}\int_b^\infty \frac{1}{r^2}dr\right\} = \frac{Q^2}{8\pi}\left\{\frac{1}{\epsilon}\left(\frac{-1}{r}\right)\Big|_a^b + \frac{1}{\epsilon_0}\left(\frac{-1}{r}\right)\Big|_b^\infty\right\}$$

$$= \frac{Q^2}{8\pi\epsilon_0}\left\{\frac{1}{(1+\chi_e)}\left(\frac{1}{a}-\frac{1}{b}\right) + \frac{1}{b}\right\} = \boxed{\frac{Q^2}{8\pi\epsilon_0(1+\chi_e)}\left(\frac{1}{a}+\frac{\chi_e}{b}\right).}$$

Problem 4.27

Using Eq. 4.55: $W = \frac{\epsilon_0}{2}\int E^2\,d\tau$. From Ex. 4.2 and Eq. 3.103,

$$\mathbf{E} = \begin{Bmatrix} \dfrac{-1}{3\epsilon_0}P\,\hat{\mathbf{z}}, & (r<R) \\ \dfrac{R^3P}{3\epsilon_0 r^3}(2\cos\theta\,\hat{\mathbf{r}} + \sin\theta\,\hat{\boldsymbol{\theta}}), & (r>R) \end{Bmatrix}, \quad \text{so}$$

$$W_{r<R} = \frac{\epsilon_0}{2}\left(\frac{P}{3\epsilon_0}\right)^2 \frac{4}{3}\pi R^3 = \frac{2\pi}{27}\frac{P^2R^3}{\epsilon_0}.$$

$$W_{r>R} = \frac{\epsilon_0}{2}\left(\frac{R^3P}{3\epsilon_0}\right)^2 \int \frac{1}{r^6}\left(4\cos^2\theta + \sin^2\theta\right) r^2 \sin\theta\,dr\,d\theta\,d\phi$$

$$= \frac{(R^3P)^2}{18\epsilon_0}2\pi\int_0^\pi (1+3\cos^2\theta)\sin\theta\,d\theta\int_R^\infty \frac{1}{r^4}\,dr = \frac{\pi(R^3P)^2}{9\epsilon_0}\left(-\cos\theta - \cos^3\theta\right)\Big|_0^\pi\left(-\frac{1}{3r^3}\right)\Big|_R^\infty$$

$$= \frac{\pi(R^3P)^2}{9\epsilon_0}\left(\frac{4}{3R^3}\right) = \frac{4\pi R^3P^2}{27\epsilon_0}.$$

$$W_{\text{tot}} = \boxed{\frac{2\pi R^3P^2}{9\epsilon_0}.}$$

This is the correct electrostatic energy of the configuration, but it is not the "total work necessary to assemble the system," because it leaves out the mechanical energy involved in polarizing the molecules.

Using Eq. 4.58: $W = \frac{1}{2}\int \mathbf{D}\cdot\mathbf{E}\,d\tau$. For $r<R$, $\mathbf{D} = \epsilon_0\mathbf{E}$, so this contribution is the same as before. For $r<R$, $\mathbf{D} = \epsilon_0\mathbf{E} + \mathbf{P} = -\frac{1}{3}\mathbf{P} + \mathbf{P} = \frac{2}{3}\mathbf{P} = -2\epsilon_0\mathbf{E}$, so $\frac{1}{2}\mathbf{D}\cdot\mathbf{E} = -2\frac{\epsilon_0}{2}E^2$, and this contribution is now $(-2)\left(\frac{2\pi}{27}\frac{P^2R^3}{\epsilon_0}\right) = -\frac{4\pi}{27}\frac{R^3P^2}{\epsilon_0}$, exactly cancelling the exterior term. *Conclusion:* $\boxed{W_{\text{tot}} = 0.}$ This is not surprising, since the derivation in Sect. 4.4.3 calculates the work done on the *free* charge, and in this problem there is no free charge in sight. Since this is a nonlinear dielectric, however, the result cannot be interpreted as the "work necessary to assemble the configuration"—the latter would depend entirely on *how* you assemble it.

Problem 4.28

First find the capacitance, as a function of h:

$$\left.\begin{aligned} &\textit{Air part: } E = \frac{2\lambda}{4\pi\epsilon_0 s} \Longrightarrow V = \frac{2\lambda}{4\pi\epsilon_0}\ln(b/a), \\ &\textit{Oil part: } D = \frac{2\lambda'}{4\pi s} \Longrightarrow E = \frac{2\lambda'}{4\pi\epsilon s} \Longrightarrow V = \frac{2\lambda'}{4\pi\epsilon}\ln(b/a), \end{aligned}\right\} \Longrightarrow \frac{\lambda}{\epsilon_0} = \frac{\lambda'}{\epsilon};\ \lambda' = \frac{\epsilon}{\epsilon_0}\lambda = \epsilon_r\lambda.$$

$Q = \lambda' h + \lambda(\ell - h) = \epsilon_r\lambda h - \lambda h + \lambda\ell = \lambda[(\epsilon_r - 1)h + \ell] = \lambda(\chi_e h + \ell)$, where ℓ is the total height.

$$C = \frac{Q}{V} = \frac{\lambda(\chi_e h + \ell)}{2\lambda\ln(b/a)}4\pi\epsilon_0 = 2\pi\epsilon_0\frac{(\chi_e h + \ell)}{\ln(b/a)}.$$

The net upward force is given by Eq. 4.64: $F = \frac{1}{2}V^2\frac{dC}{dh} = \frac{1}{2}V^2\frac{2\pi\epsilon_0\chi_e}{\ln(b/a)}$.
The gravitational force *down* is $F = mg = \rho\pi(b^2 - a^2)gh$.

$$\boxed{h = \frac{\epsilon_0\chi_e V^2}{\rho(b^2 - a^2)g\ln(b/a)}.}$$

Problem 4.29

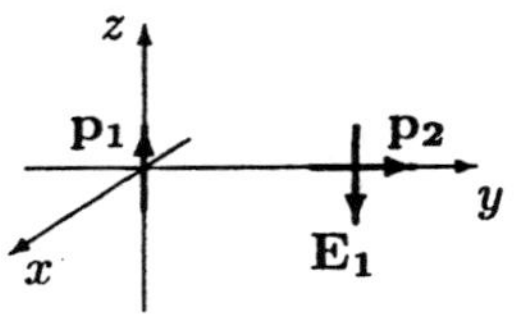

(a) Eq. 4.5 $\Rightarrow \mathbf{F}_2 = (\mathbf{p}_2 \cdot \nabla)\,\mathbf{E}_1 = p_2 \frac{\partial}{\partial y}(\mathbf{E}_1)$;

Eq. 3.103 $\Rightarrow \mathbf{E}_1 = \frac{p_1}{4\pi\epsilon_0 r^3}\,\hat{\boldsymbol{\theta}} = -\frac{p_1}{4\pi\epsilon_0 y^3}\,\hat{\mathbf{z}}$. Therefore

$$\mathbf{F}_2 = -\frac{p_1 p_2}{4\pi\epsilon_0}\left[\frac{d}{dy}\left(\frac{1}{y^3}\right)\right]\hat{\mathbf{z}} = \frac{3p_1p_2}{4\pi\epsilon_0 y^4}\,\hat{\mathbf{z}}, \text{ or } \boxed{\mathbf{F}_2 = \frac{3p_1p_2}{4\pi\epsilon_0 r^4}\,\hat{\mathbf{z}}} \text{ (upward).}$$

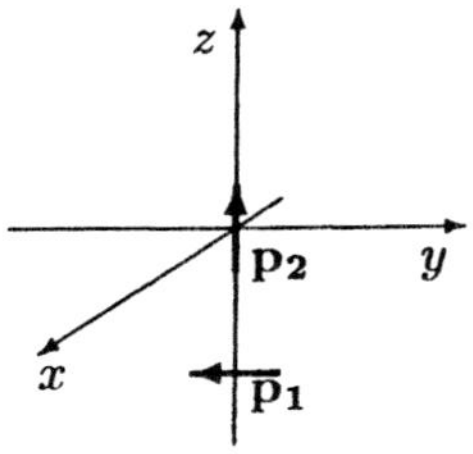

To calculate $\mathbf{F}_1$, put $\mathbf{p}_2$ at the origin, pointing in the z direction; then $\mathbf{p}_1$ is at $-r\,\hat{\mathbf{z}}$, and it points in the $-\hat{\mathbf{y}}$ direction. So $\mathbf{F}_1 = (\mathbf{p}_1 \cdot \nabla)\,\mathbf{E}_2 = -p_1 \left.\frac{\partial \mathbf{E}_2}{\partial y}\right|_{x=y=0,\,z=-r}$; we need $\mathbf{E}_2$ as a function of x, y, and z.

From Eq. 3.104: $\mathbf{E}_2 = \frac{1}{4\pi\epsilon_0}\frac{1}{r^3}\left[\frac{3(\mathbf{p}_2\cdot\mathbf{r})\mathbf{r}}{r^2} - \mathbf{p}\right]$, where $\mathbf{r} = x\,\hat{\mathbf{x}} + y\,\hat{\mathbf{y}} + z\,\hat{\mathbf{z}}$, $\mathbf{p}_2 = -p_2\,\hat{\mathbf{y}}$, and hence $\mathbf{p}_2 \cdot \mathbf{r} = -p_2 y$.

$$\mathbf{E}_2 = \frac{p_2}{4\pi\epsilon_0}\left[\frac{-3y(x\,\hat{\mathbf{x}} + y\,\hat{\mathbf{y}} + z\,\hat{\mathbf{z}}) + (x^2+y^2+z^2)\,\hat{\mathbf{y}}}{(x^2+y^2+z^2)^{5/2}}\right] = \frac{p_2}{4\pi\epsilon_0}\left[\frac{-3xy\,\hat{\mathbf{x}} + (x^2-2y^2+z^2)\,\hat{\mathbf{y}} - 3yz\,\hat{\mathbf{z}}}{(x^2+y^2+z^2)^{5/2}}\right]$$

$$\frac{\partial \mathbf{E}_2}{\partial y} = \frac{p_2}{4\pi\epsilon_0}\left\{-\frac{5}{2}\frac{1}{r^7}2y[-3xy\,\hat{\mathbf{x}} + (x^2-2y^2+z^2)\,\hat{\mathbf{y}} - 3yz\,\hat{\mathbf{z}}] + \frac{1}{r^5}(-3x\,\hat{\mathbf{x}} - 4y\,\hat{\mathbf{y}} - 3z\,\hat{\mathbf{z}})\right\};$$

$$\left.\frac{\partial \mathbf{E}_2}{\partial y}\right|_{(0,0)} = \frac{p_2}{4\pi\epsilon_0}\frac{-3z}{r^5}\,\hat{\mathbf{z}}; \quad \mathbf{F}_1 = -p_1\left(\frac{p_2}{4\pi\epsilon_0}\frac{3r}{r^5}\,\hat{\mathbf{z}}\right) = \boxed{-\frac{3p_1p_2}{4\pi\epsilon_0 r^4}\,\hat{\mathbf{z}}.}$$

These results *are* consistent with Newton's third law: $\mathbf{F}_1 = -\mathbf{F}_2$.

(b) From page 165, $\mathbf{N}_2 = (\mathbf{p}_2 \times \mathbf{E}_1) + (\mathbf{r} \times \mathbf{F}_2)$. The first term was calculated in Prob. 4.5; the second we get from (a), using $\mathbf{r} = r\,\hat{\mathbf{y}}$:

$$\mathbf{p}_2 \times \mathbf{E}_1 = \frac{p_1p_2}{4\pi\epsilon_0 r^3}(-\hat{\mathbf{x}}); \quad \mathbf{r} \times \mathbf{F}_2 = (r\,\hat{\mathbf{y}}) \times \left(\frac{3p_1p_2}{4\pi\epsilon_0 r^4}\,\hat{\mathbf{z}}\right) = \frac{3p_1p_2}{4\pi\epsilon_0 r^3}\,\hat{\mathbf{x}}; \text{ so } \boxed{\mathbf{N}_2 = \frac{2p_1p_2}{4\pi\epsilon_0 r^3}\,\hat{\mathbf{x}}.}$$

This is equal and opposite to the torque on $\mathbf{p}_1$ due to $\mathbf{p}_2$, with respect to the center of $\mathbf{p}_1$ (see Prob. 4.5).

Problem 4.30

Net force is $\boxed{\text{to the right}}$ (see diagram). Note that the field lines must bulge to the right, as shown, because $\mathbf{E}$ is perpendicular to the surface of each conductor.

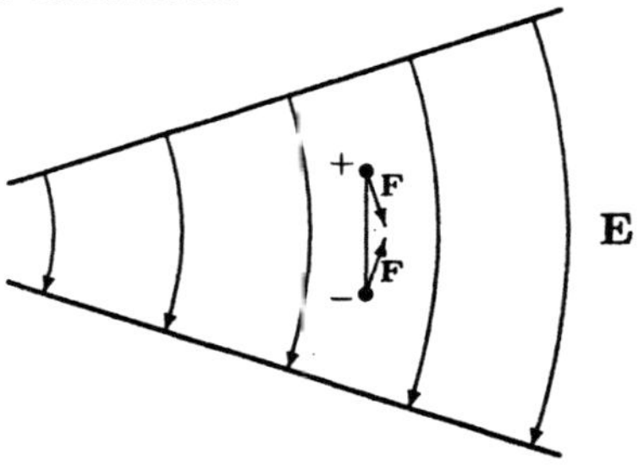

Problem 4.31

$\mathbf{P} = k\mathbf{r} = k(x\,\hat{\mathbf{x}} + y\,\hat{\mathbf{y}} + z\,\hat{\mathbf{z}}) \Longrightarrow \rho_b = -\nabla\cdot\mathbf{P} = -k(1+1+1) = \boxed{-3k.}$

Total volume bound charge: $\boxed{Q_{\text{vol}} = -3ka^3.}$

$\sigma_b = \mathbf{P}\cdot\hat{\mathbf{n}}$. At top surface, $\hat{\mathbf{n}} = \hat{\mathbf{z}}$, $z = a/2$; so $\sigma_b = ka/2$. Clearly, $\boxed{\sigma_b = ka/2}$ on all six surfaces.

Total surface bound charge: $\boxed{Q_{\text{surf}} = 6(ka/2)a^2 = 3ka^3.}$ Total bound charge is zero. ✓

Problem 4.32

$$\oint \mathbf{D}\cdot d\mathbf{a} = Q_{f_{\text{enc}}} \Rightarrow \mathbf{D} = \frac{q}{4\pi r^2}\,\hat{\mathbf{r}};\ \mathbf{E} = \frac{1}{\epsilon}\mathbf{D} = \boxed{\frac{q}{4\pi\epsilon_0(1+\chi_e)}\frac{\hat{\mathbf{r}}}{r^2};}\ \mathbf{P} = \epsilon_0\chi_e\mathbf{E} = \boxed{\frac{q\chi_e}{4\pi(1+\chi_e)}\frac{\hat{\mathbf{r}}}{r^2}.}$$

$$\rho_b = -\nabla\cdot\mathbf{P} = -\frac{q\chi_e}{4\pi(1+\chi_e)}\left(\nabla\cdot\frac{\hat{\mathbf{r}}}{r^2}\right) = \boxed{-q\frac{\chi_e}{1+\chi_e}\delta^3(\mathbf{r})}\ \text{(Eq. 1.99)};\ \sigma_b = \mathbf{P}\cdot\hat{\mathbf{r}} = \boxed{\frac{q\chi_e}{4\pi(1+\chi_e)R^2};}$$

$Q_{\text{surf}} = \sigma_b(4\pi R^2) = \boxed{q\frac{\chi_e}{1+\chi_e}.}$ The compensating negative charge is at the center:

$$\int \rho_b\,d\tau = -\frac{q\chi_e}{1+\chi_e}\int \delta^3(\mathbf{r})\,d\tau = -q\frac{\chi_e}{1+\chi_e}.$$

Problem 4.33

$\mathbf{E}^{\parallel}$ is continuous (Eq. 4.29); $D_\perp$ is continuous (Eq. 4.26, with $\sigma_f = 0$). So $E_{x_1} = E_{x_2}$, $D_{y_1} = D_{y_2} \Rightarrow \epsilon_1 E_{y_1} = \epsilon_2 E_{y_2}$, and hence

$$\frac{\tan\theta_2}{\tan\theta_1} = \frac{E_{x_2}/E_{y_2}}{E_{x_1}/E_{y_1}} = \frac{E_{y_1}}{E_{y_2}} = \frac{\epsilon_2}{\epsilon_1}.\quad \text{qed}$$

If 1 is air and 2 is dielectric, $\tan\theta_2/\tan\theta_1 = \epsilon_2/\epsilon_0 > 1$, and the field lines bend *away* from the normal. This is the opposite of light rays, so a convex "lens" would *defocus* the field lines.

Problem 4.34

In view of Eq. 4.39, the *net* dipole moment at the center is $\mathbf{p}' = \mathbf{p} - \frac{\chi_e}{1+\chi_e}\mathbf{p} = \frac{1}{1+\chi_e}\mathbf{p} = \frac{1}{\epsilon_r}\mathbf{p}$. We want the potential produced by $\mathbf{p}'$ (at the center) and σ_b (at R). Use separation of variables:

$$\left\{\begin{array}{lll} \textit{Outside:} & V(r,\theta) = \displaystyle\sum_{l=0}^{\infty}\frac{B_l}{r^{l+1}}P_l(\cos\theta) & \text{(Eq. 3.72)} \\ \textit{Inside:} & V(r,\theta) = \dfrac{1}{4\pi\epsilon_0}\dfrac{p\cos\theta}{\epsilon_r r^2} + \displaystyle\sum_{l=0}^{\infty}A_l r^l P_l(\cos\theta) & \text{(Eqs. 3.66, 3.102)} \end{array}\right\}.$$

$$V \text{ continuous at } R \Rightarrow \left\{\begin{array}{ll} \dfrac{B_l}{R^{l+1}} = A_l R^l, & \text{or}\quad B_l = R^{2l+1}A_l\ (l \neq 1) \\ \dfrac{B_1}{R^2} = \dfrac{1}{4\pi\epsilon_0}\dfrac{p}{\epsilon_r R^2} + A_1 R, & \text{or}\quad B_1 = \frac{p}{4\pi\epsilon_0\epsilon_r} + A_1 R^3 \end{array}\right\}.$$

$$\begin{aligned} \left.\frac{\partial V}{\partial r}\right|_{R+} - \left.\frac{\partial V}{\partial r}\right|_{R-} &= -\sum(l+1)\frac{B_l}{R^{l+2}}P_l(\cos\theta) + \frac{1}{4\pi\epsilon_0}\frac{2p\cos\theta}{\epsilon_r R^3} - \sum lA_lR^{l-1}P_l(\cos\theta) = -\frac{1}{\epsilon_0}\sigma_b \\ &= -\frac{1}{\epsilon_0}\mathbf{P}\cdot\hat{\mathbf{r}} = -\frac{1}{\epsilon_0}(\epsilon_0\chi_e\mathbf{E}\cdot\hat{\mathbf{r}}) = \chi_e\left.\frac{\partial V}{\partial r}\right|_{R-} = \chi_e\left\{-\frac{1}{4\pi\epsilon_0}\frac{2p\cos\theta}{\epsilon_r R^3} + \sum lA_lR^{l-1}P_l(\cos\theta)\right\}. \end{aligned}$$

$$-(l+1)\frac{B_l}{R^{l+2}} - lA_lR^{l-1} = \chi_e lA_lR^{l-1}\ (l \neq 1);\ \text{or}\ -(2l+1)A_lR^{l-1} = \chi_e lA_lR^{l-1} \Rightarrow A_l = 0\ (\ell \neq 1).$$

For $l = 1$: $-2\frac{B_1}{R^3} + \frac{1}{4\pi\epsilon_0}\frac{2p}{\epsilon_r R^3} - A_1 = \chi_e\left(-\frac{1}{4\pi\epsilon_0}\frac{2p}{\epsilon_r R^3} + A_1\right) - B_1 + \frac{p}{4\pi\epsilon_0\epsilon_r} - \frac{A_1R^3}{2} = -\frac{1}{4\pi\epsilon_0}\frac{\chi_e p}{\epsilon_r} + \chi_e\frac{A_1R^3}{2}$;

$$-\frac{p}{4\pi\epsilon_0\epsilon_r} - A_1R^3 + \frac{p}{4\pi\epsilon_0\epsilon_r} - \frac{A_1R^3}{2} = -\frac{1}{4\pi\epsilon_0}\frac{\chi_e p}{\epsilon_r} + \chi_e\frac{A_1R^3}{2} \Rightarrow \frac{A_1R^3}{2}(3+\chi_e) = \frac{1}{4\pi\epsilon_0}\frac{\chi_e p}{\epsilon_r}.$$

$$\Rightarrow A_1 = \frac{1}{4\pi\epsilon_0}\frac{2\chi_e p}{R^3\epsilon_r(3+\chi_e)} = \frac{1}{4\pi\epsilon_0}\frac{2(\epsilon_r-1)p}{R^3\epsilon_r(\epsilon_r+2)};\quad B_1 = \frac{p}{4\pi\epsilon_0\epsilon_r}\left[1 + \frac{2(\epsilon_r-1)}{(\epsilon_r+2)}\right] = \frac{p}{4\pi\epsilon_0\epsilon_r}\frac{3\epsilon_r}{\epsilon_r+2}.$$

$$\boxed{V(r,\theta) = \left(\frac{p\cos\theta}{4\pi\epsilon_0 r^2}\right)\left(\frac{3}{\epsilon_r+2}\right)\ (r \geq R).}$$

Meanwhile, for $r \leq R$, $V(r,\theta) = \frac{1}{4\pi\epsilon_0}\frac{p\cos\theta}{\epsilon_r r^2} + \frac{1}{4\pi\epsilon_0}\frac{pr\cos\theta}{R^3}\frac{2(\epsilon_r-1)}{\epsilon_r(\epsilon_r+2)}$

$$= \boxed{\frac{p\cos\theta}{4\pi\epsilon_0 r^2\epsilon_r}\left[1 + 2\left(\frac{\epsilon_r-1}{\epsilon_r+2}\right)\frac{r^3}{R^3}\right]\ (r \leq R).}$$

Problem 4.35

Given two solutions, V_1 (and $\mathbf{E}_1 = -\nabla V_1$, $\mathbf{D}_1 = \epsilon\mathbf{E}_1$) and V_2 ($\mathbf{E}_2 = -\nabla V_2$, $\mathbf{D}_2 = \epsilon\mathbf{E}_2$), define $V_3 \equiv V_2 - V_1$ ($\mathbf{E}_3 = \mathbf{E}_2 - \mathbf{E}_1$, $\mathbf{D}_3 = \mathbf{D}_2 - \mathbf{D}_1$).

$\int_\mathcal{V} \nabla\cdot(V_3\mathbf{D}_3)\,d\tau = \int_\mathcal{S} V_3\mathbf{D}_3\cdot d\mathbf{a} = 0$, ($V_3 = 0$ on $\mathcal{S}$), so $\int(\nabla V_3)\cdot\mathbf{D}_3\,d\tau + \int V_3(\nabla\cdot\mathbf{D}_3)\,d\tau = 0$.
But $\nabla\cdot\mathbf{D}_3 = \nabla\cdot\mathbf{D}_2 - \nabla\cdot\mathbf{D}_1 = \rho_f - \rho_f = 0$, and $\nabla V_3 = \nabla V_2 - \nabla V_1 = -\mathbf{E}_2 + \mathbf{E}_1 = -\mathbf{E}_3$, so $\int \mathbf{E}_3\cdot\mathbf{D}_3\,d\tau = 0$. But $\mathbf{D}_3 = \mathbf{D}_2 - \mathbf{D}_1 = \epsilon\mathbf{E}_2 - \epsilon\mathbf{E}_1 = \epsilon\mathbf{E}_3$, so $\int \epsilon(E_3)^2\,d\tau = 0$. But $\epsilon > 0$, so $\mathbf{E}_3 = 0$, so $V_2 - V_1 =$ constant. But at surface, $V_2 = V_1$, so $V_2 = V_1$ everywhere. qed

Problem 4.36

(a) Proposed potential: $\boxed{V(r) = V_0\frac{R}{r}.}$ If so, then $\boxed{\mathbf{E} = -\nabla V = V_0\frac{R}{r^2}\,\hat{\mathbf{r}},}$ in which case $\boxed{\mathbf{P} = \epsilon_0\chi_e V_0\frac{R}{r^2}\,\hat{\mathbf{r}},}$ in the region $z < 0$. ($\mathbf{P} = 0$ for $z > 0$, of course.) Then $\sigma_b = \epsilon_0\chi_e V_0\frac{R}{R^2}(\hat{\mathbf{r}}\cdot\hat{\mathbf{n}}) = \boxed{-\frac{\epsilon_0\chi_e V_0}{R}.}$ (*Note:* $\hat{\mathbf{n}}$ *points out* of dielectric $\Rightarrow \hat{\mathbf{n}} = -\hat{\mathbf{r}}$.) This σ_b is on the surface at $r = R$. The flat surface $z = 0$ carries no bound charge, since $\hat{\mathbf{n}} = \hat{\mathbf{z}} \perp \hat{\mathbf{r}}$. Nor is there any volume bound charge (Eq. 4.39). If V is to have the required spherical symmetry, the *net* charge must be uniform:
$\sigma_{tot}4\pi R^2 = Q_{tot} = 4\pi\epsilon_0 RV_0$ (since $V_0 = Q_{tot}/4\pi\epsilon_0 R$), so $\sigma_{tot} = \epsilon_0 V_0/R$. Therefore

$$\boxed{\sigma_f = \begin{cases} (\epsilon_0 V_0/R), & \text{on northern hemisphere} \\ (\epsilon_0 V_0/R)(1+\chi_e), & \text{on southern hemisphere} \end{cases}.}$$

(b) By construction, $\sigma_{tot} = \sigma_b + \sigma_f = \epsilon_0 V_0/R$ is uniform (on the northern hemisphere $\sigma_b = 0$, $\sigma_f = \epsilon_0 V_0/R$; on the southern hemisphere $\sigma_b = -\epsilon_0\chi_e V_0/R$, so $\sigma_f = \epsilon V_0/R$). The potential of a uniformly charged sphere is

$$V_0 = \frac{Q_{tot}}{4\pi\epsilon_0 r} = \frac{\sigma_{tot}(4\pi R^2)}{4\pi\epsilon_0 r} = \frac{\epsilon_0 V_0}{R}\frac{R^2}{\epsilon_0 r} = V_0\frac{R}{r}.\quad \checkmark$$

(c) Since everything is consistent, and the boundary conditions ($V = V_0$ at $r = R$, $V \to 0$ at ∞) are met, Prob. 4.35 guarantees that this is *the* solution.

(d) Figure (b) works the same way, but Fig. (a) does *not:* on the flat surface, $\mathbf{P}$ is *not* perpendicular to $\hat{\mathbf{n}}$, so we'd get bound charge on this surface, spoiling the symmetry.

Problem 4.37

$\mathbf{E}_{\text{ext}} = \frac{\lambda}{2\pi\epsilon_0 s}\,\hat{\mathbf{s}}$. Since the sphere is tiny, this is essentially constant, and hence $\mathbf{P} = \frac{\epsilon_0\chi_e}{1+\chi_e/3}\mathbf{E}_{\text{ext}}$ (Ex. 4.7).

$$\mathbf{F} = \int\left(\frac{\epsilon_0\chi_e}{1+\chi_e/3}\right)\left(\frac{\lambda}{2\pi\epsilon_0 s}\right)\frac{d}{ds}\left(\frac{\lambda}{2\pi\epsilon_0 s}\right)\hat{\mathbf{s}}\,d\tau = \left(\frac{\epsilon_0\chi_e}{1+\chi_e/3}\right)\left(\frac{\lambda}{2\pi\epsilon_0}\right)^2\left(\frac{1}{s}\right)\left(\frac{-1}{s^2}\right)\hat{\mathbf{s}}\int d\tau$$

$$= \frac{-\chi_e}{1+\chi_e/3}\left(\frac{\lambda^2}{4\pi^2\epsilon_0}\right)\frac{1}{s^3}\frac{4}{3}\pi R^3\,\hat{\mathbf{s}} = \boxed{-\left(\frac{\chi_e}{3+\chi_e}\right)\frac{\lambda^2 R^3}{\pi\epsilon_0 s^3}\,\hat{\mathbf{s}}.}$$

Problem 4.38

The density of atoms is $N = \frac{1}{(4/3)\pi R^3}$. The macroscopic field $\mathbf{E}$ is $\mathbf{E}_{\text{self}} + \mathbf{E}_{\text{else}}$, where $\mathbf{E}_{\text{self}}$ is the average field over the sphere due to the atom itself.

$$\mathbf{p} = \alpha\mathbf{E}_{\text{else}} \Rightarrow \mathbf{P} = N\alpha\mathbf{E}_{\text{else}}.$$

[Actually, it is the field at the *center*, not the average over the sphere, that belongs here, but the two are in fact equal, as we found in Prob. 3.41d.] Now

$$\mathbf{E}_{\text{self}} = -\frac{1}{4\pi\epsilon_0}\frac{\mathbf{p}}{R^3}$$

(Eq. 3.105), so

$$\mathbf{E} = -\frac{1}{4\pi\epsilon_0}\frac{\alpha}{R^3}\mathbf{E}_{\text{else}} + \mathbf{E}_{\text{else}} = \left(1 - \frac{\alpha}{4\pi\epsilon_0 R^3}\right)\mathbf{E}_{\text{else}} = \left(1 - \frac{N\alpha}{3\epsilon_0}\right)\mathbf{E}_{\text{else}}.$$

So

$$\mathbf{P} = \frac{N\alpha}{(1 - N\alpha/3\epsilon_0)}\mathbf{E} = \epsilon_0\chi_e\mathbf{E},$$

and hence

$$\chi_e = \frac{N\alpha/\epsilon_0}{(1 - N\alpha/3\epsilon_0)}.$$

Solving for α:

$$\chi_e - \frac{N\alpha}{3\epsilon_0}\chi_e = \frac{N\alpha}{\epsilon_0} \Rightarrow \frac{N\alpha}{\epsilon_0}\left(1 + \frac{\chi_e}{3}\right) = \chi_e,$$

or

$$\alpha = \frac{\epsilon_0}{N}\frac{\chi_e}{(1+\chi_e/3)} = \frac{3\epsilon_0}{N}\frac{\chi_e}{(3+\chi_e)}.\quad \text{But } \chi_e = \epsilon_r - 1, \text{ so } \alpha = \frac{3\epsilon_0}{N}\left(\frac{\epsilon_r - 1}{\epsilon_r + 2}\right).\quad \text{qed}$$

Problem 4.39

For an ideal gas, N = Avagadro's number/22.4 liters $= (6.02\times10^{23})/(22.4\times10^{-3}) = 2.7\times10^{25}$. $N\alpha/\epsilon_0 = (2.7\times10^{25})(4\pi\epsilon_0\times10^{-30})\beta/\epsilon_0 = 3.4\times10^{-4}\beta$, where β is the number listed in Table 4.1.

H:	$\beta = 0.667$,	$N\alpha/\epsilon_0 = (3.4\times10^{-4})(0.67) = 2.3\times10^{-4}$,	$\chi_e = 2.5\times10^{-4}$	agreement is quite good.
He:	$\beta = 0.205$,	$N\alpha/\epsilon_0 = (3.4\times10^{-4})(0.21) = 7.1\times10^{-5}$,	$\chi_e = 6.5\times10^{-5}$	
Ne:	$\beta = 0.396$,	$N\alpha/\epsilon_0 = (3.4\times10^{-4})(0.40) = 1.4\times10^{-4}$,	$\chi_e = 1.3\times10^{-4}$	
Ar:	$\beta = 1.64$,	$N\alpha/\epsilon_0 = (3.4\times10^{-4})(1.64) = 5.6\times10^{-4}$,	$\chi_e = 5.2\times10^{-4}$	

Problem 4.40

(a)
$$\langle u\rangle = \frac{\int_{-pE}^{pE} u e^{-u/kT}\,du}{\int_{-pE}^{pE} e^{-u/kT}\,du} = \frac{(kT)^2 e^{-u/kT}\left[-(u/kT)-1\right]\Big|_{-pE}^{pE}}{-kTe^{-u/kT}\Big|_{-pE}^{pE}}$$
$$= kT\left\{\frac{\left[e^{-pE/kT}-e^{pE/kT}\right]+\left[(pE/kT)e^{-pE/kT}+(pE/kT)e^{pE/kT}\right]}{e^{-pE/kT}-e^{pE/kT}}\right\}$$
$$= kT - pE\left[\frac{e^{pE/kT}+e^{-pE/kT}}{e^{pE/kT}-e^{-pE/kT}}\right] = kT - pE\coth\left(\frac{pE}{kT}\right).$$

$$\mathbf{P}=N\langle\mathbf{p}\rangle;\ \mathbf{p}=\langle p\cos\theta\rangle\hat{\mathbf{E}}=\langle\mathbf{P}\cdot\mathbf{E}\rangle(\hat{\mathbf{E}}/E)=-\langle u\rangle(\hat{\mathbf{E}}/E);\ P=Np\frac{-\langle u\rangle}{pE}=\boxed{Np\left\{\coth\left(\frac{pE}{kT}\right)-\frac{kT}{pE}\right\}.}$$

Let $y\equiv P/Np$, $x\equiv pE/kT$. Then $y=\coth x - 1/x$. As $x\to 0$, $y=\left(\frac{1}{x}+\frac{x}{3}-\frac{x^3}{45}+\cdots\right)-\frac{1}{x}=\frac{x}{3}-\frac{x^3}{45}+\cdots\to 0$, so the graph starts at the origin, with an initial slope of $1/3$. As $x\to\infty$, $y\to\coth(\infty)=1$, so the graph goes asymptotically to $y=1$ (see Figure).

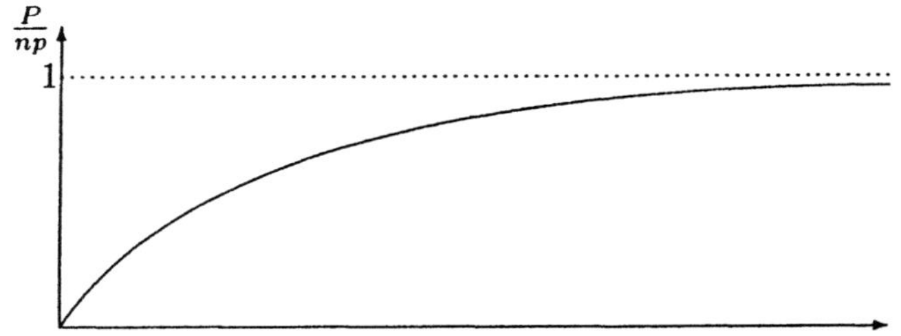

(b) For small x, $y\approx\frac{1}{3}x$, so $\frac{P}{Np}\approx\frac{pE}{3kT}$, or $P\approx\frac{Np^2}{3kT}E=\epsilon_0\chi_e E\Rightarrow P$ is proportional to E, and $\boxed{\chi_e=\frac{Np^2}{3\epsilon_0 kT}.}$

For water at $20^\circ = 293\,\text{K}$, $p=6.1\times10^{-30}\,\text{C m}$; $N=\frac{\text{molecules}}{\text{volume}}=\frac{\text{molecules}}{\text{mole}}\times\frac{\text{moles}}{\text{gram}}\times\frac{\text{grams}}{\text{volume}}$.

$N=(6.0\times10^{23})\times\left(\frac{1}{18}\right)\times(10^6)=0.33\times10^{29}$; $\chi_e=\frac{(0.33\times10^{29})(6.1\times10^{-30})^2}{(3)(8.85\times10^{-12})(1.38\times10^{-23})(293)}=\boxed{12.}$ Table 4.2 gives an experimental value of 79, so it's pretty far off.

For water vapor at $100^\circ=373\,\text{K}$, treated as an ideal gas, $\frac{\text{volume}}{\text{mole}}=(22.4\times10^{-3})\times\left(\frac{373}{293}\right)=2.85\times10^{-2}\ \text{m}^3$.

$$N=\frac{6.0\times10^{23}}{2.85\times10^{-2}}=2.11\times10^{25};\quad \chi_e=\frac{(2.11\times10^{25})(6.1\times10^{-30})^2}{(3)(8.85\times10^{-12})(1.38\times10^{-23})(373)}=\boxed{5.7\times10^{-3}.}$$

Table 4.2 gives 5.9×10^{-3}, so this time the agreement is quite good.

Chapter 5

Magnetostatics

Problem 5.1

Since $\mathbf{v} \times \mathbf{B}$ points upward, and that is also the direction of the force, q must be $\boxed{\text{positive.}}$ To find R, in terms of a and d, use the pythagorean theorem:

$$(R-d)^2 + a^2 = R^2 \Rightarrow R^2 - 2Rd + d^2 + a^2 = R^2 \Rightarrow R = \frac{a^2+d^2}{2d}.$$

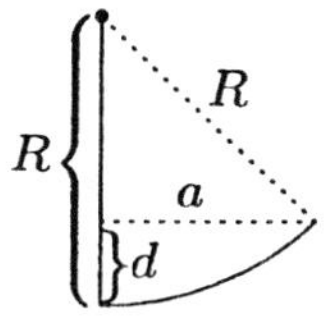

The cyclotron formula then gives

$$p = qBR = \boxed{qB\frac{(a^2+d^2)}{2d}}.$$

Problem 5.2

The general solution is (Eq. 5.6):

$$y(t) = C_1\cos(\omega t) + C_2\sin(\omega t) + \frac{E}{B}t + C_3; \quad z(t) = C_2\cos(\omega t) - C_1\sin(\omega t) + C_4.$$

(a) $y(0) = z(0) = 0$; $\dot{y}(0) = E/B$; $\dot{z}(0) = 0$. Use these to determine C_1, C_2, C_3, and C_4. $y(0) = 0 \Rightarrow C_1 + C_3 = 0$; $\dot{y}(0) = \omega C_2 + E/B = E/B \Rightarrow C_2 = 0$; $z(0) = 0 \Rightarrow C_2 + C_4 = 0 \Rightarrow C_4 = 0$; $\dot{z}(0) = 0 \Rightarrow C_1 = 0$, and hence also $C_3 = 0$. So $\boxed{y(t) = Et/B;\ z(t) = 0.}$ Does this make sense? The magnetic force is $q(\mathbf{v}\times\mathbf{B}) = -q(E/B)B\,\hat{\mathbf{z}} = -q\mathbf{E}$, which exactly cancels the electric force; since there is no net force, the particle moves in a straight line at constant speed. ✓

(b) Assuming it starts from the origin, so $C_3 = -C_1$, $C_4 = -C_2$, we have $\dot{z}(0) = 0 \Rightarrow C_1 = 0 \Rightarrow C_3 = 0$; $\dot{y}(0) = \dfrac{E}{2B} \Rightarrow C_2\omega + \dfrac{E}{B} = \dfrac{E}{2B} \Rightarrow C_2 = -\dfrac{E}{2\omega B} = -C_4$; $y(t) = -\dfrac{E}{2\omega B}\sin(\omega t) + \dfrac{E}{B}t$; $z(t) = -\dfrac{E}{2\omega B}\cos(\omega t) + \dfrac{E}{2\omega B}$, or $\boxed{y(t) = \dfrac{E}{2\omega B}\left[2\omega t - \sin(\omega t)\right];\ z(t) = \dfrac{E}{2\omega B}\left[1-\cos(\omega t)\right].}$ Let $\beta \equiv E/2\omega B$. Then $y(t) = \beta\left[2\omega t - \sin(\omega t)\right]$; $z(t) = \beta\left[1-\cos(\omega t)\right]$; $(y - 2\beta\omega t) = -\beta\sin(\omega t)$, $(z-\beta) = -\beta\cos(\omega t) \Rightarrow$ $(y-2\beta\omega t)^2 + (z-\beta)^2 = \beta^2$. This is a circle of radius β whose center moves to the right at constant speed: $y_0 = 2\beta\omega t$; $z_0 = \beta$.

(c) $\dot{z}(0) = \dot{y}(0) = \dfrac{E}{B} \Rightarrow -C_1\omega = \dfrac{E}{B} \Rightarrow C_1 = -C_3 = -\dfrac{E}{\omega B}$; $C_2\omega + \dfrac{E}{B} = \dfrac{E}{B} \Rightarrow C_2 = C_4 = 0.$

$y(t) = -\frac{E}{\omega B}\cos(\omega t) + \frac{E}{B}t + \frac{E}{\omega B}$; $z(t) = \frac{E}{\omega B}\sin(\omega t)$. $\boxed{y(t) = \frac{E}{\omega B}[1 + \omega t - \cos(\omega t)];\ z(t) = \frac{E}{\omega B}\sin(\omega t).}$

Let $\beta \equiv E/\omega B$; then $[y - \beta(1+\omega t)] = -\beta\cos(\omega t)$, $z = \beta\sin(\omega t)$; $[y - \beta(1+\omega t)]^2 + z^2 = \beta^2$. This is a circle of radius β whose center is at $y_0 = \beta(1+\omega t)$, $z_0 = 0$.

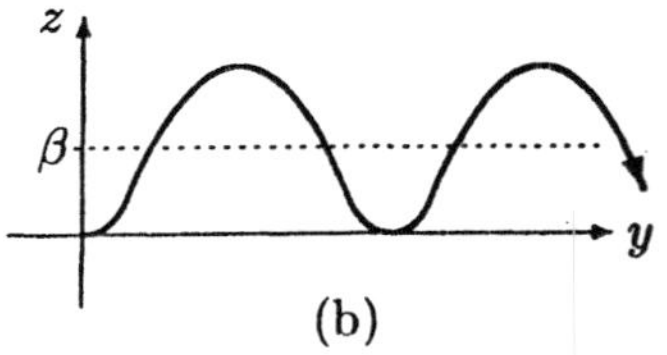

(b)

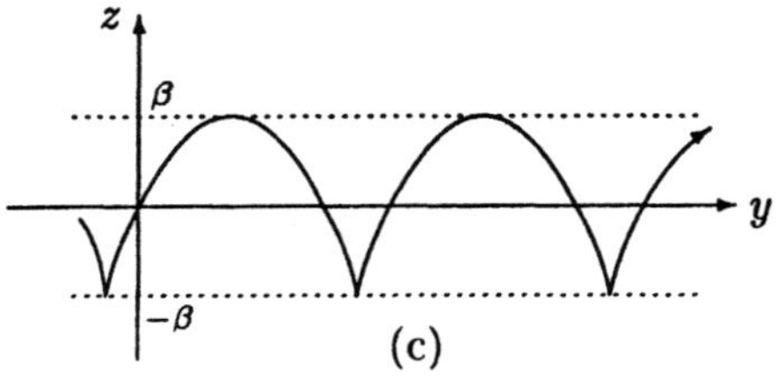

(c)

Problem 5.3

(a) From Eq. 5.2, $\mathbf{F} = q[\mathbf{E} + (\mathbf{v}\times\mathbf{B})] = 0 \Rightarrow E = vB \Rightarrow \boxed{v = \frac{E}{B}.}$

(b) From Eq. 5.3, $mv = qBR \Rightarrow \frac{q}{m} = \frac{v}{BR} = \boxed{\frac{E}{B^2R}.}$

Problem 5.4

Suppose I flows counterclockwise (if not, change the sign of the answer). The force on the left side (toward the left) cancels the force on the right side (toward the right); the force on the top is $IaB = Iak(a/2) = Ika^2/2$, (pointing upward), and the force on the bottom is $IaB = -Ika^2/2$ (also upward). So the net force is $\mathbf{F} = \boxed{Ika^2\,\hat{\mathbf{z}}.}$

Problem 5.5

(a) $\boxed{K = \frac{I}{2\pi a},}$ because the length-perpendicular-to-flow is the circumference.

(b) $J = \frac{\alpha}{s} \Rightarrow I = \int J\,da = \alpha\int\frac{1}{s}s\,ds\,d\phi = 2\pi\alpha\int ds = 2\pi\alpha a \Rightarrow \alpha = \frac{I}{2\pi a}$; $J = \boxed{\frac{I}{2\pi a s}.}$

Problem 5.6

(a) $v = \omega r$, so $\boxed{K = \sigma\omega r.}$ (b) $\mathbf{v} = \omega r\sin\theta\,\hat{\boldsymbol{\phi}} \Rightarrow \boxed{\mathbf{J} = \rho\omega r\sin\theta\,\hat{\boldsymbol{\phi}},}$ where $\rho \equiv Q/(4/3)\pi R^3$.

Problem 5.7

$\frac{d\mathbf{p}}{dt} = \frac{d}{dt}\int\rho\mathbf{r}\,d\tau = \int\left(\frac{\partial\rho}{\partial t}\right)\mathbf{r}\,d\tau = -\int(\nabla\cdot\mathbf{J})\mathbf{r}\,d\tau$ (by the continuity equation). Now product rule #5 says $\nabla\cdot(x\mathbf{J}) = x(\nabla\cdot\mathbf{J}) + \mathbf{J}\cdot(\nabla x)$. But $\nabla x = \hat{\mathbf{x}}$, so $\nabla\cdot(x\mathbf{J}) = x(\nabla\cdot\mathbf{J}) + J_x$. Thus $\int_{\mathcal{V}}(\nabla\cdot\mathbf{J})x\,d\tau = \int_{\mathcal{V}}\nabla\cdot(x\mathbf{J})\,d\tau - \int_{\mathcal{V}}J_x\,d\tau$. The first term is $\int_{\mathcal{S}}x\mathbf{J}\cdot d\mathbf{a}$ (by the divergence theorem), and since $\mathbf{J}$ is entirely *inside* $\mathcal{V}$, it is zero on the surface $\mathcal{S}$. Therefore $\int_{\mathcal{V}}(\nabla\cdot\mathbf{J})x\,d\tau = -\int_{\mathcal{V}}J_x\,d\tau$, or, combining this with the y and z components, $\int_{\mathcal{V}}(\nabla\cdot\mathbf{J})\mathbf{r}\,d\tau = -\int_{\mathcal{V}}\mathbf{J}\,d\tau$. Or, referring back to the first line, $\frac{d\mathbf{p}}{dt} = \int\mathbf{J}\,d\tau$. qed

Problem 5.8

(a) Use Eq. 5.35, with $z = R, \theta_2 = -\theta_1 = 45°$, and four sides: $B = \boxed{\frac{\sqrt{2}\mu_0 I}{\pi R}.}$

(b) $z = R$, $\theta_2 = -\theta_1 = \frac{\pi}{n}$, and n sides: $B = \boxed{\frac{n\mu_0 I}{2\pi R}\sin(\pi/n).}$

(c) For small θ, $\sin\theta \approx \theta$. So as $n \to \infty, B \to \frac{n\mu_0 I}{2\pi R}\left(\frac{\pi}{n}\right) = \boxed{\frac{\mu_0 I}{2R}}$ (same as Eq. 5.38, with $z = 0$).

Problem 5.9

(a) The straight segments produce no field at P. The two quarter-circles give $B = \boxed{\frac{\mu_0 I}{8}\left(\frac{1}{a} - \frac{1}{b}\right)}$ (out).

(b) The two half-lines are the same as one infinite line: $\frac{\mu_0 I}{2\pi R}$; the half-circle contributes $\frac{\mu_0 I}{4R}$.

So $B = \boxed{\frac{\mu_0 I}{4R}\left(1 + \frac{2}{\pi}\right)}$ (into the page).

Problem 5.10

(a) The forces on the two sides cancel. At the bottom, $B = \frac{\mu_0 I}{2\pi s} \Rightarrow F = \left(\frac{\mu_0 I}{2\pi s}\right) Ia = \frac{\mu_0 I^2 a}{2\pi s}$ (up). At the top, $B = \frac{\mu_0 I}{2\pi(s+a)} \Rightarrow F = \frac{\mu_0 I^2 a}{2\pi(s+a)}$ (down). The net force is $\boxed{\frac{\mu_0 I^2 a^2}{2\pi s(s+a)}}$ (up).

(b) The force on the bottom is the same as before, $\mu_0 I^2/2\pi$ (up). On the left side, $\mathbf{B} = \frac{\mu_0 I}{2\pi y}\,\hat{\mathbf{z}}$; $d\mathbf{F} = I(d\mathbf{l} \times \mathbf{B}) = I(dx\,\hat{\mathbf{x}} + dy\,\hat{\mathbf{y}} + dz\,\hat{\mathbf{z}}) \times \left(\frac{\mu_0 I}{2\pi y}\,\hat{\mathbf{z}}\right) = \frac{\mu_0 I^2}{2\pi y}(-dx\,\hat{\mathbf{y}} + dy\,\hat{\mathbf{x}})$. But the x component cancels the corresponding term from the right side, and $F_y = -\frac{\mu_0 I^2}{2\pi}\int_{s/\sqrt{3}}^{(s/\sqrt{3}+a/2)} \frac{1}{y}\,dx$. Here $y = \sqrt{3}x$, so

$F_y = -\frac{\mu_0 I^2}{2\sqrt{3}\pi}\ln\left(\frac{s/\sqrt{3} + a/2}{s/\sqrt{3}}\right) = -\frac{\mu_0 I^2}{2\sqrt{3}\pi}\ln\left(1 + \frac{\sqrt{3}a}{2s}\right)$. The force on the right side is the same, so the net force on the triangle is $\boxed{\frac{\mu_0 I^2}{2\pi}\left[1 - \frac{2}{\sqrt{3}}\ln\left(1 + \frac{\sqrt{3}a}{2s}\right)\right]}$.

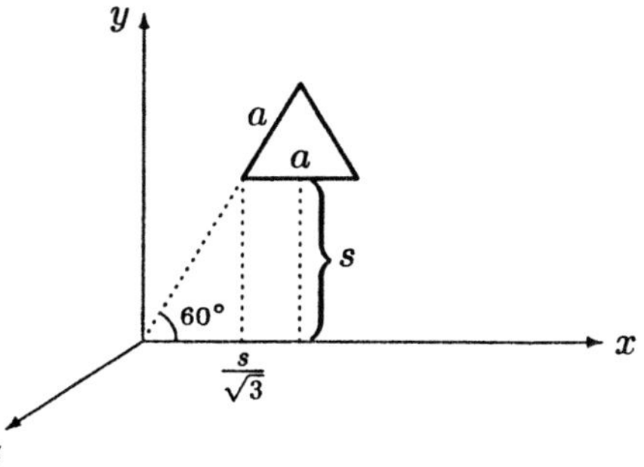

Problem 5.11

Use Eq. 5.38 for a ring of width dz, with $I \to nI\,dz$:

$B = \frac{\mu_0 nI}{2}\int \frac{a^2}{(a^2+z^2)^{3/2}}\,dz$. But $z = a\cot\theta$,

so $dz = -\frac{a}{\sin^2\theta}\,d\theta$, and $\frac{1}{(a^2+z^2)^{3/2}} = \frac{\sin^3\theta}{a^3}$.

So

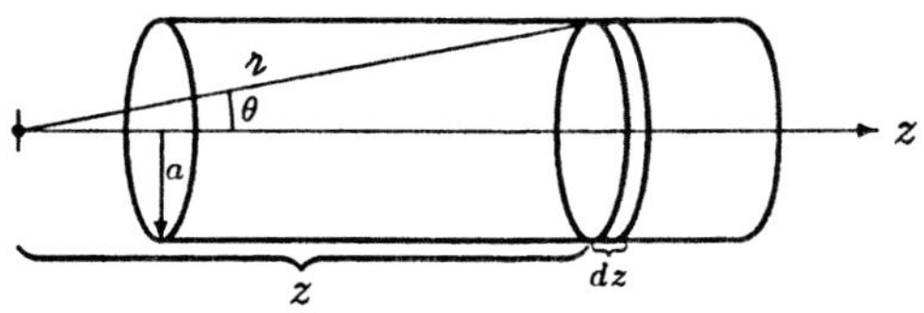

$$B = \frac{\mu_0 nI}{2}\int \frac{a^2\sin^3\theta}{a^3\sin^2\theta}(-a\,d\theta) = -\frac{\mu_0 nI}{2}\int \sin\theta\,d\theta = \frac{\mu_0 nI}{2}\cos\theta\Big|_{\theta_1}^{\theta_2} = \boxed{\frac{\mu_0 nI}{2}(\cos\theta_2 - \cos\theta_1).}$$

For an infinite solenoid, $\theta_2 = 0$, $\theta_1 = \pi$, so $(\cos\theta_2 - \cos\theta_1) = 1 - (-1) = 2$, and $B = \boxed{\mu_0 nI.}$ ✓

Problem 5.12

Magnetic attraction per unit length (Eqs. 5.37 and 5.13): $f_m = \frac{\mu_0}{2\pi}\frac{\lambda^2 v^2}{d}$.

Electric field of one wire (Eq. 2.9): $E = \frac{1}{2\pi\epsilon_0}\frac{\lambda}{s}$. Electric repulsion per unit length on the other wire: $f_e = \frac{1}{2\pi\epsilon_0}\frac{\lambda^2}{d}$. They balance when $\mu_0 v^2 = \frac{1}{\epsilon_0}$, or $\boxed{v = \frac{1}{\sqrt{\epsilon_0\mu_0}}.}$ Putting in the numbers, $v = \frac{1}{\sqrt{(8.85\times 10^{-12})(4\pi\times 10^{-7})}} = \boxed{3.00\times 10^8\ \text{m/s}.}$ This is precisely the *speed of light(!)*, so in fact you could *never* get the wires going fast enough; the electric force always dominates.

Problem 5.13

(a) $\oint \mathbf{B}\cdot d\mathbf{l} = B\,2\pi s = \mu_0 I_{\text{enc}} \Rightarrow \boxed{\mathbf{B} = \left\{\begin{array}{ll} 0, & \text{for } s < a; \\ \frac{\mu_0 I}{2\pi s}\hat{\boldsymbol{\phi}}, & \text{for } s > a. \end{array}\right\}}$

(b) $J = ks;\ I = \int_0^a J\,da = \int_0^a ks(2\pi s)\,ds = \frac{2\pi k a^3}{3} \Rightarrow k = \frac{3I}{2\pi a^3}$. $I_{\text{enc}} = \int_0^s J\,da = \int_0^s k\bar{s}(2\pi\bar{s})\,d\bar{s} = \frac{2\pi k s^3}{3} = I\frac{s^3}{a^3}$, for $s < a$; $I_{\text{enc}} = I$, for $s > a$. So $\boxed{\mathbf{B} = \left\{\begin{array}{ll} \frac{\mu_0 I s^2}{2\pi a^3}\hat{\boldsymbol{\phi}}, & \text{for } s < a; \\ \frac{\mu_0 I}{2\pi s}\hat{\boldsymbol{\phi}}, & \text{for } s > a. \end{array}\right\}}$

Problem 5.14

By the right-hand-rule, the field points in the $-\hat{\mathbf{y}}$ direction for $z > 0$, and in the $+\hat{\mathbf{y}}$ direction for $z < 0$. At $z = 0, B = 0$. Use the amperian loop shown:

$\oint \mathbf{B}\cdot d\mathbf{l} = Bl = \mu_0 I_{\text{enc}} = \mu_0 l z J \Rightarrow \boxed{\mathbf{B} = -\mu_0 J z\,\hat{\mathbf{y}}}$ $(-a < z < a)$. If $z > a, I_{\text{enc}} = \mu_0 l a J$,

so $\boxed{\mathbf{B} = \left\{\begin{array}{ll} -\mu_0 J a\,\hat{\mathbf{y}}, & \text{for } z > +a; \\ +\mu_0 J a\,\hat{\mathbf{y}}, & \text{for } z > -a. \end{array}\right\}}$

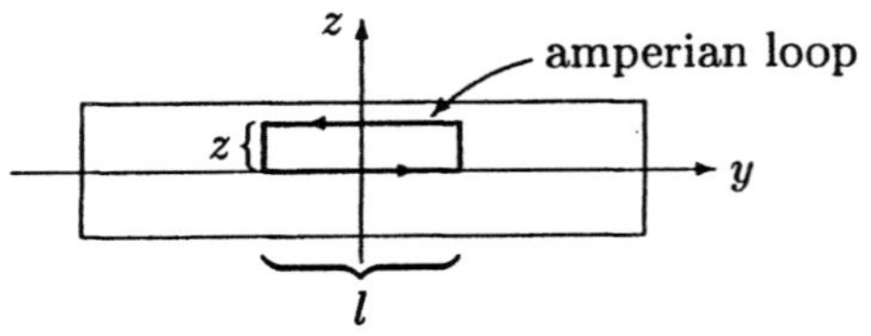

Problem 5.15

The field inside a solenoid is $\mu_0 n I$, and outside it is zero. The outer solenoid's field points to the left $(-\hat{\mathbf{z}})$, whereas the inner one points to the right $(+\hat{\mathbf{z}})$. So: (i) $\boxed{\mathbf{B} = \mu_0 I(n_1 - n_2)\,\hat{\mathbf{z}},}$ (ii) $\boxed{\mathbf{B} = -\mu_0 I n_2\,\hat{\mathbf{z}},}$ (iii) $\boxed{\mathbf{B} = 0.}$

Problem 5.16

From Ex. 5.8, the top plate produces a field $\mu_0 K/2$ (aiming *out of the page*, for points above it, and *into the page*, for points below). The bottom plate produces a field $\mu_0 K/2$ (aiming *into the page*, for points above it, and *out of the page*, for points below). Above and below *both* plates the two fields cancel; *between* the plates they add up to $\mu_0 K$, pointing *in*.

(a) $\boxed{B = \mu_0 \sigma v \text{ (in)}}$ betweem the plates, $\boxed{B = 0}$ elsewhere.

(b) The Lorentz force law says $\mathbf{F} = \int(\mathbf{K}\times\mathbf{B})\,da$, so the force *per unit area* is $\mathbf{f} = \mathbf{K}\times\mathbf{B}$. Here $K = \sigma v$, to the right, and $\mathbf{B}$ (the field of the lower plate) is $\mu_0\sigma v/2$, into the page. So $\boxed{f_m = \mu_0\sigma^2 v^2/2 \text{ (up).}}$

(c) The electric field of the lower plate is $\sigma/2\epsilon_0$; the electric force per unit area on the upper plate is $\boxed{f_e = \sigma^2/2\epsilon_0 \text{ (down)}.}$ They balance if $\mu_0 v^2 = 1/\epsilon_0$, or $\boxed{v = 1/\sqrt{\epsilon_0\mu_0} = c}$ (the speed of light), as in Prob. 5.12.

Problem 5.17

We might as well orient the axes so the field point $\mathbf{r}$ lies on the y axis: $\mathbf{r} = (0, y, 0)$. Consider a source point at (x', y', z') on loop #1:

$$\boldsymbol{\mathscr{r}} = -x'\,\hat{\mathbf{x}} + (y - y')\,\hat{\mathbf{y}} - z'\,\hat{\mathbf{z}};\quad d\mathbf{l}' = dx'\,\hat{\mathbf{x}} + dy'\,\hat{\mathbf{y}};$$

$$d\mathbf{l}' \times \boldsymbol{\mathscr{r}} = \begin{vmatrix} \hat{\mathbf{x}} & \hat{\mathbf{y}} & \hat{\mathbf{z}} \\ dx' & dy' & 0 \\ -x' & (y-y') & -z' \end{vmatrix} = (-z'\,dy')\,\hat{\mathbf{x}} + (z'\,dx')\,\hat{\mathbf{y}} + [(y-y')\,dx' + x'\,dy']\,\hat{\mathbf{z}}.$$

$$d\mathbf{B}_1 = \frac{\mu_0 I}{4\pi}\frac{d\mathbf{l}' \times \boldsymbol{\mathscr{r}}}{\mathscr{r}^3} = \frac{\mu_0 I}{4\pi}\frac{(-z'\,dy')\,\hat{\mathbf{x}} + (z'\,dx')\,\hat{\mathbf{y}} + [(y-y')\,dx' + x'\,dy']\,\hat{\mathbf{z}}}{[(x')^2 + (y-y')^2 + (z')^2]^{3/2}}.$$

Now consider the symmetrically placed source element on loop #2, at $(x', y', -z')$. Since z' changes sign, while everything else is the same, the $\hat{\mathbf{x}}$ and $\hat{\mathbf{y}}$ components from $d\mathbf{B}_1$ and $d\mathbf{B}_2$ cancel, leaving only a $\hat{\mathbf{z}}$ component. qed

With this, Ampére's law yields immediately:

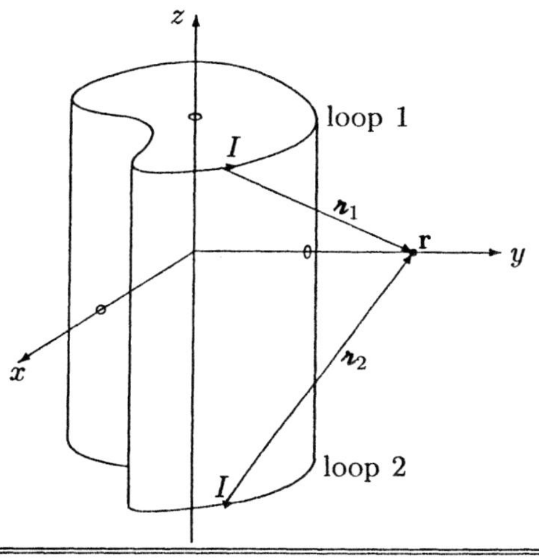

$$\boxed{\mathbf{B} = \begin{cases} \mu_0 n I\,\hat{\mathbf{z}}, & \text{inside the solenoid;} \\ 0, & \text{outside} \end{cases}}$$

(the same as for a circular solenoid—Ex. 5.9).

For the toroid, $N/2\pi s = n$ (the number of turns per unit length), so Eq. 5.58 yields $B = \mu_0 n I$ inside, and zero outside, consistent with the solenoid. [*Note:* $N/2\pi s = n$ applies only if the toroid is large in circumference, so that s is essentially constant over the cross-section.]

Problem 5.18

$\boxed{\text{It doesn't matter.}}$ According to Theorem 2, in Sect. 1.6.2, $\int \mathbf{J}\cdot d\mathbf{a}$ is independent of surface, for any given boundary line, provided that $\mathbf{J}$ is divergenceless, which it *is*, for steady currents (Eq. 5.31).

Problem 5.19

(a) $\rho = \dfrac{\text{charge}}{\text{volume}} = \dfrac{\text{charge}}{\text{atom}}\cdot\dfrac{\text{atoms}}{\text{mole}}\cdot\dfrac{\text{moles}}{\text{gram}}\cdot\dfrac{\text{grams}}{\text{volume}} = (e)(N)\left(\dfrac{1}{M}\right)(d)$, where

$$\begin{aligned} e &= \text{charge of electron} &&= 1.6\times10^{-19}\,\text{C}, \\ N &= \text{Avogadro's number} &&= 6.0\times10^{23}\,\text{mole}, \\ M &= \text{atomic mass of copper} &&= 64\,\text{gm/mole}, \\ d &= \text{density of copper} &&= 9.0\,\text{gm/cm}^3. \end{aligned}$$

$$\rho = (1.6\times10^{-19})(6.0\times10^{23})\left(\frac{9.0}{64}\right) = \boxed{1.4\times10^{4}\ \text{C/cm}^3.}$$

(b) $J = \dfrac{I}{\pi s^2} = \rho v \Rightarrow v = \dfrac{I}{\pi s^2 \rho} = \dfrac{1}{\pi(2.5\times10^{-3})(1.4\times10^4)} = \boxed{9.1\times10^{-3}\ \text{cm/s},}$ or about 33 cm/hr. This is astonishingly small—literally slower than a snail's pace.

(c) From Eq. 5.37, $f_m = \dfrac{\mu_0}{2\pi}\left(\dfrac{I_1 I_2}{d}\right) = \dfrac{(4\pi\times10^{-7})}{2\pi} = \boxed{2\times10^{-7}\ \text{N/cm}.}$

(d) $E = \frac{1}{2\pi\epsilon_0}\frac{\lambda}{d}$; $f_e = \frac{1}{2\pi\epsilon_0}\left(\frac{\lambda_1\lambda_2}{d}\right) = \frac{1}{v^2}\frac{1}{2\pi\epsilon_0}\left(\frac{I_1I_2}{d}\right) = \left(\frac{c^2}{v^2}\right)\frac{\mu_0}{2\pi}\left(\frac{I_1I_2}{d}\right) = \frac{c^2}{v^2}f_m$, where $c \equiv 1/\sqrt{\epsilon_0\mu_0} = 3.00 \times 10^8$ m/s. Here $\frac{f_e}{f_m} = \frac{c^2}{v^2} = \left(\frac{3.0\times10^{10}}{9.1\times10^{-3}}\right)^2 = \boxed{1.1\times10^{25}.}$

$f_e = (1.1\times10^{25})(2\times10^{-7}) = \boxed{2\times10^{18}\,\text{N/cm.}}$

Problem 5.20

Ampére's law says $\nabla\times\mathbf{B} = \mu_0\mathbf{J}$. Together with the continuity equation (5.29) this gives $\nabla\cdot(\nabla\times\mathbf{B}) = \mu_0\nabla\cdot\mathbf{J} = -\mu_0\partial\rho/\partial t$, which is *inconsistent* with div(curl)=0 unless ρ is constant (magnetostatics). The other Maxwell equations are OK: $\nabla\times\mathbf{E} = 0 \Rightarrow \nabla\cdot(\nabla\times\mathbf{E}) = 0$ (✓), and as for the two divergence equations, there is no relevant vanishing second derivative (the other one is curl(grad), which doesn't involve the divergence).

Problem 5.21

At this stage I'd expect no changes in Gauss's law or Ampére's law. The divergence of **B** would take the form $\boxed{\nabla\cdot\mathbf{B} = \alpha_0\rho_m,}$ where ρ_m is the density of magnetic charge, and α_0 is some constant (analogous to ϵ_0 and μ_0). The curl of **E** becomes $\boxed{\nabla\times\mathbf{E} = \beta_0\mathbf{J}_m,}$ where $\mathbf{J}_m$ is the magnetic current density (representing the flow of magnetic charge), and β_0 is another constant. Presumably magnetic charge is conserved, so ρ_m and $\mathbf{J}_m$ satisfy a continuity equation: $\nabla\cdot\mathbf{J}_m = -\partial\rho_m/\partial t$.

As for the Lorentz force law, one might guess something of the form $q_m[\mathbf{B} + (\mathbf{v}\times\mathbf{E})]$ (where q_m is the magnetic charge). But this is dimensionally impossible, since E has the same units as vB. Evidently we need to divide $(\mathbf{v}\times\mathbf{E})$ by something with the dimensions of velocity-squared. The natural candidate is $c^2 = 1/\epsilon_0\mu_0$: $\boxed{\mathbf{F} = q_e[\mathbf{E} + (\mathbf{v}\times\mathbf{B})] + q_m\left[\mathbf{B} - \frac{1}{c^2}(\mathbf{v}\times\mathbf{E})\right].}$ In this form the magnetic analog to Coulomb's law reads $\mathbf{F} = \frac{\alpha_0}{4\pi}\frac{q_{m_1}q_{m_2}}{r^2}\hat{\mathbf{r}}$, so to determine α_0 we would first introduce (arbitrarily) a unit of magnetic charge, then measure the force between unit charges at a given separation. [For further details, and an explanation of the minus sign in the force law, see Prob. 7.35.]

Problem 5.22

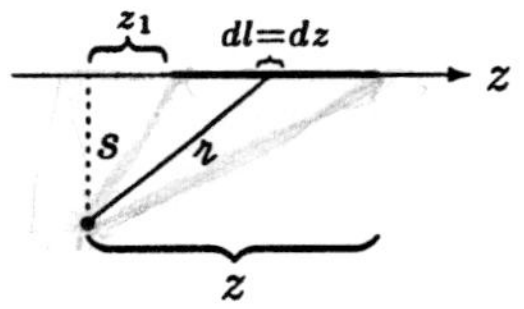

$$\mathbf{A} = \frac{\mu_0}{4\pi}\int\frac{I\,\hat{\mathbf{z}}}{\mathscr{r}}\,dz = \frac{\mu_0 I}{4\pi}\hat{\mathbf{z}}\int_{z_1}^{z_2}\frac{dz}{\sqrt{z^2+s^2}}$$

$$= \frac{\mu_0 I}{4\pi}\hat{\mathbf{z}}\left[\ln\left(z+\sqrt{z^2+s^2}\right)\right]\Big|_{z_1}^{z_2} = \boxed{\frac{\mu_0 I}{4\pi}\ln\left[\frac{z_2+\sqrt{(z_2)^2+s^2}}{z_1+\sqrt{(z_1)^2+s^2}}\right]\hat{\mathbf{z}}}$$

$$\mathbf{B} = \nabla\times\mathbf{A} = -\frac{\partial A}{\partial s}\hat{\boldsymbol{\phi}} = -\frac{\mu_0 I}{4\pi}\left[\frac{1}{z_2+\sqrt{(z_2)^2+s^2}}\frac{s}{\sqrt{(z_2)^2+s^2}} - \frac{1}{z_1+\sqrt{(z_1)^2+s^2}}\frac{s}{\sqrt{(z_1)^2+s^2}}\right]\hat{\boldsymbol{\phi}}$$

$$= -\frac{\mu_0 Is}{4\pi}\left[\frac{z_2-\sqrt{(z_2)^2+s^2}}{(z_2)^2-[(z_2)^2+s^2]}\frac{1}{\sqrt{(z_2)^2+s^2}} - \frac{z_1-\sqrt{(z_1)^2+s^2}}{z_1^2-[(z_1)^2+s^2]}\frac{1}{\sqrt{(z_1)^2+s^2}}\right]\hat{\boldsymbol{\phi}}$$

$$= -\frac{\mu_0 Is}{4\pi}\left(-\frac{1}{s^2}\right)\left[\frac{z_2}{\sqrt{(z_2)^2+s^2}} - 1 - \frac{z_1}{\sqrt{(z_1)^2+s^2}} + 1\right]\hat{\boldsymbol{\phi}} = \frac{\mu_0 I}{4\pi s}\left[\frac{z_2}{\sqrt{(z_2)^2+s^2}} - \frac{z_1}{\sqrt{(z_1)^2+s^2}}\right]\hat{\boldsymbol{\phi}},$$

or, since $\sin\theta_1 = \frac{z_1}{\sqrt{(z_1)^2+s^2}}$ and $\sin\theta_2 = \frac{z_2}{\sqrt{(z_2)^2+s^2}}$,

$$= \boxed{\frac{\mu_0 I}{4\pi s}(\sin\theta_2 - \sin\theta_1)\,\hat{\boldsymbol{\phi}}}\ \text{(as in Eq. 5.35).}$$

Problem 5.23

$$A_\phi = k \Rightarrow \mathbf{B} = \nabla \times \mathbf{A} = \frac{1}{s}\frac{\partial}{\partial s}(sk)\,\hat{\mathbf{z}} = \frac{k}{s}\,\hat{\mathbf{z}};\ \mathbf{J} = \frac{1}{\mu_0}(\nabla \times \mathbf{B}) = \frac{1}{\mu_0}\left[-\frac{\partial}{\partial s}\left(\frac{k}{s}\right)\right]\hat{\boldsymbol{\phi}} = \boxed{\frac{k}{\mu_0 s^2}\,\hat{\boldsymbol{\phi}}.}$$

Problem 5.24

$\nabla \cdot \mathbf{A} = -\frac{1}{2}\nabla \cdot (\mathbf{r} \times \mathbf{B}) = -\frac{1}{2}[\mathbf{B} \cdot (\nabla \times \mathbf{r}) - \mathbf{r} \cdot (\nabla \times \mathbf{B})] = 0$, since $\nabla \times \mathbf{B} = 0$ ($\mathbf{B}$ is uniform) and $\nabla \times \mathbf{r} = 0$ (Prob. 1.62). $\nabla \times \mathbf{A} = -\frac{1}{2}\nabla \times (\mathbf{r} \times \mathbf{B}) = -\frac{1}{2}[(\mathbf{B} \cdot \nabla)\mathbf{r} - (\mathbf{r} \cdot \nabla)\mathbf{B} + \mathbf{r}(\nabla \cdot \mathbf{B}) - \mathbf{B}(\nabla \cdot \mathbf{r})]$. But $(\mathbf{r} \cdot \nabla)\mathbf{B} = 0$ and $\nabla \cdot \mathbf{B} = 0$ (since $\mathbf{B}$ is uniform), and $\nabla \cdot \mathbf{r} = \frac{\partial x}{\partial x} + \frac{\partial y}{\partial y} + \frac{\partial z}{\partial z} = 1 + 1 + 1 = 3$. Finally, $(\mathbf{B} \cdot \nabla)\mathbf{r} = \left(B_x\frac{\partial}{\partial x} + B_y\frac{\partial}{\partial y} + B_z\frac{\partial}{\partial z}\right)(x\,\hat{\mathbf{x}} + y\,\hat{\mathbf{y}} + z\,\hat{\mathbf{z}}) = B_x\,\hat{\mathbf{x}} + B_y\,\hat{\mathbf{y}} + B_z\,\hat{\mathbf{z}} = \mathbf{B}$. So $\nabla \times \mathbf{A} = -\frac{1}{2}(\mathbf{B} - 3\mathbf{B}) = \mathbf{B}$. qed

Problem 5.25

(a) $\mathbf{A}$ points in the same direction as $\mathbf{I}$, and is a function only of s (the distance from the wire). In cylindrical coordinates, then, $\mathbf{A} = A(s)\,\hat{\mathbf{z}}$, so $\mathbf{B} = \nabla \times \mathbf{A} = -\frac{\partial A}{\partial s}\,\hat{\boldsymbol{\phi}} = \frac{\mu_0 I}{2\pi s}\,\hat{\boldsymbol{\phi}}$ (the field of an infinite wire). Therefore $\frac{\partial A}{\partial s} = -\frac{\mu_0 I}{2\pi s}$, and $\boxed{\mathbf{A}(\mathbf{r}) = -\frac{\mu_0 I}{2\pi}\ln(s/a)\,\hat{\mathbf{z}}}$ (the constant a is arbitrary; you could use 1, but then the units look fishy). $\nabla \cdot \mathbf{A} = \frac{\partial A_z}{\partial z} = 0.\ \checkmark$ $\nabla \times \mathbf{A} = -\frac{\partial A_z}{\partial s}\,\hat{\boldsymbol{\phi}} = \frac{\mu_0 I}{2\pi s}\,\hat{\boldsymbol{\phi}} = \mathbf{B}.\ \checkmark$

(b) Here Ampére's law gives $\oint \mathbf{B} \cdot d\mathbf{l} = B\,2\pi s = \mu_0 I_{\text{enc}} = \mu_0 J\,\pi s^2 = \mu_0\frac{I}{\pi R^2}\pi s^2 = \frac{\mu_0 I s^2}{R^2}$. $\mathbf{B} = \frac{\mu_0}{2\pi}\frac{Is}{R^2}\,\hat{\boldsymbol{\phi}}$. $\frac{\partial A}{\partial s} = -\frac{\mu_0 I}{2\pi}\frac{s}{R^2} \Rightarrow \mathbf{A} = -\frac{\mu_0 I}{4\pi R^2}(s^2 - b^2)\,\hat{\mathbf{z}}$. Here b is again arbitrary, except that since $\mathbf{A}$ must be continuous at R, $-\frac{\mu_0 I}{2\pi}\ln(R/a) = -\frac{\mu_0 I}{4\pi R^2}(R^2 - b^2)$, which means that we must pick a and b such that $2\ln(R/b) = 1 - (b/R)^2$. I'll use $a = b = R$. Then

$$\boxed{\mathbf{A} = \begin{cases} -\dfrac{\mu_0 I}{4\pi R^2}(s^2 - R^2)\,\hat{\mathbf{z}}, & \text{for } s \le R; \\ -\dfrac{\mu_0 I}{2\pi}\ln(s/R)\,\hat{\mathbf{z}}, & \text{for } s \ge R. \end{cases}}$$

Problem 5.26

$\mathbf{K} = K\,\hat{\mathbf{x}} \Rightarrow \mathbf{B} = \pm\frac{\mu_0 K}{2}\,\hat{\mathbf{y}}$ (plus for $z < 0$, minus for $z > 0$).

$\mathbf{A}$ is parallel to $\mathbf{K}$, and depends only on z, so $\mathbf{A} = A(z)\,\hat{\mathbf{x}}$.

$$\mathbf{B} = \nabla \times \mathbf{A} = \begin{vmatrix} \hat{\mathbf{x}} & \hat{\mathbf{y}} & \hat{\mathbf{z}} \\ \partial/\partial x & \partial/\partial y & \partial/\partial z \\ A(z) & 0 & 0 \end{vmatrix} = \frac{\partial A}{\partial z}\,\hat{\mathbf{y}} = \pm\frac{\mu_0 K}{2}\,\hat{\mathbf{y}}.$$

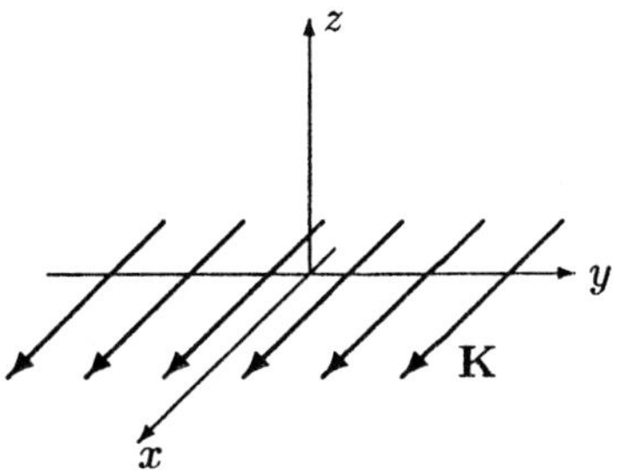

$\boxed{\mathbf{A} = -\frac{\mu_0 K}{2}|z|\,\hat{\mathbf{x}}}$ will do the job—or this plus any constant.

Problem 5.27

(a) $\nabla \cdot \mathbf{A} = \frac{\mu_0}{4\pi}\int \nabla \cdot \left(\frac{\mathbf{J}}{\mathscr{r}}\right) d\tau'$. $\nabla \cdot \left(\frac{\mathbf{J}}{\mathscr{r}}\right) = \frac{1}{\mathscr{r}}(\nabla \cdot \mathbf{J}) + \mathbf{J} \cdot \nabla\left(\frac{1}{\mathscr{r}}\right)$. But the first term is zero, because $\mathbf{J}(\mathbf{r}')$ is a function of the *source* coordinates, not the *field* coordinates. And since $\boldsymbol{\mathscr{r}} = \mathbf{r} - \mathbf{r}'$, $\nabla\left(\frac{1}{\mathscr{r}}\right) = -\nabla'\left(\frac{1}{\mathscr{r}}\right)$. So

$\nabla\cdot\left(\frac{\mathbf{J}}{\eta}\right) = -\mathbf{J}\cdot\nabla'\left(\frac{1}{\eta}\right)$. But $\nabla'\cdot\left(\frac{\mathbf{J}}{\eta}\right) = \frac{1}{\eta}(\nabla'\cdot\mathbf{J}) + \mathbf{J}\cdot\nabla'\left(\frac{1}{\eta}\right)$, and $\nabla'\cdot\mathbf{J} = 0$ in magnetostatics (Eq. 5.31). So $\nabla\cdot\left(\frac{\mathbf{J}}{\eta}\right) = -\nabla'\cdot\left(\frac{\mathbf{J}}{\eta}\right)$, and hence, by the divergence theorem, $\nabla\cdot\mathbf{A} = -\frac{\mu_0}{4\pi}\int\nabla'\cdot\left(\frac{\mathbf{J}}{\eta}\right)d\tau' = -\frac{\mu_0}{4\pi}\oint\frac{\mathbf{J}}{\eta}\cdot d\mathbf{a}'$, where the integral is now over the *surface* surrounding all the currents. But $\mathbf{J} = 0$ on this surface, so $\nabla\cdot\mathbf{A} = 0$. ✓

(b) $\nabla\times\mathbf{A} = \frac{\mu_0}{4\pi}\int\nabla\times\left(\frac{\mathbf{J}}{\eta}\right)d\tau' = \frac{\mu_0}{4\pi}\int\left[\frac{1}{\eta}(\nabla\times\mathbf{J}) - \mathbf{J}\times\nabla\left(\frac{1}{\eta}\right)\right]d\tau'$. But $\nabla\times\mathbf{J} = 0$ (since $\mathbf{J}$ is not a function of $\mathbf{r}$), and $\nabla\left(\frac{1}{\eta}\right) = -\frac{\hat{\boldsymbol{\eta}}}{\eta^2}$ (Eq. 1.101), so $\nabla\times\mathbf{A} = \frac{\mu_0}{4\pi}\int\frac{\mathbf{J}\times\hat{\boldsymbol{\eta}}}{\eta^2}d\tau' = \mathbf{B}$. ✓

(c) $\nabla^2\mathbf{A} = \frac{\mu_0}{4\pi}\int\nabla^2\left(\frac{\mathbf{J}}{\eta}\right)d\tau'$. But $\nabla^2\left(\frac{\mathbf{J}}{\eta}\right) = \mathbf{J}\nabla^2\left(\frac{1}{\eta}\right)$ (once again, $\mathbf{J}$ is a *constant*, as far as differentiation with respect to $\mathbf{r}$ is concerned), and $\nabla^2\left(\frac{1}{\eta}\right) = -4\pi\delta^3(\boldsymbol{\eta})$ (Eq. 1.102).

So $\nabla^2\mathbf{A} = \frac{\mu_0}{4\pi}\int\mathbf{J}(\mathbf{r}')\left[-4\pi\delta^3(\boldsymbol{\eta})\right]d\tau' = -\mu_0\mathbf{J}(\mathbf{r})$. ✓

Problem 5.28

$\mu_0 I = \oint\mathbf{B}\cdot d\mathbf{l} = -\int_{\mathbf{a}}^{\mathbf{b}}\nabla U\cdot d\mathbf{l} = -[U(\mathbf{b}) - U(\mathbf{a})]$ (by the gradient theorem), so $U(\mathbf{b}) \neq U(\mathbf{a})$. qed

For an infinite straight wire, $\mathbf{B} = \frac{\mu_0 I}{2\pi s}\hat{\boldsymbol{\phi}}$. $\boxed{U = -\frac{\mu_0 I\phi}{2\pi}}$ would do the job, in the sense that $-\nabla U = \frac{\mu_0 I}{2\pi}\nabla(\phi) = \frac{\mu_0 I}{2\pi}\frac{1}{s}\frac{\partial\phi}{\partial\phi}\hat{\boldsymbol{\phi}} = \mathbf{B}$. But when ϕ advances by 2π, this function does *not* return to its initial value; it works (say) for $0 \le \phi < 2\pi$, but at 2π it "jumps" back to zero.

Problem 5.29

Use Eq. 5.67, with $R \to \bar{r}$ and $\sigma \to \rho\, d\bar{r}$:

$$\begin{aligned}
\mathbf{A} &= \frac{\mu_0\omega\rho}{3}\frac{\sin\theta}{r^2}\hat{\boldsymbol{\phi}}\int_0^r \bar{r}^4\,d\bar{r} + \frac{\mu_0\omega\rho}{3}r\sin\theta\,\hat{\boldsymbol{\phi}}\int_r^R \bar{r}\,d\bar{r} \\
&= \left(\frac{\mu_0\omega\rho}{3}\right)\sin\theta\left[\frac{1}{r^2}\left(\frac{r^5}{5}\right) + \frac{r}{2}(R^2 - r^2)\right]\hat{\boldsymbol{\phi}} = \frac{\mu_0\omega\rho}{2}r\sin\theta\left(\frac{R^2}{3} - \frac{r^2}{5}\right)\hat{\boldsymbol{\phi}}. \\
\mathbf{B} &= \nabla\times\mathbf{A} = \frac{\mu_0\omega\rho}{2}\left\{\frac{1}{r\sin\theta}\frac{\partial}{\partial\theta}\left[\sin\theta\, r\sin\theta\left(\frac{R^2}{3} - \frac{r^2}{5}\right)\right]\hat{\mathbf{r}} - \frac{1}{r}\frac{\partial}{\partial r}\left[r^2\sin\theta\left(\frac{R^2}{3} - \frac{r^2}{5}\right)\right]\hat{\boldsymbol{\theta}}\right\} \\
&= \mu_0\omega\rho\left[\left(\frac{R^2}{3} - \frac{r^2}{5}\right)\cos\theta\,\hat{\mathbf{r}} - \left(\frac{R^2}{3} - \frac{2r^2}{5}\right)\sin\theta\,\hat{\boldsymbol{\theta}}\right]. \text{ But } \rho = \frac{Q}{(4/3)\pi R^3}, \text{ so} \\
&= \boxed{\frac{\mu_0\omega Q}{4\pi R}\left[\left(1 - \frac{3r^2}{5R^2}\right)\cos\theta\,\hat{\mathbf{r}} - \left(1 - \frac{6r^2}{5R^2}\right)\sin\theta\,\hat{\boldsymbol{\theta}}\right]}.
\end{aligned}$$

Problem 5.30

(a)
$$\left\{\begin{aligned}
-\frac{\partial W_z}{\partial x} &= F_y \Rightarrow W_z(x,y,z) = -\int_0^x F_y(x',y,z)\,dx' + C_1(y,z). \\
\frac{\partial W_y}{\partial x} &= F_z \Rightarrow W_y(x,y,z) = +\int_0^x F_z(x',y,z)\,dx' + C_2(y,z).
\end{aligned}\right\}$$

These satisfy (ii) and (iii), for *any* C_1 and C_2; it remains to choose these functions so as to satisfy (i):

$-\int_0^x \frac{\partial F_y(x',y,z)}{\partial y}\,dx' + \frac{\partial C_1}{\partial y} - \int_0^x \frac{\partial F_z(x',y,z)}{\partial z}\,dx' - \frac{\partial C_2}{\partial z} = F_x(x,y,z)$. But $\frac{\partial F_x}{\partial x} + \frac{\partial F_y}{\partial y} + \frac{\partial F_z}{\partial z} = 0$, so $\int_0^x \frac{\partial F_x(x',y,z)}{\partial x'}\,dx' + \frac{\partial C_1}{\partial y} - \frac{\partial C_2}{\partial z} = F_x(x,y,z)$. Now $\int_0^x \frac{\partial F_x(x',y,z)}{\partial x'}\,dx' = F_x(x,y,z) - F_x(0,y,z)$, so $\frac{\partial C_1}{\partial y} - \frac{\partial C_2}{\partial z} = F_x(0,y,z)$. We may as well pick $C_2 = 0$, $C_1(y,z) = \int_0^y F_x(0,y',z)\,dy'$, and we're done, with

$$W_x = 0;\quad W_y = \int_0^x F_z(x',y,z)\,dx';\quad W_z = \int_0^y F_x(0,y',z)\,dy' - \int_0^x F_y(x',y,z)\,dx'.$$

(b) $\nabla \times \mathbf{W} = \left(\frac{\partial W_z}{\partial y} - \frac{\partial W_y}{\partial z}\right)\hat{\mathbf{x}} + \left(\frac{\partial W_x}{\partial z} - \frac{\partial W_z}{\partial x}\right)\hat{\mathbf{y}} + \left(\frac{\partial W_y}{\partial x} - \frac{\partial W_x}{\partial y}\right)\hat{\mathbf{z}}$

$= \left[F_x(0,y,z) - \int_0^x \frac{\partial F_y(x',y,z)}{\partial y}\,dx' - \int_0^x \frac{\partial F_z(x',y,z)}{\partial z}\,dx'\right]\hat{\mathbf{x}} + [0 + F_y(x,y,z)]\,\hat{\mathbf{y}} + [F_z(x,y,z) - 0]\,\hat{\mathbf{z}}.$

But $\nabla \cdot \mathbf{F} = 0$, so the $\hat{\mathbf{x}}$ term is $\left[F_x(0,y,z) + \int_0^x \frac{\partial F_x(x',y,z)}{\partial x'}\,dx'\right] = F_x(0,y,z) + F_x(x,y,z) - F_x(0,y,z)$, so $\nabla \times \mathbf{W} = \mathbf{F}$. ✓

$\nabla \cdot \mathbf{W} = \frac{\partial W_x}{\partial x} + \frac{\partial W_y}{\partial y} + \frac{\partial W_z}{\partial z} = 0 + \int_0^x \frac{\partial F_z(x',y,z)}{\partial y}\,dx' + \int_0^y \frac{\partial F_x(0,y',z)}{\partial z}\,dy' - \int_0^x \frac{\partial F_y(x',y,z)}{\partial z}\,dx' \neq 0$, in general.

(c) $W_y = \int_0^x x'\,dx' = \frac{x^2}{2}$; $W_z = \int_0^y y'\,dy' - \int_0^x z\,dx' = \frac{y^2}{2} - zx$.

$$\boxed{\mathbf{W} = \frac{x^2}{2}\,\hat{\mathbf{y}} + \left(\frac{y^2}{2} - zx\right)\hat{\mathbf{z}}.}\qquad \nabla \times \mathbf{W} = \begin{vmatrix} \hat{\mathbf{x}} & \hat{\mathbf{y}} & \hat{\mathbf{z}} \\ \partial/\partial x & \partial/\partial y & \partial/\partial z \\ 0 & x^2/2 & (y^2/2 - zx) \end{vmatrix} = y\,\hat{\mathbf{x}} + z\,\hat{\mathbf{y}} + x\,\hat{\mathbf{z}} = \mathbf{F}. ✓$$

Problem 5.31

(a) At the surface of the solenoid, $\mathbf{B}_{\text{above}} = 0$, $\mathbf{B}_{\text{below}} = \mu_0 n I\,\hat{\mathbf{z}} = \mu_0 K\,\hat{\mathbf{z}}$; $\hat{\mathbf{n}} = \hat{\mathbf{s}}$; so $\mathbf{K} \times \hat{\mathbf{n}} = -K\,\hat{\mathbf{z}}$. Evidently Eq. 5.74 holds. ✓

(b) In Eq. 5.67, both expressions reduce to $(\mu_0 R^2 \omega\sigma/3)\sin\theta\,\hat{\boldsymbol{\phi}}$ at the surface, so Eq. 5.75 is satisfied. $\left.\frac{\partial \mathbf{A}}{\partial r}\right|_{R^+} = \frac{\mu_0 R^4 \omega\sigma}{3}\left(-\frac{2\sin\theta}{r^3}\right)\hat{\boldsymbol{\phi}}\Big|_R = -\frac{2\mu_0 R\omega\sigma}{3}\sin\theta\,\hat{\boldsymbol{\phi}}$; $\left.\frac{\partial \mathbf{A}}{\partial r}\right|_{R^-} = \frac{\mu_0 R\omega\sigma}{3}\sin\theta\,\hat{\boldsymbol{\phi}}$. So the left side of Eq. 5.76 is $-\mu_0 R\omega\sigma\sin\theta\,\hat{\boldsymbol{\phi}}$. Meanwhile $\mathbf{K} = \sigma\mathbf{v} = \sigma(\boldsymbol{\omega} \times \mathbf{r}) = \sigma\omega R\sin\theta\,\hat{\boldsymbol{\phi}}$, so the right side of Eq. 5.76 is $-\mu_0\sigma\omega R\sin\theta\,\hat{\boldsymbol{\phi}}$, and the equation is satisfied.

Problem 5.32

Because $\mathbf{A}_{\text{above}} = \mathbf{A}_{\text{below}}$ at every point on the surface, it follows that $\frac{\partial \mathbf{A}}{\partial x}$ and $\frac{\partial \mathbf{A}}{\partial y}$ are the same above and below; any discontinuity is confined to the normal derivative.

$\mathbf{B}_{\text{above}} - \mathbf{B}_{\text{below}} = \left(-\frac{\partial A_{y_{\text{above}}}}{\partial z} + \frac{\partial A_{y_{\text{below}}}}{\partial z}\right)\hat{\mathbf{x}} + \left(\frac{\partial A_{x_{\text{above}}}}{\partial z} - \frac{\partial A_{x_{\text{below}}}}{\partial z}\right)\hat{\mathbf{y}}$. But Eq. 5.74 says this equals $\mu_0 K(-\hat{\mathbf{y}})$. So $\frac{\partial A_{y_{\text{above}}}}{\partial z} = \frac{\partial A_{y_{\text{below}}}}{\partial z}$, and $\frac{\partial A_{x_{\text{above}}}}{\partial z} - \frac{\partial A_{x_{\text{below}}}}{\partial z} = -\mu_0 K$. Thus the *normal* derivative of the *component of* $\mathbf{A}$ *parallel to* $\mathbf{K}$ suffers a discontinuity $-\mu_0 K$, or, more compactly: $\boxed{\frac{\partial \mathbf{A}_{\text{above}}}{\partial n} - \frac{\partial \mathbf{A}_{\text{below}}}{\partial n} = -\mu_0 \mathbf{K}.}$

Problem 5.33

(Same idea as Prob. 3.33.) Write $\mathbf{m} = (\mathbf{m} \cdot \hat{\mathbf{r}})\,\hat{\mathbf{r}} + (\mathbf{m} \cdot \hat{\boldsymbol{\theta}})\,\hat{\boldsymbol{\theta}} = m\cos\theta\,\hat{\mathbf{r}} - m\sin\theta\,\hat{\boldsymbol{\theta}}$ (Fig. 5.54). Then $3(\mathbf{m} \cdot \hat{\mathbf{r}})\,\hat{\mathbf{r}} - \mathbf{m} = 3m\cos\theta\,\hat{\mathbf{r}} - m\cos\theta\,\hat{\mathbf{r}} + m\sin\theta\,\hat{\boldsymbol{\theta}} = 2m\cos\theta\,\hat{\mathbf{r}} + m\sin\theta\,\hat{\boldsymbol{\theta}}$, and Eq. 5.87 ⇔ Eq. 5.86. qed

Problem 5.34

(a) $\mathbf{m} = I\mathbf{a} = \boxed{I\pi R^2\,\hat{\mathbf{z}}.}$

(b) $\mathbf{B} \approx \boxed{\dfrac{\mu_0}{4\pi}\dfrac{I\pi R^2}{r^3}\left(2\cos\theta\,\hat{\mathbf{r}} + \sin\theta\,\hat{\boldsymbol{\theta}}\right).}$

(c) On the z axis, $\theta = 0$, $r = z$, $\hat{\mathbf{r}} = \hat{\mathbf{z}}$ (for $z > 0$), so $\boxed{\mathbf{B} \approx \dfrac{\mu_0 I R^2}{2z^3}\,\hat{\mathbf{z}}}$ (for $z < 0$, $\theta = \pi$, $\hat{\mathbf{r}} = -\hat{\mathbf{z}}$, so the field is the same, with $|z|^3$ in place of z^3). The exact answer (Eq. 5.38) reduces (for $z \gg R$) to $B \approx \mu_0 I R^2/2|z|^3$, so they agree.

Problem 5.35

For a ring, $m = I\pi r^2$. Here $I \to \sigma v\,dr = \sigma\omega r\,dr$, so $m = \int_0^R \pi r^2 \sigma\omega r\,dr = \boxed{\pi\sigma\omega R^4/4.}$

Problem 5.36

The total charge on the shaded ring is $dq = \sigma(2\pi R\sin\theta)R\,d\theta$. The time for one revolution is $dt = 2\pi/\omega$. So the current in the ring is $I = \dfrac{dq}{dt} = \sigma\omega R^2\sin\theta\,d\theta$. The area of the ring is $\pi(R\sin\theta)^2$, so the magnetic moment of the ring is $dm = (\sigma\omega R^2\sin\theta\,d\theta)\pi R^2\sin^2\theta$, and the total dipole moment of the shell is

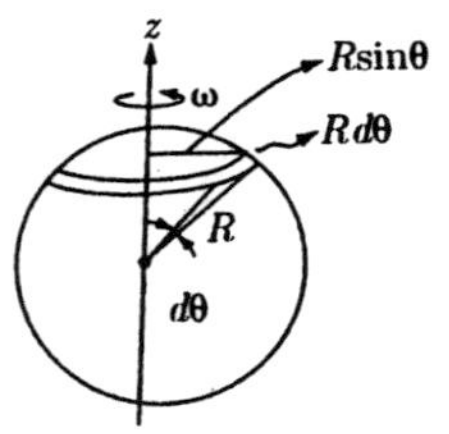

$m = \sigma\omega\pi R^4\int_0^\pi \sin^3\theta\,d\theta = (4/3)\sigma\omega\pi R^4$, or $\boxed{\mathbf{m} = \dfrac{4\pi}{3}\sigma\omega R^4\,\hat{\mathbf{z}}.}$

The dipole term in the multipole expansion for $\mathbf{A}$ is therefore $\mathbf{A}_{\text{dip}} = \dfrac{\mu_0}{4\pi}\dfrac{4\pi}{3}\sigma\omega R^4\dfrac{\sin\theta}{r^2}\,\hat{\boldsymbol{\phi}} = \dfrac{\mu_0\sigma\omega R^4}{3}\dfrac{\sin\theta}{r^2}\,\hat{\boldsymbol{\phi}}$, which is also the *exact* potential (Eq. 5.67); evidently a spinning sphere produces a perfect dipole field, with no higher multipole contributions.

Problem 5.37

The field of one side is given by Eq. 5.35, with $s \to \sqrt{z^2 + (w/2)^2}$ and $\sin\theta_2 = -\sin\theta_1 = \dfrac{(w/2)}{\sqrt{z^2 + w^2/2}}$;

$B = \dfrac{\mu_0 I}{4\pi}\dfrac{w}{\sqrt{z^2 + (w^2/4)}\sqrt{z^2 + (w^2/2)}}$. To pick off the vertical component, multiply by $\sin\phi = \dfrac{(w/2)}{\sqrt{z^2 + (w/2)^2}}$; for all four sides, multiply by 4: $\boxed{\mathbf{B} = \dfrac{\mu_0 I}{2\pi}\dfrac{w^2}{(z^2 + w^2/4)\sqrt{z^2 + w^2/2}}\,\hat{\mathbf{z}}.}$ For $z \gg w$, $\mathbf{B} \approx \dfrac{\mu_0 I w^2}{2\pi z^3}\,\hat{\mathbf{z}}$. The field of a dipole $\boxed{m = Iw^2,}$ for points on the z axis (Eq. 5.86, with $r \to z$, $\hat{\mathbf{r}} \to \hat{\mathbf{z}}$, $\theta = 0$) is $\mathbf{B} = \dfrac{\mu_0}{2\pi}\dfrac{m}{z^3}\,\hat{\mathbf{z}}$. ✓

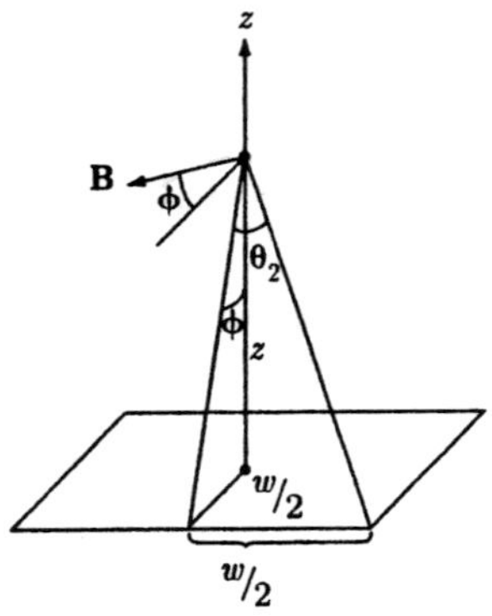

Problem 5.38

The mobile charges *do* pull in toward the axis, but the resulting concentration of (negative) charge sets up an *electric* field that repels away further accumulation. Equilibrium is reached when the electric repulsion on a mobile charge q balances the magnetic attraction: $\mathbf{F} = q[\mathbf{E} + (\mathbf{v}\times\mathbf{B})] = 0 \Rightarrow \mathbf{E} = -(\mathbf{v}\times\mathbf{B})$. Say the current is in the z direction: $\mathbf{J} = \rho_- v\,\hat{\mathbf{z}}$ (where ρ_- and v are both negative).

$$\oint \mathbf{B}\cdot d\mathbf{l} = B\,2\pi s = \mu_0 J\pi s^2 \Rightarrow \mathbf{B} = \frac{\mu_0\rho_- v s}{2}\,\hat{\boldsymbol{\phi}};$$

$$\int \mathbf{E}\cdot d\mathbf{a} = E\,2\pi s l = \frac{1}{\epsilon_0}(\rho_+ + \rho_-)\pi s^2 l \Rightarrow \mathbf{E} = \frac{1}{2\epsilon_0}(\rho_+ + \rho_-)s\,\hat{\mathbf{s}}.$$

$$\frac{1}{2\epsilon_0}(\rho_+ + \rho_-)s\,\hat{\mathbf{s}} = -\left[(v\,\hat{\mathbf{z}})\times\left(\frac{\mu_0\rho_- v s}{2}\,\hat{\boldsymbol{\phi}}\right)\right] = \frac{\mu_0}{2}\rho_- v^2 s\,\hat{\mathbf{s}} \Rightarrow \rho_+ + \rho_- = \rho_-(\epsilon_0\mu_0 v^2) = \rho_-\left(\frac{v^2}{c^2}\right).$$

Evidently $\rho_+ = -\rho_-\left(1-\frac{v^2}{c^2}\right) = \frac{\rho_-}{\gamma^2}$, or $\rho_- = -\gamma^2\rho_+$. In this naive model, the mobile negative charges fill a smaller inner cylinder, leaving a shell of positive (stationary) charge at the outside. But since $v \ll c$, the effect is extremely small.

Problem 5.39

(a) If *positive* charges flow to the *right*, they are deflected $\boxed{\text{down,}}$ and the bottom plate acquires a *positive* charge.

(b) $qvB = qE \Rightarrow E = vB \Rightarrow V = Et = \boxed{vBt,}$ with the *bottom* at higher potential.

(c) If *negative* charges flow to the *left*, they are *also* deflected down, and the bottom plate acquires a *negative* charge. The potential difference is still the same, but this time the *top* plate is at the higher potential.

Problem 5.40

From Eq. 5.17, $\mathbf{F} = I\int(d\mathbf{l}\times\mathbf{B})$. But $\mathbf{B}$ is constant, in this case, so it comes outside the integral: $\mathbf{F} = I\left(\int d\mathbf{l}\right)\times\mathbf{B}$, and $\int d\mathbf{l} = \mathbf{w}$, the vector displacement from the point at which the wire first enters the field to the point where it leaves. Since $\mathbf{w}$ and $\mathbf{B}$ are perpendicular, $F = IBw$, and $\mathbf{F}$ is perpendicular to $\mathbf{w}$.

Problem 5.41

The angular momentum acquired by the particle as it moves out from the center to the edge is

$$\mathbf{L} = \int\frac{d\mathbf{L}}{dt}\,dt = \int\mathbf{N}\,dt = \int(\mathbf{r}\times\mathbf{F})\,dt = \int\mathbf{r}\times q(\mathbf{v}\times\mathbf{B})\,dt = q\int\mathbf{r}\times(d\mathbf{l}\times\mathbf{B}) = q\left[\int(\mathbf{r}\cdot\mathbf{B})\,d\mathbf{l} - \int\mathbf{B}(\mathbf{r}\cdot d\mathbf{l})\right].$$

But $\mathbf{r}$ is perpendicular to $\mathbf{B}$, so $\mathbf{r}\cdot\mathbf{B} = 0$, and $\mathbf{r}\cdot d\mathbf{l} = \mathbf{r}\cdot d\mathbf{r} = \frac{1}{2}d(\mathbf{r}\cdot\mathbf{r}) = \frac{1}{2}d(r^2) = r\,dr = (1/2\pi)(2\pi r\,dr)$. So $\mathbf{L} = -\frac{q}{2\pi}\int_0^R \mathbf{B}\,2\pi r\,dr = -\frac{q}{2\pi}\int\mathbf{B}\,da$. It follows that $\boxed{L = -\frac{q}{2\pi}\Phi,}$ where $\Phi = \int B\,da$ is the total flux. In particular, if $\Phi = 0$, then $L = 0$, and the charge emerges with zero angular momentum, which means it is going along a radial line. qed

---

Problem 5.42

From Eq. 5.24, $\mathbf{F} = \int(\mathbf{K}\times\mathbf{B}_{\text{ave}})\,da$. Here $\mathbf{K} = \sigma\mathbf{v}$, $\mathbf{v} = \omega R\sin\theta\,\hat{\boldsymbol{\phi}}$, $da = R^2\sin\theta\,d\theta\,d\phi$, and $\mathbf{B}_{\text{ave}} = \frac{1}{2}(\mathbf{B}_{\text{in}} + \mathbf{B}_{\text{out}})$. From Eq. 5.68,

$$\mathbf{B}_{\text{in}} = \frac{2}{3}\mu_0\sigma R\omega\,\hat{\mathbf{z}} = \frac{2}{3}\mu_0\sigma R\omega(\cos\theta\,\hat{\mathbf{r}} - \sin\theta\,\hat{\boldsymbol{\theta}}).\ \text{From Eq. 5.67,}$$

$$\mathbf{B}_{\text{out}} = \nabla\times\mathbf{A} = \nabla\times\left(\frac{\mu_0 R^4\omega\sigma}{3}\frac{\sin\theta}{r^2}\hat{\boldsymbol{\phi}}\right) = \frac{\mu_0 R^4\omega\sigma}{3}\left[\frac{1}{r\sin\theta}\frac{\partial}{\partial\theta}\left(\frac{\sin^2\theta}{r^2}\right)\hat{\mathbf{r}} - \frac{1}{r}\frac{\partial}{\partial r}\left(\frac{\sin\theta}{r}\right)\hat{\boldsymbol{\theta}}\right]$$

$$= \frac{\mu_0 R^4\omega\sigma}{3r^3}(2\cos\theta\,\hat{\mathbf{r}} + \sin\theta\,\hat{\boldsymbol{\theta}}) = \frac{\mu_0 R\omega\sigma}{3}(2\cos\theta\,\hat{\mathbf{r}} + \sin\theta\,\hat{\boldsymbol{\theta}})\ \text{(since } r = R).$$

$$\mathbf{B}_{\text{ave}} = \frac{\mu_0 R\omega\sigma}{6}(4\cos\theta\,\hat{\mathbf{r}} - \sin\theta\,\hat{\boldsymbol{\theta}}).$$

$$\mathbf{K}\times\mathbf{B}_{\text{ave}} = (\sigma\omega R\sin\theta)\left(\frac{\mu_0 R\omega\sigma}{6}\right)\left[\hat{\boldsymbol{\phi}}\times(4\cos\theta\,\hat{\mathbf{r}} - \sin\theta\,\hat{\boldsymbol{\theta}})\right] = \frac{\mu_0}{6}(\sigma\omega R)^2(4\cos\theta\,\hat{\boldsymbol{\theta}} + \sin\theta\,\hat{\mathbf{r}})\sin\theta.$$

Picking out the z component of $\hat{\boldsymbol{\theta}}$ (namely, $-\sin\theta$) and of $\hat{\mathbf{r}}$ (namely, $\cos\theta$), we have $(\mathbf{K}\times\mathbf{B}_{\text{ave}})_z = -\frac{\mu_0}{2}(\sigma\omega R)^2\sin^2\theta\cos\theta$, so

$$F_z = -\frac{\mu_0}{2}(\sigma\omega R)^2 R^2\int\sin^3\theta\cos\theta\,d\theta\,d\phi = -\frac{\mu_0}{2}(\sigma\omega R^2)^2\,2\pi\left(\frac{\sin^4\theta}{4}\right)\Bigg|_0^{\pi/2},\ \text{or}\ \boxed{\mathbf{F} = -\frac{\mu_0\pi}{4}(\sigma\omega R^2)^2\,\hat{\mathbf{z}}.}$$

Problem 5.43

(a) $\mathbf{F} = m\mathbf{a} = q_e(\mathbf{v}\times\mathbf{B}) = \frac{\mu_0}{4\pi}\frac{q_e q_m}{r^2}(\mathbf{v}\times\hat{\mathbf{r}})$; $\boxed{\mathbf{a} = \frac{\mu_0}{4\pi}\frac{q_e q_m}{m r^3}(\mathbf{v}\times\mathbf{r}).}$

(b) Because $\mathbf{a}\perp\mathbf{v}$, $\mathbf{a}\cdot\mathbf{v} = 0$. But $\mathbf{a}\cdot\mathbf{v} = \frac{1}{2}\frac{d}{dt}(\mathbf{v}\cdot\mathbf{v}) = \frac{1}{2}\frac{d}{dt}(v^2) = v\frac{dv}{dt}$. So $\frac{dv}{dt} = 0$. qed

(c) $\frac{d\mathbf{Q}}{dt} = m(\mathbf{v}\times\mathbf{v}) + m(\mathbf{r}\times\mathbf{a}) - \frac{\mu_0 q_e q_m}{4\pi}\frac{d}{dt}\left(\frac{\mathbf{r}}{r}\right) = 0 + \frac{\mu_0 q_e q_m}{4\pi r^3}[\mathbf{r}\times(\mathbf{v}\times\mathbf{r})] - \frac{\mu_0 q_e q_m}{4\pi}\left(\frac{\mathbf{v}}{r} - \frac{\mathbf{r}}{r^2}\frac{dr}{dt}\right)$

$$= \frac{\mu_0 q_e q_m}{4\pi}\left\{\frac{1}{r^3}[r^2\mathbf{v} - (\mathbf{r}\cdot\mathbf{v})\mathbf{r}] - \frac{\mathbf{v}}{r} + \frac{\mathbf{r}}{r^2}\frac{d}{dt}(\sqrt{\mathbf{r}\cdot\mathbf{r}})\right\} = \frac{\mu_0 q_e q_m}{4\pi}\left[\frac{\mathbf{v}}{r} - \frac{(\hat{\mathbf{r}}\cdot\mathbf{v})}{r}\hat{\mathbf{r}} - \frac{\mathbf{v}}{r} + \frac{\hat{\mathbf{r}}}{2r}\frac{2(\mathbf{r}\cdot\mathbf{v})}{r}\right] = 0.\ \checkmark$$

(d) (i) $\mathbf{Q}\cdot\hat{\boldsymbol{\phi}} = Q(\hat{\mathbf{z}}\cdot\hat{\boldsymbol{\phi}}) = m(\mathbf{r}\times\mathbf{v})\cdot\hat{\boldsymbol{\phi}} - \frac{\mu_0 q_e q_m}{4\pi}(\hat{\mathbf{r}}\cdot\hat{\boldsymbol{\phi}})$. But $\hat{\mathbf{z}}\cdot\hat{\boldsymbol{\phi}} = \hat{\mathbf{r}}\cdot\hat{\boldsymbol{\phi}} = 0$, so $(\mathbf{r}\times\mathbf{v})\cdot\hat{\boldsymbol{\phi}} = 0$. But $\mathbf{r} = r\,\hat{\mathbf{r}}$, and $\mathbf{v} = \frac{d\mathbf{l}}{dt} = \dot{r}\,\hat{\mathbf{r}} + r\dot{\theta}\,\hat{\boldsymbol{\theta}} + r\sin\theta\dot{\phi}\,\hat{\boldsymbol{\phi}}$ (where dots denote differentiation with respect to time), so

$$\mathbf{r}\times\mathbf{v} = \begin{vmatrix}\hat{\mathbf{r}} & \hat{\boldsymbol{\theta}} & \hat{\boldsymbol{\phi}}\\ r & 0 & 0\\ \dot{r} & r\dot{\theta} & r\sin\theta\dot{\phi}\end{vmatrix} = (-r^2\sin\theta\dot{\phi})\,\hat{\boldsymbol{\theta}} + (r^2\dot{\theta})\,\hat{\boldsymbol{\phi}}.$$

Therefore $(\mathbf{r}\times\mathbf{v})\cdot\hat{\boldsymbol{\phi}} = r^2\dot{\theta} = 0$, so θ is constant. qed

(ii) $\mathbf{Q}\cdot\hat{\mathbf{r}} = Q(\hat{\mathbf{z}}\cdot\hat{\mathbf{r}}) = m(\mathbf{r}\times\mathbf{v})\cdot\hat{\mathbf{r}} - \frac{\mu_0 q_e q_m}{4\pi}(\hat{\mathbf{r}}\cdot\hat{\mathbf{r}})$. But $\hat{\mathbf{z}}\cdot\hat{\mathbf{r}} = \cos\theta$, and $(\mathbf{r}\times\mathbf{v})\perp\mathbf{r}\Rightarrow(\mathbf{r}\times\mathbf{v})\cdot\hat{\mathbf{r}} = 0$, so $Q\cos\theta = -\frac{\mu_0 q_e q_m}{4\pi}$, or $Q = -\frac{\mu_0 q_e q_m}{4\pi\cos\theta}$. And since θ is constant, so too is Q. qed

(iii) $\mathbf{Q}\cdot\hat{\boldsymbol{\theta}} = Q(\hat{\mathbf{z}}\cdot\hat{\boldsymbol{\theta}}) = m(\mathbf{r}\times\mathbf{v})\cdot\hat{\boldsymbol{\theta}} - \frac{\mu_0 q_e q_m}{4\pi}(\hat{\mathbf{r}}\cdot\hat{\boldsymbol{\theta}})$. But $\hat{\mathbf{z}}\cdot\hat{\boldsymbol{\theta}} = -\sin\theta$, $\hat{\mathbf{r}}\cdot\hat{\boldsymbol{\theta}} = 0$, and $(\mathbf{r}\times\mathbf{v})\cdot\hat{\boldsymbol{\theta}} = -r^2\sin\theta\dot{\phi}$ (from (i)), so $-Q\sin\theta = -mr^2\sin\theta\dot{\phi}\Rightarrow\dot{\phi} = \frac{Q}{mr^2} = \frac{k}{r^2}$, with $\boxed{k\equiv\frac{Q}{m} = -\frac{\mu_0 q_e q_m}{4\pi m\cos\theta}.}$

(e) $v^2 = \dot{r}^2 + r^2\dot{\theta}^2 + r^2\sin^2\theta\dot{\phi}^2$, but $\dot{\theta} = 0$ and $\dot{\phi} = \frac{k}{r^2}$, so $\dot{r}^2 = v^2 - r^2\sin^2\theta\frac{k^2}{r^4} = v^2 - \frac{k^2\sin^2\theta}{r^2}$.

$$\left(\frac{dr}{d\phi}\right)^2 = \frac{\dot{r}^2}{\dot{\phi}^2} = \frac{v^2 - (k\sin\theta/r)^2}{(k^2/r^4)} = r^2\left[\left(\frac{vr}{k}\right)^2 - \sin^2\theta\right]; \quad \boxed{\frac{dr}{d\phi} = r\sqrt{\left(\frac{vr}{k}\right)^2 - \sin^2\theta.}}$$

(f) $\displaystyle\int \frac{dr}{r\sqrt{(vr/k)^2 - \sin^2\theta}} = \int d\phi \Rightarrow \phi - \phi_0 = \frac{1}{\sin\theta}\sec^{-1}\left(\frac{vr}{k\sin\theta}\right)$; $\sec[(\phi-\phi_0)\sin\theta] = \dfrac{vr}{k\sin\theta}$, or

$$\boxed{r(\phi) = \frac{A}{\cos[(\phi-\phi_0)\sin\theta]},} \text{ where } A \equiv -\frac{\mu_0 q_e q_m \tan\theta}{4\pi m v}.$$

Problem 5.44

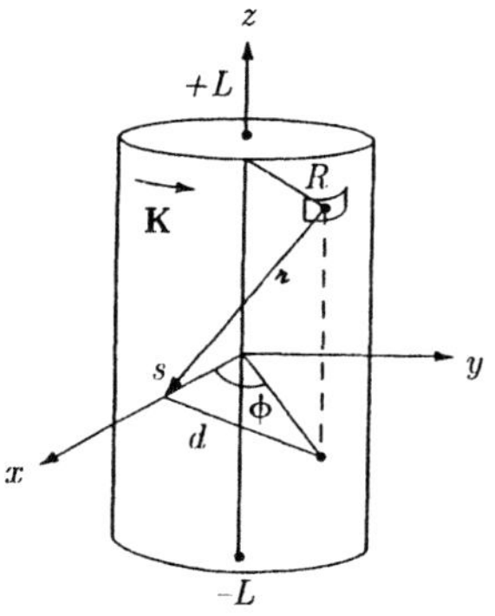

Put the field point on the x axis, so $\mathbf{r} = (s, 0, 0)$. Then $\mathbf{B} = \dfrac{\mu_0}{4\pi}\displaystyle\int \frac{(\mathbf{K}\times\hat{\boldsymbol{\imath}})}{\imath^2}\,da$; $da = R\,d\phi\,dz$; $\mathbf{K} = K\hat{\boldsymbol{\phi}} = K(-\sin\phi\,\hat{\mathbf{x}} + \cos\phi\,\hat{\mathbf{y}})$; $\boldsymbol{\imath} = (s - R\cos\phi)\,\hat{\mathbf{x}} - R\sin\phi\,\hat{\mathbf{y}} - z\,\hat{\mathbf{z}}$.

$$\mathbf{K}\times\boldsymbol{\imath} = K\begin{vmatrix} \hat{\mathbf{x}} & \hat{\mathbf{y}} & \hat{\mathbf{z}} \\ -\sin\phi & \cos\phi & 0 \\ (s - R\cos\phi) & (-R\sin\phi) & (-z)\end{vmatrix} =$$

$K\left[(-z\cos\phi)\,\hat{\mathbf{x}} + (-z\sin\phi)\,\hat{\mathbf{y}} + (R - s\cos\phi)\,\hat{\mathbf{z}}\right]$; $\imath^2 = z^2 + R^2 + s^2 - 2Rs\cos\phi$. The x and y components integrate to zero (z integrand is odd, as in Prob. 5.17).

$$\begin{aligned}
B_z &= \frac{\mu_0}{4\pi}KR\int \frac{(R - s\cos\phi)}{(z^2 + R^2 + s^2 - 2Rs\cos\phi)^{3/2}}\,d\phi\,dz \\
&= \frac{\mu_0 KR}{4\pi}\int_0^{2\pi}(R - s\cos\phi)\left\{\int_{-\infty}^{\infty}\frac{dz}{(z^2 + d^2)^{3/2}}\right\}d\phi,
\end{aligned}$$

where $d^2 \equiv R^2 + s^2 - 2Rs\cos\phi$. Now $\displaystyle\int_{-\infty}^{\infty}\frac{dz}{(z^2+d^2)^{3/2}} = \left.\frac{2z}{d^2\sqrt{z^2+d^2}}\right|_0^{\infty} = \frac{2}{d^2}$.

$$\begin{aligned}
&= \frac{\mu_0 KR}{2\pi}\int_0^{2\pi}\frac{(R - s\cos\phi)}{(R^2 + s^2 - 2Rs\cos\phi)}\,d\phi; \quad (R - s\cos\phi) = \frac{1}{2R}\left[(R^2 - s^2) + (R^2 + s^2 - 2Rs\cos\phi)\right]. \\
&= \frac{\mu_0 K}{4\pi}\left[(R^2 - s^2)\int_0^{2\pi}\frac{d\phi}{(R^2 + s^2 - 2Rs\cos\phi)} + \int_0^{2\pi}d\phi\right].
\end{aligned}$$

$$\begin{aligned}
\int_0^{2\pi}\frac{d\phi}{a + b\cos\phi} &= 2\int_0^{\pi}\frac{d\phi}{a + b\cos\phi} = \frac{4}{\sqrt{a^2 - b^2}}\tan^{-1}\left[\frac{\sqrt{a^2 - b^2}\tan(\phi/2)}{a + b}\right]\Bigg|_0^{\pi} \\
&= \frac{4}{\sqrt{a^2 - b^2}}\tan^{-1}\left[\frac{\sqrt{a^2 - b^2}\tan(\pi/2)}{a + b}\right] = \frac{4}{\sqrt{a^2 - b^2}}\left(\frac{\pi}{2}\right) = \frac{2\pi}{\sqrt{a^2 - b^2}}.
\end{aligned}$$

Here $a = R^2 + s^2$, $b = -2Rs$, so $a^2 - b^2 = R^4 + 2R^2s^2 + s^4 - 4R^2s^2 = R^4 - 2R^2s^2 + s^4 = (R^2 - s^2)^2$; $\sqrt{a^2 - b^2} = |R^2 - d^2|$.

$$B_z = \frac{\mu_0 K}{4\pi}\left[\frac{(R^2 - s^2)}{|R^2 - s^2|}2\pi + 2\pi\right] = \frac{\mu_0 K}{2}\left(\frac{R^2 - s^2}{|R^2 - s^2|} + 1\right).$$

Inside the solenoid, $s < R$, so $B_z = \dfrac{\mu_0 K}{2}(1+1) = \mu_0 K$. *Outside* the solenoid, $s > R$, so $B_z = \dfrac{\mu_0 K}{2}(-1+1) = 0$. Here $K = nI$, so $\boxed{\mathbf{B} = \mu_0 nI\,\hat{\mathbf{z}}\text{(inside), and 0(outside)}}$ (as we found more easily using Ampére's law, in Ex. 5.9).

Problem 5.45

Let the source point be $\mathbf{r}' = R\cos\phi\,\hat{\mathbf{x}} - R\sin\phi\,\hat{\mathbf{y}}$, and the field point be $\mathbf{r} = R\cos\theta\,\hat{\mathbf{x}} + R\sin\theta\,\hat{\mathbf{y}}$; then $\boldsymbol{\mathscr{r}} = R\left[(\cos\theta - \cos\phi)\,\hat{\mathbf{x}} + (\sin\theta + \sin\phi)\,\hat{\mathbf{y}}\right]$ and $d\mathbf{l} = R\sin\phi\,d\phi\,\hat{\mathbf{x}} + R\cos\phi\,d\phi\,\hat{\mathbf{y}} = R\,d\phi(\sin\phi\,\hat{\mathbf{x}} + \cos\phi\,\hat{\mathbf{y}})$.

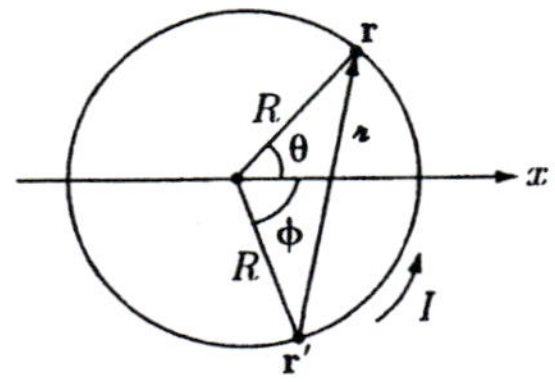

$$
\begin{aligned}
d\mathbf{l}\times\boldsymbol{\mathscr{r}} &= R^2\,d\phi\begin{vmatrix}\hat{\mathbf{x}} & \hat{\mathbf{y}} & \hat{\mathbf{z}}\\ \sin\phi & \cos\phi & 0\\ (\cos\theta-\cos\phi) & (\sin\theta+\sin\phi) & 0\end{vmatrix}\\
&= R^2(\sin\phi\sin\theta + \sin^2\phi - \cos\theta\cos\phi + \cos^2\phi)\,d\phi\,\hat{\mathbf{z}}\\
&= R^2(1+\sin\theta\sin\phi - \cos\theta\cos\phi)\,d\phi\,\hat{\mathbf{z}} = R^2\left[1-\cos(\theta+\phi)\right]\,d\phi\,\hat{\mathbf{z}}.\\
\mathbf{B} &= \frac{\mu_0 I}{4\pi}\int\frac{d\mathbf{l}\times\boldsymbol{\mathscr{r}}}{\mathscr{r}^3} = \frac{\mu_0 I}{4\pi}R^2\,\hat{\mathbf{z}}\int_0^{\pi}\frac{[1-\cos(\theta+\phi)]}{[2R^2-2R^2\cos(\theta+\phi)]^{3/2}}\,d\phi = \frac{\mu_0 I R^2}{4\pi(2R^2)^{3/2}}\,\hat{\mathbf{z}}\int_0^{\pi}\frac{d\phi}{\sqrt{1-\cos(\theta+\phi)}}\\
&= \frac{\mu_0 I}{8\sqrt{2}\pi R}\,\hat{\mathbf{z}}\int_0^{\pi}\frac{d\phi}{\sqrt{2}\sin[(\theta+\phi)/2]} = \frac{\mu_0 I}{16\pi R}\,\hat{\mathbf{z}}\left\{2\ln\left[\tan\left(\frac{\theta+\phi}{4}\right)\right]\right\}\Bigg|_0^{\pi} = \boxed{\frac{\mu_0 I}{8\pi R}\ln\left[\frac{\tan\left(\frac{\theta+\pi}{4}\right)}{\tan\left(\frac{\theta}{4}\right)}\right]\hat{\mathbf{z}}.}
\end{aligned}
$$

Problem 5.46

(a) From Eq. 5.38, $\boxed{\mathbf{B} = \frac{\mu_0 I R^2}{2}\left\{\frac{1}{[R^2+(d/2+z)^2]^{3/2}} + \frac{1}{[R^2+(d/2-z)^2]^{3/2}}\right\}.}$

$$
\begin{aligned}
\frac{\partial B}{\partial z} &= \frac{\mu_0 I R^2}{2}\left\{\frac{(-3/2)2(d/2+z)}{[R^2+(d/2+z)^2]^{5/2}} + \frac{(-3/2)2(d/2-z)(-1)}{[R^2+(d/2-z)^2]^{5/2}}\right\}\\
&= \frac{3\mu_0 I R^2}{2}\left\{\frac{-(d/2+z)}{[R^2+(d/2+z)^2]^{5/2}} + \frac{(d/2-z)}{[R^2+(d/2-z)^2]^{5/2}}\right\}.\\
\left.\frac{\partial B}{\partial z}\right|_{z=0} &= \frac{3\mu_0 I R^2}{2}\left\{\frac{-d/2}{[R^2+(d/2)^2]^{5/2}} + \frac{d/2}{[R^2+(d/2)^2]^{5/2}}\right\} = 0.\ \checkmark
\end{aligned}
$$

(b) Differentiating again:

$$
\begin{aligned}
\frac{\partial^2 B}{\partial z^2} &= \frac{3\mu_0 I R^2}{2}\left\{\frac{-1}{[R^2+(d/2+z)^2]^{5/2}} + \frac{-(d/2+z)(-5/2)2(d/2+z)}{[R^2+(d/2+z)^2]^{7/2}}\right.\\
&\qquad\left.+\frac{-1}{[R^2+(d/2-z)^2]^{5/2}} + \frac{(d/2-z)(-5/2)2(d/2-z)(-1)}{[R^2+(d/2-z)^2]^{7/2}}\right\}.\\
\left.\frac{\partial^2 B}{\partial z^2}\right|_{z=0} &= \frac{3\mu_0 I R^2}{2}\left\{\frac{-2}{[R^2+(d/2)^2]^{5/2}} + \frac{2(5/2)2(d/2)^2 2}{[R^2+(d/2)^2]^{7/2}}\right\} = \frac{3\mu_0 I R^2}{[R^2+(d/2)^2]^{7/2}}\left(-R^2-\frac{d^2}{4}+\frac{5d^2}{4}\right)\\
&= \frac{3\mu_0 I R^2}{[R^2+(d/2)^2]^{7/2}}\left(d^2-R^2\right).\quad \text{Zero if } \boxed{d=R,} \text{ in which case}\\
B(0) &= \frac{\mu_0 I R^2}{2}\left\{\frac{1}{[R^2+(R/2)^2]^{3/2}} + \frac{1}{[R^2+(R/2)^2]^{3/2}}\right\} = \mu_0 I R^2\frac{1}{(5R^2/4)^{3/2}} = \boxed{\frac{8\mu_0 I}{5^{3/2}R}.}
\end{aligned}
$$

Problem 5.47

(a) The total charge on the shaded ring is $dq = \sigma(2\pi r)\,dr$. The time for one revolution is $dt = 2\pi/\omega$. So the current in the ring is $I = \dfrac{dq}{dt} = \sigma\omega r\,dr$. From Eq. 5.38, the magnetic field of this ring (for points on the axis) is $d\mathbf{B} = \dfrac{\mu_0}{2}\sigma\omega r \dfrac{r^2}{(r^2+z^2)^{3/2}}\,dr\,\hat{\mathbf{z}}$, and the total field of the disk is

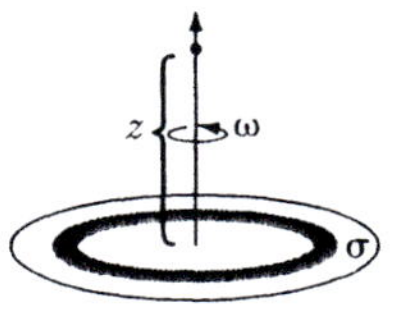

$$\begin{aligned}
\mathbf{B} &= \frac{\mu_0\sigma\omega}{2}\int_0^R \frac{r^3\,dr}{(r^2+z^2)^{3/2}}\,\hat{\mathbf{z}}. \quad \text{Let } u \equiv r^2, \text{ so } du = 2r\,dr. \quad \text{Then} \\
&= \frac{\mu_0\sigma\omega}{4}\int_0^{R^2} \frac{u\,du}{(u+z^2)^{3/2}} = \frac{\mu_0\sigma\omega}{4}\left[2\left(\frac{u+2z^2}{\sqrt{u+z^2}}\right)\right]\Bigg|_0^{R^2} = \boxed{\frac{\mu_0\sigma\omega}{2}\left[\frac{(R^2+2z^2)}{\sqrt{R^2+z^2}} - 2z\right]\hat{\mathbf{z}}.}
\end{aligned}$$

(b) Slice the sphere into slabs of thickness t, and use (a). Here $t = |d(R\cos\theta)| = R\sin\theta\,d\theta$; $\sigma \to \rho t = \rho R\sin\theta\,d\theta; R \to R\sin\theta;\ z \to z - R\cos\theta$. First rewrite the term in square brackets:

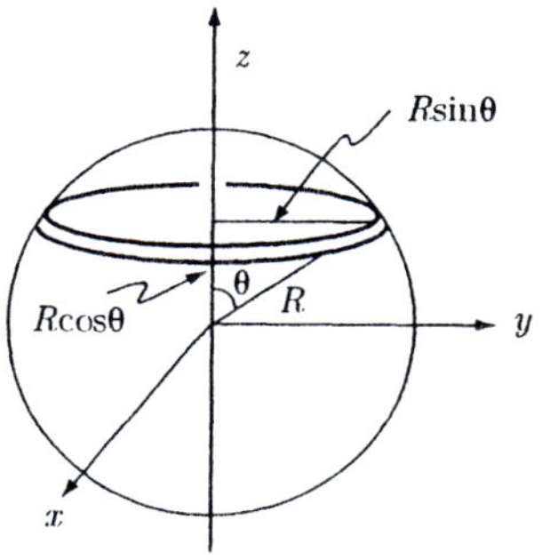

$$\begin{aligned}
\left[\frac{(R^2+2z^2)}{\sqrt{R^2+z^2}} - 2z\right] &= \frac{2(R^2+z^2)}{\sqrt{R^2+z^2}} - \frac{R^2}{\sqrt{R^2+z^2}} - 2z \\
&= 2\left[\sqrt{R^2+z^2} - \frac{R^2/2}{\sqrt{R^2+z^2}} - z\right].
\end{aligned}$$

But $R^2+z^2 \to R^2\sin^2\theta + (z^2 - 2Rz\cos\theta + R^2\cos^2\theta) = R^2 + z^2 - 2Rz\cos\theta$. So

$$\begin{aligned}
B_z &= \frac{\mu_0\rho R\omega}{2}2\int_0^\pi \sin\theta\,d\theta\left[\sqrt{R^2+z^2-2Rz\cos\theta} - \frac{(R^2/2)\sin^2\theta}{\sqrt{R^2+z^2-2Rz\cos\theta}} - (z - R\cos\theta)\right].
\end{aligned}$$

Let $u \equiv \cos\theta$, so $du = -\sin\theta\,d\theta$; $\theta: 0 \to \pi \Rightarrow u: 1 \to -1$; $\sin^2\theta = 1 - u^2$.

$$\begin{aligned}
&= \mu_0\rho R\omega\int_{-1}^1\left[\sqrt{R^2+z^2-2Rzu} - \frac{(R^2/2)(1-u^2)}{\sqrt{R^2+z^2-2Rzu}} - z + Ru\right]du \\
&= \mu_0\rho R\omega\left[I_1 - \frac{R^2}{2}(I_2 - I_3) - I_4 + I_5\right].
\end{aligned}$$

$$\begin{aligned}
I_1 &= \int_{-1}^1 \sqrt{R^2+z^2-2Rzu}\,du = -\frac{1}{3Rz}\left(R^2+z^2-2Rzu\right)^{3/2}\Big|_{-1}^1 \\
&= -\frac{1}{3Rz}\left[\left(R^2+z^2-2Rz\right)^{3/2} - \left(R^2+z^2+2Rz\right)^{3/2}\right] = -\frac{1}{3Rz}\left[(z-R)^3 - (z+R)^3\right] \\
&= -\frac{1}{3Rz}\left(z^3 - 3z^2R + 3zR^2 - R^3 - z^3 - 3z^2R - 3zR^2 - R^3\right) = \frac{2}{3z}(3z^2 + R^2). \\
I_2 &= \int_{-1}^1 \frac{1}{\sqrt{R^2+z^2-2Rzu}}\,du = -\frac{1}{Rz}\sqrt{R^2+z^2-2Rzu}\Big|_{-1}^1 = -\frac{1}{Rz}\left[(z-R) - (z+R)\right] = \frac{2}{z}.
\end{aligned}$$

$$\begin{aligned}
I_3 &= \int_{-1}^{1} \frac{u^2}{\sqrt{R^2+z^2-2Rzu}}\,du \\
&= -\frac{1}{60R^3z^3}\left[8(R^2+z^2)^2+4(R^2+z^2)2Rzu+3(2Rz)^2u^2\right]\sqrt{R^2+z^2-2Rzu}\Big|_{-1}^{1} \\
&= -\frac{1}{60R^3z^3}\Big\{\left[8(R^2+z^2)^2+8Rz(R^2+z^2)+12R^2z^2\right](z-R) \\
&\quad - \left[8(R^2+z^2)^2-8Rz(R^2+z^2)+12R^2z^2\right](z+R)\Big\} \\
&= -\frac{1}{60R^3z^3}\left\{z\left[16Rz(R^2+z^2)\right]-R\left[16(R^2+z^2)^2+24R^2z^2\right]\right\} \\
&= -\frac{1}{60R^3z^3}16R\left(R^2z^2+z^4-R^4-2R^2z^2-z^4-\frac{3}{2}R^2z^2\right) \\
&= -\frac{4}{15R^2z^3}\left(-\frac{5}{2}R^2z^2-R^4\right)=\frac{4}{15z^3}\left(R^2+\frac{5}{2}z^2\right). \quad I_4 = z\int_{-1}^{1}du = 2z; \quad I_5 = R\int_{-1}^{1}u\,du = 0.
\end{aligned}$$

$$\begin{aligned}
B_z &= \mu_0 R\rho\omega\left[\frac{2}{3z}(3z^2+R^2)-\frac{R^2}{2}\frac{2}{z}+\frac{R^2}{2}\frac{4}{15z^3}\left(R^2+\frac{5}{2}z^2\right)-2z\right] \\
&= \mu_0 R\rho\omega\left(2z+\frac{2R^2}{3z}-\frac{R^2}{z}+\frac{2R^4}{15z^3}+\frac{R^2}{3z}-2z\right) \\
&= \mu_0\rho\omega\frac{2R^5}{15z^3}. \quad \text{But } \rho=\frac{Q}{(4/3)\pi R^3}, \text{ so } \boxed{\mathbf{B}=\frac{\mu_0 Q\omega R^2}{10\pi z^3}\,\hat{\mathbf{z}}.}
\end{aligned}$$

Problem 5.48

$\mathbf{B} = \dfrac{\mu_0 I}{4\pi}\displaystyle\int \frac{d\mathbf{l}'\times \boldsymbol{\mathscr{r}}}{\mathscr{r}^3}$. $\boldsymbol{\mathscr{r}} = -R\cos\phi\,\hat{\mathbf{x}} + (y - R\sin\phi)\,\hat{\mathbf{y}} + z\,\hat{\mathbf{z}}$. (For simplicity I'll drop the prime on ϕ.) $\mathscr{r}^2 = R^2\cos^2\phi + y^2 - 2Ry\sin\phi + R^2\sin^2\phi + z^2 = R^2+y^2+z^2-2Ry\sin\phi$. The source coordinates (x',y',z') satisfy $x' = R\cos\phi \Rightarrow dx' = -R\sin\phi\,d\phi$; $y' = R\sin\phi \Rightarrow dy' = R\cos\phi\,d\phi$; $z' = 0 \Rightarrow dz' = 0$. So $d\mathbf{l}' = -R\sin\phi\,d\phi\,\hat{\mathbf{x}} + R\cos\phi\,d\phi\,\hat{\mathbf{y}}$.

$$d\mathbf{l}'\times\boldsymbol{\mathscr{r}} = \begin{vmatrix} \hat{\mathbf{x}} & \hat{\mathbf{y}} & \hat{\mathbf{z}} \\ -R\sin\phi\,d\phi & R\cos\phi\,d\phi & 0 \\ -R\cos\phi & (y-R\sin\phi) & z \end{vmatrix} = (Rz\cos\phi\,d\phi)\,\hat{\mathbf{x}} + (Rz\sin\phi\,d\phi)\,\hat{\mathbf{y}} + (-Ry\sin\phi\,d\phi + R^2\,d\phi)\,\hat{\mathbf{z}}.$$

$$B_x = \frac{\mu_0 IRz}{4\pi}\int_0^{2\pi}\frac{\cos\phi\,d\phi}{(R^2+y^2+z^2-2Ry\sin\phi)^{3/2}} = \frac{\mu_0 IRz}{4\pi}\frac{1}{Ry}\frac{1}{\sqrt{R^2+y^2+z^2-2Ry\sin\phi}}\Bigg|_0^{2\pi} = 0,$$

since $\sin\phi = 0$ at both limits. The y and z components are elliptic integrals, and cannot be expressed in terms of elementary functions.

$$\boxed{B_x = 0;\quad B_y = \frac{\mu_0 IRz}{4\pi}\int_0^{2\pi}\frac{\sin\phi\,d\phi}{(R^2+y^2+z^2-2Ry\sin\phi)^{3/2}};\; B_z = \frac{\mu_0 IR}{4\pi}\int_0^{2\pi}\frac{(R-y\sin\phi)\,d\phi}{(R^2+y^2+z^2-2Ry\sin\phi)^{3/2}}.}$$

Problem 5.49

From the Biot-Savart law, the field of loop #1 is $\mathbf{B} = \dfrac{\mu_0 I_1}{4\pi}\displaystyle\oint_1 \frac{d\mathbf{l}_1\times\hat{\boldsymbol{\mathscr{r}}}}{\mathscr{r}^2}$; the force on loop #2 is

$\mathbf{F} = I_2\displaystyle\oint_2 d\mathbf{l}_2\times\mathbf{B} = \frac{\mu_0}{4\pi}I_1I_2\oint_1\oint_2\frac{d\mathbf{l}_2\times(d\mathbf{l}_1\times\hat{\boldsymbol{\mathscr{r}}})}{\mathscr{r}^2}$. Now $d\mathbf{l}_2\times(d\mathbf{l}_1\times\hat{\boldsymbol{\mathscr{r}}}) = d\mathbf{l}_1(d\mathbf{l}_2\cdot\hat{\boldsymbol{\mathscr{r}}}) - \hat{\boldsymbol{\mathscr{r}}}(d\mathbf{l}_1\cdot d\mathbf{l}_2)$, so

$$\mathbf{F} = -\frac{\mu_0}{4\pi} I_1 I_2 \left\{ \oint \oint \frac{\hat{\boldsymbol{\imath}}}{\imath^2}(d\mathbf{l}_1 \cdot d\mathbf{l}_2) - \oint d\mathbf{l}_1 \oint \frac{(d\mathbf{l}_2 \cdot \hat{\boldsymbol{\imath}})}{\imath^2} \right\}$$

The first term is what we want. It remains to show that the second term is zero:

$\boldsymbol{\imath} = (x_2 - x_1)\,\hat{\mathbf{x}} + (y_2 - y_1)\,\hat{\mathbf{y}} + (z_2 - z_1)\,\hat{\mathbf{z}}$, so $\nabla_2(1/\imath) = \frac{\partial}{\partial x_2}\left[(x_2 - x_1)^2 + (y_2 - y_1)^2 + (z_2 - z_1)^2\right]^{-1/2} \hat{\mathbf{x}}$ $+ \frac{\partial}{\partial y_2}\left[(x_2 - x_1)^2 + (y_2 - y_1)^2 + (z_2 - z_1)^2\right]^{-1/2} \hat{\mathbf{y}} + \frac{\partial}{\partial z_2}\left[(x_2 - x_1)^2 + (y_2 - y_1)^2 + (z_2 - z_1)^2\right]^{-1/2} \hat{\mathbf{z}}$ $= -\frac{(x_2 - x_1)}{\imath^3}\hat{\mathbf{x}} - \frac{(y_2 - y_1)}{\imath^3}\hat{\mathbf{y}} - \frac{(z_2 - z_1)}{\imath^3}\hat{\mathbf{z}} = -\frac{\boldsymbol{\imath}}{\imath^3} = -\frac{\hat{\boldsymbol{\imath}}}{\imath^2}$. So $\oint \frac{\hat{\boldsymbol{\imath}}}{\imath^2} \cdot d\mathbf{l}_2 = -\oint \nabla_2\left(\frac{1}{\imath}\right) \cdot d\mathbf{l}_2 = 0$ (by Corollary 2 in Sect. 1.3.3). qed

Problem 5.50

Poisson's equation (Eq. 2.24) says $\nabla^2 V = -\frac{1}{\epsilon_0}\rho$. For *dielectrics* (with no *free* charge), $\rho_b = -\nabla \cdot \mathbf{P}$ (Eq. 4.12), and the resulting potential is $V(\mathbf{r}) = \frac{1}{4\pi\epsilon_0}\int \frac{\mathbf{P}(\mathbf{r}') \cdot \hat{\boldsymbol{\imath}}}{\imath^2}\, d\tau'$. In *general*, $\rho = \epsilon_0 \nabla \cdot \mathbf{E}$ (Gauss's law), so the analogy is $\mathbf{P} \to -\epsilon_0 \mathbf{E}$, and hence $V(\mathbf{r}) = -\frac{1}{4\pi}\int \frac{\mathbf{E}(\mathbf{r}') \cdot \hat{\boldsymbol{\imath}}}{\imath^2}\, d\tau'$. qed

[There are many other ways to obtain this result. For example, using Eq. 1.100:

$$\nabla \cdot \left(\frac{\hat{\boldsymbol{\imath}}}{\imath^2}\right) = -\nabla' \cdot \left(\frac{\hat{\boldsymbol{\imath}}}{\imath^2}\right) = 4\pi\delta^3(\boldsymbol{\imath}) = 4\pi\delta^3(\mathbf{r} - \mathbf{r}'),$$

$$V(\mathbf{r}) = \int V(\mathbf{r}')\delta^3(\mathbf{r} - \mathbf{r}')\, d\tau' = -\frac{1}{4\pi}\int V(\mathbf{r}')\nabla' \cdot \left(\frac{\hat{\boldsymbol{\imath}}}{\imath^2}\right) d\tau' = \frac{1}{4\pi}\int \frac{\hat{\boldsymbol{\imath}}}{\imath^2} \cdot [\nabla' V(\mathbf{r}')]\, d\tau' - \frac{1}{4\pi}\oint V(\mathbf{r}')\frac{\hat{\boldsymbol{\imath}}}{\imath^2} \cdot d\mathbf{a}'$$

(Eq. 1.59). But $\nabla' V(\mathbf{r}') = -\mathbf{E}(\mathbf{r}')$, and the surface integral $\to 0$ at ∞, so $V(\mathbf{r}) = -\frac{1}{4\pi}\int \frac{\mathbf{E}(\mathbf{r}') \cdot \hat{\boldsymbol{\imath}}}{\imath^2}\, d\tau'$, as before. You can also *check* the result, by computing its gradient—but it's not easy.]

Problem 5.51

(a) For uniform $\mathbf{B}$, $\int_0^{\mathbf{r}} (\mathbf{B} \times d\mathbf{l}) = \mathbf{B} \times \int_0^{\mathbf{r}} d\mathbf{l} = \boxed{\mathbf{B} \times \mathbf{r}} \neq \mathbf{A} = -\frac{1}{2}(\mathbf{B} \times \mathbf{r})$.

(b) $\mathbf{B} = \frac{\mu_0 I}{2\pi s}\hat{\boldsymbol{\phi}}$, so $\oint \mathbf{B} \times d\mathbf{l} = \left(\frac{\mu_0 I}{2\pi a}\hat{\mathbf{s}} - \frac{\mu_0 I}{2\pi b}\hat{\mathbf{s}}\right) w = \boxed{\frac{\mu_0 I w}{2\pi}\left(\frac{1}{a} - \frac{1}{b}\right)\hat{\mathbf{s}} \neq 0.}$

(c) $\mathbf{A} = -\mathbf{r} \times \mathbf{B} \int_0^1 \lambda\, d\lambda = \boxed{-\frac{1}{2}(\mathbf{r} \times \mathbf{B}).}$

(d) $\mathbf{B} = \frac{\mu_0 I}{2\pi s}\hat{\boldsymbol{\phi}}$; $\mathbf{B}(\lambda\mathbf{r}) = \frac{\mu_0 I}{2\pi \lambda s}\hat{\boldsymbol{\phi}}$; $\mathbf{A} = -\frac{\mu_0 I}{2\pi s}(\mathbf{r} \times \hat{\boldsymbol{\phi}})\int_0^1 \lambda \frac{1}{\lambda}\, d\lambda = -\frac{\mu_0 I}{2\pi s}(\mathbf{r} \times \hat{\boldsymbol{\phi}})$. But $\mathbf{r}$ here is the vector from the origin—in cylindrical coordinates $\mathbf{r} = s\,\hat{\mathbf{s}} + z\,\hat{\mathbf{z}}$. So $\mathbf{A} = -\frac{\mu_0 I}{2\pi s}\left[s(\hat{\mathbf{s}} \times \hat{\boldsymbol{\phi}}) + z(\hat{\mathbf{z}} \times \hat{\boldsymbol{\phi}})\right]$, and $(\hat{\mathbf{s}} \times \hat{\boldsymbol{\phi}}) = \hat{\mathbf{z}}$, $(\hat{\mathbf{z}} \times \hat{\boldsymbol{\phi}}) = -\hat{\mathbf{s}}$. So $\boxed{\mathbf{A} = \frac{\mu_0 I}{2\pi s}(z\,\hat{\mathbf{s}} - s\,\hat{\mathbf{z}}).}$

The examples in (c) and (d) happen to be divergenceless, but this is not the case in general. For (letting $\mathbf{L} \equiv \int_0^1 \lambda \mathbf{B}(\lambda\mathbf{r})\, d\lambda$, for short) $\nabla \cdot \mathbf{A} = -\nabla \cdot (\mathbf{r} \times \mathbf{L}) = -[\mathbf{L} \cdot (\nabla \times \mathbf{r}) - \mathbf{r} \cdot (\nabla \times \mathbf{L})] = \mathbf{r} \cdot (\nabla \times \mathbf{L})$, and $\nabla \times \mathbf{L} = \int_0^1 \lambda[\nabla \times \mathbf{B}(\lambda\mathbf{r})]\, d\lambda = \int_0^1 \lambda^2[\nabla_\lambda \times \mathbf{B}(\lambda\mathbf{r})]\, d\lambda = \mu_0 \int_0^1 \lambda^2 \mathbf{J}(\lambda\mathbf{r})\, d\lambda$, so $\nabla \cdot \mathbf{A} = \mu_0 \mathbf{r} \cdot \int_0^1 \lambda^2 \mathbf{J}(\lambda\mathbf{r})\, d\lambda$, and it vanishes in regions where $\mathbf{J} = 0$ (which is why the examples in (c) and (d) were divergenceless). To construct an explicit counterexample, we need the field at a point where $\mathbf{J} \neq 0$—say, *inside* a wire with uniform current.

Here Ampére's law gives $B\,2\pi s = \mu_0 I_{\text{enc}} = \mu_0 J \pi s^2 \Rightarrow \mathbf{B} = \frac{\mu_0 J}{2} s\,\hat{\boldsymbol{\phi}}$, so

$$\mathbf{A} = -\mathbf{r} \times \int_0^1 \lambda \left(\frac{\mu_0 J}{2}\right) \lambda s\,\hat{\boldsymbol{\phi}}\,d\lambda = -\frac{\mu_0 J}{6} s(\mathbf{r} \times \hat{\boldsymbol{\phi}}) = \frac{\mu_0 J s}{6}(z\,\hat{\mathbf{s}} - s\hat{\mathbf{z}}).$$

$$\boldsymbol{\nabla}\cdot\mathbf{A} = \frac{\mu_0 J}{6}\left[\frac{1}{s}\frac{\partial}{\partial s}(s^2 z) + \frac{\partial}{\partial z}(-s^2)\right] = \frac{\mu_0 J}{6}\left(\frac{1}{s}2sz\right) = \frac{\mu_0 J z}{3} \neq 0.$$

Conclusion: (ii) does *not* automatically yield $\boldsymbol{\nabla}\cdot\mathbf{A} = 0.$

Problem 5.52

(a) Exploit the analogy with the electrical case:

$$\mathbf{E} = \frac{1}{4\pi\epsilon_0}\frac{1}{r^3}[3(\mathbf{p}\cdot\hat{\mathbf{r}})\,\hat{\mathbf{r}} - \mathbf{p}] \quad \text{(Eq. 3.104)} \quad = -\boldsymbol{\nabla}V, \quad \text{with } V = \frac{1}{4\pi\epsilon_0}\frac{\mathbf{p}\cdot\hat{\mathbf{r}}}{r^2} \quad \text{(Eq. 3.102).}$$

$$\mathbf{B} = \frac{\mu_0}{4\pi}\frac{1}{r^3}[3(\mathbf{m}\cdot\hat{\mathbf{r}})\,\hat{\mathbf{r}} - \mathbf{m}] \quad \text{(Eq. 5.87)} \quad = -\boldsymbol{\nabla}U, \quad \text{(Eq. 5.65).}$$

Evidently the prescription is $\mathbf{p}/\epsilon_0 \to \mu_0\mathbf{m}$: $U(\mathbf{r}) = \frac{\mu_0}{4\pi}\frac{\mathbf{m}\cdot\hat{\mathbf{r}}}{r^2}.$

(b) Comparing Eqs. 5.67 and 5.85, the dipole moment of the shell is $\mathbf{m} = (4\pi/3)\omega\sigma R^4\,\hat{\mathbf{z}}$ (which we also got in Prob. 5.36). Using the result of (a), then, $U(\mathbf{r}) = \frac{\mu_0\omega\sigma R^4}{3}\frac{\cos\theta}{r^2}$ for $r > R$.

Inside the shell, the field is uniform (Eq. 5.38): $\mathbf{B} = \frac{2}{3}\mu_0\sigma\omega R\,\hat{\mathbf{z}}$, so $U(\mathbf{r}) = -\frac{2}{3}\mu_0\sigma\omega R z + \text{constant}$. We may as well pick the constant to be zero, so $U(\mathbf{r}) = -\frac{2}{3}\mu_0\sigma\omega R r\cos\theta$ for $r < R$.

[Notice that $U(\mathbf{r})$ is *not continuous* at the surface ($r = R$): $U_{\text{in}}(R) = -\frac{2}{3}\mu_0\sigma\omega R^2\cos\theta \neq U_{\text{out}}(R) = \frac{1}{3}\mu_0\sigma\omega R^2\cos\theta$. As I warned you on p. 236: if you insist on using magnetic scalar potentials, keep away from places where there is current!]

(c)

$$\mathbf{B} = \frac{\mu_0\omega Q}{4\pi R}\left[\left(1 - \frac{3r^2}{5R^2}\right)\cos\theta\,\hat{\mathbf{r}} - \left(1 - \frac{6r^2}{5R^2}\right)\sin\theta\,\hat{\boldsymbol{\theta}}\right] = -\boldsymbol{\nabla}U = -\frac{\partial U}{\partial r}\hat{\mathbf{r}} - \frac{1}{r}\frac{\partial U}{\partial\theta}\hat{\boldsymbol{\theta}} - \frac{1}{r\sin\theta}\frac{\partial U}{\partial\phi}\hat{\boldsymbol{\phi}}.$$

$$\frac{\partial U}{\partial\phi} = 0 \Rightarrow U(r,\theta,\phi) = U(r,\theta).$$

$$\frac{1}{r}\frac{\partial U}{\partial\theta} = \left(\frac{\mu_0\omega Q}{4\pi R}\right)\left(1 - \frac{6r^2}{5R^2}\right)\sin\theta \Rightarrow U(r,\theta) = -\left(\frac{\mu_0\omega Q}{4\pi R}\right)\left(1 - \frac{6r^2}{5R^2}\right)r\cos\theta + f(r).$$

$$\frac{\partial U}{\partial r} = -\left(\frac{\mu_0\omega Q}{4\pi R}\right)\left(1 - \frac{3r^2}{5R^2}\right)\cos\theta \Rightarrow U(r,\theta) = -\left(\frac{\mu_0\omega Q}{4\pi R}\right)\left(r - \frac{r^3}{5R^2}\right)\cos\theta + g(\theta).$$

Equating the two expressions:

$$-\left(\frac{\mu_0\omega Q}{4\pi R}\right)\left(1 - \frac{6r^2}{5R^2}\right)r\cos\theta + f(r) = -\left(\frac{\mu_0\omega Q}{4\pi R}\right)\left(1 - \frac{r^2}{5R^2}\right)r\cos\theta + g(\theta),$$

or

$$\left(\frac{\mu_0\omega Q}{4\pi R^3}\right)r^3\cos\theta + f(r) = g(\theta).$$

But there is no way to write $r^3\cos\theta$ as the sum of a function of θ and a function of r, so we're stuck. The reason is that you can't *have* a scalar magnetic potential in a region where the current is nonzero.

Problem 5.53

(a) $\nabla\cdot\mathbf{B}=0$, $\nabla\times\mathbf{B}=\mu_0\mathbf{J}$, and $\nabla\cdot\mathbf{A}=0$, $\nabla\times\mathbf{A}=\mathbf{B}\Rightarrow\mathbf{A}=\frac{\mu_0}{4\pi}\int\frac{\mathbf{J}}{\mathscr{r}}\,d\tau'$, so

$\nabla\cdot\mathbf{A}=0$, $\nabla\times\mathbf{A}=\mathbf{B}$, and $\nabla\cdot\mathbf{W}=0$ (we'll choose it so), $\nabla\times\mathbf{W}=\mathbf{A}\Rightarrow\boxed{\mathbf{W}=\frac{1}{4\pi}\int\frac{\mathbf{B}}{\mathscr{r}}\,d\tau'.}$

(b) $\mathbf{W}$ will be proportional to $\mathbf{B}$ and to two factors of $\mathbf{r}$ (since differentiating *twice* must recover $\mathbf{B}$), so I'll try something of the form $\mathbf{W}=\alpha\mathbf{r}(\mathbf{r}\cdot\mathbf{B})+\beta r^2\mathbf{B}$, and see if I can pick the constants α and β in such a way that $\nabla\cdot\mathbf{W}=0$ and $\nabla\times\mathbf{W}=\mathbf{A}$.

$\nabla\cdot\mathbf{W}=\alpha\left[(\mathbf{r}\cdot\mathbf{B})(\nabla\cdot\mathbf{r})+\mathbf{r}\cdot\nabla(\mathbf{r}\cdot\mathbf{B})\right]+\beta\left[r^2(\nabla\cdot\mathbf{B})+\mathbf{B}\cdot\nabla(r^2)\right]$. $\nabla\mathbf{r}=\frac{\partial x}{\partial x}+\frac{\partial y}{\partial y}+\frac{\partial z}{\partial z}=1+1+1=3$;

$\nabla(\mathbf{r}\cdot\mathbf{B})=\mathbf{r}\times(\nabla\times\mathbf{B})+\mathbf{B}\times(\nabla\times\mathbf{r})+(\mathbf{r}\cdot\nabla)\mathbf{B}+(\mathbf{B}\cdot\nabla)\mathbf{r}$; but $\mathbf{B}$ is constant, so all derivatives of $\mathbf{B}$ vanish, and $\nabla\times\mathbf{r}=0$ (Prob. 1.62), so

$\nabla(\mathbf{r}\cdot\mathbf{B})=(\mathbf{B}\cdot\nabla)\mathbf{r}=\left(B_x\frac{\partial}{\partial x}+B_y\frac{\partial}{\partial y}+B_z\frac{\partial}{\partial z}\right)(x\,\hat{\mathbf{x}}+y\,\hat{\mathbf{y}}+z\,\hat{\mathbf{z}})=B_x\,\hat{\mathbf{x}}+B_y\,\hat{\mathbf{y}}+B_z\,\hat{\mathbf{z}}=\mathbf{B}$;

$\nabla(r^2)=\left(\hat{\mathbf{x}}\frac{\partial}{\partial x}+\hat{\mathbf{y}}\frac{\partial}{\partial y}+\hat{\mathbf{z}}\frac{\partial}{\partial z}\right)(x^2+y^2+z^2)=2x\,\hat{\mathbf{x}}+2y\,\hat{\mathbf{y}}+2z\,\hat{\mathbf{z}}=2\mathbf{r}$. So

$\nabla\cdot\mathbf{W}=\alpha\left[3(\mathbf{r}\cdot\mathbf{B})+(\mathbf{r}\cdot\mathbf{B})\right]+\beta\left[0+2(\mathbf{r}\cdot\mathbf{B})\right]=2(\mathbf{r}\cdot\mathbf{B})(2\alpha+\beta)$, which is zero if $2\alpha+\beta=0$.

$\nabla\times\mathbf{W}=\alpha\left[(\mathbf{r}\cdot\mathbf{B})(\nabla\times\mathbf{r})-\mathbf{r}\times\nabla(\mathbf{r}\cdot\mathbf{B})\right]+\beta\left[r^2(\nabla\times\mathbf{B})-\mathbf{B}\times\nabla(r^2)\right]=\alpha\left[0-(\mathbf{r}\times\mathbf{B})\right]+\beta\left[0-2(\mathbf{B}\times\mathbf{r})\right]$

$=-(\mathbf{r}\times\mathbf{B})(\alpha-2\beta)=-\frac{1}{2}(\mathbf{r}\times\mathbf{B})$ (Prob. 5.24). So we want $\alpha-2\beta=1/2$. Evidently $\alpha-2(-2\alpha)=5\alpha=1/2$,

or $\alpha=1/10;\beta=-2\alpha=-1/5$. *Conclusion:* $\boxed{\mathbf{W}=\frac{1}{10}\left[\mathbf{r}(\mathbf{r}\cdot\mathbf{B})-2r^2\mathbf{B}\right].}$ (But this is certainly not unique.)

(c) $\nabla\times\mathbf{W}=\mathbf{A}\Rightarrow\int(\nabla\times\mathbf{W})\cdot d\mathbf{a}=\int\mathbf{A}\cdot d\mathbf{a}$. Or $\oint\mathbf{W}\cdot d\mathbf{l}=\int\mathbf{A}\cdot d\mathbf{a}$. Integrate around the amperian loop shown, taking $\mathbf{W}$ to point parallel to the axis, and choosing $\mathbf{W}=0$ on the axis:

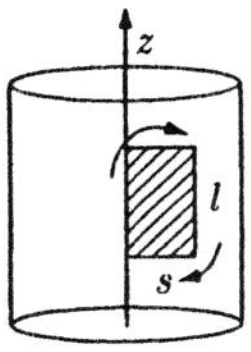

$$-Wl=\int_0^s\left(\frac{\mu_0 nI}{2}\right)l\bar{s}\,d\bar{s}=\frac{\mu_0 nI}{2}\frac{s^2l}{2}\text{ (using Eq. 5.70 for }\mathbf{A}).$$

$$\boxed{\mathbf{W}=-\frac{\mu_0 nIs^2}{4}\,\hat{\mathbf{z}}}\ (s<R).$$

For $s>R$, $-Wl=\frac{\mu_0 nIR^2l}{4}+\int_R^s\left(\frac{\mu_0 nI}{2}\right)\frac{R^2}{\bar{s}}l\,d\bar{s}=\frac{\mu_0 nIR^2l}{4}+\frac{\mu_0 nIR^2l}{2}\ln(s/R)$;

$$\boxed{\mathbf{W}=-\frac{\mu_0 nIR^2}{4}\left[1+2\ln(s/R)\right]\hat{\mathbf{z}}}\ (s>R).$$

Problem 5.54

Apply the divergence theorem to the function $[\mathbf{U}\times(\nabla\times\mathbf{V})]$, noting (from the product rule) that $\nabla\cdot[\mathbf{U}\times(\nabla\times\mathbf{V})]=(\nabla\times\mathbf{V})\cdot(\nabla\times\mathbf{U})-\mathbf{U}\cdot[\nabla\times(\nabla\times\mathbf{V})]$:

$$\int\nabla\cdot[\mathbf{U}\times(\nabla\times\mathbf{V})]\,d\tau=\int\{(\nabla\times\mathbf{V})\cdot(\nabla\times\mathbf{U})-\mathbf{U}\cdot[\nabla\times(\nabla\times\mathbf{V})]\}\,d\tau=\oint[\mathbf{U}\times(\nabla\times\mathbf{V})]\cdot d\mathbf{a}.$$

As always, suppose we have *two* solutions, $\mathbf{B}_1$ (and $\mathbf{A}_1$) and $\mathbf{B}_2$ (and $\mathbf{A}_2$). Define $\mathbf{B}_3\equiv\mathbf{B}_2-\mathbf{B}_1$ (and $\mathbf{A}_3\equiv\mathbf{A}_2-\mathbf{A}_1$), so that $\nabla\times\mathbf{A}_3=\mathbf{B}_3$ and $\nabla\times\mathbf{B}_3=\nabla\times\mathbf{B}_1-\nabla\times\mathbf{B}_2=\mu_0\mathbf{J}-\mu_0\mathbf{J}=0$. Set $\mathbf{U}=\mathbf{V}=\mathbf{A}_3$ in the above identity:

$$\int \{(\nabla \times \mathbf{A}_3)\cdot(\nabla \times \mathbf{A}_3) - \mathbf{A}_3 \cdot [\nabla \times (\nabla \times \mathbf{A}_3)]\}\, d\tau = \int \{(\mathbf{B}_3)\cdot(\mathbf{B}_3) - \mathbf{A}_3 \cdot [\nabla \times \mathbf{B}_3]\}\, d\tau = \int (B_3)^2\, d\tau$$

$= \oint [\mathbf{A}_3 \times (\nabla \times \mathbf{A}_3)] \cdot d\mathbf{a} = \oint (\mathbf{A}_3 \times \mathbf{B}_3) \cdot d\mathbf{a}$. But either $\mathbf{A}$ is specified (in which case $\mathbf{A}_3 = 0$), or else $\mathbf{B}$ is specified (in which case $\mathbf{B}_3 = 0$), at the surface. In either case $\oint(\mathbf{A}_3 \times \mathbf{B}_3) \cdot d\mathbf{a} = 0$. So $\int (B_3)^2\, d\tau = 0$, and hence $\mathbf{B}_1 = \mathbf{B}_2$. qed

Problem 5.55

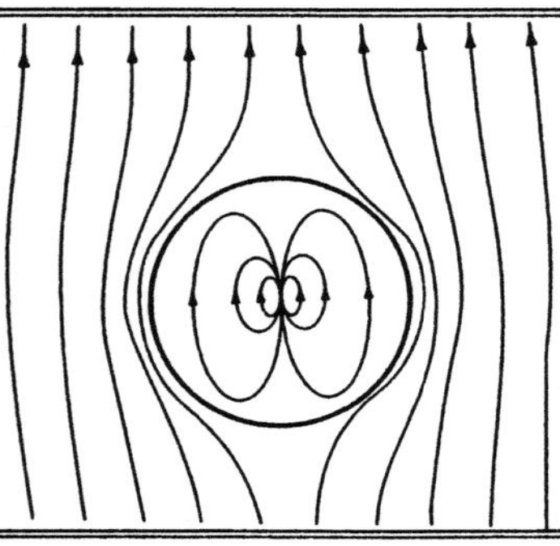

From Eq. 5.86, $\mathbf{B}_{\text{tot}} = B_0\,\hat{\mathbf{z}} - \frac{\mu_0 m_0}{4\pi r^3}(2\cos\theta\,\hat{\mathbf{r}} + \sin\theta\,\hat{\boldsymbol{\theta}})$. Therefore $\mathbf{B}\cdot\hat{\mathbf{r}} = B_0(\hat{\mathbf{z}}\cdot\hat{\mathbf{r}}) - \frac{\mu_0 m_0}{4\pi r^3}2\cos\theta = \left(B_0 - \frac{\mu_0 m_0}{2\pi r^3}\right)\cos\theta$. This is zero, for all θ, when $r = R$, given by $B_0 = \frac{\mu_0 m_0}{2\pi R^3}$, or

$\boxed{R = \left(\frac{\mu_0 m_0}{2\pi B_0}\right)^{1/3}.}$ Evidently no field lines cross this sphere.

Problem 5.56

(a) $I = \frac{Q}{(2\pi/\omega)} = \frac{Q\omega}{2\pi}$; $a = \pi R^2$; $\mathbf{m} = \frac{Q\omega}{2\pi}\pi R^2\,\hat{\mathbf{z}} = \frac{Q}{2}\omega R^2\,\hat{\mathbf{z}}$. $L = RMv = M\omega R^2; \mathbf{L} = M\omega R^2\,\hat{\mathbf{z}}$.

$\frac{m}{L} = \frac{Q}{2}\frac{\omega R^2}{M\omega R^2} = \frac{Q}{2M}$. $\boxed{\mathbf{m} = \left(\frac{Q}{2M}\right)\mathbf{L},}$ and the gyromagnetic ratio is $\boxed{g = \frac{Q}{2M}.}$

(b) Because g is *independent of* R, the same ratio applies to all "donuts", and hence to the entire sphere (or any other figure of revolution): $\boxed{g = \frac{Q}{2M}.}$

(c) $m = \frac{e}{2m}\frac{\hbar}{2} = \frac{e\hbar}{4m} = \frac{(1.60 \times 10^{-19})(1.05 \times 10^{-34})}{4(9.11 \times 10^{-31})} = \boxed{4.61 \times 10^{-24}\,\mathrm{A\,m^2}.}$

Problem 5.57

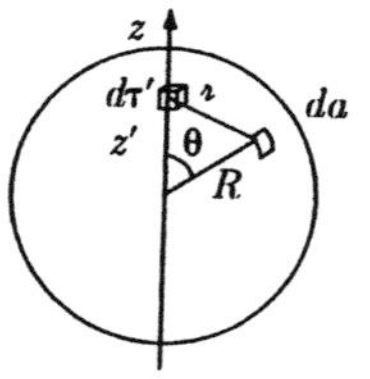

(a) $\mathbf{B}_{\text{ave}} = \frac{1}{(3/4)\pi R^3}\int \mathbf{B}\, d\tau = \frac{3}{4\pi R^3}\int (\nabla \times \mathbf{A})\, d\tau = -\frac{3}{4\pi R^3}\oint \mathbf{A} \times d\mathbf{a} = -\frac{3}{4\pi R^3}\frac{\mu_0}{4\pi}\oint\left\{\int \frac{\mathbf{J}}{\mathscr{r}}\, d\tau'\right\} \times d\mathbf{a} = -\frac{3\mu_0}{(4\pi)^2 R^3}\int \mathbf{J} \times \left\{\oint \frac{1}{\mathscr{r}}\, d\mathbf{a}\right\} d\tau'$. Note that $\mathbf{J}$ depends on the source point $\mathbf{r}'$, not on the field point $\mathbf{r}$. To do the surface integral, choose the (x, y, z) coordinates so that $\mathbf{r}'$ lies on the z axis (see diagram). Then $\mathscr{r} = \sqrt{R^2 + (z')^2 - 2Rz'\cos\theta}$, while $d\mathbf{a} = R^2 \sin\theta\, d\theta\, d\phi\,\hat{\mathbf{r}}$. By symmetry, the x and y components must integrate to zero; since the z component of $\hat{\mathbf{r}}$ is $\cos\theta$, we have

$$\oint \frac{1}{\mathscr{r}}\,d\mathbf{a} = \hat{\mathbf{z}}\int \frac{\cos\theta}{\sqrt{R^2+(z')^2-2Rz'\cos\theta}}R^2\sin\theta\,d\theta\,d\phi = 2\pi R^2\,\hat{\mathbf{z}}\int_0^\pi \frac{\cos\theta\sin\theta}{\sqrt{R^2+(z')^2-2Rz'\cos\theta}}\,d\theta.$$

Let $u \equiv \cos\theta$, so $du = -\sin\theta\,d\theta$.

$$= 2\pi R^2\,\hat{\mathbf{z}}\int_{-1}^{1}\frac{u}{\sqrt{R^2+(z')^2-2Rz'u}}\,du$$

$$= 2\pi R^2\,\hat{\mathbf{z}}\left\{-\frac{2\left[2(R^2+(z')^2)+2Rz'u\right]}{3(2Rz')^2}\sqrt{R^2+(z')^2-2Rz'u}\right\}\Bigg|_{-1}^{1}$$

$$= -\frac{2\pi R^2\,\hat{\mathbf{z}}}{3(Rz')^2}\left\{\left[R^2+(z')^2+Rz'\right]\sqrt{R^2+(z')^2-2Rz'} - \left[R^2+(z')^2-Rz'\right]\sqrt{R^2+(z')^2+2Rz'}\right\}$$

$$= -\left[\frac{2\pi}{3(z')^2}\,\hat{\mathbf{z}}\right]\left\{\left[R^2+(z')^2+Rz'\right]|R-z'| - \left[R^2+(z')^2-Rz'\right](R+z')\right\}$$

$$= \left\{\begin{array}{ll} \dfrac{4\pi}{3}z'\,\hat{\mathbf{z}} = \dfrac{4\pi}{3}\mathbf{r}', & (r' < R); \\ \dfrac{4\pi R^3}{3(z')^2}\,\hat{\mathbf{z}} = \dfrac{4\pi}{3}\dfrac{R^3}{(r')^3}\mathbf{r}', & (r' > R). \end{array}\right\}$$

For now we want $r' < R$, so $\mathbf{B}_{\text{ave}} = -\dfrac{3\mu_0}{(4\pi)^2R^3}\dfrac{4\pi}{3}\displaystyle\int(\mathbf{J}\times\mathbf{r}')\,d\tau' = -\dfrac{\mu_0}{4\pi R^3}\displaystyle\int(\mathbf{J}\times\mathbf{r}')\,d\tau'$. Now $\mathbf{m} = \frac{1}{2}\int(\mathbf{r}\times\mathbf{J})\,d\tau$ (Eq. 5.91), so $\mathbf{B}_{\text{ave}} = \dfrac{\mu_0}{4\pi}\dfrac{2\mathbf{m}}{R^3}$. qed

(b) This time $r' > R$, so $\mathbf{B}_{\text{ave}} = -\dfrac{3\mu_0}{(4\pi)^2R^3}\dfrac{4\pi}{3}R^3\displaystyle\int\left(\mathbf{J}\times\frac{\mathbf{r}'}{(r')^3}\right)d\tau' = \dfrac{\mu_0}{4\pi}\displaystyle\int\frac{\mathbf{J}\times\hat{\mathscr{r}}}{\mathscr{r}^2}\,d\tau'$, where $\mathscr{r}$ now goes from the source point to the *center* ($\boldsymbol{\mathscr{r}} = -\mathbf{r}'$). Thus $\mathbf{B}_{\text{ave}} = \mathbf{B}_{\text{cen}}$. qed

Problem 5.58

(a) Problem 5.51 gives the dipole moment of a shell: $\mathbf{m} = \dfrac{4\pi}{3}\sigma\omega R^4\,\hat{\mathbf{z}}$. Let $R \to r, \sigma \to \rho\,dr$, and integrate:

$$\mathbf{m} = \frac{4\pi}{3}\omega\rho\,\hat{\mathbf{z}}\int_0^R r^4\,dr = \frac{4\pi}{3}\omega\rho\frac{R^5}{5}\,\hat{\mathbf{z}}.\quad \text{But } \rho = \frac{Q}{(4/3)\pi R^3},\ \text{so}\ \boxed{\mathbf{m} = \frac{1}{5}Q\omega R^2\,\hat{\mathbf{z}}.}$$

(b) $\mathbf{B}_{\text{ave}} = \dfrac{\mu_0}{4\pi}\dfrac{2\mathbf{m}}{R^3} = \boxed{\dfrac{\mu_0}{4\pi}\dfrac{2Q\omega}{5R}\,\hat{\mathbf{z}}.}$

(c) $\mathbf{A} \cong \dfrac{\mu_0}{4\pi}\dfrac{m\sin\theta}{r^2}\,\hat{\boldsymbol{\phi}} = \boxed{\dfrac{\mu_0}{4\pi}\dfrac{Q\omega R^2}{5}\dfrac{\sin\theta}{r^2}\,\hat{\boldsymbol{\phi}}.}$

(d) Use Eq. 5.67, with $R \to \bar{r}, \sigma \to \rho\,d\bar{r}$, and integrate:

$$\mathbf{A} = \frac{\mu_0\omega\rho}{3}\frac{\sin\theta}{r^2}\,\hat{\boldsymbol{\phi}}\int_0^R \bar{r}^4\,d\bar{r} = \frac{\mu_0\omega}{3}\frac{3Q}{4\pi R^3}\frac{\sin\theta}{r^2}\frac{R^5}{5}\,\hat{\boldsymbol{\phi}} = \boxed{\frac{\mu_0}{4\pi}\frac{Q\omega R^2}{5}\frac{\sin\theta}{r^2}\,\hat{\boldsymbol{\phi}}.}$$

This is *identical* to (c); evidently the field is *pure* dipole, for points outside the sphere.

(e) According to Prob. 5.29, the field is $\mathbf{B} = \dfrac{\mu_0\omega Q}{4\pi R}\left[\left(1-\dfrac{3r^2}{5R^2}\right)\cos\theta\,\hat{\mathbf{r}} - \left(1-\dfrac{6r^2}{5R^2}\right)\sin\theta\,\hat{\boldsymbol{\theta}}\right]$. The average obviously points in the z direction, so take the z component of $\hat{\mathbf{r}}$ ($\cos\theta$) and $\hat{\boldsymbol{\theta}}$ ($-\sin\theta$):

$$\begin{aligned}
B_{\text{ave}} &= \frac{\mu_0\omega Q}{4\pi R}\frac{1}{(4/3)\pi R^3}\int\left[\left(1-\frac{3r^2}{5R^2}\right)\cos^2\theta+\left(1-\frac{6r^2}{5R^2}\right)\sin^2\theta\right]r^2\sin\theta\,dr\,d\theta\,d\phi \\
&= \frac{3\mu_0\omega Q}{(4\pi R^2)^2}2\pi\int_0^\pi\left[\left(\frac{r^3}{3}-\frac{3}{5}\frac{R^5}{5R^2}\right)\cos^2\theta+\left(\frac{R^3}{3}-\frac{6}{5}\frac{R^5}{5R^2}\right)\sin^2\theta\right]\sin\theta\,d\theta \\
&= \frac{3\mu_0\omega Q}{8\pi R^4}R^3\int_0^\pi\left(\frac{16}{75}\cos^2\theta+\frac{7}{75}\sin^2\theta\right)\sin\theta\,d\theta = \frac{3\mu_0\omega Q}{8\pi R}\frac{1}{75}\int_0^\pi(7+9\cos^2\theta)\sin\theta\,d\theta \\
&= \frac{\mu_0\omega Q}{200\pi R}\left(-7\cos\theta-3\cos^3\theta\right)\Big|_0^\pi = \frac{\mu_0\omega Q}{200\pi R}(20) = \frac{\mu_0\omega Q}{10\pi R}\ \text{(same as (b)).}\ \checkmark
\end{aligned}$$

Problem 5.59

The issue (and the integral) is identical to the one in Prob. 3.42. The resolution (as before) is to regard Eq. 5.87 as correct outside an infinitesimal sphere centered at the dipole. *Inside* this sphere the field is a delta-function, $\mathbf{A}\delta^3(\mathbf{r})$, with $\mathbf{A}$ selected so as to make the average field consistent with Prob. 5.57:

$$\mathbf{B}_{\text{ave}} = \frac{1}{(4/3)\pi R^3}\int\mathbf{A}\delta^3(\mathbf{r})\,d\tau = \frac{3}{4\pi R^3}\mathbf{A} = \frac{\mu_0}{4\pi}\frac{2\mathbf{m}}{R^3} \Rightarrow \mathbf{A} = \frac{2\mu_0\mathbf{m}}{3}.\ \text{The added term is}\ \boxed{\frac{2\mu_0}{3}\mathbf{m}\delta^3(\mathbf{r}).}$$

Problem 5.60

(a) $I\,d\mathbf{l}\to\mathbf{J}\,d\tau$, so $\boxed{\mathbf{A} = \frac{\mu_0}{4\pi}\sum_{n=0}^{\infty}\frac{1}{r^{n+1}}\int(r')^nP_n(\cos\theta)\mathbf{J}\,d\tau.}$

(b) $\mathbf{A}_{\text{mon}} = \frac{\mu_0}{4\pi r}\int\mathbf{J}\,d\tau = \frac{\mu_0}{4\pi r}\frac{d\mathbf{p}}{dt}$ (Prob. 5.7), where $\mathbf{p}$ is the total electric dipole moment. In magnetostatics, $\mathbf{p}$ is constant, so $d\mathbf{p}/dt = 0$, and hence $\mathbf{A}_{\text{mon}} = 0$. qed

(c) $\mathbf{m} = I\mathbf{a} = \frac{1}{2}I\oint(\mathbf{r}\times d\mathbf{l})\to\mathbf{m} = \frac{1}{2}\int(\mathbf{r}\times\mathbf{J})\,d\tau$. qed

---

Problem 5.61

For a dipole at the origin and a field point in the xz plane ($\phi = 0$), we have

$$\begin{aligned}
\mathbf{B} &= \frac{\mu_0}{4\pi}\frac{m}{r^3}(2\cos\theta\,\hat{\mathbf{r}}+\sin\theta\,\hat{\boldsymbol{\theta}}) = \frac{\mu_0}{4\pi}\frac{m}{r^3}[2\cos\theta(\sin\theta\,\hat{\mathbf{x}}+\cos\theta\,\hat{\mathbf{z}})+\sin\theta(\cos\theta\,\hat{\mathbf{x}}-\sin\theta\,\hat{\mathbf{z}})] \\
&= \frac{\mu_0}{4\pi}\frac{m}{r^3}[3\sin\theta\cos\theta\,\hat{\mathbf{x}}+(2\cos^2\theta-\sin^2\theta)\,\hat{\mathbf{z}}].
\end{aligned}$$

Here we have a *stack* of such dipoles, running from $z = -L/2$ to $z = +L/2$. Put the field point at s on the x axis. The $\hat{\mathbf{x}}$ components cancel (because of symmetrically placed dipoles above and below $z = 0$), leaving $\mathbf{B} = \frac{\mu_0}{4\pi}2\mathcal{M}\,\hat{\mathbf{z}}\int_0^{L/2}\frac{(3\cos^2\theta-1)}{r^3}\,dz$, where $\mathcal{M}$ is the dipole moment per unit length: $m = I\pi R^2 = (\sigma vh)\pi R^2 = \sigma\omega R\pi R^2h \Rightarrow \mathcal{M} = \frac{m}{h} = \pi\sigma\omega R^3$. Now $\sin\theta = \frac{s}{r}$, so $\frac{1}{r^3} = \frac{\sin^3\theta}{s^3}$; $z = -s\cot\theta \Rightarrow dz = \frac{s}{\sin^2\theta}\,d\theta$. Therefore

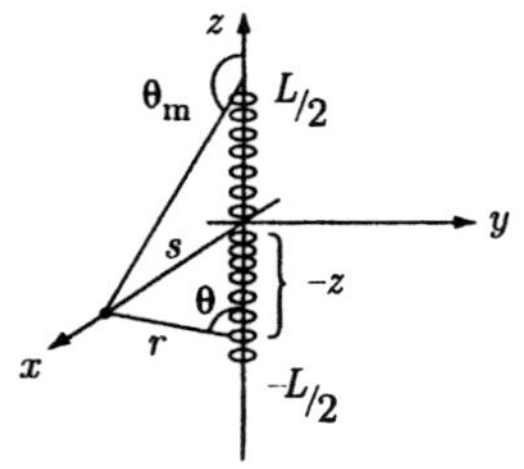

$$\begin{aligned}
\mathbf{B} &= \frac{\mu_0}{2\pi}(\pi\sigma\omega R^3)\,\hat{\mathbf{z}}\int_{\pi/2}^{\theta_m}(3\cos^2\theta-1)\frac{\sin^3\theta}{s^3}\frac{s}{\sin^2\theta}\,d\theta = \frac{\mu_0\sigma\omega R^3}{2s^2}\,\hat{\mathbf{z}}\int_{\pi/2}^{\theta_m}(3\cos^2\theta-1)\sin\theta\,d\theta \\
&= \frac{\mu_0\sigma\omega R^3}{2s^2}\,\hat{\mathbf{z}}\left(-\cos^3\theta+\cos\theta\right)\Big|_{\pi/2}^{\theta_m} = \frac{\mu_0\sigma\omega R^3}{2s^2}\cos\theta_m\left(1-\cos^2\theta_m\right)\hat{\mathbf{z}} = \frac{\mu_0\sigma\omega R^3}{2s^2}\cos\theta_m\sin^2\theta_m\,\hat{\mathbf{z}}.
\end{aligned}$$

But $\sin\theta_m = \frac{s}{\sqrt{s^2+(L/2)^2}}$, and $\cos\theta_m = \frac{-(L/2)}{\sqrt{s^2+(L/2)^2}}$, so $\boxed{\mathbf{B} = -\frac{\mu_0\sigma\omega R^3 L}{4[s^2+(L/2)^2]^{3/2}}\,\hat{\mathbf{z}}.}$

Chapter 6

Magnetostatic Fields in Matter

Problem 6.1

$$\mathbf{N} = \mathbf{m}_2 \times \mathbf{B}_1;\ \mathbf{B}_1 = \frac{\mu_0}{4\pi}\frac{1}{r^3}\left[3(\mathbf{m}_1\cdot\hat{\mathbf{r}})\hat{\mathbf{r}} - \mathbf{m}_1\right];\ \hat{\mathbf{r}} = \hat{\mathbf{y}};\mathbf{m}_1 = m_1\hat{\mathbf{z}};\ \mathbf{m}_2 = m_2\hat{\mathbf{y}}.\quad \mathbf{B}_1 = -\frac{\mu_0}{4\pi}\frac{m_1}{r^3}\hat{\mathbf{z}}.$$

$\mathbf{N} = -\frac{\mu_0}{4\pi}\frac{m_1 m_2}{r^3}(\hat{\mathbf{y}}\times\hat{\mathbf{z}}) = -\frac{\mu_0}{4\pi}\frac{m_1 m_2}{r^3}\hat{\mathbf{x}}$. Here $m_1 = \pi a^2 I$, $m_2 = b^2 I$. So $\boxed{\mathbf{N} = -\frac{\mu_0}{4}\frac{(abI)^2}{r^3}\hat{\mathbf{x}}.}$ Final orientation : $\boxed{\text{downward}}$ $(-\hat{\mathbf{z}})$.

Problem 6.2

$d\mathbf{F} = I\,d\mathbf{l}\times\mathbf{B}$; $d\mathbf{N} = \mathbf{r}\times d\mathbf{F} = I\mathbf{r}\times(d\mathbf{l}\times\mathbf{B})$. Now (Prob. 1.6): $\mathbf{r}\times(d\mathbf{l}\times\mathbf{B}) + d\mathbf{l}\times(\mathbf{B}\times\mathbf{r}) + \mathbf{B}\times(\mathbf{r}\times d\mathbf{l}) = 0$. But $d[\mathbf{r}\times(\mathbf{r}\times\mathbf{B})] = d\mathbf{r}\times(\mathbf{r}\times\mathbf{B}) + \mathbf{r}\times(d\mathbf{r}\times\mathbf{B})$ (since $\mathbf{B}$ is constant), and $d\mathbf{r} = d\mathbf{l}$, so $d\mathbf{l}\times(\mathbf{B}\times\mathbf{r}) = \mathbf{r}\times(d\mathbf{l}\times\mathbf{B}) - d[\mathbf{r}\times(\mathbf{r}\times\mathbf{B})]$. Hence $2\mathbf{r}\times(d\mathbf{l}\times\mathbf{B}) = d[\mathbf{r}\times(\mathbf{r}\times\mathbf{B})] - \mathbf{B}\times(\mathbf{r}\times d\mathbf{l})$. $d\mathbf{N} = \frac{1}{2}I\left\{d[\mathbf{r}\times(\mathbf{r}\times\mathbf{B})] - \mathbf{B}\times(\mathbf{r}\times d\mathbf{l})\right\}$. $\therefore \mathbf{N} = \frac{1}{2}I\left\{\oint d[\mathbf{r}\times(\mathbf{r}\times\mathbf{B})] - \mathbf{B}\times\oint(\mathbf{r}\times d\mathbf{l})\right\}$. But the first term is zero ($\oint d(\cdots) = 0$), and the second integral is $2\mathbf{a}$ (Eq. 1.107). So $\mathbf{N} = -I(\mathbf{B}\times\mathbf{a}) = \mathbf{m}\times\mathbf{B}$. qed

Problem 6.3

(a)

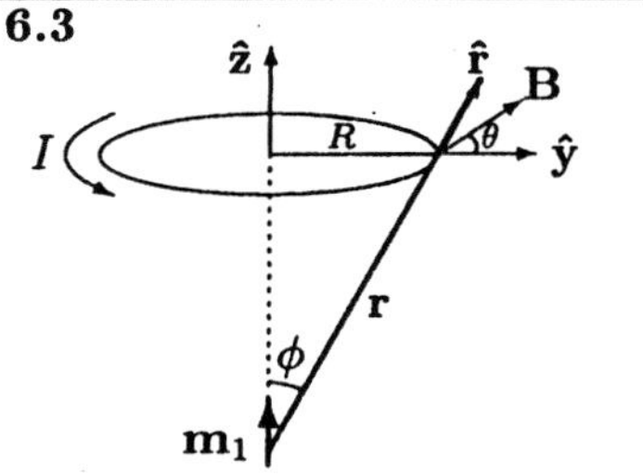

According to Eq. 6.2, $F = 2\pi IRB\cos\theta$. But $\mathbf{B} = \frac{\mu_0}{4\pi}\frac{[3(\mathbf{m}_1\cdot\hat{\mathbf{r}})\hat{\mathbf{r}} - \mathbf{m}_1]}{r^3}$, and $B\cos\theta = \mathbf{B}\cdot\hat{\mathbf{y}}$, so $B\cos\theta = \frac{\mu_0}{4\pi}\frac{1}{r^3}[3(\mathbf{m}_1\cdot\hat{\mathbf{r}})(\hat{\mathbf{r}}\cdot\hat{\mathbf{y}}) - (\mathbf{m}_1\cdot\hat{\mathbf{y}})]$. But $\mathbf{m}_1\cdot\hat{\mathbf{y}} = 0$ and $\hat{\mathbf{r}}\cdot\hat{\mathbf{y}} = \sin\phi$, while $\mathbf{m}_1\cdot\hat{\mathbf{r}} = m_1\cos\theta$. $\therefore\ B\cos\theta = \frac{\mu_0}{4\pi}\frac{1}{r^3}3m_1\sin\phi\cos\phi$.

$F = 2\pi IR\frac{\mu_0}{4\pi}\frac{1}{r^3}3m_1\sin\phi\cos\phi$. Now $\sin\phi = \frac{R}{r}$, $\cos\phi = \sqrt{r^2 - R^2}/r$, so $F = 3\frac{\mu_0}{2}m_1 IR^2\frac{\sqrt{r^2-R^2}}{r^5}$.
But $IR^2\pi = m_2$, so $F = \frac{3\mu_0}{2\pi}m_1 m_2\frac{\sqrt{r^2-R^2}}{r^5}$, while for a dipole, $R \ll r$, so $\boxed{F = \frac{3\mu_0}{2\pi}\frac{m_1 m_2}{r^4}.}$

(b) $\mathbf{F} = \boldsymbol{\nabla}(\mathbf{m}_2\cdot\mathbf{B}) = (\mathbf{m}_2\cdot\boldsymbol{\nabla})\mathbf{B} = \left(m_2\frac{d}{dz}\right)\left[\frac{\mu_0}{4\pi}\frac{1}{z^3}\underbrace{(3(\mathbf{m}_1\cdot\hat{\mathbf{z}})\hat{\mathbf{z}} - \mathbf{m}_1)}_{2\mathbf{m}_1}\right] = \frac{\mu_0}{2\pi}m_1 m_2\hat{\mathbf{z}}\underbrace{\frac{d}{dz}\left(\frac{1}{z^3}\right)}_{-3\frac{1}{z^4}}$,

or, since $z = r$: $\boxed{\mathbf{F} = -\frac{3\mu_0}{2\pi}\frac{m_1 m_2}{r^4}\hat{\mathbf{z}}.}$

Problem 6.4

$$d\mathbf{F} = I\left\{(dy\,\hat{\mathbf{y}}) \times \mathbf{B}(0,y,0) + (dz\,\hat{\mathbf{z}}) \times \mathbf{B}(0,\epsilon,z) - (dy\,\hat{\mathbf{y}}) \times \mathbf{B}(0,y,\epsilon) - (dz\,\hat{\mathbf{z}}) \times \mathbf{B}(0,0,z)\right\}$$
$$= I\Big\{-(dy\,\hat{\mathbf{y}}) \times \underbrace{[\mathbf{B}(0,y,\epsilon) - \mathbf{B}(0,y,0)]}_{\approx\,\epsilon\frac{\partial\mathbf{B}}{\partial z}} + (dz\,\hat{\mathbf{z}}) \times \underbrace{[\mathbf{B}(0,\epsilon,z) - \mathbf{B}(0,0,z)]}_{\approx\,\epsilon\frac{\partial\mathbf{B}}{\partial y}}\Big\}$$
$$\Rightarrow I\epsilon^2\left\{\hat{\mathbf{z}} \times \frac{\partial\mathbf{B}}{\partial y} - \hat{\mathbf{y}} \times \frac{\partial\mathbf{B}}{\partial z}\right\}. \quad \left[\text{Note that } \int dy\, \frac{\partial\mathbf{B}}{\partial z}\Big|_{0,y,0} \approx \epsilon\,\frac{\partial\mathbf{B}}{\partial z}\Big|_{0,0,0} \text{ and } \int dz\, \frac{\partial\mathbf{B}}{\partial y}\Big|_{0,0,z} \approx \epsilon\,\frac{\partial\mathbf{B}}{\partial y}\Big|_{0,0,0}.\right]$$

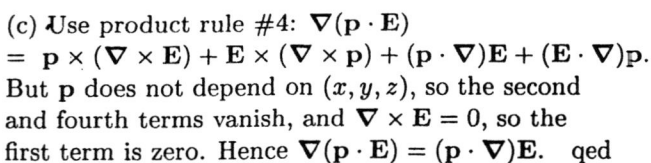

$$\mathbf{F} = m\left\{\begin{vmatrix} \hat{\mathbf{x}} & \hat{\mathbf{y}} & \hat{\mathbf{z}} \\ 0 & 0 & 1 \\ \frac{\partial B_x}{\partial y} & \frac{\partial B_y}{\partial y} & \frac{\partial B_z}{\partial y} \end{vmatrix} - \begin{vmatrix} \hat{\mathbf{x}} & \hat{\mathbf{y}} & \hat{\mathbf{z}} \\ 0 & 1 & 0 \\ \frac{\partial B_x}{\partial z} & \frac{\partial B_y}{\partial z} & \frac{\partial B_z}{\partial z} \end{vmatrix}\right\} = m\left\{\hat{\mathbf{y}}\frac{\partial B_x}{\partial y} - \hat{\mathbf{x}}\frac{\partial B_y}{\partial y} - \hat{\mathbf{x}}\frac{\partial B_z}{\partial z} - \hat{\mathbf{z}}\frac{\partial B_x}{\partial z}\right\}$$
$$= m\left[\hat{\mathbf{x}}\frac{\partial B_x}{\partial x} + \hat{\mathbf{y}}\frac{\partial B_x}{\partial y} + \hat{\mathbf{z}}\frac{\partial B_x}{\partial z}\right] \quad \left(\text{using } \nabla\cdot\mathbf{B} = 0 \text{ to write } \frac{\partial B_y}{\partial y} + \frac{\partial B_z}{\partial z} = -\frac{\partial B_x}{\partial x}\right).$$

But $\mathbf{m}\cdot\mathbf{B} = mB_x$ (since $\mathbf{m} = m\hat{\mathbf{x}}$, here), so $\nabla(\mathbf{m}\cdot\mathbf{B}) = m\nabla(B_x) = m\left(\frac{\partial B_x}{\partial x}\hat{\mathbf{x}} + \frac{\partial B_x}{\partial y}\hat{\mathbf{y}} + \frac{\partial B_x}{\partial z}\hat{\mathbf{z}}\right)$.
Therefore $\mathbf{F} = \nabla(\mathbf{m}\cdot\mathbf{B})$. qed

Problem 6.5

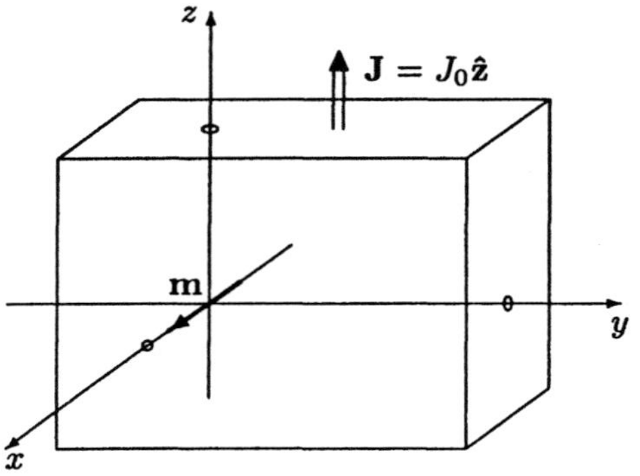

(a) $\mathbf{B} = \mu_0 J_0 x\hat{\mathbf{y}}$ (Prob. 5.14).
$\mathbf{m}\cdot\mathbf{B} = 0$, so Eq. 6.3 says $\boxed{\mathbf{F} = 0.}$

(b) $\mathbf{m}\cdot\mathbf{B} = m_0\mu_0 J_0 x$, so $\boxed{\mathbf{F} = m_0\mu_0 J_0\hat{\mathbf{x}}.}$

(c) Use product rule #4: $\nabla(\mathbf{p}\cdot\mathbf{E})$
$= \mathbf{p}\times(\nabla\times\mathbf{E}) + \mathbf{E}\times(\nabla\times\mathbf{p}) + (\mathbf{p}\cdot\nabla)\mathbf{E} + (\mathbf{E}\cdot\nabla)\mathbf{p}$.
But $\mathbf{p}$ does not depend on (x, y, z), so the second and fourth terms vanish, and $\nabla\times\mathbf{E} = 0$, so the first term is zero. Hence $\nabla(\mathbf{p}\cdot\mathbf{E}) = (\mathbf{p}\cdot\nabla)\mathbf{E}$. qed

This argument does *not* apply to the magnetic analog, since $\nabla\times\mathbf{B} \neq 0$. In fact, $\nabla(\mathbf{m}\cdot\mathbf{B}) = (\mathbf{m}\cdot\nabla)\mathbf{B} + \mu_0(\mathbf{m}\times\mathbf{J})$.
$(\mathbf{m}\cdot\nabla)\mathbf{B}_a = m_0\frac{\partial}{\partial x}(\mathbf{B}) = m_0\mu_0 J_0\hat{\mathbf{y}}$, $(\mathbf{m}\cdot\nabla)\mathbf{B}_b = m_0\frac{\partial}{\partial y}(\mu_0 J_0 x\hat{\mathbf{y}}) = 0$.

Problem 6.6

Aluminum, copper, copper chloride, and sodium all have an *odd* number of electrons, so we expect them to be paramagnetic. The rest (having an even number) should be diamagnetic.

Problem 6.7

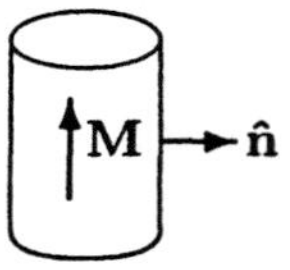

$\mathbf{J}_b = \nabla\times\mathbf{M} = 0$; $\mathbf{K}_b = \mathbf{M}\times\hat{\mathbf{n}} = M\hat{\boldsymbol{\phi}}$.
The field is that of a surface current $\mathbf{K}_b = M\hat{\boldsymbol{\phi}}$, but that's just a solenoid, so the field $\boxed{\text{outside is zero,}}$ and inside $B = \mu_0 K_b = \mu_0 M$. Moreover, it points upward (in the drawing), so $\boxed{\mathbf{B} = \mu_0\mathbf{M}.}$

Problem 6.8

$$\nabla\times\mathbf{M}=\mathbf{J}_b=\frac{1}{s}\frac{\partial}{\partial s}(s\,ks^2)\hat{\mathbf{z}}=\frac{1}{s}(3ks^2)\hat{\mathbf{z}}=3ks\hat{\mathbf{z}},\quad \mathbf{K}_b=\mathbf{M}\times\hat{\mathbf{n}}=ks^2(\hat{\boldsymbol{\phi}}\times\hat{\mathbf{s}})=-kR^2\hat{\mathbf{z}}.$$

So the bound current flows up the cylinder, and returns down the surface. [Incidentally, the *total* current should be zero ... *is* it? *Yes,* for $\int J_b\,da=\int_0^R(3ks)(2\pi s\,ds)=2\pi kR^3$, while $\int K_b\,dl=(-kR^2)(2\pi R)=-2\pi kR^3$.] Since these currents have cylindrical symmetry, we can get the field by Ampère's law:

$$B\cdot 2\pi s=\mu_0 I_{\text{enc}}=\mu_0\int_0^s J_b\,da=2\pi k\mu_0 s^3\Rightarrow\boxed{\mathbf{B}=\mu_0 ks^2\hat{\boldsymbol{\phi}}}=\mu_0\mathbf{M}.$$

Outside the cylinder $I_{\text{enc}}=0$, so $\boxed{\mathbf{B}=0.}$

Problem 6.9

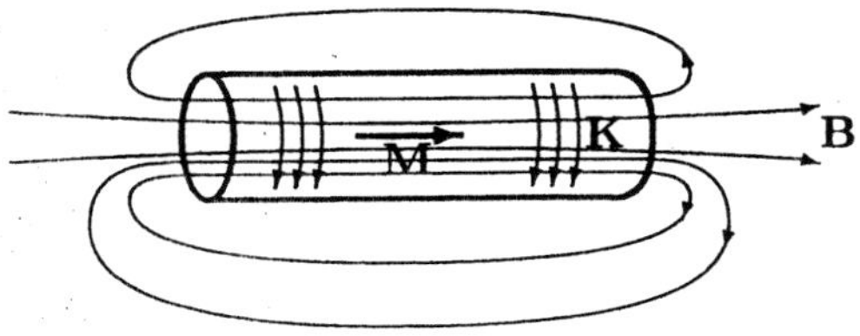

$\boxed{\mathbf{K}_b=\mathbf{M}\times\hat{\mathbf{n}}=M\hat{\boldsymbol{\phi}}.}$
(Essentially a long solenoid)

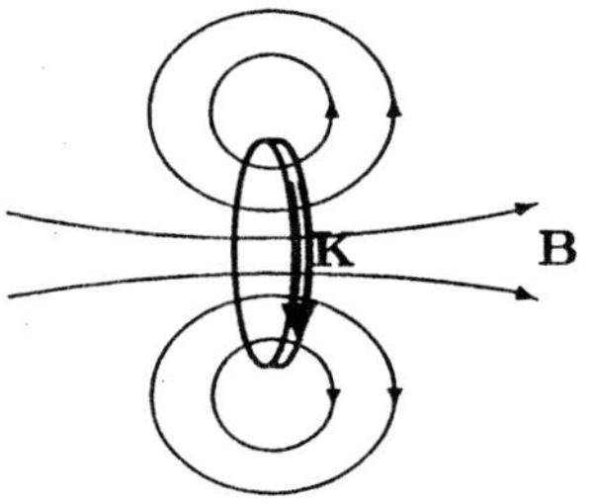

(Essentially a physical dipole)

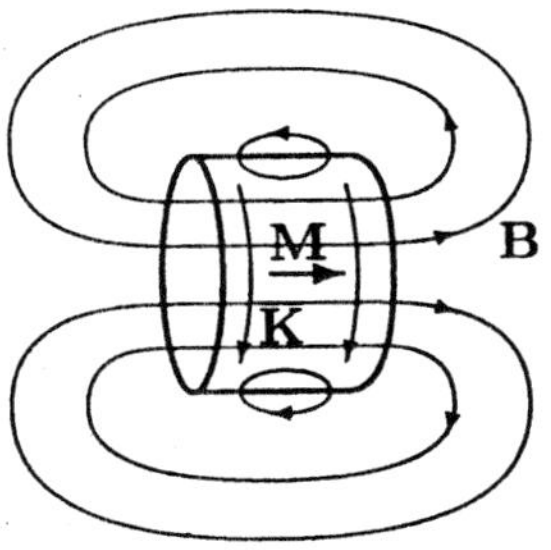

(Intermediate case)
[The external fields are the same as in the electrical case; the *internal* fields (inside the bar) are completely different—in fact, opposite in direction.]

Problem 6.10

$K_b=M$, so the field inside a *complete* ring would be $\mu_0 M$. The field of a square loop, at the center, is given by Prob. 5.8: $B_{\text{sq}}=\sqrt{2}\,\mu_0 I/\pi R$. Here $I=Mw$, and $R=a/2$, so

$$B_{\text{sq}}=\frac{\sqrt{2}\,\mu_0 Mw}{\pi(a/2)}=\frac{2\sqrt{2}\,\mu_0 Mw}{\pi a};\quad\text{net field in gap : }\boxed{\mathbf{B}=\mu_0\mathbf{M}\left(1-\frac{2\sqrt{2}\,w}{\pi a}\right).}$$

Problem 6.11

As in Sec. 4.2.3, we want the average of $\mathbf{B} = \mathbf{B}_{\text{out}} + \mathbf{B}_{\text{in}}$, where $\mathbf{B}_{\text{out}}$ is due to molecules *outside* a small sphere around point P, and $\mathbf{B}_{\text{in}}$ is due to molecules *inside* the sphere. The average of $\mathbf{B}_{\text{out}}$ is same as field at center (Prob. 5.57b), and for this it is OK to use Eq. 6.10, since the center is "far" from all the molecules in question:

$$\mathbf{A}_{\text{out}} = \frac{\mu_0}{4\pi} \int\limits_{\text{outside}} \frac{\mathbf{M} \times \hat{\boldsymbol{\imath}}}{\imath^2} d\tau$$

The average of $\mathbf{B}_{\text{in}}$ is $\frac{\mu_0}{4\pi}\left(\frac{2\mathbf{m}}{R^3}\right)$—Eq. 5.89—where $\mathbf{m} = \frac{4}{3}\pi R^3 \mathbf{M}$. Thus the average $\mathbf{B}_{\text{in}}$ is $2\mu_0\mathbf{M}/3$. But what is *left out* of the integral $\mathbf{A}_{\text{out}}$ is the contribution of a uniformly magnetized sphere, to wit: $2\mu_0\mathbf{M}/3$ (Eq. 6.16), and this is precisely what $\mathbf{B}_{\text{in}}$ puts back in. So we'll get the correct macroscopic field using Eq. 6.10. qed

Problem 6.12

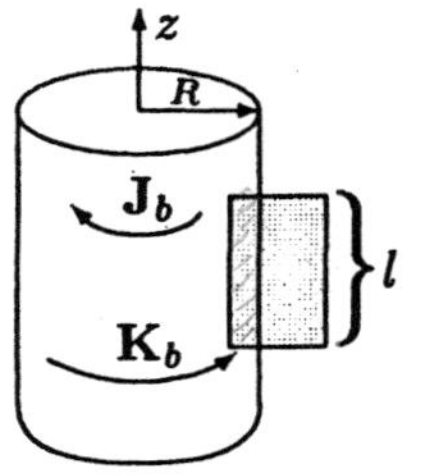

(a) $\mathbf{M} = ks\hat{\mathbf{z}}$; $\mathbf{J}_b = \boldsymbol{\nabla}\times\mathbf{M} = -k\hat{\boldsymbol{\phi}}$; $\mathbf{K}_b = \mathbf{M} \times \hat{\mathbf{n}} = kR\hat{\boldsymbol{\phi}}$.

$\mathbf{B}$ is in the z direction (this is essentially a superposition of solenoids). So $\boxed{\mathbf{B} = 0 \text{ outside.}}$ Use the amperian loop shown (shaded)—inner side at radius s:

$\oint \mathbf{B} \cdot d\mathbf{l} = Bl = \mu_0 I_{\text{enc}} = \mu_0\left[\int J_b\, da + K_b l\right] = \mu_0\left[-kl(R-s) + kRl\right] = \mu_0 kls.$

$\therefore \boxed{\mathbf{B} = \mu_0 ks\hat{\mathbf{z}} \text{ inside.}}$

(b) By symmetry, $\mathbf{H}$ points in the z direction. That same amperian loop gives $\oint \mathbf{H} \cdot d\mathbf{l} = Hl = \mu_0 I_{f_{\text{enc}}} = 0$, since there *is* no free current here. So $\boxed{\mathbf{H} = 0}$, and hence $\boxed{\mathbf{B} = \mu_0\mathbf{M}.}$ *Outside* $\mathbf{M} = 0$, so $\mathbf{B} = 0$; *inside* $\mathbf{M} = ks\hat{\mathbf{z}}$, so $\mathbf{B} = \mu_0 ks\hat{\mathbf{z}}$.

Problem 6.13

(a) The field of a magnetized sphere is $\frac{2}{3}\mu_0\mathbf{M}$ (Eq. 6.16), so $\boxed{\mathbf{B} = \mathbf{B}_0 - \frac{2}{3}\mu_0\mathbf{M},}$ with the sphere removed.

In the cavity, $\mathbf{H} = \frac{1}{\mu_0}\mathbf{B}$, so $\mathbf{H} = \frac{1}{\mu_0}\left(\mathbf{B}_0 - \frac{2}{3}\mu_0\mathbf{M}\right) = \mathbf{H}_0 + \mathbf{M} - \frac{2}{3}\mathbf{M} \Rightarrow \boxed{\mathbf{H} = \mathbf{H}_0 + \frac{1}{3}\mathbf{M}.}$

(b)

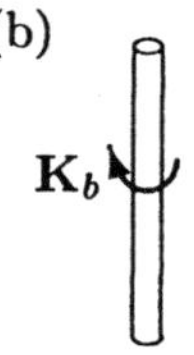

The field inside a long solenoid is $\mu_0 K$. Here $K = M$, so the field of the bound current on the inside surface of the cavity is $\mu_0 M$, pointing *down*. Therefore

$$\boxed{\mathbf{B} = \mathbf{B}_0 - \mu_0\mathbf{M};}$$

$$\mathbf{H} = \frac{1}{\mu_0}(\mathbf{B}_0 - \mu_0\mathbf{M}) = \frac{1}{\mu_0}\mathbf{B}_0 - \mathbf{M} \Rightarrow \boxed{\mathbf{H} = \mathbf{H}_0.}$$

(c)

This time the bound currents are small, and far away from the center, so $\boxed{\mathbf{B} = \mathbf{B}_0,}$ while $\mathbf{H} = \frac{1}{\mu_0}\mathbf{B}_0 = \mathbf{H}_0 + \mathbf{M} \Rightarrow \boxed{\mathbf{H} = \mathbf{H}_0 + \mathbf{M}.}$

[*Comment*: In the wafer, $\mathbf{B}$ is the field in the medium; in the needle, $\mathbf{H}$ is the $\mathbf{H}$ in the medium; in the sphere (intermediate case) both $\mathbf{B}$ and $\mathbf{H}$ are modified.]

Problem 6.14

M: 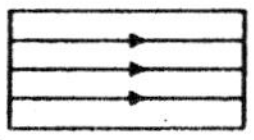 ; **B is the same as the field of a short solenoid;** $\mathbf{H} = \frac{1}{\mu_0}\mathbf{B} - \mathbf{M}$.

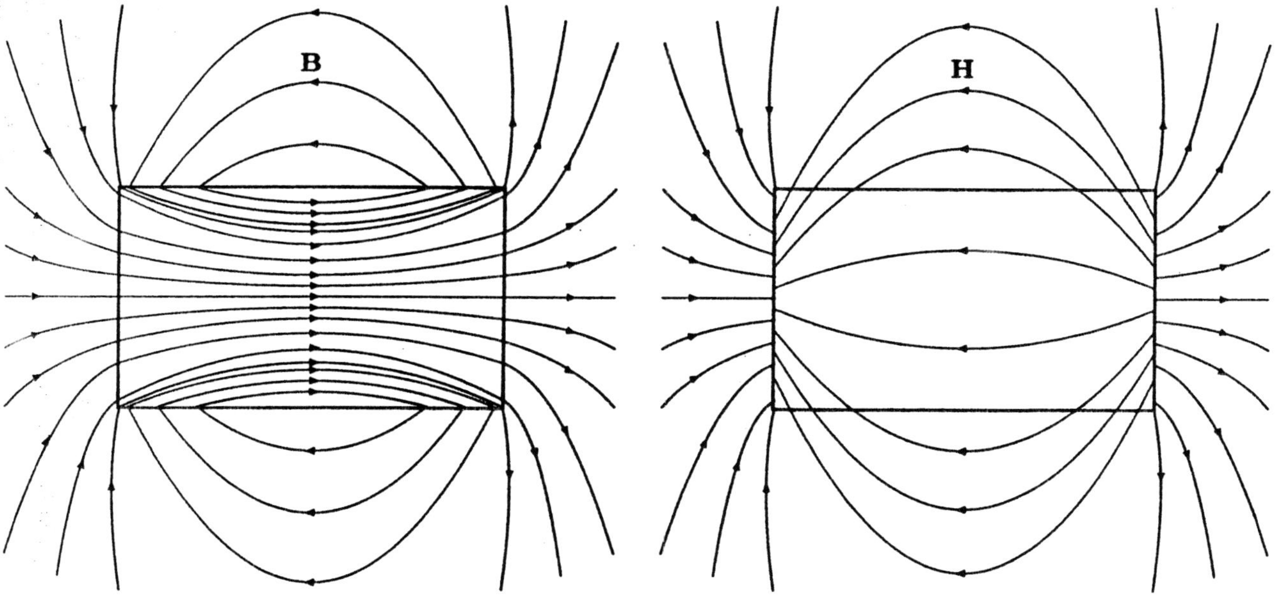

Problem 6.15

"Potentials":

$$\begin{cases} W_{\text{in}}(r,\theta) &= \sum A_l r^l P_l(\cos\theta), \quad (r < R); \\ W_{\text{out}}(r,\theta) &= \sum \frac{B_l}{r^{l+1}} P_l(\cos\theta), \quad (r > R). \end{cases}$$

Boundary Conditions:

$$\begin{cases} \text{(i)} \quad W_{\text{in}}(R,\theta) = W_{\text{out}}(R,\theta), \\ \text{(ii)} \quad -\left.\frac{\partial W_{\text{out}}}{\partial r}\right|_R + \left.\frac{\partial W_{\text{in}}}{\partial r}\right|_R = M^{\perp} = M\,\hat{\mathbf{z}}\cdot\hat{\mathbf{r}} = M\cos\theta. \end{cases}$$

(The continuity of W follows from the gradient theorem: $W(\mathbf{b}) - W(\mathbf{a}) = \int_{\mathbf{a}}^{\mathbf{b}} \nabla W \cdot d\mathbf{l} = -\int_{\mathbf{a}}^{\mathbf{b}} \mathbf{H}\cdot d\mathbf{l}$; if the two points are infinitesimally separated, this last integral $\to 0$.)

$$\begin{cases} \text{(i)} \quad \Rightarrow \quad A_l R^l = \frac{B_l}{R^{l+1}} \quad \Rightarrow \quad B_l = R^{2l+1} A_l, \\ \text{(ii)} \quad \Rightarrow \quad \sum (l+1)\frac{B_l}{R^{l+2}} P_l(\cos\theta) + \sum l A_l R^{l-2} P_l(\cos\theta) = M\cos\theta. \end{cases}$$

Combining these:

$$\sum (2l+1) R^{l-1} A_l P_l(\cos\theta) = M\cos\theta, \text{ so } A_l = 0 \ (l \neq 1), \text{ and } 3A_1 = M \Rightarrow A_1 = \frac{M}{3}.$$

Thus $W_{\text{in}}(r,\theta) = \frac{M}{3} r\cos\theta = \frac{M}{3} z$, and hence $\mathbf{H}_{\text{in}} = -\nabla W_{\text{in}} = -\frac{M}{3}\hat{\mathbf{z}} = -\frac{1}{3}\mathbf{M}$, so

$$\mathbf{B} = \mu_0(\mathbf{H} + \mathbf{M}) = \mu_0\left(-\frac{1}{3}\mathbf{M} + \mathbf{M}\right) = \boxed{\frac{2}{3}\mu_0\mathbf{M}.} \ \checkmark$$

Problem 6.16

$\oint \mathbf{H}\cdot d\mathbf{l} = I_{f_{\rm enc}} = I$, so $\mathbf{H} = \frac{I}{2\pi s}\hat{\boldsymbol{\phi}}$. $\mathbf{B} = \mu_0(1+\chi_m)\mathbf{H} = \boxed{\mu_0(1+\chi_m)\frac{I}{2\pi s}\hat{\boldsymbol{\phi}}.}$ $\mathbf{M} = \chi_m\mathbf{H} = \boxed{\frac{\chi_m I}{2\pi s}\hat{\boldsymbol{\phi}}.}$

$$\mathbf{J}_b = \nabla\times\mathbf{M} = \frac{1}{s}\frac{\partial}{\partial s}\left(s\frac{\chi_m I}{2\pi s}\right)\hat{\mathbf{z}} = \boxed{0.} \quad \mathbf{K}_b = \mathbf{M}\times\hat{\mathbf{n}} = \boxed{\begin{cases} \frac{\chi_m I}{2\pi a}\hat{\mathbf{z}}, & \text{at } s = a; \\ -\frac{\chi_m I}{2\pi b}\hat{\mathbf{z}}, & \text{at } s = b. \end{cases}}$$

Total enclosed current, for an amperian loop between the cylinders:

$$I + \frac{\chi_m I}{2\pi a}2\pi a = (1+\chi_m)I, \quad \text{so} \quad \oint \mathbf{B}\cdot d\mathbf{l} = \mu_0 I_{\rm enc} = \mu_0(1+\chi_m)I \Rightarrow \mathbf{B} = \frac{\mu_0(1+\chi_m)I}{2\pi s}\hat{\boldsymbol{\phi}}. \checkmark$$

Problem 6.17

From Eq. 6.20: $\oint \mathbf{H}\cdot d\mathbf{l} = H(2\pi s) = I_{f_{\rm enc}} = \begin{cases} I(s^2/a^2), & (s<a); \\ I & (s>a). \end{cases}$

$$H = \left\{ \begin{array}{ll} \frac{Is}{2\pi a^2}, & (s<a) \\ \frac{I}{2\pi s}, & (s>a) \end{array} \right\}, \quad \text{so } B = \mu H = \boxed{\begin{cases} \frac{\mu_0(1+\chi_m)Is}{2\pi a^2}, & (s<a); \\ \frac{\mu_0 I}{2\pi s}, & (s>a). \end{cases}}$$

$J_b = \nabla\times M = \nabla\times(\chi_m H) = \chi_m \nabla\times H = \chi_m J_f$

$\mathbf{J}_b = \chi_m\mathbf{J}_f$ (Eq. 6.33), and $J_f = \frac{I}{\pi a^2}$, so $\boxed{J_b = \frac{\chi_m I}{\pi a^2}}$ (same direction as I).

$\mathbf{K}_b = \mathbf{M}\times\hat{\mathbf{n}} = \chi_m\mathbf{H}\times\hat{\mathbf{n}} \Rightarrow \boxed{K_b = \frac{\chi_m I}{2\pi a}}$ (opposite direction to I).

$I_b = J_b(\pi a^2) + K_b(2\pi a) = \chi_m I - \chi_m I = \boxed{0}$ (as it *should* be, of course).

Problem 6.18

By the method of Prob. 6.15:

For large r, we want $\mathbf{B}(r,\theta) \to \mathbf{B}_0 = B_0\,\hat{\mathbf{z}}$, so $\mathbf{H} = \frac{1}{\mu_0}\mathbf{B} \to \frac{1}{\mu_0}B_0\,\hat{\mathbf{z}}$, and hence $W \to -\frac{1}{\mu_0}B_0 z = -\frac{1}{\mu_0}B_0 r\cos\theta$.

"Potentials":

$$\begin{cases} W_{\rm in}(r,\theta) &= \sum A_l r^l P_l(\cos\theta), & (r<R); \\ W_{\rm out}(r,\theta) &= -\frac{1}{\mu_0}B_0 r\cos\theta + \sum \frac{B_l}{r^{l+1}}P_l(\cos\theta), & (r>R). \end{cases}$$

Boundary Conditions:

$$\begin{cases} \text{(i)} & W_{\rm in}(R,\theta) = W_{\rm out}(R,\theta), \\ \text{(ii)} & -\mu_0\frac{\partial W_{\rm out}}{\partial r}\Big|_R + \mu\frac{\partial W_{\rm in}}{\partial r}\Big|_R = 0. \end{cases}$$

(The latter follows from Eq. 6.26.)

$$\text{(ii)} \Rightarrow \mu_0\left[\frac{1}{\mu_0}B_0\cos\theta + \sum(l+1)\frac{B_l}{R^{l+2}}P_l(\cos\theta)\right] + \mu\sum lA_lR^{l-1}P_l(\cos\theta) = 0.$$

For $l \neq 1$, (i) $\Rightarrow B_l = R^{2l+1}A_l$, so $[\mu_0(l+1)+\mu l]A_lR^{l-1} = 0$, and hence $A_l = 0$.
For $l = 1$, (i) $\Rightarrow A_1R = -\frac{1}{\mu_0}B_0R + B_1/R^2$, and (ii) $\Rightarrow B_0 + 2\mu_0B_1/R^3 + \mu A_1 = 0$, so $A_1 = -3B_0/(2\mu_0+\mu)$.

$$W_{\rm in}(r,\theta) = -\frac{3B_0}{(2\mu_0+\mu)}r\cos\theta = -\frac{3B_0 z}{(2\mu_0+\mu)}. \quad \mathbf{H}_{\rm in} = -\nabla W_{\rm in} = \frac{3B_0}{(2\mu_0+\mu)}\hat{\mathbf{z}} = \frac{3\mathbf{B}_0}{(2\mu_0+\mu)}.$$

$$\mathbf{B} = \mu\mathbf{H} = \frac{3\mu\mathbf{B}_0}{(2\mu_0+\mu)} = \boxed{\left(\frac{1+\chi_m}{1+\chi_m/3}\right)\mathbf{B}_0.}$$

By the method of Prob. 4.23:

Step 1: $\mathbf{B}_0$ magnetizes the sphere: $\mathbf{M}_0 = \chi_m \mathbf{H}_0 = \frac{\chi_m}{\mu_0(1+\chi_m)}\mathbf{B}_0$. This magnetization sets up a field within the sphere given by Eq. 6.16:

$$\mathbf{B}_1 = \frac{2}{3}\mu_0 \mathbf{M}_0 = \frac{2}{3}\frac{\chi_m}{1+\chi_m}\mathbf{B}_0 = \frac{2}{3}\kappa \mathbf{B}_0 \quad (\text{where } \kappa \equiv \tfrac{\chi_m}{1+\chi_m}).$$

Step 2: $\mathbf{B}_1$ magnetizes the sphere an additional amount $\mathbf{M}_1 = \frac{\kappa}{\mu_0}\mathbf{B}_1$. This sets up an additional field in the sphere:

$$\mathbf{B}_2 = \frac{2}{3}\mu_0 \mathbf{M}_1 = \frac{2}{3}\kappa \mathbf{B}_1 = \left(\frac{2\kappa}{3}\right)^2 \mathbf{B}_0, \quad \text{etc.}$$

The *total* field is:

$$\mathbf{B} = \mathbf{B}_0 + \mathbf{B}_1 + \mathbf{B}_2 + \cdots = \mathbf{B}_0 + (2\kappa/3)\mathbf{B}_0 + (2\kappa/3)^2\mathbf{B}_0 + \cdots = \left[1 + (2\kappa/3) + (2\kappa/3)^2 + \cdots\right]\mathbf{B}_0 = \frac{\mathbf{B}_0}{(1-2\kappa/3)}.$$

$$\frac{1}{1-2\kappa/3} = \frac{3}{3 - 2\chi_m/(1+\chi_m)} = \frac{3+3\chi_m}{3+3\chi_m-2\chi_m} = \frac{3(1+\chi_m)}{3+\chi_m}, \text{ so } \boxed{\mathbf{B} = \left(\frac{1+\chi_m}{1+\chi_m/3}\right)\mathbf{B}_0.}$$

Problem 6.19

$\Delta \mathbf{m} = -\frac{e^2r^2}{4m_e}\mathbf{B}$; $\mathbf{M} = \frac{\Delta \mathbf{m}}{V} = -\frac{e^2r^2}{4m_eV}\mathbf{B}$, where V is the volume per electron. $\mathbf{M} = \chi_m \mathbf{H}$ (Eq. 6.29) $= \frac{\chi_m}{\mu_0(1+\chi_m)}\mathbf{B}$ (Eq. 6.30). So $\chi_m = -\frac{e^2r^2}{4m_eV}\mu_0$. [*Note*: $\chi_m \ll 1$, so I won't worry about the $(1+\chi_m)$ term; for the same reason we need not distinguish $\mathbf{B}$ from $\mathbf{B}_{\text{else}}$, as we did in deriving the Clausius-Mossotti equation in Prob. 4.38.] Let's say $V = \frac{4}{3}\pi r^3$. Then $\chi_m = -\frac{\mu_0}{4\pi}\left(\frac{3e^2}{4m_e r}\right)$. I'll use 1 Å$= 10^{-10}$ m for r. Then $\chi_m = -(10^{-7})\left(\frac{3(1.6\times10^{-19})^2}{4(9.1\times10^{-31})(10^{-10})}\right) = \boxed{-2\times10^{-5},}$ which is not bad—Table 6.1 says $\chi_m = -1\times10^{-5}$. However, I used only ***one electron*** per atom (copper has 29) and a very crude value for r. Since the orbital radius is smaller for the inner electrons, they count for less ($\Delta m \sim r^2$). I have also neglected competing paramagnetic effects. But never mind ... this is in the right ball park.

Problem 6.20

Place the object in a region of zero magnetic field, and heat it above the Curie point—or simply drop it on a hard surface. If it's delicate (a watch, say), place it between the poles of an electromagnet, and magnetize it back and forth many times; each time you reverse the direction, reduce the field slightly.

Problem 6.21

(a) Identical to Prob. 4.7, only starting with Eqs. 6.1 and 6.3 instead of Eqs. 4.4 and 4.5.

(b) Identical to Prob. 4.8, but starting with Eq. 5.87 instead of 3.104.

(c) $U = -\frac{\mu_0}{4\pi}\frac{1}{r^3}[3\cos\theta_1\cos\theta_2 - \cos(\theta_2-\theta_1)]m_1m_2$. Or, using $\cos(\theta_2-\theta_1) = \cos\theta_1\cos\theta_2 - \sin\theta_1\sin\theta_2$,

$$\boxed{U = \frac{\mu_0}{4\pi}\frac{m_1m_2}{r^3}\left(\sin\theta_1\sin\theta_2 - 2\cos\theta_1\cos\theta_2\right).}$$

Stable position occurs at minimum energy: $\frac{\partial U}{\partial\theta_1} = \frac{\partial U}{\partial\theta_2} = 0$

$$\begin{cases} \frac{\partial U}{\partial\theta_1} = \frac{\mu_0 m_1 m_2}{4\pi r^3}(\cos\theta_1\sin\theta_2 + 2\sin\theta_1\cos\theta_2) = 0 \Rightarrow 2\sin\theta_1\cos\theta_2 = -\cos\theta_1\sin\theta_2; \\ \frac{\partial U}{\partial\theta_2} = \frac{\mu_0 m_1 m_2}{4\pi r^3}(\sin\theta_1\cos\theta_2 + 2\cos\theta_1\sin\theta_2) = 0 \Rightarrow 2\sin\theta_1\cos\theta_2 = -4\cos\theta_1\sin\theta_2. \end{cases}$$

Thus $\sin\theta_1\cos\theta_2 = \sin\theta_2\cos\theta_1 = 0.$ $\begin{cases} \text{Either } \sin\theta_1 = \sin\theta_2 = 0: & \overset{①}{\longrightarrow\longrightarrow} \text{ or } \overset{②}{\longrightarrow\longleftarrow} \\ \text{or } \cos\theta_1 = \cos\theta_2 = 0: & \underset{③}{\uparrow\uparrow} \text{ or } \underset{④}{\uparrow\downarrow} \end{cases}$

Which of these is the *stable* minimum? Certainly not ② or ③—for these $\mathbf{m}_2$ is not parallel to $\mathbf{B}_1$, whereas we know $\mathbf{m}_2$ will line up along $\mathbf{B}_1$. It remains to compare ① (with $\theta_1 = \theta_2 = 0$) and ④ (with $\theta_1 = \pi/2, \theta_2 = -\pi/2$): $U_1 = \frac{\mu_0 m_1 m_2}{4\pi r^3}(-2)$; $U_2 = \frac{\mu_0 m_1 m_2}{4\pi r^3}(-1)$. U_1 is the lower energy, hence the more stable configuration.

Conclusion: They line up parallel, along the line joining them: $\longrightarrow\longrightarrow$

(d) They'd line up the same way: $\longrightarrow\longrightarrow\longrightarrow\longrightarrow\longrightarrow\longrightarrow$

Problem 6.22

$$\mathbf{F} = I\oint d\mathbf{l}\times\mathbf{B} = I\left(\oint d\mathbf{l}\right)\times\mathbf{B}_0 + I\oint d\mathbf{l}\times[(\mathbf{r}\cdot\boldsymbol{\nabla}_0)\mathbf{B}_0] - I\left(\oint d\mathbf{l}\right)\times[(\mathbf{r}_0\cdot\boldsymbol{\nabla}_0)\mathbf{B}_0] = I\oint d\mathbf{l}\times[(\mathbf{r}\cdot\boldsymbol{\nabla}_0)\mathbf{B}_0]$$

(because $\oint d\mathbf{l} = 0$). Now

$$(d\mathbf{l}\times\mathbf{B}_0)_i = \sum_{j,k}\epsilon_{ijk}\,dl_j(B_0)_k, \quad \text{and } (\mathbf{r}\cdot\boldsymbol{\nabla}_0) = \sum_l r_l(\nabla_0)_l, \text{ so}$$

$$\begin{aligned} F_i &= I\sum_{j,k,l}\epsilon_{ijk}\left[\oint r_l\,dl_j\right][(\nabla_0)_l(B_0)_k] \quad \left\{\text{Lemma 1: } \oint r_l\,dl_j = \sum_m \epsilon_{ljm}a_m \text{ (proof below).}\right\} \\ &= I\sum_{j,k,l,m}\epsilon_{ijk}\epsilon_{ljm}a_m(\nabla_0)_l(B_0)_k \quad \left\{\text{Lemma 2: } \sum_j \epsilon_{ijk}\epsilon_{ljm} = \delta_{il}\delta_{km} - \delta_{im}\delta_{kl} \text{ (proof below).}\right\} \\ &= I\sum_{k,l,m}(\delta_{il}\delta_{km} - \delta_{im}\delta_{kl})\,a_m(\nabla_0)_l(B_0)_k = I\sum_k[a_k(\nabla_0)_i(B_0)_k - a_i(\nabla_0)_k(B_0)_k] \\ &= I\left[(\nabla_0)_i(\mathbf{a}\cdot\mathbf{B}_0) - a_i(\boldsymbol{\nabla}_0\cdot\mathbf{B}_0)\right]. \end{aligned}$$

But $\boldsymbol{\nabla}_0\cdot\mathbf{B}_0 = 0$ (Eq. 5.48), and $\mathbf{m} = I\mathbf{a}$ (Eq. 5.84), so $\mathbf{F} = \boldsymbol{\nabla}_0(\mathbf{m}\cdot\mathbf{B}_0)$ (the subscript just reminds us to take the derivatives at the point where $\mathbf{m}$ is located). qed

Proof of Lemma 1:
Eq. 1.108 says $\oint(\mathbf{c}\cdot\mathbf{r})\,d\mathbf{l} = \mathbf{a}\times\mathbf{c} = -\mathbf{c}\times\mathbf{a}$. The jth component is $\sum_p\oint c_p r_p\,dl_j = -\sum_{p,m}\epsilon_{jpm}c_p a_m$. Pick $c_p = \delta_{pl}$ (i.e. 1 for the lth component, zero for the others). Then $\oint r_l\,dl_j = -\sum_m\epsilon_{jlm}a_m = \sum_m\epsilon_{ljm}a_m$. qed

Proof of Lemma 2:
$\epsilon_{ijk}\epsilon_{ljm} = 0$ unless ijk and ljm are both permutations of 123. In particular, i must either be l or m, and k must be the other, so

$$\sum_j\epsilon_{ijk}\epsilon_{ljm} = A\delta_{il}\delta_{km} + B\delta_{im}\delta_{kl}.$$

To determine the constant A, pick $i = l = 1$, $k = m = 3$; the only contribution comes from $j = 2$:

$$\epsilon_{123}\epsilon_{123} = 1 = A\delta_{11}\delta_{33} + B\delta_{13}\delta_{31} = A \Rightarrow A = 1.$$

To determine B, pick $i = m = 1$, $k = l = 3$:

$$\epsilon_{123}\epsilon_{321} = -1 = A\delta_{13}\delta_{31} + B\delta_{11}\delta_{33} = B \Rightarrow B = -1.$$

So

$$\sum_j\epsilon_{ijk}\epsilon_{ljm} = \delta_{il}\delta_{km} - \delta_{im}\delta_{kl}. \quad \text{qed}$$

Problem 6.23

(a) The *electric* field inside a uniformly *polarized* sphere, $\mathbf{E} = -\frac{1}{3\epsilon_0}\mathbf{P}$ (Eq. 4.14) translates to $\mathbf{H} = -\frac{1}{3\mu_0}(\mu_0\mathbf{M}) = -\frac{1}{3}\mathbf{M}$. But $\mathbf{B} = \mu_0(\mathbf{H}+\mathbf{M})$. So the *magnetic* field inside a uniformly *magnetized* sphere is $\mathbf{B} = \mu_0(-\frac{1}{3}\mathbf{M}+\mathbf{M}) = \boxed{\frac{2}{3}\mu_0\mathbf{M}}$ (same as Eq. 6.16).

(b) The *electric* field inside a sphere of linear *dielectric* in an otherwise uniform *electric* field is $\mathbf{E} = \frac{1}{1+\chi_e/3}\mathbf{E}_0$ (Eq. 4.49). Now χ_e translates to χ_m, for then Eq. 4.30 ($\mathbf{P} = \epsilon_0\chi_e\mathbf{E}$) goes to $\mu_0\mathbf{M} = \mu_0\chi_m\mathbf{H}$, or $\mathbf{M} = \chi_m\mathbf{H}$ (Eq. 6.29). So Eq. 4.49 $\Rightarrow$ $\mathbf{H} = \frac{1}{1+\chi_m/3}\mathbf{H}_0$. But $\mathbf{B} = \mu_0(1+\chi_m)\mathbf{H}$, and $\mathbf{B}_0 = \mu_0\mathbf{H}_0$ (Eqs. 6.31 and 6.32), so the *magnetic* field inside a sphere of linear *magnetic* material in an otherwise uniform *magnetic* field is

$$\frac{\mathbf{B}}{\mu_0(1+\chi_m)} = \frac{1}{(1+\chi_m/3)}\frac{\mathbf{B}_0}{\mu_0}, \text{ or } \boxed{\mathbf{B} = \left(\frac{1+\chi_m}{1+\chi_m/3}\right)\mathbf{B}_0} \text{ (as in Prob. 6.18).}$$

(c) The average *electric* field over a sphere, due to charges within, is $\mathbf{E}_{\text{ave}} = -\frac{1}{4\pi\epsilon_0}\frac{\mathbf{p}}{R^3}$. Let's pretend the charges are all due to the frozen-in polarization of some medium (whatever ρ might be, we can solve $\nabla\cdot\mathbf{P} = -\rho$ to find the appropriate $\mathbf{P}$). In this case there are *no* free charges, and $\mathbf{p} = \int\mathbf{P}\,d\tau$, so $\mathbf{E}_{\text{ave}} = -\frac{1}{4\pi\epsilon_0}\frac{1}{R^3}\int\mathbf{P}\,d\tau$, which translates to

$$\mathbf{H}_{\text{ave}} = -\frac{1}{4\pi\mu_0}\frac{1}{R^3}\int\mu_0\mathbf{M}\,d\tau = -\frac{1}{4\pi R^3}\mathbf{m}.$$

But $\mathbf{B} = \mu_0(\mathbf{H}+\mathbf{M})$, so $\mathbf{B}_{\text{ave}} = -\frac{\mu_0}{4\pi}\frac{\mathbf{m}}{R^3} + \mu_0\mathbf{M}_{\text{ave}}$, and $\mathbf{M}_{\text{ave}} = \frac{\mathbf{m}}{\frac{4}{3}\pi R^3}$, so $\boxed{\mathbf{B}_{\text{ave}} = \frac{\mu_0}{4\pi}\frac{2\mathbf{m}}{R^3}}$, in agreement with Eq. 5.89. (We must assume for this argument that all the currents are *bound*, but again it doesn't really matter, since we can model any current configuration by an appropriate frozen-in magnetization. See G. H. Goedecke, *Am. J. Phys.* **66**, 1010 (1998).)

Problem 6.24

$$\begin{aligned}
\text{Eq. 2.15: } \quad \mathbf{E} &= \rho\left\{\frac{1}{4\pi\epsilon_0}\int_{\mathcal{V}}\frac{\hat{\boldsymbol{\mathscr{r}}}}{\mathscr{r}^2}\,d\tau'\right\} && \text{(for uniform charge density);}\\
\text{Eq. 4.9: } \quad V &= \mathbf{P}\cdot\left\{\frac{1}{4\pi\epsilon_0}\int_{\mathcal{V}}\frac{\hat{\boldsymbol{\mathscr{r}}}}{\mathscr{r}^2}\,d\tau'\right\} && \text{(for uniform polarization);}\\
\text{Eq. 6.11: } \quad \mathbf{A} &= \mu_0\epsilon_0\mathbf{M}\times\left\{\frac{1}{4\pi\epsilon_0}\int_{\mathcal{V}}\frac{\hat{\boldsymbol{\mathscr{r}}}}{\mathscr{r}^2}\,d\tau'\right\} && \text{(for uniform magnetization).}
\end{aligned}$$

For a uniformly charged sphere (radius R): $\begin{cases}\mathbf{E}_{\text{in}} = \rho\left(\frac{1}{3\epsilon_0}\mathbf{r}\right) & \text{(Prob. 2.12),}\\ \mathbf{E}_{\text{out}} = \rho\left(\frac{1}{3\epsilon_0}\frac{R^3}{r^2}\hat{\mathbf{r}}\right) & \text{(Ex. 2.2).}\end{cases}$

So the scalar potential of a uniformly polarized sphere is: $\begin{cases}V_{\text{in}} = \frac{1}{3\epsilon_0}(\mathbf{P}\cdot\mathbf{r}),\\ V_{\text{out}} = \frac{1}{3\epsilon_0}\frac{R^3}{r^2}(\mathbf{P}\cdot\hat{\mathbf{r}}),\end{cases}$

and the vector potential of a uniformly magnetized sphere is: $\begin{cases}\mathbf{A}_{\text{in}} = \frac{\mu_0}{3}(\mathbf{M}\times\mathbf{r}),\\ \mathbf{A}_{\text{out}} = \frac{\mu_0}{3}\frac{R^3}{r^2}(\mathbf{M}\times\hat{\mathbf{r}}),\end{cases}$

(confirming the results of Ex. 4.2 and of Exs. 6.1 and 5.11).

Problem 6.25

(a) $\mathbf{B}_1 = \frac{\mu_0}{4\pi}\frac{2m}{z^3}\hat{\mathbf{z}}$ (Eq. 5.86, with $\theta = 0$). So $\mathbf{m}_2\cdot\mathbf{B}_1 = -\frac{\mu_0}{2\pi}\frac{m^2}{z^3}$. $\mathbf{F} = \nabla(\mathbf{m}\cdot\mathbf{B})$ (Eq. 6.3) $\Rightarrow$ $\mathbf{F} = \frac{\partial}{\partial z}\left[-\frac{\mu_0}{2\pi}\frac{m^2}{z^3}\right]\hat{\mathbf{z}} = \frac{3\mu_0 m^2}{2\pi z^4}\hat{\mathbf{z}}$. This is the magnetic force *upward* (on the upper magnet); it balances the gravitational force downward ($-m_d g\hat{\mathbf{z}}$):

$$\frac{3\mu_0 m^2}{2\pi z^4} - m_d g = 0 \Rightarrow \boxed{z = \left[\frac{3\mu_0 m^2}{2\pi m_d g}\right]^{1/4}}.$$

(b) The middle magnet is repelled *upward* by lower magnet and *downward* by upper magnet:

$$\frac{3\mu_0 m^2}{2\pi x^4} - \frac{3\mu_0 m^2}{2\pi y^4} - m_d g = 0.$$

The top magnet is repelled *upward* by middle magnet, and attracted *downward* by lower magnet:

$$\frac{3\mu_0 m^2}{2\pi y^4} - \frac{3\mu_0 m^2}{2\pi (x+y)^4} - m_d g = 0.$$

Subtracting: $\frac{3\mu_0 m^2}{2\pi}\left[\frac{1}{x^4} - \frac{1}{y^4} - \frac{1}{y^4} + \frac{1}{(x+y)^4}\right] - m_d g + m_d g = 0$, or $\frac{1}{x^4} - \frac{2}{y^4} + \frac{1}{(x+y)^4} = 0$, so: $2 = \frac{1}{(x/y)^4} + \frac{1}{(x/y+1)^4}$.
Let $\alpha \equiv x/y$; then $2 = \frac{1}{\alpha^4} + \frac{1}{(\alpha+1)^4}$. Mathematica gives the numerical solution $\alpha = \boxed{x/y = 0.850115\ldots}$

Problem 6.26

At the interface, the perpendicular component of $\mathbf{B}$ is continuous (Eq. 6.26), and the parallel component of $\mathbf{H}$ is continuous (Eq. 6.25 with $\mathbf{K}_f = 0$). So $B_1^{\perp} = B_2^{\perp}$, $\mathbf{H}_1^{\parallel} = \mathbf{H}_2^{\parallel}$. But $\mathbf{B} = \mu\mathbf{H}$ (Eq. 6.31), so $\frac{1}{\mu_1}\mathbf{B}_1^{\parallel} = \frac{1}{\mu_2}\mathbf{B}_2^{\parallel}$. Now $\tan\theta_1 = B_1^{\parallel}/B_1^{\perp}$, and $\tan\theta_2 = B_2^{\parallel}/B_2^{\perp}$, so

$$\frac{\tan\theta_2}{\tan\theta_1} = \frac{B_2^{\parallel}}{B_2^{\perp}}\frac{B_1^{\perp}}{B_1^{\parallel}} = \frac{B_2^{\parallel}}{B_1^{\parallel}} = \frac{\mu_2}{\mu_1}$$

(the same form, though for different reasons, as Eq. 4.68).

Problem 6.27

In view of Eq. 6.33, there is a *bound* dipole at the center: $\mathbf{m}_b = \chi_m\mathbf{m}$. So the *net* dipole moment at the center is $\mathbf{m}_{\text{center}} = \mathbf{m} + \mathbf{m}_b = (1+\chi_m)\mathbf{m} = \frac{\mu}{\mu_0}\mathbf{m}$. This produces a field given by Eq. 5.87:

$$\mathbf{B}_{\substack{\text{center}\\\text{dipole}}} = \frac{\mu}{4\pi}\frac{1}{r^3}\left[3(\mathbf{m}\cdot\hat{\mathbf{r}})\hat{\mathbf{r}} - \mathbf{m}\right].$$

This accounts for the *first* term in the field. The remainder must be due to the bound surface current ($\mathbf{K}_b$) at $r = R$ (since there can be no volume bound current, according to Eq. 6.33). Let us make an educated guess (based either on the answer provided or on the analogous electrical Prob. 4.34) that the field due to the surface bound current is (for interior points) of the form $\mathbf{B}_{\substack{\text{surface}\\\text{current}}} = A\mathbf{m}$ (i.e. a constant, proportional to $\mathbf{m}$). In that case the magnetization will be:

$$\mathbf{M} = \chi_m\mathbf{H} = \frac{\chi_m}{\mu}\mathbf{B} = \frac{\chi_m}{4\pi}\frac{1}{r^3}\left[3(\mathbf{m}\cdot\hat{\mathbf{r}})\hat{\mathbf{r}} - \mathbf{m}\right] + \frac{\chi_m}{\mu}A\mathbf{m}.$$

This will produce bound currents $\mathbf{J}_b = \nabla\times\mathbf{M} = 0$, as it should, for $0 < r < R$ (no need to *calculate* this curl—the second term is *constant*, and the first is essentially the field of a dipole, which we know is curl-less, except at $r = 0$), and

$$\mathbf{K}_b = \mathbf{M}(R)\times\hat{\mathbf{r}} = \frac{\chi_m}{4\pi R^3}(-\mathbf{m}\times\hat{\mathbf{r}}) + \frac{\chi_m A}{\mu}(\mathbf{m}\times\hat{\mathbf{r}}) = \chi_m m\left(-\frac{1}{4\pi R^3} + \frac{A}{\mu}\right)\sin\theta\,\hat{\boldsymbol{\phi}}.$$

But this is exactly the surface current produced by a spinning sphere: $\mathbf{K} = \sigma\mathbf{v} = \sigma\omega R\sin\theta\,\hat{\boldsymbol{\phi}}$, with $(\sigma\omega R) \leftrightarrow \chi_m m\left(\frac{A}{\mu} - \frac{1}{4\pi R^3}\right)$. So the field it produces (for points inside) is (Eq. 5.68):

$$\mathbf{B}_{\substack{\text{surface}\\\text{current}}} = \frac{2}{3}\mu_0(\sigma\omega R) = \frac{2}{3}\mu_0\chi_m\mathbf{m}\left(\frac{A}{\mu} - \frac{1}{4\pi R^3}\right).$$

Everything is consistent, therefore, provided $A = \frac{2}{3}\mu_0\chi_m\left(\frac{A}{\mu} - \frac{1}{4\pi R^3}\right)$, or $A\left(1 - \frac{2\mu_0}{3\mu}\chi_m\right) = -\frac{2}{3}\frac{\mu_0\chi_m}{4\pi R^3}$. But $\chi_m = \left(\frac{\mu}{\mu_0}\right) - 1$, so $A\left(1 - \frac{2}{3} + \frac{2}{3}\frac{\mu_0}{\mu}\right) = -\frac{2}{3}\frac{(\mu-\mu_0)}{4\pi R^3}$, or $A\left(1 + \frac{2\mu_0}{\mu}\right) = 2\frac{(\mu_0-\mu)}{4\pi R^3}$; $A = \frac{\mu}{4\pi}\frac{2(\mu_0-\mu)}{R^3(2\mu_0+\mu)}$, and hence

$$\mathbf{B} = \frac{\mu}{4\pi}\left\{\frac{1}{r^3}[3(\mathbf{m}\cdot\hat{\mathbf{r}})\hat{\mathbf{r}} - \mathbf{m}] + \frac{2(\mu_0 - \mu)\mathbf{m}}{R^3(2\mu_0 + \mu)}\right\}. \quad \text{qed}$$

The *exterior* field is that of the central dipole plus that of the surface current, which, according to Prob. 5.36, is *also* a perfect dipole field, of dipole moment

$$\mathbf{m}_{\substack{\text{surface}\\\text{current}}} = \frac{4}{3}\pi R^3(\sigma\omega R) = \frac{4}{3}\pi R^3\left(\frac{3}{2\mu_0}\mathbf{B}_{\substack{\text{surface}\\\text{current}}}\right) = \frac{2\pi R^3}{\mu_0}\frac{\mu}{4\pi}\frac{2(\mu_0 - \mu)\mathbf{m}}{R^3(2\mu_0 + \mu)} = \frac{\mu(\mu_0 - \mu)\mathbf{m}}{\mu_0(2\mu_0 + \mu)}.$$

So the *total* dipole moment is:

$$\mathbf{m}_{\text{tot}} = \frac{\mu}{\mu_0}\mathbf{m} + \frac{\mu}{\mu_0}\mathbf{m}\frac{(\mu_0 - \mu)}{(2\mu_0 + \mu)} = \frac{3\mu\mathbf{m}}{(2\mu_0 + \mu)},$$

and hence the field (for $r > R$) is

$$\boxed{\mathbf{B} = \frac{\mu_0}{4\pi}\left(\frac{3\mu}{2\mu_0 + \mu}\right)\frac{1}{r^3}[3(\mathbf{m}\cdot\hat{\mathbf{r}})\hat{\mathbf{r}} - \mathbf{m}].}$$

Problem 6.28

The problem is that the field inside a *cavity* is not the same as the field in the material itself.

(a) *Ampére type.* The field deep inside the magnet is that of a long solenoid, $\mathbf{B}_0 \approx \mu_0\mathbf{M}$. From Prob. 6.13:

$$\begin{cases} \text{Sphere}: & \mathbf{B} = \mathbf{B}_0 - \frac{2}{3}\mu_0\mathbf{M} = \frac{1}{3}\mu_0\mathbf{M}; \\ \text{Needle}: & \mathbf{B} = \mathbf{B}_0 - \mu_0\mathbf{M} = 0; \\ \text{Wafer}: & \mathbf{B} = \mu_0\mathbf{M}. \end{cases}$$

(b) *Gilbert type.* This is analogous to the *electric* case. The field at the center is approximately that midway between two distant point charges, $\mathbf{B}_0 \approx 0$. From Prob. 4.16 (with $\mathbf{E} \to \mathbf{B}$, $1/\epsilon_0 \to \mu_0$, $\mathbf{P} \to \mathbf{M}$):

$$\begin{cases} \text{Sphere}: & \mathbf{B} = \mathbf{B}_0 + \frac{\mu_0}{3}\mathbf{M} = \frac{1}{3}\mu_0\mathbf{M}; \\ \text{Needle}: & \mathbf{B} = \mathbf{B}_0 = 0; \\ \text{Wafer}: & \mathbf{B} = \mathbf{B}_0 + \mu_0\mathbf{M} = \mu_0\mathbf{M}. \end{cases}$$

In the cavities, then, the fields are the *same* for the two models, and this will be no test at all. $\boxed{\text{Yes.}}$ Fund it with \$1 M from the Office of Alternative Medicine.

Chapter 7

Electrodynamics

Problem 7.1

(a) Let Q be the charge on the inner shell. Then $\mathbf{E} = \frac{1}{4\pi\epsilon_0}\frac{Q}{r^2}\hat{\mathbf{r}}$ in the space between them, and $(V_a - V_b) = -\int_b^a \mathbf{E}\cdot d\mathbf{r} = -\frac{1}{4\pi\epsilon_0}Q\int_b^a \frac{1}{r^2}dr = \frac{Q}{4\pi\epsilon_0}\left(\frac{1}{a}-\frac{1}{b}\right)$.

$$I = \int \mathbf{J}\cdot d\mathbf{a} = \sigma\int \mathbf{E}\cdot d\mathbf{a} = \sigma\frac{Q}{\epsilon_0} = \frac{\sigma}{\epsilon_0}\frac{4\pi\epsilon_0(V_a - V_b)}{(1/a - 1/b)} = \boxed{4\pi\sigma\frac{(V_a - V_b)}{(1/a - 1/b)}.}$$

(b) $R = \frac{V_a - V_b}{I} = \boxed{\frac{1}{4\pi\sigma}\left(\frac{1}{a}-\frac{1}{b}\right).}$

(c) For large b ($b \gg a$), the second term is negligible, and $R = 1/4\pi\sigma a$. Essentially all of the resistance is in the region right around the inner sphere. Successive shells, as you go out, contribute less and less, because the cross-sectional area ($4\pi r^2$) gets larger and larger. For the two submerged spheres, $R = \frac{2}{4\pi\sigma a} = \frac{1}{2\pi\sigma a}$ (one R as the current leaves the first, one R as it converges on the second). Therefore $I = V/R = \boxed{2\pi\sigma a V.}$

Problem 7.2

(a) $V = Q/C = IR$. Because positive I means the charge on the capacitor is *decreasing*, $\frac{dQ}{dt} = -I = -\frac{1}{RC}Q$, so $Q(t) = Q_0 e^{-t/RC}$. But $Q_0 = Q(0) = CV_0$, so $\boxed{Q(t) = CV_0 e^{-t/RC}.}$

Hence $I(t) = -\frac{dQ}{dt} = CV_0\frac{1}{RC}e^{-t/RC} = \boxed{\frac{V_0}{R}e^{-t/RC}.}$

(b) $W = \boxed{\frac{1}{2}cV_0^2.}$ The energy delivered to the resistor is $\int_0^\infty P\,dt = \int_0^\infty I^2 R\,dt = \frac{V_0^2}{R}\int_0^\infty e^{-2t/RC}dt = \frac{V_0^2}{R}\left(-\frac{RC}{2}e^{-2t/RC}\right)\Big|_0^\infty = \frac{1}{2}CV_0^2.$ ✓

(c) $V_0 = Q/C + IR$. This time positive I means Q is *increasing*: $\frac{dQ}{dt} = I = \frac{1}{RC}(CV_0 - Q) \Rightarrow \frac{dQ}{Q - CV_0} = -\frac{1}{RC}dt \Rightarrow \ln(Q - CV_0) = -\frac{1}{RC}t + \text{constant} \Rightarrow Q(t) = CV_0 + ke^{-t/RC}$. But $Q(0) = 0 \Rightarrow k = -CV_0$, so $\boxed{Q(t) = CV_0\left(1 - e^{-t/RC}\right).}$ $I(t) = \frac{dQ}{dt} = CV_0\left(\frac{1}{RC}e^{-t/RC}\right) = \boxed{\frac{V_0}{R}e^{-t/RC}.}$

(d) Energy from battery: $\int_0^\infty V_0 I\,dt = \frac{V_0^2}{R}\int_0^\infty e^{-t/RC}dt = \frac{V_0^2}{R}\left(-RCe^{-t/RC}\right)\Big|_0^\infty = \frac{V_0^2}{R}RC = \boxed{CV_0^2.}$

Since $I(t)$ is the same as in (a), the energy delivered to the resistor is again $\boxed{\frac{1}{2}CV_0^2.}$ The final energy in the capacitor is also $\boxed{\frac{1}{2}CV_0^2,}$ so $\boxed{\text{half}}$ the energy from the battery goes to the capacitor, and the other half to the resistor.

Problem 7.3

(a) $I = \int \mathbf{J}\cdot d\mathbf{a}$, where the integral is taken over a surface enclosing the positively charged conductor. But $\mathbf{J} = \sigma\mathbf{E}$, and Gauss's law says $\int \mathbf{E}\cdot d\mathbf{a} = \frac{1}{\epsilon_0}Q$, so $I = \sigma\int \mathbf{E}\cdot d\mathbf{a} = \frac{\sigma}{\epsilon_0}Q$. But $Q = CV$, and $V = IR$, so $I = \frac{\sigma}{\epsilon_0}CIR$, or $\boxed{R = \frac{\epsilon_0}{\sigma C}.}$ qed

(b) $Q = CV = CIR \Rightarrow \frac{dQ}{dt} = -I = -\frac{1}{RC}Q \Rightarrow \boxed{Q(t) = Q_0e^{-t/RC}}$, or, since $V = Q/C$, $V(t) = V_0e^{-t/RC}$. The time constant is $\tau = RC = \boxed{\epsilon_0/\sigma.}$

Problem 7.4

$I = J(s)\,2\pi sL \;\Rightarrow\; J(s) = I/2\pi sL.\quad E = J/\sigma = I/2\pi s\sigma L = I/2\pi kL.$

$V = -\int_b^a \mathbf{E}\cdot d\mathbf{l} = -\frac{I}{2\pi kL}(a-b).$ So $\boxed{R = \frac{b-a}{2\pi kL}.}$

Problem 7.5

$$I = \frac{\mathcal{E}}{r+R};\; P = I^2R = \frac{\mathcal{E}^2R}{(r+R)^2};\; \frac{dP}{dR} = \mathcal{E}^2\left[\frac{1}{(r+R)^2} - \frac{2R}{(r+R)^3}\right] = 0 \Rightarrow r+R = 2R \Rightarrow \boxed{R = r.}$$

Problem 7.6

$\mathcal{E} = \oint \mathbf{E}\cdot d\mathbf{l} = \boxed{\text{zero}}$ for *all* electrostatic fields. It *looks* as though $\mathcal{E} = \oint \mathbf{E}\cdot d\mathbf{l} = (\sigma/\epsilon_0)h$, as would indeed be the case if the field were really just σ/ϵ_0 inside and zero outside. But in fact there is always a "fringing field" at the edges (Fig. 4.31), and this is evidently just right to kill off the contribution from the left end of the loop. The current is $\boxed{\text{zero.}}$

Problem 7.7

(a) $\mathcal{E} = -\frac{d\Phi}{dt} = -Bl\frac{dx}{dt} = -Blv$; $\mathcal{E} = IR \Rightarrow \boxed{I = \frac{Blv}{R}.}$ (Never mind the minus sign—it just tells you the *direction* of flow: $(\mathbf{v}\times\mathbf{B})$ is *upward*, in the bar, so *downward* through the resistor.)

(b) $F = IlB = \boxed{\frac{B^2l^2v}{R},}$ to the $\boxed{\text{left.}}$

(c) $F = ma = m\frac{dv}{dt} = -\frac{B^2l^2}{R}v \Rightarrow \frac{dv}{dt} = -(\frac{B^2l^2}{Rm})v \Rightarrow \boxed{v = v_0e^{-\frac{B^2l^2}{mR}t}.}$

(d) The energy goes into heat in the resistor. The power delivered to resistor is I^2R, so

$$\frac{dW}{dt} = I^2R = \frac{B^2l^2v^2}{R^2}R = \frac{B^2l^2}{R}v_0^2e^{-2\alpha t},\text{ where } \alpha \equiv \frac{B^2l^2}{mR};\quad \frac{dW}{dt} = \alpha mv_0^2e^{-2\alpha t}.$$

The total energy delivered to the resistor is $W = \alpha mv_0^2\int_0^\infty e^{-2\alpha t}dt = \alpha mv_0^2\left.\frac{e^{-2\alpha t}}{-2\alpha}\right|_0^\infty = \alpha mv_0^2\frac{1}{2\alpha} = \frac{1}{2}mv_0^2.$ ✓

ln(s+a) - ln(s)

Problem 7.8

(a) The field of long wire is $\mathbf{B} = \frac{\mu_0 I}{2\pi s}\hat{\boldsymbol{\phi}}$, so $\Phi = \int \mathbf{B}\cdot d\mathbf{a} = \frac{\mu_0 I}{2\pi}\int_s^{s+a} \frac{1}{s}(a\,ds) = \boxed{\frac{\mu_0 I a}{2\pi}\ln\left(\frac{s+a}{s}\right)}.$

(b) $\mathcal{E} = -\frac{d\Phi}{dt} = -\frac{\mu_0 I a}{2\pi}\frac{d}{dt}\ln\left(\frac{s+a}{s}\right)$, and $\frac{ds}{dt} = v$, so $-\frac{\mu_0 I a}{2\pi}\left(\frac{1}{s+a}\frac{ds}{dt} - \frac{1}{s}\frac{ds}{dt}\right) = \boxed{\frac{\mu_0 I a^2 v}{2\pi s(s+a)}}.$

The field points *out* of the page, so the force on a charge in the nearby side of the square is *to the right*. In the far side it's also to the right, but here the field is weaker, so the current flows $\boxed{\text{counterclockwise.}}$

(c) This time the flux is *constant*, so $\boxed{\mathcal{E} = 0.}$

Problem 7.9

Since $\nabla\cdot\mathbf{B} = 0$, Theorem 2(c) (Sect. 1.6.2) guarantees that $\int \mathbf{B}\cdot d\mathbf{a}$ is the same for *all* surfaces with a given boundary line.

Problem 7.10

$\Phi = \mathbf{B}\cdot\mathbf{a} = Ba^2\cos\theta$

Here $\theta = \omega t$, so

$\mathcal{E} = -\frac{d\Phi}{dt} = -Ba^2(-\sin\omega t)\omega;$

$\boxed{\mathcal{E} = B\omega a^2\sin\omega t.}$

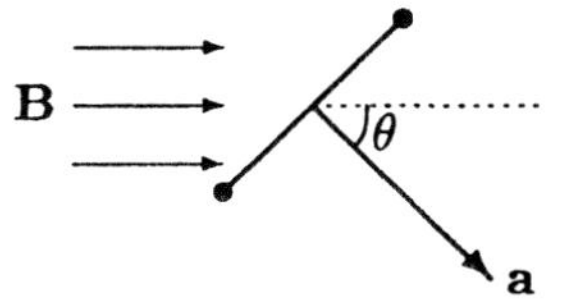

(view from above)

Problem 7.11

$\mathcal{E} = Blv = IR \Rightarrow I = \frac{Bl}{R}v \Rightarrow$ upward magnetic force $= IlB = \frac{B^2l^2}{R}v$. This opposes the gravitational force downward:

$$mg - \frac{B^2l^2}{R}v = m\frac{dv}{dt};\quad \frac{dv}{dt} = g - \alpha v,\ \text{where } \alpha \equiv \frac{B^2l^2}{mR}.\quad g - \alpha v_t = 0 \Rightarrow v_t = \frac{g}{\alpha} = \boxed{\frac{mgR}{B^2l^2}}.$$

$$\frac{dv}{g - \alpha v} = dt \Rightarrow -\frac{1}{\alpha}\ln(g - \alpha v) = t + \text{const.} \Rightarrow g - \alpha v = Ae^{-\alpha t};\ \text{at } t = 0,\ v = 0,\ \text{so } A = g.$$

$$\alpha v = g(1 - e^{-\alpha t});\quad v = \frac{g}{\alpha}(1 - e^{-\alpha t}) = \boxed{v_t(1 - e^{-\alpha t}).}$$

At 90% of terminal velocity, $v/v_t = 0.9 = 1 - e^{-\alpha t} \Rightarrow e^{-\alpha t} = 1 - 0.9 = 0.1;\ \ln(0.1) = -\alpha t;\ \ln 10 = \alpha t;$

$t = \frac{1}{\alpha}\ln 10$, or $\boxed{t_{90\%} = \frac{v_t}{g}\ln 10.}$

Now the numbers: $m = 4\eta A l$, where η is the mass density of aluminum, A is the cross-sectional area, and l is the length of a side. $R = 4l/A\sigma$, where σ is the conductivity of aluminum. So

$$v_t = \frac{4\eta A l g 4l}{A\sigma B^2 l^2} = \frac{16\eta g}{\sigma B^2} = \frac{16 g\eta\rho}{B^2},\ \text{and}\ \left\{\begin{array}{l}\rho = 2.8\times10^{-8}\,\Omega\,\text{m}\\ g = 9.8\,\text{m/s}^2\\ \eta = 2.7\times10^3\,\text{kg/m}^3\\ B = 1\,\text{T}\end{array}\right\}.$$

So $v_t = \frac{(16)(9.8)(2.7\times10^3)(2.8\times10^{-8})}{1} = \boxed{1.2\text{ cm/s};}\ t_{90\%} = \frac{1.2\times10^{-2}}{9.8}\ln(10) = \boxed{2.8\text{ ms.}}$

$\boxed{\text{If the loop were cut, it would fall freely, with acceleration } g.}$

Problem 7.12

$$\Phi = \pi\left(\frac{a}{2}\right)^2 B = \frac{\pi a^2}{4} B_0\cos(\omega t);\ \mathcal{E} = -\frac{d\Phi}{dt} = \frac{\pi a^2}{4} B_0\omega\sin(\omega t).\quad I(t) = \frac{\mathcal{E}}{R} = \boxed{\frac{\pi a^2\omega}{4R} B_0\sin(\omega t).}$$

Problem 7.13

$$\Phi = \int B\,dx\,dy = kt^2\int_0^a dx\int_0^a y^3\,dy = \frac{1}{4}kt^2a^5.\quad \mathcal{E} = -\frac{d\Phi}{dt} = -\boxed{\tfrac{1}{2}kta^5.}$$

Problem 7.14

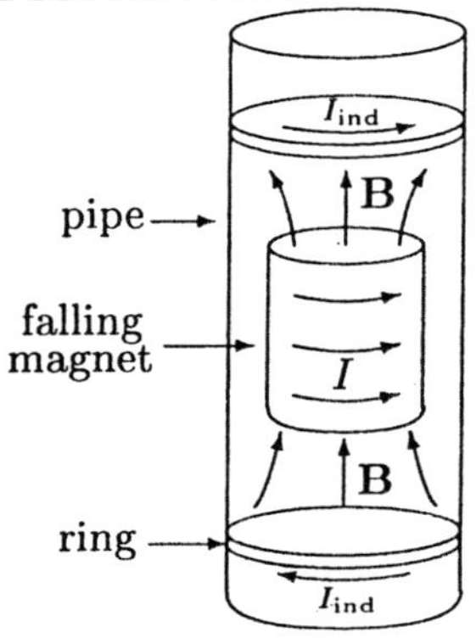

Suppose the current (I) in the magnet flows counterclockwise (viewed from above), as shown, so its field, near the ends, points *upward.* A ring of pipe *below* the magnet experiences an increasing upward flux, as the magnet approaches, and hence (by Lenz's law) a current (I_{ind}) will be induced in it such as to produce a *downward* flux. Thus I_{ind} must flow *clockwise*, which is *opposite* to the current in the magnet. Since opposite currents repel, the force on the magnet is *upward.* Meanwhile, a ring *above* the magnet experiences a *decreasing* (upward) flux, so *its* induced current is *parallel* to I, and it *attracts* the magnet upward. And the flux through rings *next to* the magnet is constant, so *no* current is induced in them. *Conclusion:* the delay is due to forces exerted on the magnet by induced eddy currents in the pipe.

Problem 7.15

In the quasistatic approximation, $\mathbf{B} = \begin{cases} \mu_0 nI\,\hat{\mathbf{z}}, & (s < a); \\ 0, & (s > a). \end{cases}$

Inside: for an "amperian loop" of radius $s < a$,

$$\Phi = B\pi s^2 = \mu_0 nI\pi s^2;\ \oint \mathbf{E}\cdot d\mathbf{l} = E\,2\pi s = -\frac{d\Phi}{dt} = -\mu_0 n\pi s^2\frac{dI}{dt};\quad \boxed{\mathbf{E} = -\frac{\mu_0 ns}{2}\frac{dI}{dt}\,\hat{\boldsymbol{\phi}}.}$$

Outside: for an "amperian loop" of radius $s > a$:

$$\Phi = B\pi a^2 = \mu_0 nI\pi a^2;\ E\,2\pi s = -\mu_0 n\pi a^2\frac{dI}{dt};\quad \boxed{\mathbf{E} = -\frac{\mu_0 na^2}{2s}\frac{dI}{dt}\,\hat{\boldsymbol{\phi}}.}$$

Problem 7.16

(a) The magnetic field (in the quasistatic approximation) is "circumferential". This is analogous to the current in a solenoid, and hence the field is $\boxed{\text{longitudinal.}}$

(b) Use the "amperian loop" shown.
Outside, $\mathbf{B} = 0$, so here $\mathbf{E} = 0$ (like $\mathbf{B}$ outside a solenoid).
So $\oint \mathbf{E}\cdot d\mathbf{l} = El = -\frac{d\Phi}{dt} = -\frac{d}{dt}\int \mathbf{B}\cdot d\mathbf{a} = -\frac{d}{dt}\int_s^a \frac{\mu_0 I}{2\pi s'}l\,ds'$
$\therefore E = -\frac{\mu_0}{2\pi}\frac{dI}{dt}\ln\left(\frac{a}{s}\right)$. But $\frac{dI}{dt} = -I_0\omega\sin\omega t$,

so $\boxed{\mathbf{E} = \frac{\mu_0 I_0\omega}{2\pi}\sin(\omega t)\ln\left(\frac{a}{s}\right)\hat{\mathbf{z}}.}$

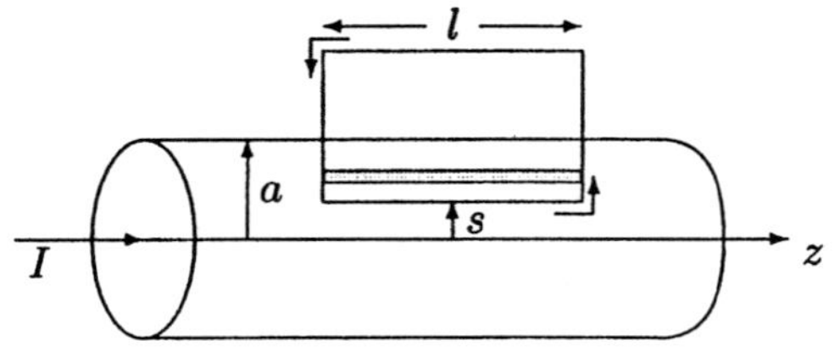

Problem 7.17

(a) The field inside the solenoid is $B = \mu_0 n I$. So $\Phi = \pi a^2 \mu_0 n I \Rightarrow \mathcal{E} = -\pi a^2 \mu_0 n (dI/dt)$.
In magnitude, then, $\mathcal{E} = \pi a^2 \mu_0 n k$. Now $\mathcal{E} = I_r R$, so $\boxed{I_{\text{resistor}} = \dfrac{\pi a^2 \mu_0 n k}{R}.}$
B is to the right and increasing, so the field of the loop is to the *left*, so the current is counterclockwise, or $\boxed{\text{to the right,}}$ through the resistor.

(b) $\Delta\Phi = 2\pi a^2 \mu_0 n I$; $I = \dfrac{dQ}{dt} = \dfrac{\mathcal{E}}{R} = -\dfrac{1}{R}\dfrac{d\Phi}{dt} \Rightarrow \Delta Q = \dfrac{1}{R}\Delta\Phi$, in magnitude. So $\boxed{\Delta Q = \dfrac{2\pi a^2 \mu_0 n I}{R}.}$

Problem 7.18

$$\Phi = \int \mathbf{B}\cdot d\mathbf{a};\ \mathbf{B} = \frac{\mu_0 I}{2\pi s}\hat{\boldsymbol{\phi}};\ \Phi = \frac{\mu_0 I a}{2\pi}\int_a^{2a}\frac{ds}{s} = \frac{\mu_0 I a \ln 2}{2\pi};\ \mathcal{E} = I_{\text{loop}}R = \frac{dQ}{dt}R = -\frac{d\Phi}{dt} = -\frac{\mu_0 a \ln 2}{2\pi}\frac{dI}{dt}.$$

$$dQ = -\frac{\mu_0 a \ln 2}{2\pi R}dI \Rightarrow \boxed{Q = \frac{I\mu_0 a \ln 2}{2\pi R}.}$$

The field of the wire, at the square loop, is *out of the page*, and *decreasing*, so the field of the induced current must point out of page, within the loop, and hence the induced current flows $\boxed{\text{counterclockwise.}}$

Problem 7.19

In the quasistatic approximation, $\mathbf{B} = \begin{cases} \frac{\mu_0 NI}{2\pi s}\hat{\boldsymbol{\phi}}, & \text{(inside toroid);} \\ 0, & \text{(outside toroid)} \end{cases}$
(Eq. 5.58). The flux around the toroid is therefore

$$\Phi = \frac{\mu_0 NI}{2\pi}\int_a^{a+w}\frac{1}{s}h\,ds = \frac{\mu_0 NIh}{2\pi}\ln\left(1+\frac{w}{a}\right) \approx \frac{\mu_0 Nhw}{2\pi a}I.\quad \frac{d\Phi}{dt} = \frac{\mu_0 Nhw}{2\pi a}\frac{dI}{dt} = \frac{\mu_0 Nhwk}{2\pi a}.$$

The electric field is the same as the *magnetic* field of a circular current (Eq. 5.38):

$$\mathbf{B} = \frac{\mu_0 I}{2}\frac{a^2}{(a^2+z^2)^{3/2}}\hat{\mathbf{z}},$$

with (Eq. 7.18)

$$I \to -\frac{1}{\mu_0}\frac{d\Phi}{dt} = -\frac{Nhwk}{2\pi a}.\quad \text{So } \mathbf{E} = \frac{\mu_0}{2}\left(-\frac{Nhwk}{2\pi a}\right)\frac{a^2}{(a^2+z^2)^{3/2}}\hat{\mathbf{z}} = \boxed{-\frac{\mu_0}{4\pi}\frac{Nhwka}{(a^2+z^2)^{3/2}}\hat{\mathbf{z}}.}$$

Problem 7.20

(a) From Eq. 5.38, the field (on the axis) is $\mathbf{B} = \frac{\mu_0 I}{2}\frac{b^2}{(b^2+z^2)^{3/2}}\hat{\mathbf{z}}$, so the flux through the little loop (area πa^2) is $\boxed{\Phi = \dfrac{\mu_0 \pi I a^2 b^2}{2(b^2+z^2)^{3/2}}.}$

(b) The field (Eq. 5.86) is $\mathbf{B} = \frac{\mu_0}{4\pi}\frac{m}{r^3}(2\cos\theta\,\hat{\mathbf{r}} + \sin\theta\,\hat{\boldsymbol{\theta}})$, where $m = I\pi a^2$. Integrating over the spherical "cap" (bounded by the big loop and centered at the little loop):

$$\Phi = \int \mathbf{B}\cdot d\mathbf{a} = \frac{\mu_0}{4\pi}\frac{I\pi a^2}{r^3}\int(2\cos\theta)(r^2\sin\theta\,d\theta\,d\phi) = \frac{\mu_0 I a^2}{2r}2\pi\int_0^{\bar{\theta}}\cos\theta\sin\theta\,d\theta$$

da = r sinθ dφ r dθ r̂

where $r = \sqrt{b^2 + z^2}$ and $\sin\bar{\theta} = b/r$. Evidently $\Phi = \frac{\mu_0 I \pi a^2}{r} \left.\frac{\sin^2\theta}{2}\right|_0^{\bar{\theta}} = \boxed{\frac{\mu_0 \pi I a^2 b^2}{2(b^2+z^2)^{3/2}},}$ the same as in (a)!!

(c) Dividing off I ($\Phi_1 = M_{12}I_2$, $\Phi_2 = M_{21}I_1$): $\boxed{M_{12} = M_{21} = \frac{\mu_0 \pi a^2 b^2}{2(b^2+z^2)^{3/2}}.}$

Problem 7.21

$$\mathcal{E} = -\frac{d\Phi}{dt} = -M\frac{dI}{dt} = -Mk.$$

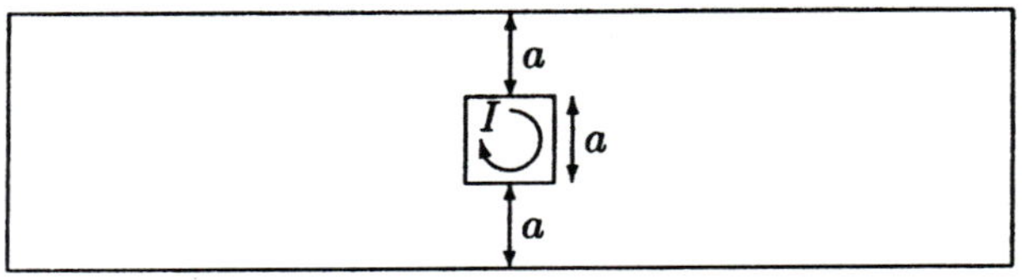

It's hard to calculate M using a current in the little loop, so, exploiting the equality of the mutual inductances, I'll find the flux through the *little* loop when a current I flows in the *big* loop: $\Phi = MI$. The field of *one* long wire is $B = \frac{\mu_0 I}{2\pi s} \Rightarrow \Phi_1 = \frac{\mu_0 I}{2\pi}\int_a^{2a} \frac{1}{s} a\, ds = \frac{\mu_0 I a}{2\pi}\ln 2$, so the *total* flux is

$$\Phi = 2\Phi_1 = \frac{\mu_0 I a \ln 2}{\pi} \Rightarrow M = \frac{\mu_0 a \ln 2}{\pi} \Rightarrow \boxed{\mathcal{E} = \frac{\mu_0 k a \ln 2}{\pi},} \quad \text{in magnitude.}$$

Direction: The net flux (through the big loop), due to I in the little loop, is ***into the page.*** (Why? Field lines point *in*, for the inside of the little loop, and ***out*** everywhere outside the little loop. The big loop encloses *all* of the former, and only *part* of the latter, so ***net*** flux is ***inward.***) This flux is ***increasing,*** so the induced current in the big loop is such that ***its*** field points ***out*** of the page: it flows $\boxed{\text{counterclockwise.}}$

Problem 7.22

$B = \mu_0 n I \Rightarrow \Phi_1 = \mu_0 n I \pi R^2$ (flux through a single turn). In a length l there are nl such turns, so the total flux is $\Phi = \mu_0 n^2 \pi R^2 I l$. The self-inductance is given by $\Phi = LI$, so the self-inductance per unit length is $\boxed{\mathcal{L} = \mu_0 n^2 \pi R^2.}$

Problem 7.23

The field of one wire is $B_1 = \frac{\mu_0}{2\pi}\frac{I}{s}$, so $\Phi = 2\cdot\frac{\mu_0 I}{2\pi}\cdot l \int_\epsilon^{d-\epsilon}\frac{ds}{s} = \frac{\mu_0 I l}{\pi}\ln\left(\frac{d-\epsilon}{\epsilon}\right)$. The ϵ in the numerator is negligible (compared to d), but in the denominator we *cannot* let $\epsilon \to 0$, else the flux is *infinite.*

$\boxed{L = \frac{\mu_0 l}{\pi}\ln(d/\epsilon)}$. Evidently the size of the wire itself is critical in determining L.

Problem 7.24

(a) In the quasistatic approximation $\mathbf{B} = \frac{\mu_0}{2\pi s}\hat{\boldsymbol{\phi}}$. So $\Phi_1 = \frac{\mu_0 I}{2\pi}\int_a^b \frac{1}{s} h\, ds = \frac{\mu_0 I h}{2\pi}\ln(b/a)$.

This is the flux through *one* turn; the *total* flux is N times Φ_1: $\Phi = \frac{\mu_0 N h}{2\pi}\ln(b/a) I_0 \cos(\omega t)$. So

$$\mathcal{E} = -\frac{d\Phi}{dt} = \frac{\mu_0 N h}{2\pi}\ln(b/a) I_0 \omega \sin(\omega t) = \frac{(4\pi\times10^{-7})(10^3)(10^{-2})}{2\pi}\ln(2)(0.5)(2\pi\, 60)\sin(\omega t)$$

$= \boxed{2.61\times10^{-4}\sin(\omega t)}$ (in volts), where $\omega = 2\pi\, 60 = 377/\text{s}$. $I_r = \frac{\mathcal{E}}{R} = \frac{2.61\times10^{-4}}{500}\sin(\omega t)$

$= \boxed{5.22\times10^{-7}\sin(\omega t)}$ (amperes).

(b) $\mathcal{E}_b = -L\frac{dI_r}{dt}$; where (Eq. 7.27) $L = \frac{\mu_0 N^2 h}{2\pi}\ln(b/a) = \frac{(4\pi\times10^{-7})(10^6)(10^{-2})}{2\pi}\ln(2) = 1.39\times10^{-3}$ (henries).

Therefore $\mathcal{E}_b = -(1.39\times10^{-3})(5.22\times10^{-7}\omega)\cos(\omega t) = \boxed{-2.74\times10^{-7}\cos(\omega t)}$ (volts).

Ratio of amplitudes: $\dfrac{2.74\times 10^{-7}}{2.61\times 10^{-4}} = \boxed{1.05\times 10^{-3}} = \dfrac{\mu_0 N^2 h\omega}{2\pi R}\ln(b/a).$

Problem 7.25

With I positive clockwise, $\mathcal{E} = -L\frac{dI}{dt} = Q/C$, where Q is the charge on the capacitor; $I = \frac{dQ}{dt}$, so $\frac{d^2Q}{dt^2} = -\frac{1}{LC}Q = -\omega^2 Q$, where $\omega = \frac{1}{\sqrt{LC}}$. The general solution is $Q(t) = A\cos\omega t + B\sin\omega t$. At $t=0$, $Q = CV$, so $A = CV$; $I(t) = \frac{dQ}{dt} = -A\omega\sin\omega t + B\omega\sin\omega t$. At $t=0$, $I=0$, so $B=0$, and

$$I(t) = -CV\omega\sin\omega t = \boxed{-V\sqrt{\frac{C}{L}}\sin\left(\frac{t}{\sqrt{LC}}\right).}$$

If you put in a resistor, the oscillation is "damped". This time $-L\frac{dI}{dt} = \frac{Q}{C} + IR$, so $L\frac{d^2Q}{dt^2} + R\frac{dQ}{dt} + \frac{1}{C}Q = 0$. For an analysis of this case, see Purcell's *Electricity and Magnetism* (Ch. 8) or any book on oscillations and waves.

Problem 7.26

(a) $W = \frac{1}{2}LI^2$. $L = \mu_0 n^2\pi R^2 l$ (Prob. 7.22) $\boxed{W = \frac{1}{2}\mu_0 n^2\pi R^2 l I^2}.$

(b) $W = \frac{1}{2}\oint(\mathbf{A}\cdot\mathbf{I})dl$. $\mathbf{A} = (\mu_0 nI/2)R\,\hat{\boldsymbol{\phi}}$, at the surface (Eq. 5.70 or 5.71). So $W_1 = \frac{1}{2}\frac{\mu_0 nI}{2}RI\cdot 2\pi R$, for one turn. There are nl such turns in length l, so $W = \frac{1}{2}\mu_0 n^2\pi R^2 l I^2$. ✓

(c) $W = \frac{1}{2\mu_0}\int B^2\,d\tau$. $B = \mu_0 nI$, inside, and zero outside; $\int d\tau = \pi R^2 l$, so $W = \frac{1}{2\mu_0}\mu_0^2 n^2 I^2\pi R^2 l = \frac{1}{2}\mu_0 n^2\pi R^2 l I^2$. ✓

(d) $W = \frac{1}{2\mu_0}\left[\int B^2\,d\tau - \oint(\mathbf{A}\times\mathbf{B})\cdot d\mathbf{a}\right]$. This time $\int B^2\,d\tau = \mu_0^2 n^2 I^2\pi(R^2-a^2)l$. Meanwhile, $\mathbf{A}\times\mathbf{B} = 0$ outside (at $s=b$). Inside, $\mathbf{A} = \frac{\mu_0 nI}{2}a\,\hat{\boldsymbol{\phi}}$ (at $s=a$), while $\mathbf{B} = \mu_0 nI\,\hat{\mathbf{z}}$.

$\mathbf{A}\times\mathbf{B} = \frac{1}{2}\mu_0^2 n^2 I^2 a\underbrace{(\hat{\boldsymbol{\phi}}\times\hat{\mathbf{z}})}_{\hat{\mathbf{s}}}$

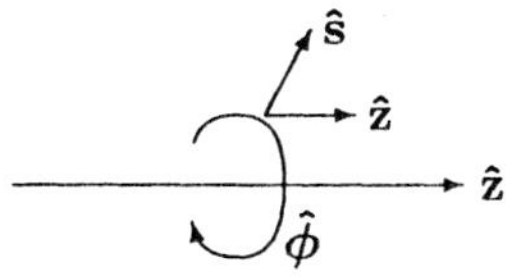

points *inward* ("out" of the volume)

$\oint(\mathbf{A}\times\mathbf{B})\cdot d\mathbf{a} = \int(\frac{1}{2}\mu_0^2 n^2 I^2 a\hat{\mathbf{s}})\cdot[a\,d\phi\,dz(-\hat{\mathbf{s}})] = -\frac{1}{2}\mu_0^2 n^2 I^2 a^2 2\pi l.$

$W = \frac{1}{2\mu_0}\left[\mu_0^2 n^2 I^2\pi(R^2-a^2)l + \mu_0^2 n^2 I^2\pi a^2 l\right] = \frac{1}{2}\mu_0 n^2 I^2 R^2\pi l.$ ✓

Problem 7.27

$$B = \frac{\mu_0 nI}{2\pi s};\ W = \frac{1}{2\mu_0}\int B^2\,d\tau = \frac{1}{2\mu_0}\frac{\mu_0^2 n^2 I^2}{4\pi^2}\int\frac{1}{s^2}hr\,d\phi\,ds = \frac{\mu_0 n^2 I^2}{8\pi^2}h2\pi\ln\left(\frac{b}{a}\right) = \boxed{\frac{1}{4\pi}\mu_0 n^2 I^2 h\ln(b/a).}$$

$\boxed{L = \frac{\mu_0}{2\pi}n^2 h\ln(b/a)}$ (same as Eq. 7.27).

Problem 7.28

$$\oint\mathbf{B}\cdot d\mathbf{l} = B(2\pi s) = \mu_0 I_{\text{enc}} = \mu_0 I(s^2/R^2) \Rightarrow B = \frac{\mu_0 Is}{2\pi R^2}.$$

$$W = \frac{1}{2\mu_0}\int B^2\,d\tau = \frac{1}{2\mu_0}\frac{\mu_0^2 I^2}{4\pi^2 R^4}\int_0^R s^2(2\pi s)l\,ds = \frac{\mu_0 I^2 l}{4\pi R^4}\left(\frac{s^4}{4}\right)\Big|_0^R = \frac{\mu_0 l}{16\pi}I^2 = \frac{1}{2}LI^2.$$

So $L = \frac{\mu_0}{8\pi}l$, and $\mathcal{L} = L/l = \boxed{\mu_0/8\pi,}$ independent of R!

Problem 7.29

(a) Initial current: $I_0 = \mathcal{E}_0/R$. So $-L\frac{dI}{dt} = IR \Rightarrow \frac{dI}{dt} = -\frac{R}{L}I \Rightarrow I = I_0 e^{-Rt/L}$, or $\boxed{I(t) = \frac{\mathcal{E}_0}{R}e^{-Rt/L}.}$

(b) $P = I^2 R = (\mathcal{E}_0/R)^2 e^{-2Rt/L}R = \frac{\mathcal{E}_0^2}{R}e^{-2Rt/L} = \frac{dW}{dt}$.

$$W = \frac{\mathcal{E}_0^2}{R}\int_0^\infty e^{-2Rt/L}\,dt = \frac{\mathcal{E}_0^2}{R}\left(-\frac{L}{2R}e^{-2Rt/L}\right)\Big|_0^\infty = \frac{\mathcal{E}_0^2}{R}(0 + L/2R) = \boxed{\frac{1}{2}L(\mathcal{E}_0/R)^2.}$$

(c) $W_0 = \frac{1}{2}LI_0^2 = \frac{1}{2}\left(\mathcal{E}_0/R\right)^2$. ✓

Problem 7.30

(a) $\mathbf{B}_1 = \frac{\mu_0}{4\pi}\frac{1}{r^3}I_1[3(\mathbf{a}_1\cdot\hat{\boldsymbol{\imath}})\hat{\boldsymbol{\imath}} - \mathbf{a}_1]$, since $\mathbf{m}_1 = I_1\mathbf{a}_1$. The flux through loop 2 is then

$$\Phi_2 = \mathbf{B}_1\cdot\mathbf{a}_2 = \frac{\mu_0}{4\pi}\frac{1}{\imath^3}I_1[3(\mathbf{a}_1\cdot\hat{\boldsymbol{\imath}})(\mathbf{a}_2\cdot\hat{\boldsymbol{\imath}}) - \mathbf{a}_1\cdot\mathbf{a}_2] = MI_1. \quad \boxed{M = \frac{\mu_0}{4\pi\imath^3}[3(\mathbf{a}_1\cdot\hat{\boldsymbol{\imath}})(\mathbf{a}_2\cdot\hat{\boldsymbol{\imath}}) - \mathbf{a}_1\cdot\mathbf{a}_2].}$$

(b) $\mathcal{E}_1 = -M\frac{dI_2}{dt}$, $\left.\frac{dW}{dt}\right|_1 = -\mathcal{E}_1 I_1 = MI_1\frac{dI_2}{dt}$. (This is the work done per unit time *against* the mutual emf in loop 1—hence the minus sign.) So (since I_1 is constant) $W_1 = MI_1I_2$, where I_2 is the final current in loop 2:

$$\boxed{W = \frac{\mu_0}{4\pi\imath^3}[3(\mathbf{m}_1\cdot\hat{\boldsymbol{\imath}})(\mathbf{m}_2\cdot\hat{\boldsymbol{\imath}}) - \mathbf{m}_1\cdot\mathbf{m}_2].}$$

Notice that this is *opposite in sign* to Eq. 6.35. In Prob. 6.21 we assumed that the magnitudes of the dipole moments were *fixed*, and we did not worry about the energy necessary to sustain the currents themselves—only the energy required to move them into position and rotate them into their final orientations. But in *this* problem we are including it *all*, and it is a curious fact that this merely changes the sign of the answer. For commentary on this subtle issue see R. H. Young, *Am. J. Phys.* **66**, 1043 (1998), and the references cited there.

Problem 7.31

The displacement current density (Sect. 7.3.2) is $\mathbf{J}_d = \epsilon_0\frac{\partial\mathbf{E}}{\partial t} = \frac{I}{A} = \frac{I}{\pi a^2}\,\hat{\mathbf{z}}$. Drawing an "amperian loop" at radius s,

$$\oint \mathbf{B}\cdot d\mathbf{l} = B\cdot 2\pi s = \mu_0 I_{d_{\rm enc}} = \mu_0\frac{I}{\pi a^2}\cdot\pi s^2 = \mu_0 I\frac{s^2}{a^2} \Rightarrow B = \frac{\mu_0 I s^2}{2\pi s a^2}; \quad \boxed{\mathbf{B} = \frac{\mu_0 I s}{2\pi a^2}\,\hat{\boldsymbol{\phi}}.}$$

Problem 7.32

(a) $\mathbf{E} = \dfrac{\sigma(t)}{\epsilon_0}\,\hat{\mathbf{z}};\quad \sigma(t) = \dfrac{Q(t)}{\pi a^2} = \dfrac{It}{\pi a^2};\quad \boxed{\dfrac{It}{\pi\epsilon_0 a^2}\,\hat{\mathbf{z}}.}$

(b) $I_{d_{\rm enc}} = J_d\pi s^2 = \epsilon_0\dfrac{dE}{dt}\pi s^2 = \boxed{I\dfrac{s^2}{a^2}.}\quad \oint\mathbf{B}\cdot d\mathbf{l} = \mu_0 I_{d_{\rm enc}} \Rightarrow B\,2\pi s = \mu_0 I\dfrac{s^2}{a^2} \Rightarrow \mathbf{B} = \boxed{\dfrac{\mu_0 I}{2\pi a^2}s\,\hat{\boldsymbol{\phi}}.}$

(c) A surface current flows radially outward over the left plate; let $I(s)$ be the total current crossing a circle of radius s. The charge density (at time t) is

$$\sigma(t) = \frac{[I - I(s)]t}{\pi s^2}.$$

Since we are told this is independent of s, it must be that $I - I(s) = \beta s^2$, for some constant β. But $I(a) = 0$, so $\beta a^2 = I$, or $\beta = I/a^2$. Therefore $I(s) = I(1 - s^2/a^2)$.

$$B\,2\pi s = \mu_0 I_{\rm enc} = \mu_0[I - I(s)] = \mu_0\frac{s^2}{a^2} \Rightarrow \boxed{\mathbf{B} = \frac{\mu_0}{2\pi a^2}s\,\hat{\boldsymbol{\phi}}.}\ ✓$$

Problem 7.33

(a) $\mathbf{J}_d = \epsilon_0\dfrac{\mu_0 I_0\omega^2}{2\pi}\cos(\omega t)\ln(a/s)\,\hat{\mathbf{z}}$. But $I_0\cos(\omega t) = I$. So $\boxed{\mathbf{J}_d = \dfrac{\mu_0\epsilon_0}{2\pi}\omega^2 I\ln(a/s)\,\hat{\mathbf{z}}.}$

(b) $I_d = \displaystyle\int\mathbf{J}_d\cdot d\mathbf{a} = \frac{\mu_0\epsilon_0\omega^2 I}{2\pi}\int_0^a \ln(a/s)(2\pi s\,ds) = \mu_0\epsilon_0\omega^2 I\int_0^a(s\ln a - s\ln s)ds$

$$= \mu_0\epsilon_0\omega^2 I\left[(\ln a)\frac{s^2}{2} - \frac{s^2}{2}\ln s + \frac{s^2}{4}\right]\Bigg|_0^a = \mu_0\epsilon_0\omega^2 I\left[\frac{a^2}{2}\ln a - \frac{a^2}{2}\ln a + \frac{a^2}{4}\right] = \boxed{\frac{\mu_0\epsilon_0\omega^2 I a^2}{4}.}$$

(c) $\boxed{\dfrac{I_d}{I} = \dfrac{\mu_0\epsilon_0\omega^2a^2}{4}.}$ Since $\mu_0\epsilon_0 = 1/c^2$, $I_d/I = (\omega a/2c)^2$. If $a = 10^{-3}\,\text{m}$, and $\frac{I_d}{I} = \frac{1}{100}$, so that $\frac{\omega a}{2c} = \frac{1}{10}$, $\omega = \frac{2c}{10a} = \frac{3\times10^8\,\text{m/s}}{5\times10^{-3}\,\text{m}}$, or $\omega = 0.6\times10^{11}/\text{s} = \boxed{6\times10^{10}/\text{s};}$ $\nu = \frac{\omega}{2\pi} \approx 10^{10}\,\text{Hz}$, or 10^4 megahertz. (This is the *microwave* region, *way* above radio frequencies.)

Problem 7.34

Physically, this is the field of a point charge $-q$ at the origin, out to an expanding spherical shell of radius vt; outside this shell the field is zero. Evidently the shell carries the opposite charge, $+q$. *Mathematically*, using product rule #5 and Eq. 1.99:

$$\nabla\cdot\mathbf{E} = \theta(vt-r)\nabla\cdot\left(-\frac{1}{4\pi\epsilon_0}\frac{q}{r^2}\hat{\mathbf{r}}\right) - \frac{1}{4\pi\epsilon_0}\frac{q}{r^2}\hat{\mathbf{r}}\cdot\nabla[\theta(vt-r)] = -\frac{q}{\epsilon_0}\delta^3(\mathbf{r})\theta(vt-r) - \frac{1}{4\pi\epsilon_0}\frac{q}{r^2}(\hat{\mathbf{r}}\cdot\hat{\mathbf{r}})\frac{\partial}{\partial r}\theta(vt-r).$$

But $\delta^3(\mathbf{r})\theta(vt-r) = \delta^3(\mathbf{r})\theta(t)$, and $\frac{\partial}{\partial r}\theta(vt-r) = -\delta(vt-r)$ (Prob. 1.45), so

$$\rho = \epsilon_0\nabla\cdot\mathbf{E} = \boxed{-q\delta^3(\mathbf{r})\theta(t) + \frac{q}{4\pi r^2}\delta(vt-r).}$$

(For $t<0$ the field and the charge density are zero everywhere.)

Clearly $\nabla\cdot\mathbf{B} = 0$, and $\nabla\times\mathbf{E} = 0$ (since $\mathbf{E}$ has only an r component, and it is independent of θ and ϕ). There remains only the Ampére/Maxwell law, $\nabla\times\mathbf{B} = 0 = \mu_0\mathbf{J} + \mu_0\epsilon_0\partial\mathbf{E}/\partial t$. Evidently

$$\mathbf{J} = -\epsilon_0\frac{\partial\mathbf{E}}{\partial t} = -\epsilon_0\left\{-\frac{q}{4\pi\epsilon_0 r^2}\frac{\partial}{\partial t}[\theta(vt-r)]\right\}\hat{\mathbf{r}} = \boxed{\frac{q}{4\pi r^2}v\delta(vt-r)\,\hat{\mathbf{r}}.}$$

(The stationary charge at the origin does not contribute to $\mathbf{J}$, of course; for the expanding shell we have $\mathbf{J} = \rho\mathbf{v}$, as expected—Eq. 5.26.)

Problem 7.35

From $\nabla\cdot\mathbf{B} = \mu_0\rho_m$ it follows that the field of a point monopole is $\mathbf{B} = \frac{\mu_0}{4\pi}\frac{q_m}{\imath^2}\hat{\boldsymbol{\imath}}$. The force law has the form $\mathbf{F}\propto q_m\left(\mathbf{B} - \frac{1}{c^2}\mathbf{v}\times\mathbf{E}\right)$ (see Prob. 5.21—the c^2 is needed on dimensional grounds). The proportionality constant must be 1 to reproduce "Coulomb's law" for point charges at rest. So $\boxed{\mathbf{F} = q_m\left(\mathbf{B} - \dfrac{1}{c^2}\mathbf{v}\times\mathbf{E}\right).}$

Problem 7.36

Integrate the "generalized Faraday law" (Eq. 7.43iii), $\nabla\times\mathbf{E} = -\mu_0\mathbf{J}_m - \frac{\partial\mathbf{B}}{\partial t}$, over the surface of the loop:

$$\int(\nabla\times\mathbf{E})\cdot d\mathbf{a} = \oint\mathbf{E}\cdot d\mathbf{l} = \mathcal{E} = -\mu_0\int\mathbf{J}_m\cdot d\mathbf{a} - \frac{d}{dt}\int\mathbf{B}\cdot d\mathbf{a} = -\mu_0 I_{m_{\text{enc}}} - \frac{d\Phi}{dt}.$$

But $\mathcal{E} = -L\dfrac{dI}{dt}$, so $\dfrac{dI}{dt} = \dfrac{\mu_0}{L}I_{m_{\text{enc}}} + \dfrac{1}{L}\dfrac{d\Phi}{dt}$, or $I = \dfrac{\mu_0}{L}\Delta Q_m + \dfrac{1}{L}\Delta\Phi$, where ΔQ_m is the total magnetic charge passing through the surface, and $\Delta\Phi$ is the change in flux through the surface. If we use the *flat* surface, then $\Delta Q_m = q_m$ and $\Delta\Phi = 0$ (when the monopole is far away, $\Phi = 0$; the flux builds up to $\mu_0 q_m/2$ just before it passes through the loop; then it abruptly drops to $-\mu_0 q_m/2$, and rises back up to zero as the monopole disappears into the distance). If we use a huge balloon-shaped surface, so that q_m remains *inside* it on the far side, then $\Delta Q_m = 0$, but Φ rises monotonically from 0 to $\mu_0 q_m$. In either case,

$$\boxed{I = \frac{\mu_0 q_m}{L}.}$$

Problem 7.37

$$E = \frac{V}{d} \Rightarrow J_c = \sigma E = \frac{1}{\rho}E = \frac{V}{\rho d}.\ J_d = \frac{\partial D}{\partial t} = \frac{\partial}{\partial t}(\epsilon E) = \epsilon\frac{\partial}{\partial t}\left[\frac{V_0\cos(2\pi\nu t)}{d}\right] = \frac{\epsilon V_0}{d}[-2\pi\nu\sin(2\pi\nu t)].$$

The ratio of the amplitudes is therefore:

$$\frac{J_c}{J_d} = \frac{V_0}{\rho d}\frac{d}{2\pi\nu\epsilon V_0} = \frac{1}{2\pi\nu\epsilon\rho} = \left[2\pi(4\times10^8)(81)(8.85\times10^{-12})(0.23)\right]^{-1} = \boxed{2.41.}$$

Problem 7.38

The potential and field in this configuration are identical to those in the upper half of Ex. 3.8. Therefore:

$$I = \int \mathbf{J}\cdot d\mathbf{a} = \sigma\int \mathbf{E}\cdot d\mathbf{a}$$

where the integral is over the hemispherical surface just outside the sphere.

But I can with impunity *close* this surface: (because $E = 0$ down there anyway—inside a conductor).

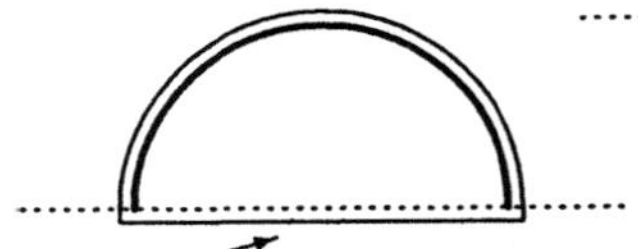

So $I = \sigma\int\mathbf{E}\cdot d\mathbf{a} = \frac{\sigma}{\epsilon_0}Q_{\text{enc}} = \frac{\sigma}{\epsilon_0}\int\sigma_e\,da$, where σ_e is the electric charge density on the surface of the hemisphere—to wit (Eq. 3.77) $\sigma_e = 3\epsilon_0E_0\cos\theta$.

$$I = \frac{\sigma}{\epsilon_0}3\epsilon_0E_0\int\cos\theta\,a^2\sin\theta\,d\theta\,d\phi = 3\sigma E_0a^2 2\pi\underbrace{\int_0^{\pi/2}\sin\theta\cos\theta\,d\theta}_{\left.\frac{\sin^2\theta}{2}\right|_0^{\pi/2} = \frac{1}{2}} = 3\sigma E_0\pi a^2.$$

But in this case $E_0 = V_0/d$, so $\boxed{I = \dfrac{3\sigma\pi V_0a^2}{d}.}$

Problem 7.39

Begin with a different problem: two parallel wires carrying charges $+\lambda$ and $-\lambda$ as shown.

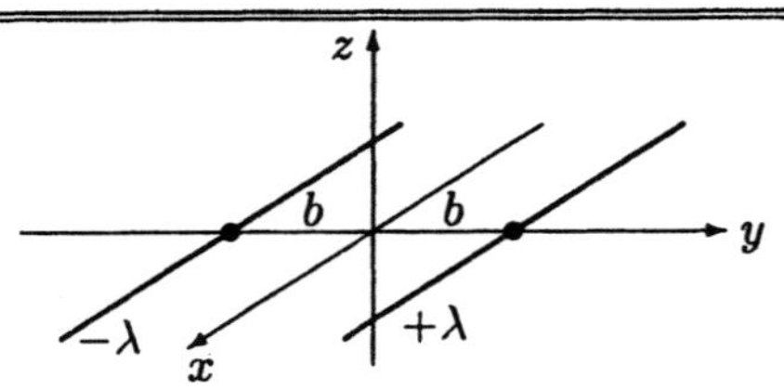

Field of one wire: $\mathbf{E} = \frac{\lambda}{2\pi\epsilon_0 s}\hat{\mathbf{s}}$; potential: $V = -\frac{\lambda}{2\pi\epsilon_0}\ln(s/a)$.

Potential of combination: $V = \frac{\lambda}{2\pi\epsilon_0}\ln(s_-/s_+)$,

or $V(y,z) = \frac{\lambda}{4\pi\epsilon_0}\ln\left\{\frac{(y+b)^2+z^2}{(y-b)^2+z^2}\right\}$.

Find the locus of points of fixed V (i.e. equipotential surfaces):

$$e^{4\pi\epsilon_0V/\lambda} \equiv \mu = \frac{(y+b)^2+z^2}{(y-b)^2+z^2} \Longrightarrow \mu(y^2-2yb+b^2+z^2) = y^2+2yb+b^2+z^2;$$

$$y^2(\mu-1)+b^2(\mu-1)+z^2(\mu-1)-2yb(\mu+1) = 0 \Longrightarrow y^2+z^2+b^2-2yb\beta = 0 \quad \left(\beta\equiv\frac{\mu+1}{\mu-1}\right);$$

$$(y-b\beta)^2+z^2+b^2-b^2\beta^2 = 0 \Longrightarrow (y-b\beta^2)+z^2 = b^2(\beta^2-1).$$

This is a *circle*, with center at $y_0 = b\beta = b\left(\frac{\mu+1}{\mu-1}\right)$ and radius $= b\sqrt{\beta^2-1} = b\sqrt{\frac{(\mu^2+2\mu+1)-(\mu^2-2\mu+1)}{(\mu-1)^2}} = \frac{2b\sqrt{\mu}}{\mu-1}$.

This suggests an image solution to the problem at hand. We want $y_0 = d$, radius $= a$, and $V = V_0$. These determine the parameters b, μ, and λ of the image solution:

$$\frac{d}{a} = \frac{y_0}{\text{radius}} = \frac{b\left(\frac{\mu+1}{\mu-1}\right)}{\frac{2b\sqrt{\mu}}{\mu-1}} = \frac{\mu+1}{2\sqrt{\mu}}. \quad \text{Call } \frac{d}{a} \equiv \alpha.$$

$$4\alpha^2\mu = (\mu+1)^2 = \mu^2 + 2\mu + 1 \Longrightarrow \mu^2 + (2-4\alpha^2)\mu + 1 = 0;$$

$$\mu = \frac{4\alpha^2 - 2 \pm \sqrt{4(1-2\alpha^2)^2 - 4}}{2} = 2\alpha^2 - 1 \pm \sqrt{1 - 4\alpha^2 + 4\alpha^4 - 1} = 2\alpha^2 - 1 \pm 2\alpha\sqrt{\alpha^2-1};$$

$$\frac{4\pi\epsilon_0 V_0}{\lambda} = \ln\mu \Longrightarrow \lambda = \frac{4\pi\epsilon_0 V_0}{\ln\left(2\alpha^2 - 1 \pm 2\alpha\sqrt{\alpha^2-1}\right)}. \quad \text{That's the line charge in the image problem.}$$

$$I = \int \mathbf{J}\cdot d\mathbf{a} = \sigma\int \mathbf{E}\cdot d\mathbf{a} = \sigma\frac{1}{\epsilon_0}Q_{\text{enc}} = \frac{\sigma}{\epsilon_0}\lambda l.$$

The current per unit length is $i = \dfrac{I}{l} = \dfrac{\sigma\lambda}{\epsilon_0} = \dfrac{4\pi\sigma V_0}{\ln\left(2\alpha^2 - 1 \pm 2\alpha\sqrt{\alpha^2-1}\right)}$. Which sign do we want? Suppose the cylinders are far apart, $d \gg a$, so that $\alpha \gg 1$.

$$(\;) = 2\alpha^2 - 1 \pm 2\alpha^2\sqrt{1 - 1/\alpha^2} = 2\alpha^2 - 1 \pm 2\alpha^2\left[1 - \frac{1}{2\alpha^2} - \frac{1}{8\alpha^4} + \cdots\right]$$

$$= 2\alpha^2(1\pm 1) - (1\pm 1) \mp \frac{1}{4\alpha^2} \pm \cdots = \begin{cases} 4\alpha^2 - 2 - 1/2\alpha^2 + \cdots \approx 4\alpha^2 & (+\text{ sign}), \\ -1/4\alpha^2 & (-\text{ sign}). \end{cases}$$

The current must surely ***decrease*** with increasing α, so evidently the + sign is correct:

$$\boxed{i = \frac{4\pi\sigma V_0}{\ln\left(2\alpha^2 - 1 + 2\alpha\sqrt{\alpha^2-1}\right)}, \quad \text{where } \alpha = \frac{d}{a}.}$$

Problem 7.40

(a) The resistance of *one* disk (Ex. 7.1) is $dR = \frac{dz}{\sigma A} = \frac{\rho}{\pi r^2}\,dz$, where $r = \left(\frac{b-a}{L}\right)z + a$ is the radius of the disk. The total resistance is

$$R = \frac{\rho}{\pi}\int_0^L \frac{1}{\left[\left(\frac{b-a}{L}\right)z + a\right]^2}\,dz = \frac{\rho}{\pi}\left(\frac{L}{b-a}\right)\left\{\frac{-1}{\left[\left(\frac{b-a}{L}\right)z + a\right]}\right\}\Bigg|_0^L = \frac{\rho L}{\pi(b-a)}\left[-\frac{1}{(b-a+a)} + \frac{1}{a}\right]$$

$$= \frac{\rho L}{\pi(b-a)}\left(\frac{b-a}{ab}\right) = \boxed{\frac{\rho L}{\pi ab}}.$$

(b) In Ex. 7.1 the current was parallel to the axis; here it certainly is *not*. (Nor is it radial with respect to the apex of the cone, since the ends are *flat*. This is not an easy configuration to solve exactly.)

(c) This time the flow *is* radial, and we can add the resistances of nested spherical shells: $dR = \dfrac{\rho}{A}\,dr$, where

$$A = \int_0^\theta r^2\sin\theta\,d\theta\,d\phi = 2\pi r^2(-\cos\theta)\Big|_0^\theta = 2\pi r^2(1-\cos\theta).$$

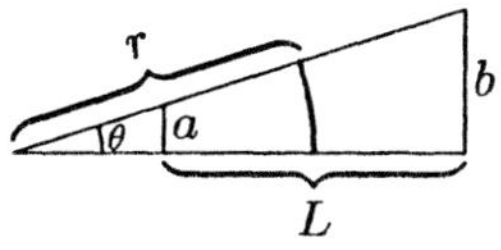

$$\begin{aligned}R &= \frac{\rho}{2\pi(1-\cos\theta)}\int_{r_a}^{r_b}\frac{1}{r^2}\,dr = \frac{\rho}{2\pi(1-\cos\theta)}\left(\frac{r_b-r_a}{r_a r_b}\right).\ \text{Now } \frac{a}{r_a}=\frac{b}{r_b}=\sin\theta.\\ &= \frac{\rho(b-a)}{2\pi ab}\frac{\sin\theta}{(1-\cos\theta)}.\ \text{But } \sin\theta = \frac{b-a}{\sqrt{L^2+(b-a)^2}} \text{ and } \cos\theta = \frac{L}{\sqrt{L^2+(b-a)^2}}.\\ &= \boxed{\frac{\rho(b-a)^2}{2\pi ab}\frac{1}{\left[\sqrt{L^2+(b-a)^2}-L\right]}.}\end{aligned}$$

[Note that if $b-a \ll L$, then $\sqrt{L^2+(b-a)^2} \cong L\left[1+\frac{1}{2}\frac{(b-a)^2}{L^2}\right]$, and $R \cong \frac{\rho(b-a)^2}{2\pi ab}\frac{1}{(b-a)^2/2L} = \frac{\rho L}{\pi ab}$, as in (a).]

Problem 7.41

From Prob. 3.23, $\begin{cases} V_{\text{in}}(s,\phi) = \sum_{k=1}^{\infty} s^k b_k \sin(k\phi), & (s<a);\\ V_{\text{out}}(s,\phi) = \sum_{k=1}^{\infty} s^{-k} d_k \sin(k\phi), & (s>a).\end{cases}$

(We don't need the cosine terms, because V is clearly an *odd* function of ϕ.) At $s=a$, $V_{\text{in}} = V_{\text{out}} = V_0\phi/2\pi$. Let's start with V_{in}, and use Fourier's trick to determine b_k:

$\sum_{k=1}^{\infty} a^k b_k \sin(k\phi) = \frac{V_0\phi}{2\pi} \Rightarrow \sum_{k=1}^{\infty} a^k b_k \int_{-\pi}^{\pi}\sin(k\phi)\sin(k'\phi)\,d\phi = \frac{V_0}{2\pi}\int_{-\pi}^{\pi}\phi\sin(k'\phi)\,d\phi$. But

$\int_{-\pi}^{\pi}\sin(k\phi)\sin(k'\phi)\,d\phi = \pi\delta_{kk'}$, and

$\int_{-\pi}^{\pi}\phi\sin(k'\phi)\,d\phi = \left[\frac{1}{(k')^2}\sin(k'\phi) - \frac{\phi}{k'}\cos(k'\phi)\right]\Big|_{-\pi}^{\pi} = -\frac{2\pi}{k'}\cos(k'\phi) = -\frac{2\pi}{k'}(-1)^{k'}$. So

$\pi a^k b_k = \frac{V_0}{2\pi}\left[-\frac{2\pi}{k}(-1)^k\right]$, or $b_k = -\frac{V_0}{\pi k}\left(-\frac{1}{a}\right)^k$, and hence $V_{\text{in}}(s,\phi) = -\frac{V_0}{\pi}\sum_{k=1}^{\infty}\frac{1}{k}\left(-\frac{s}{a}\right)^k\sin(k\phi)$.

Similarly, $V_{\text{out}}(s,\phi) = -\frac{V_0}{\pi}\sum_{k=1}^{\infty}\frac{1}{k}\left(-\frac{a}{s}\right)^k\sin(k\phi)$. Both sums are of the form $S \equiv \sum_{k=1}^{\infty}\frac{1}{k}(-x)^k\sin(k\phi)$ (with $x=s/a$ for $r<a$ and $x=a/s$ for $r>a$). This series can be summed explicitly, using Euler's formula ($e^{i\theta} = \cos\theta + i\sin\theta$): $S = \text{Im}\sum_{k=1}^{\infty}\frac{1}{k}(-x)^k e^{ik\phi} = \text{Im}\sum_{k=1}^{\infty}\frac{1}{k}\left(-xe^{i\phi}\right)^k$.

But $\ln(1+w) = w - \frac{1}{2}w^2 + \frac{1}{3}w^3 - \frac{1}{4}w^4\cdots = -\sum_{k=1}^{\infty}\frac{1}{k}(-w)^k$, so $S = -\text{Im}\left[\ln\left(1+xe^{i\phi}\right)\right]$.

Now $\ln\left(Re^{i\theta}\right) = \ln R + i\theta$, so $S = -\theta$, where

$$\tan\theta = \frac{\text{Im}\left(1+xe^{i\phi}\right)}{\text{Re}\left(1+xe^{i\phi}\right)} = \frac{\frac{1}{2i}\left[(1+xe^{i\phi})-(1+xe^{-i\phi})\right]}{\frac{1}{2}\left[(1+xe^{i\phi})+(1+xe^{-i\phi})\right]} = \frac{x\left(e^{i\phi}-e^{-i\phi}\right)}{i\left[2+x\left(e^{i\phi}+e^{-i\phi}\right)\right]} = \frac{x\sin\phi}{1+x\cos\phi}.$$

$$\text{Conclusion: } \boxed{\begin{cases} V_{\text{in}}(s,\phi) &= \dfrac{V_0}{\pi}\tan^{-1}\left(\dfrac{s\sin\phi}{a+s\cos\phi}\right), \quad (s<a); \\ V_{\text{out}}(s,\phi) &= \dfrac{V_0}{\pi}\tan^{-1}\left(\dfrac{a\sin\phi}{s+a\cos\phi}\right), \quad (s>a). \end{cases}}$$

(b) From Eq. 2.36, $\sigma(\phi) = -\epsilon_0\left\{\left.\dfrac{\partial V_{\text{out}}}{\partial s}\right|_{s=a} - \left.\dfrac{\partial V_{\text{in}}}{\partial s}\right|_{s=a}\right\}$.

$$\begin{aligned} \frac{\partial V_{\text{out}}}{\partial s} &= \frac{V_0}{\pi}\left\{\frac{1}{\left[1+\left(\frac{a\sin\phi}{s+a\cos\phi}\right)^2\right]}\frac{(-a\sin\phi)}{(s+a\cos\phi)^2}\right\} = -\frac{V_0}{\pi}\left[\frac{a\sin\phi}{(s+a\cos\phi)^2+(a\sin\phi)^2}\right] \\ &= -\frac{V_0}{\pi}\left(\frac{a\sin\phi}{s^2+2as\cos\phi+a^2}\right); \\ \frac{\partial V_{\text{in}}}{\partial s} &= \frac{V_0}{\pi}\left\{\frac{1}{\left[1+\left(\frac{s\sin\phi}{a+s\cos\phi}\right)^2\right]}\frac{[(a+s\cos\phi)\sin\phi - s\sin\phi\cos\phi]}{(a+s\cos\phi)^2}\right\} = \frac{V_0}{\pi}\left[\frac{a\sin\phi}{(a+s\cos\phi)^2+(s\sin\phi)^2}\right] \\ &= \frac{V_0}{\pi}\left(\frac{a\sin\phi}{s^2+2as\cos\phi+a^2}\right). \end{aligned}$$

$$\left.\frac{\partial V_{\text{in}}}{\partial s}\right|_{s=a} = -\left.\frac{\partial V_{\text{out}}}{\partial s}\right|_{s=a} = \frac{V_0}{2\pi a}\left(\frac{\sin\phi}{1+\cos\phi}\right), \text{ so } \sigma(\phi) = \frac{\epsilon_0 V_0}{\pi a}\frac{\sin\phi}{(1+\cos\phi)} = \boxed{\frac{\epsilon_0 V_0}{\pi a}\tan(\phi/2).}$$

Problem 7.42

(a) Faraday's law says $\nabla\times\mathbf{E} = -\frac{\partial\mathbf{B}}{\partial t}$, so $\mathbf{E}=0 \Rightarrow \frac{\partial\mathbf{B}}{\partial t}=0 \Rightarrow \mathbf{B}(\mathbf{r})$ is independent of t.

(b) Faraday's law in integral form (Eq. 7.18) says $\oint\mathbf{E}\cdot d\mathbf{l} = -d\Phi/dt$. In the wire itself $\mathbf{E}=0$, so Φ through the loop is constant.

(c) Ampère-Maxwell$\Rightarrow$ $\nabla\times\mathbf{B} = \mu_0\mathbf{J} + \mu_0\epsilon_0\frac{\partial\mathbf{E}}{\partial t}$, so $\mathbf{E}=0$, $\mathbf{B}=0 \Rightarrow \mathbf{J}=0$, and hence any current must be at the surface.

(d) From Eq. 5.68, a rotating shell produces a uniform magnetic field (inside): $\mathbf{B} = \frac{2}{3}\mu_0\sigma\omega a\hat{\mathbf{z}}$. So to *cancel* such a field, we need $\sigma\omega a = -\dfrac{3}{2}\dfrac{B_0}{\mu_0}$. Now $\mathbf{K} = \sigma\mathbf{v} = \sigma\omega a\sin\theta\,\hat{\boldsymbol{\phi}}$, so $\boxed{\mathbf{K} = -\dfrac{3B_0}{2\mu_0}\sin\theta\,\hat{\boldsymbol{\phi}}.}$

Problem 7.43

(a) To make the field *parallel* to the plane, we need image monopoles of the *same* sign (compare Figs. 2.13 and 2.14), so the image dipole points $\boxed{\text{down } (-z).}$

(b) From Prob. 6.3 (with $r \to 2z$):

$$F = \boxed{\frac{3\mu_0}{2\pi}\frac{m^2}{(2z)^4}}. \quad \frac{3\mu_0}{2\pi}\frac{m^2}{(2h)^4} = Mg \Rightarrow h = \boxed{\frac{1}{2}\left(\frac{3\mu_0 m^2}{2\pi Mg}\right)^{1/4}}.$$

(c) Using Eq. 5.87, and referring to the figure:

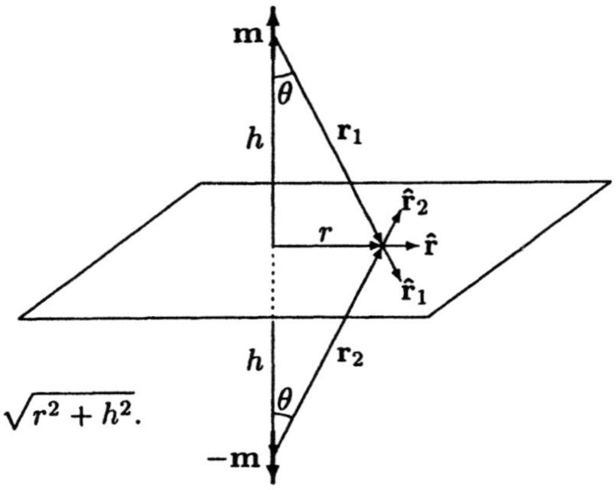

$$\mathbf{B} = \frac{\mu_0}{4\pi}\frac{1}{(r_1)^3}\left\{[3(m\,\hat{\mathbf{z}}\cdot\hat{\mathbf{r}}_1)\,\hat{\mathbf{r}}_1 - m\hat{\mathbf{z}}] + [3(-m\,\hat{\mathbf{z}}\cdot\hat{\mathbf{r}}_2)\,\hat{\mathbf{r}}_2 + m\hat{\mathbf{z}}]\right\}$$
$$= \frac{3\mu_0 m}{4\pi(r_1)^3}\left[(\hat{\mathbf{z}}\cdot\hat{\mathbf{r}}_1)\,\hat{\mathbf{r}}_1 - (\hat{\mathbf{z}}\cdot\hat{\mathbf{r}}_2)\,\hat{\mathbf{r}}_2\right]. \quad \text{But } \hat{\mathbf{z}}\cdot\hat{\mathbf{r}}_1 = -\hat{\mathbf{z}}\cdot\hat{\mathbf{r}}_2 = \cos\theta.$$
$$= -\frac{3\mu_0 m}{4\pi(r_1)^3}\cos\theta(\hat{\mathbf{r}}_1 + \hat{\mathbf{r}}_2). \quad \text{But } \hat{\mathbf{r}}_1 + \hat{\mathbf{r}}_2 = 2\sin\theta\,\hat{\mathbf{r}}.$$
$$= -\frac{3\mu_0 m}{2\pi(r_1)^3}\sin\theta\cos\theta\,\hat{\mathbf{r}}. \quad \text{But } \sin\theta = \frac{r}{r_1},\ \cos\theta = \frac{h}{r_1},\ \text{and } r_1 = \sqrt{r^2+h^2}.$$
$$= -\frac{3\mu_0 mh}{2\pi}\frac{r}{(r^2+h^2)^{5/2}}\,\hat{\mathbf{r}}.$$

Now $\mathbf{B} = \mu_0(\mathbf{K}\times\hat{\mathbf{z}}) \Rightarrow \hat{\mathbf{z}}\times\mathbf{B} = \mu_0\,\hat{\mathbf{z}}\times(\mathbf{K}\times\hat{\mathbf{z}}) = \mu_0\,[\mathbf{K} - \hat{\mathbf{z}}(\mathbf{K}\cdot\hat{\mathbf{z}})] = \mu_0\mathbf{K}$. (I used the BAC-CAB rule, and noted that $\mathbf{K}\cdot\hat{\mathbf{z}} = 0$, because the surface current is in the xy plane.)

$$\mathbf{K} = \frac{1}{\mu_0}(\hat{\mathbf{z}}\times\mathbf{B}) = -\frac{3mh}{2\pi}\frac{r}{(r^2+h^2)^{5/2}}(\hat{\mathbf{z}}\times\hat{\mathbf{r}}) = -\frac{3mh}{2\pi}\frac{r}{(r^2+h^2)^{5/2}}\,\hat{\boldsymbol{\phi}}. \quad \text{qed}$$

Problem 7.44

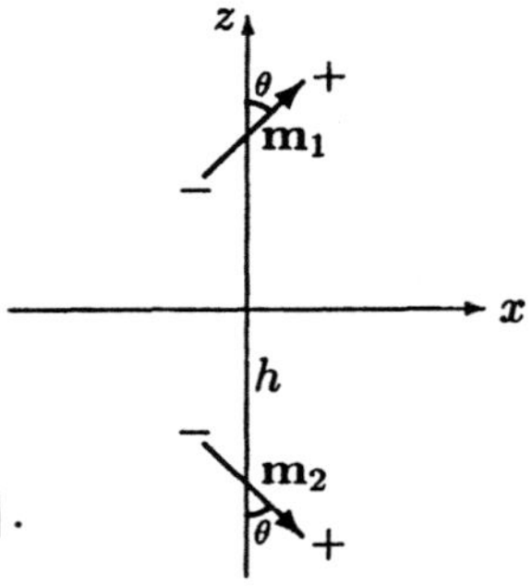

Say the angle between the dipole ($\mathbf{m}_1$) and the z axis is θ (see diagram).

The field of the image dipole ($\mathbf{m}_2$) is

$$\mathbf{B}(z) = \frac{\mu_0}{4\pi}\frac{1}{(h+z)^3}\left[3(\mathbf{m}_2\cdot\hat{\mathbf{z}})\,\hat{\mathbf{z}} - \mathbf{m}_2\right]$$

for points on the z axis (Eq. 5.87). The torque on $\mathbf{m}_1$ is (Eq. 6.1)

$$\mathbf{N} = \mathbf{m}_1\times\mathbf{B} = \frac{\mu_0}{4\pi(2h)^3}\left[3(\mathbf{m}_2\cdot\hat{\mathbf{z}})(\mathbf{m}_1\times\hat{\mathbf{z}}) - (\mathbf{m}_1\times\mathbf{m}_2)\right].$$

But $\mathbf{m}_1 = m(\sin\theta\,\hat{\mathbf{x}} + \cos\theta\hat{\mathbf{z}})$, $\mathbf{m}_2 = m(\sin\theta\,\hat{\mathbf{x}} - \cos\theta\hat{\mathbf{z}})$, so $\mathbf{m}_2\cdot\hat{\mathbf{z}} = -m\cos\theta$, $\mathbf{m}_1\times\hat{\mathbf{z}} = -m\sin\theta\,\hat{\mathbf{y}}$, and $\mathbf{m}_1\times\mathbf{m}_2 = 2m^2\sin\theta\cos\theta\hat{\mathbf{y}}$.

$$\mathbf{N} = \frac{\mu_0}{4\pi(2h)^3}\left[3m^2\sin\theta\cos\theta\,\hat{\mathbf{y}} - 2m^2\sin\theta\cos\theta\,\hat{\mathbf{y}})\right] = \frac{\mu_0 m^2}{4\pi(2h)^3}\sin\theta\cos\theta\,\hat{\mathbf{y}}.$$

Evidently the torque is zero for $\theta = 0,\ \pi/2$, or π. But 0 and π are clearly unstable, since the nearby ends of the dipoles (minus, in the figure) dominate, and they repel. The stable configuration is $\theta = \pi/2$: parallel to the surface (contrast Prob. 4.6).

In this orientation, $\mathbf{B}(z) = -\frac{\mu_0 m}{4\pi(h+z)^3}\,\hat{\mathbf{x}}$, and the force on $\mathbf{m}_1$ is (Eq. 6.3):

$$\mathbf{F} = \boldsymbol{\nabla}\left[-\frac{\mu_0 m^2}{4\pi(h+z)^3}\right]\Bigg|_{z=h} = \frac{3\mu_0 m^2}{4\pi(h+z)^4}\,\hat{\mathbf{z}}\Bigg|_{z=h} = \frac{3\mu_0 m^2}{4\pi(2h)^4}\,\hat{\mathbf{z}}.$$

At equilibrium this force upward balances the weight Mg:

$$\frac{3\mu_0 m^2}{4\pi(2h)^4} = Mg \Rightarrow h = \boxed{\frac{1}{2}\left(\frac{3\mu_0 m^2}{4\pi Mg}\right)^{1/4}}.$$

Incidentally, this is $(1/2)^{1/4} = 0.84$ times the height it would adopt in the orientation *perpendicular* to the plane (Prob. 7.43b).

Problem 7.45

$\mathbf{f} = \mathbf{v} \times \mathbf{B}$; $\mathbf{v} = \omega a \sin\theta\, \hat{\boldsymbol{\phi}}$; $\mathbf{f} = \omega a B_0 \sin\theta (\hat{\boldsymbol{\phi}} \times \hat{\mathbf{z}})$. $\mathcal{E} = \int \mathbf{f} \cdot d\mathbf{l}$, and $d\mathbf{l} = a\, d\theta\, \hat{\boldsymbol{\theta}}$.

So $\mathcal{E} = \omega a^2 B_0 \int_0^{\pi/2} \sin\theta (\hat{\boldsymbol{\phi}} \times \hat{\mathbf{z}}) \cdot \hat{\boldsymbol{\theta}}\, d\theta$. But $\hat{\boldsymbol{\theta}} \cdot (\hat{\boldsymbol{\phi}} \times \hat{\mathbf{z}}) = \hat{\mathbf{z}} \cdot (\hat{\boldsymbol{\theta}} \times \hat{\boldsymbol{\phi}}) = \hat{\mathbf{z}} \cdot \hat{\mathbf{r}} = \cos\theta$.

$$\mathcal{E} = \omega a^2 B_0 \int_0^{\pi/2} \sin\theta \cos\theta\, d\theta = \omega a^2 B_0 \left[\frac{\sin^2\theta}{2}\right]\Big|_0^{\pi/2} = \boxed{\frac{1}{2}\omega a^2 B_0}$$ (same as the rotating disk in Ex. 7.4).

Problem 7.46

(a) In the "square" orientation (□), it falls at terminal velocity $\boxed{v_{\text{square}} = \dfrac{mgR}{B^2 l^2}}$ (Prob. 7.11). In the "diamond" orientation (◇), the magnetic force upward is $F = IBd$ (Prob. 5.40).

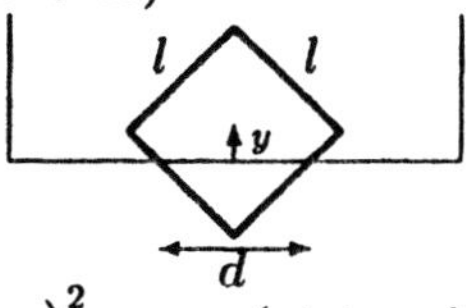

The flux is $\Phi = B\left[l^2 - (d/2)^2\right]$, and $d/2 = l/\sqrt{2} - y$,
so $\Phi = B\left[l^2 - (l/\sqrt{2} - y)^2\right]$.
$\mathcal{E} = -\frac{d\Phi}{dt} = -2B\left(l/\sqrt{2} - y\right)\frac{dy}{dt}$. But $\frac{dy}{dt} = -v$.
So $\mathcal{E} = 2Bv\left(l/\sqrt{2} - y\right) = IR \Rightarrow I = \frac{2Bv}{R}\left(l/\sqrt{2} - y\right)$; $F = 2 \cdot \frac{2B^2 v}{R}\left(l/\sqrt{2} - y\right)^2 = mg$ (at terminal velocity).

$$\boxed{v_{\text{diamond}} = \frac{mgR}{4B^2\left(l/\sqrt{2} - y\right)^2}.}$$ (This works for negative y as well as positive, if you replace y by $|y|$.)

Thus $\dfrac{v_{\text{square}}}{v_{\text{diamond}}} = \left(\dfrac{mgR}{B^2 l^2}\right)\dfrac{4B^2\left(l/\sqrt{2} - y\right)^2}{mgR} = \boxed{\left(\sqrt{2} - 2y/l\right)^2.}$ At first ($y \sim l/\sqrt{2}$) the "diamond" falls faster; toward the halfway mark ($y \sim 0$), the "square" falls twice as fast; then the diamond again takes over. The *total* time it takes for the square to fall is:

$$t_{\text{square}} = \frac{l}{v_{\text{square}}} = \boxed{\frac{B^2 l^3}{mgR}}$$

(assuming it always goes at the terminal velocity, which—as we found in Prob. 7.11—is close to the truth, if the field is strong). For the diamond, t is

$$-\int \frac{dy}{v_{\text{diamond}}} = -\frac{8B^2}{mgR}\int_{l/\sqrt{2}}^{0} -\left(l/\sqrt{2} - y\right)^2 dy = \frac{8B^2}{mgR}\left[\frac{1}{3}\left(l/\sqrt{2} - y\right)^3\right]\Bigg|_{l/\sqrt{2}}^{0} = \frac{8B^2}{mgR}\frac{1}{3}\frac{l^3}{2\sqrt{2}} = \boxed{\frac{2\sqrt{2}}{3}\frac{B^2 l^3}{mgR}.}$$

So $t_{\text{square}}/t_{\text{diamond}} = 3/2\sqrt{2} = 1.06$. The ***"square" falls faster, overall.*** If free to rotate, it would start out in the "diamond" orientation, switch to "square" for the middle portion, and then switch back to diamond, always trying to present the minimum *chord* at the field's edge.

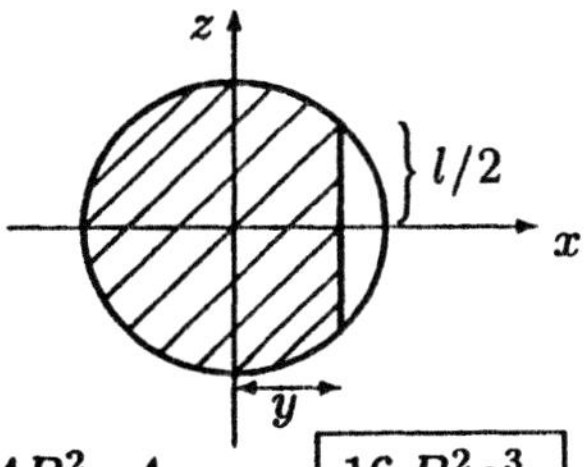

(b) $F = IBl$; $\Phi = 2B\int_{-a}^{y}\sqrt{a^2 - x^2}\, dx$ (a = radius of circle).
$\mathcal{E} = -\frac{d\Phi}{dt} = -2B\sqrt{a^2 - y^2}\frac{dy}{dt} = 2Bv\sqrt{a^2 - y^2} = IR$.
$I = \frac{2Bv}{R}\sqrt{a^2 - y^2}$; $l/2 = \sqrt{a^2 - y^2}$. So $F = \frac{4B^2 v}{R}(a^2 - y^2) = mg$.

$$v_{\text{circle}} = \frac{mgR}{4B^2(a^2 - y^2)};$$

$$t_{\text{circle}} = \int_{+a}^{-a} -\frac{dy}{v} = \frac{4B^2}{mgR}\int_{-a}^{a}(a^2 - y^2)dy = \frac{4B^2}{mgR}\left(a^2 y - \frac{1}{3}y^3\right)\Big|_{-a}^{a} = \frac{4B^2}{mgR}\left(\frac{4}{3}a^3\right) = \boxed{\frac{16}{3}\frac{B^2 a^3}{mgR}.}$$

Problem 7.47

(a) In magnetostatics

$$\nabla \cdot \mathbf{B} = 0,\ \nabla \times \mathbf{B} = \mu_0 \mathbf{J} \ \Rightarrow\ \mathbf{B}(\mathbf{r}) = \frac{\mu_0}{4\pi}\int \frac{\mathbf{J}(\mathbf{r}') \times \hat{\boldsymbol{\mathscr{r}}}}{\mathscr{r}^2}\, d\tau'.$$

For Faraday electric fields (with $\rho = 0$), therefore,

$$\nabla \cdot \mathbf{E} = 0,\ \nabla \times \mathbf{E} = -\frac{\partial \mathbf{B}}{\partial t} \ \Rightarrow\ \mathbf{E}(\mathbf{r}, t) = -\frac{1}{4\pi}\frac{\partial}{\partial t}\int \frac{\mathbf{B}(\mathbf{r}', t) \times \hat{\boldsymbol{\mathscr{r}}}}{\mathscr{r}^2}\, d\tau'$$

(with the substitution $\mathbf{J} \to -\frac{1}{\mu_0}\frac{\partial \mathbf{B}}{\partial t}$.)

(b) From Prob. 5.50a,

$$\mathbf{A}(\mathbf{r}, t) = \frac{1}{4\pi}\int \frac{\mathbf{B}(\mathbf{r}', t) \times \hat{\boldsymbol{\mathscr{r}}}}{\mathscr{r}^2}\, d\tau', \text{ so } \mathbf{E} = -\frac{\partial \mathbf{A}}{\partial t}.$$

[*Check:* $\nabla \times \mathbf{E} = -\frac{\partial}{\partial t}(\nabla \times \mathbf{A}) = -\frac{\partial \mathbf{B}}{\partial t}$, and we recover Faraday's law.]

(c) The Coulomb field is zero inside and $\frac{1}{4\pi\epsilon_0}\frac{Q}{r^2}\,\hat{\mathbf{r}} = \frac{1}{4\pi\epsilon_0}\frac{\sigma 4\pi R^2}{r^2}\,\hat{\mathbf{r}} = \frac{\sigma R^2}{\epsilon_0 r^2}\,\hat{\mathbf{r}}$ outside. The Faraday field is $-\frac{\partial \mathbf{A}}{\partial t}$, where $\mathbf{A}$ is given (in the quasistatic approximation) by Eq. 5.67, with ω a function of time. Letting $\dot{\omega} \equiv d\omega/dt$,

$$\boxed{\mathbf{E}(r, \theta, \phi, t) = \begin{cases} \dfrac{\mu_0 R \dot{\omega} \sigma}{3} r \sin\theta\, \hat{\boldsymbol{\phi}} & (r < R), \\[2ex] \dfrac{\sigma R^2}{\epsilon_0 r^2}\,\hat{\mathbf{r}} + \dfrac{\mu_0 R^4 \dot{\omega} \sigma}{3}\dfrac{\sin\theta}{r^2}\,\hat{\boldsymbol{\phi}} & (r > R). \end{cases}}$$

Problem 7.48

$qBR = mv$ (Eq. 5.3). If R is to stay fixed, then $qR\dfrac{dB}{dt} = m\dfrac{dv}{dt} = ma = F = qE$, or $E = R\dfrac{dB}{dt}$. But $\oint \mathbf{E} \cdot d\mathbf{l} = -\dfrac{d\Phi}{dt}$, so $E\, 2\pi R = -\dfrac{d\Phi}{dt}$, so $-\dfrac{1}{2\pi R}\dfrac{d\Phi}{dt} = R\dfrac{dB}{dt}$, or $B = -\dfrac{1}{2}\left(\dfrac{1}{\pi R^2}\Phi\right)$ + constant. If at time $t = 0$ the field is off, then the constant is zero, and $B(R) = \dfrac{1}{2}\left(\dfrac{1}{\pi R^2}\Phi\right)$ (in magnitude). Evidently the field at R must be *half* the *average* field over the cross-section of the orbit. qed

Problem 7.49

Initially, $\frac{mv^2}{r} = \frac{1}{4\pi\epsilon_0}\frac{qQ}{r^2} \Rightarrow T = \frac{1}{2}mv^2 = \frac{1}{2}\frac{1}{4\pi\epsilon_0}\frac{qQ}{r}$. After the magnetic field is on, the electron circles in a new orbit, of radius r_1 and velocity v_1:

$$\frac{mv_1^2}{r_1} = \frac{1}{4\pi\epsilon_0}\frac{qQ}{r_1^2} + qv_1 B \Rightarrow T_1 = \frac{1}{2}mv_1^2 = \frac{1}{2}\frac{1}{4\pi\epsilon_0}\frac{qQ}{r_1} + \frac{1}{2}qv_1 r_1 B.$$

But $r_1 = r + dr$, so $(r_1)^{-1} = r^{-1}\left(1 + \frac{dr}{r}\right)^{-1} \cong r^{-1}\left(1 - \frac{dr}{r}\right)$, while $v_1 = v + dv$, $B = dB$. To first order, then,

$$T_1 = \frac{1}{2}\frac{1}{4\pi\epsilon_0}\frac{qQ}{r}\left(1 - \frac{dr}{r}\right) + \frac{1}{2}q(vr)\,dB, \text{ and hence } dT = T_1 - T = \frac{qvr}{2}\,dB - \frac{1}{2}\frac{1}{4\pi\epsilon_0}\frac{qQ}{r^2}\,dr.$$

Now, the induced electric field is $E = \frac{r}{2}\frac{dB}{dt}$ (Ex. 7.7), so $m\frac{dv}{dt} = qE = \frac{qr}{2}\frac{dB}{dt}$, or $m\,dv = \frac{qr}{2}dB$. The increase in kinetic energy is therefore $dT = d(\frac{1}{2}mv^2) = mv\,dv = \frac{qvr}{2}dB$. Comparing the two expressions, I conclude that $dr = 0$. qed

Problem 7.50

$\mathcal{E} = -\frac{d\Phi}{dt} = -\alpha$. So the current in R_1 and R_2 is $I = \frac{\alpha}{R_1 + R_2}$; by Lenz's law, it flows counterclockwise. Now the voltage across R_1 (which voltmeter #1 measures) is $V_1 = IR_1 = \boxed{\frac{\alpha R_1}{R_1 + R_2}}$ (V_b is the *higher* potential), and $V_2 = -IR_2 = \boxed{\frac{-\alpha R^2}{R_1 + R_2}}$ (V_b is *lower*).

Problem 7.51

$$\mathcal{E} = vBh = -L\frac{dI}{dt};\ F = IhB = m\frac{dv}{dt};\ \frac{d^2v}{dt^2} = \frac{hB}{m}\frac{dI}{dt} = -\frac{hB}{m}\left(\frac{hB}{L}\right)v,\ \boxed{\frac{d^2v}{dt^2} = -\omega^2 v,}\ \text{with}\ \boxed{\omega = \frac{hB}{\sqrt{mL}}.}$$

Problem 7.52

A point on the upper loop: $\mathbf{r}_2 = (a\cos\phi_2, a\sin\phi_2, z)$; a point on the lower loop: $\mathbf{r}_1 = (b\cos\phi_1, b\sin\phi_1, 0)$.

$$\begin{aligned}
\mathscr{r}^2 &= (\mathbf{r}_2 - \mathbf{r}_1)^2 = (a\cos\phi_2 - b\cos\phi_1)^2 + (a\sin\phi_2 - b\sin\phi_1)^2 + z^2 \\
&= a^2\cos^2\phi_2 - 2ab\cos\phi_2\cos\phi_1 + b^2\cos^2\phi_1 + a^2\sin^2\phi_2 - 2ab\sin\phi_1\sin\phi_2 + b^2\sin^2\phi_1 + z^2 \\
&= a^2 + b^2 + z^2 - 2ab(\cos\phi_2\cos\phi_1 + \sin\phi_2\sin\phi_1) = a^2 + b^2 + z^2 - 2ab\cos(\phi_2 - \phi_1) \\
&= (a^2 + b^2 + z^2)\left[1 - 2\beta\cos(\phi_2 - \phi_1)\right] = \frac{ab}{\beta}\left[1 - 2\beta\cos(\phi_2 - \phi_1)\right].
\end{aligned}$$

$d\mathbf{l}_1 = b\,d\phi_1\,\hat{\boldsymbol{\phi}}_1 = b\,d\phi_1[-\sin\phi_1\,\hat{\mathbf{x}} + \cos\phi_1\,\hat{\mathbf{y}}]$; $d\mathbf{l}_2 = a\,d\phi_2\,\hat{\boldsymbol{\phi}}_2 = a\,d\phi_2[-\sin\phi_2\,\hat{\mathbf{x}} + \cos\phi_2\,\hat{\mathbf{y}}]$, so
$d\mathbf{l}_1 \cdot d\mathbf{l}_2 = ab\,d\phi_1\,d\phi_2[\sin\phi_1\sin\phi_2 + \cos\phi_1\cos\phi_2] = ab\cos(\phi_2 - \phi_1)\,d\phi_1\,d\phi_2$.

$$M = \frac{\mu_0}{4\pi}\oint\oint\frac{d\mathbf{l}_1 \cdot d\mathbf{l}_2}{\mathscr{r}} = \frac{\mu_0}{4\pi}\frac{ab}{\sqrt{ab/\beta}}\int\int\frac{\cos(\phi_2 - \phi_1)}{\sqrt{1 - 2\beta\cos(\phi_2 - \phi_1)}}\,d\phi_2\,d\phi_1.$$

Both integrals run from 0 to 2π. Do the ϕ_2 integral first, letting $u \equiv \phi_2 - \phi_1$:

$$\int_{-\phi_1}^{2\pi-\phi_1}\frac{\cos u}{\sqrt{1 - 2\beta\cos u}}\,du = \int_0^{2\pi}\frac{\cos u}{\sqrt{1 - 2\beta\cos u}}\,du$$

(since the integral runs over a complete cycle of $\cos u$, we may as well change the limits to $0 \to 2\pi$). Then the ϕ_1 integral is just 2π, and

$$M = \frac{\mu_0}{4\pi}\sqrt{ab\beta}\,2\pi\int_0^{2\pi}\frac{\cos u}{\sqrt{1 - 2\beta\cos u}}\,du = \frac{\mu_0}{2}\sqrt{ab\beta}\int_0^{2\pi}\frac{\cos u}{\sqrt{1 - 2\beta\cos u}}\,du.$$

(a) If a is small, then $\beta \ll 1$, so (using the binomial theorem)

$$\frac{1}{\sqrt{1 - 2\beta\cos u}} \cong 1 + \beta\cos u,\ \text{and}\ \int_0^{2\pi}\frac{\cos u}{\sqrt{1 - 2\beta\cos u}}\,du \cong \int_0^{2\pi}\cos u\,du + \beta\int_0^{2\pi}\cos^2 u\,du = 0 + \beta\pi,$$

and hence $M = (\mu_0\pi/2)\sqrt{ab\beta^3}$. Moreover, $\beta \cong ab/(b^2 + z^2)$, so $M \cong \boxed{\frac{\mu_0\pi a^2 b^2}{2(b^2 + z^2)^{3/2}}}$ (same as in Prob. 7.20).

(b) More generally,

$$(1+\epsilon)^{-1/2} = 1 - \frac{1}{2}\epsilon + \frac{3}{8}\epsilon^2 - \frac{5}{16}\epsilon^3 + \cdots \Longrightarrow \frac{1}{\sqrt{1-2\beta\cos u}} = 1 + \beta\cos u + \frac{3}{2}\beta^2\cos^2 u + \frac{5}{2}\beta^3\cos^3 u + \cdots,$$

so

$$M = \frac{\mu_0}{2}\sqrt{ab\beta}\left\{\int_0^{2\pi}\cos u\,du + \beta\int_0^{2\pi}\cos^2 u\,du + \frac{3}{2}\beta^2\int_0^{2\pi}\cos^3 u\,du + \frac{5}{2}\beta^3\int_0^{2\pi}\cos^4 u\,du + \cdots\right\}$$

$$= \frac{\mu_0}{2}\sqrt{ab\beta}\left[0 + \beta(\pi) + \frac{3}{2}\beta^2(0) + \frac{5}{2}\beta^3(\frac{3}{4}\pi) + \cdots\right] = \boxed{\frac{\mu_0\pi}{2}\sqrt{ab\beta^3}\left(1 + \frac{15}{8}\beta^2 + (\;)\beta^4 + \cdots\right).} \quad \text{qed}$$

Problem 7.53

Let Φ be the flux of $\mathbf{B}$ through a *single* loop of either coil, so that $\Phi_1 = N_1\Phi$ and $\Phi_2 = N_2\Phi$. Then

$$\mathcal{E}_1 = -N_1\frac{d\Phi}{dt}, \quad \mathcal{E}_2 = -N_2\frac{d\Phi}{dt}, \text{ so } \frac{\mathcal{E}_2}{\mathcal{E}_1} = \frac{N_2}{N_1}. \quad \text{qed}$$

Problem 7.54

(a) Suppose current I_1 flows in coil 1, and I_2 in coil 2. Then (if Φ is the flux through *one* turn):

$$\Phi_1 = I_1L_1 + MI_2 = N_1\Phi; \quad \Phi_2 = I_2L_2 + MI_1 = N_2\Phi, \quad \text{or } \Phi = I_1\frac{L_1}{N_1} + I_2\frac{M}{N_1} = I_2\frac{L_2}{N_2} + I_1\frac{M}{N_2}.$$

In case $I_1 = 0$, we have $\frac{M}{N_1} = \frac{L_2}{N_2}$; if $I_2 = 0$, we have $\frac{L_1}{N_1} = \frac{M}{N_2}$. Dividing: $\frac{M}{L_1} = \frac{L_2}{M}$, or $L_1L_2 = M^2$. qed

(b) $-\mathcal{E}_1 = \frac{d\Phi_1}{dt} = L_1\frac{dI_1}{dt} + M\frac{dI_2}{dt} = V_1\cos(\omega t)$; $-\mathcal{E}_2 = \frac{d\Phi_2}{dt} = L_2\frac{dI_2}{dt} + M\frac{dI_1}{dt} = -I_2R$. qed

(c) Multiply the first equation by L_2: $L_1L_2\frac{dI_1}{dt} + L_2\frac{dI_2}{dt}M = L_2V_1\cos\omega t$. Plug in $L_2\frac{dI_2}{dt} = -I_2R - M\frac{dI_1}{dt}$.

$$M^2\frac{dI_1}{dt} - MRI_2 - M^2\frac{dI_1}{dt} = L_2V_1\cos\omega t \Rightarrow \boxed{I_2(t) = -\frac{L_2V_1}{MR}\cos\omega t.} \; L_1\frac{dI_1}{dt} + M\left(\frac{L_2V_1}{MR}\omega\sin\omega t\right) = V_1\cos\omega t.$$

$$\frac{dI_1}{dt} = \frac{V_1}{L_1}\left(\cos\omega t - \frac{L_2}{R}\omega\sin\omega t\right) \Rightarrow \boxed{I_1(t) = \frac{V_1}{L_1}\left(\frac{1}{\omega}\sin\omega t + \frac{L_2}{R}\cos\omega t\right).}$$

(d) $\displaystyle\frac{V_{\text{out}}}{V_{\text{in}}} = \frac{I_2R}{V_1\cos\omega t} = \frac{-\frac{L_2V_1}{MR}\cos\omega t\,R}{V_1\cos\omega t} = -\frac{L_2}{M} = -\frac{N_2}{N_1}$. The ratio of the amplitudes is $\displaystyle\frac{N_2}{N_1}$. qed

(e) $$P_{\text{in}} = V_{\text{in}}I_1 = (V_1\cos\omega t)\left(\frac{V_1}{L_1}\right)\left(\frac{1}{\omega}\sin\omega t + \frac{L_2}{R}\cos\omega t\right) = \boxed{\frac{(V_1)^2}{L_1}\left(\frac{1}{\omega}\sin\omega t\cos\omega t + \frac{L_2}{R}\cos^2\omega t\right).}$$

$$P_{\text{out}} = V_{\text{out}}I_2 = (I_2)^2R = \boxed{\frac{(L_2V_1)^2}{M^2R}\cos^2\omega t.}$$ Average of $\cos^2\omega t$ is 1/2; average of $\sin\omega t\cos\omega t$ is zero.

$$\text{So } \langle P_{\text{in}}\rangle = \frac{1}{2}(V_1)^2\left(\frac{L_2}{L_1R}\right); \; \langle P_{\text{out}}\rangle = \frac{1}{2}(V_1)^2\left[\frac{(L_2)^2}{M^2R}\right] = \frac{1}{2}(V_1)^2\left[\frac{(L_2)^2}{L_1L_2R}\right]; \; \boxed{\langle P_{\text{in}}\rangle = \langle P_{\text{out}}\rangle = \frac{(V_1)^2L_2}{2L_1R}.}$$

Problem 7.55

(a) The continuity equation says $\frac{\partial\rho}{\partial t} = -\,\boldsymbol{\nabla}\cdot\mathbf{J}$. Here the right side is independent of t, so we can integrate: $\rho(t) = (-\,\boldsymbol{\nabla}\cdot\mathbf{J})t+$ constant. The "constant" may be a function of $\mathbf{r}$—it's only constant with respect to t. So, putting in the $\mathbf{r}$ dependence explicitly, and noting that $\boldsymbol{\nabla}\cdot\mathbf{J} = -\dot\rho(\mathbf{r},0)$, $\rho(\mathbf{r},t) = \dot\rho(\mathbf{r},0)t + \rho(\mathbf{r},0)$. qed

(b) Suppose $\mathbf{E} = \frac{1}{4\pi\epsilon_0}\int \frac{\rho\hat{\mathscr{r}}}{\mathscr{r}^2}\,d\tau$ and $\mathbf{B} = \frac{\mu_0}{4\pi}\int \frac{\mathbf{J}\times\hat{\mathscr{r}}}{\mathscr{r}^2}\,d\tau$. We want to show that $\nabla\cdot\mathbf{B} = 0$, $\nabla\times\mathbf{B} = \mu_0\mathbf{J}+\mu_0\epsilon_0\frac{\partial\mathbf{E}}{\partial t}$; $\nabla\cdot\mathbf{E} = \frac{1}{\epsilon_0}\rho$, and $\nabla\times\mathbf{E} = -\frac{\partial\mathbf{B}}{\partial t}$, provided that $\mathbf{J}$ is independent of t.

We know from Ch. 2 that Coulomb's law $\left(\mathbf{E} = \frac{1}{4\pi\epsilon_0}\int\frac{\rho\hat{\mathscr{r}}}{\mathscr{r}^2}\,d\tau\right)$ satisfies $\nabla\cdot\mathbf{E} = \frac{1}{\epsilon_0}\rho$ and $\nabla\times\mathbf{E} = 0$. Since $\mathbf{B}$ is constant (in time), the $\nabla\cdot\mathbf{E}$ and $\nabla\times\mathbf{E}$ equations are satisfied. From Chapter 5 (specifically, Eqs. 5.45-5.48) we know that the Biot-Savart law satisfies $\nabla\cdot\mathbf{B} = 0$. It remains only to check $\nabla\times\mathbf{B}$. The argument in Sect. 5.3.2 carries through until the equation following Eq. 5.52, where I invoked $\nabla'\cdot\mathbf{J} = 0$. In its place we now put $\nabla'\cdot\mathbf{J} = -\dot{\rho}$:

$$\nabla\times\mathbf{B} = \mu_0\mathbf{J} - \frac{\mu_0}{4\pi}\int \underbrace{(\mathbf{J}\cdot\nabla)\frac{\hat{\mathscr{r}}}{\mathscr{r}^2}}_{(-\mathbf{J}tp\nabla')\frac{\hat{\mathscr{r}}}{\mathscr{r}^2}\ \text{(Eq. 5.52)}}\,d\tau \qquad \text{(Eqs. 5.49-5.51)}$$

Integration by parts yields two terms, one of which becomes a surface integral, and goes to zero. The other is $\frac{\mathscr{r}}{\mathscr{r}^3}\nabla'\cdot\mathbf{J} = \frac{\hat{\mathscr{r}}}{\mathscr{r}^2}(-\dot{\rho})$. So:

$$\nabla\times\mathbf{B} = \mu_0\mathbf{J} - \frac{\mu_0}{4\pi}\int\frac{\hat{\mathscr{r}}}{\mathscr{r}^2}(-\dot{\rho})d\tau = \mu_0\mathbf{J} + \mu_0\epsilon_0\frac{\partial}{\partial t}\left\{\frac{1}{4\pi\epsilon_0}\int\frac{\rho\hat{\mathscr{r}}}{\mathscr{r}^3}\,d\tau\right\} = \mu_0\mathbf{J} + \mu_0\epsilon_0\frac{\partial\mathbf{E}}{\partial t}. \quad \text{qed}$$

Problem 7.56

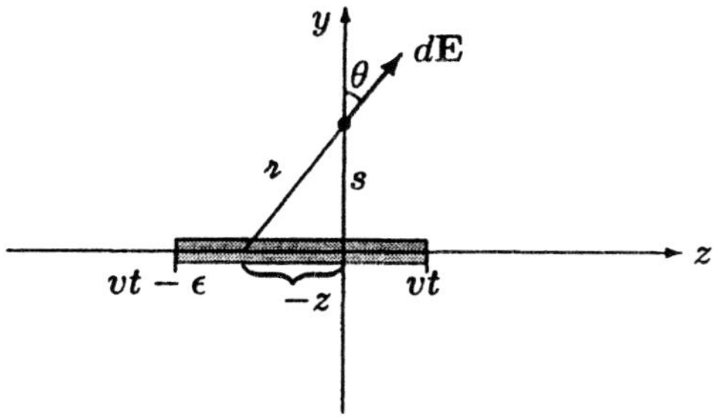

(a) $dE_z = \frac{1}{4\pi\epsilon_0}\frac{(-\lambda)dz}{\mathscr{r}^2}\sin\theta$

$\sin\theta = \frac{-z}{\mathscr{r}}$; $\mathscr{r} = \sqrt{z^2+s^2}$

$$E_z = \frac{\lambda}{4\pi\epsilon_0}\int\frac{z\,dz}{(z^2+s^2)^{3/2}} = \frac{\lambda}{4\pi\epsilon_0}\left[\frac{-1}{\sqrt{z^2+s^2}}\right]\Bigg|_{vt-\epsilon}^{vt}$$

$$\boxed{E_z = \frac{\lambda}{4\pi\epsilon_0}\left\{\frac{1}{\sqrt{(vt-\epsilon)^2+s^2}} - \frac{1}{\sqrt{(vt)^2+s^2}}\right\}.}$$

(b)

$$\Phi_E = \frac{\lambda}{4\pi\epsilon_0}\int_0^a\left\{\frac{1}{\sqrt{(vt-\epsilon)^2+s^2}} - \frac{1}{\sqrt{(vt)^2+s^2}}\right\}2\pi s\,ds = \frac{\lambda}{2\epsilon_0}\left[\sqrt{(vt-\epsilon)^2+s^2} - \sqrt{(vt)^2+s^2}\right]\Bigg|_0^a$$

$$= \boxed{\frac{\lambda}{2\epsilon_0}\left[\sqrt{(vt-\epsilon)^2+a^2} - \sqrt{(vt)^2+a^2} - (\epsilon - vt) + (vt)\right].}$$

(c) $$I_d = \epsilon_0\frac{d\Phi_E}{dt} = \boxed{\frac{\lambda}{2}\left\{\frac{v(vt-\epsilon)}{\sqrt{(vt-\epsilon)^2+a^2}} - \frac{v(vt)}{\sqrt{(vt)^2+a^2}} + 2v\right\}.}$$

As $\epsilon \to 0$, $vt < \epsilon$ also $\to 0$, so $I_d \to \frac{\lambda}{2}(2v) = \lambda v = I$. With an infinitesimal gap we attribute the magnetic field to *displacement* current, instead of real current, but we get the same answer. qed

---

Problem 7.57

(a) $\nabla^2 V = \frac{1}{s}\frac{\partial}{\partial s}\left(s\frac{\partial(zf)}{\partial s}\right) + \frac{\partial^2(zf)}{\partial z^2} = \frac{z}{s}\frac{d}{ds}\left(s\frac{df}{ds}\right) = 0 \Rightarrow \frac{d}{ds}\left(s\frac{df}{ds}\right) = 0 \Rightarrow s\frac{df}{ds} = A$ (a constant) $\Rightarrow$ $A\frac{ds}{s} = df \Rightarrow f = A\ln(s/s_0)$ (s_0 another constant). But (ii) $\Rightarrow f(b) = 0$, so $\ln(b/s_0) = 0$, so $s_0 = b$, and

$V(s,z) = Az\ln(s/b)$. But (i) $\Rightarrow Az\ln(a/b) = -(I\rho z)/(\pi a^2)$, so $A = -\dfrac{I\rho}{\pi a^2}\dfrac{1}{\ln(a/b)}$; $\boxed{V(s,z) = -\dfrac{I\rho z}{\pi a^2}\dfrac{\ln(s/b)}{\ln(a/b)}.}$

(b) $\mathbf{E} = -\nabla V = -\dfrac{\partial V}{\partial s}\hat{\mathbf{s}} - \dfrac{\partial V}{\partial z}\hat{\mathbf{z}} = \dfrac{I\rho z}{\pi a^2}\dfrac{1}{s\ln(a/b)}\hat{\mathbf{s}} + \dfrac{I\rho}{\pi a^2}\dfrac{\ln(s/b)}{\ln(a/b)}\hat{\mathbf{z}} = \boxed{\dfrac{I\rho}{\pi a^2\ln(a/b)}\left(\dfrac{z}{s}\hat{\mathbf{s}} + \ln\left(\dfrac{s}{b}\right)\hat{\mathbf{z}}\right).}$

(c) $\sigma(z) = \epsilon_0\left[E_s(a^+) - E_s(a^-)\right] = \epsilon_0\left[\dfrac{I\rho}{\pi a^2\ln(a/b)}\left(\dfrac{z}{a}\right) - 0\right] = \boxed{\dfrac{\epsilon_0 I\rho z}{\pi a^3\ln(a/b)}.}$

Problem 7.58

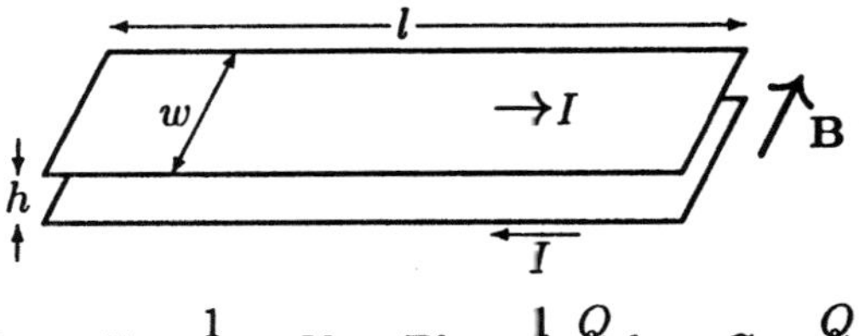

(a) Parallel-plate capacitor: $E = \dfrac{1}{\epsilon_0}\sigma$; $V = Eh = \dfrac{1}{\epsilon_0}\dfrac{Q}{wl}h \Rightarrow C = \dfrac{Q}{V} = \dfrac{\epsilon_0 wl}{h} \Rightarrow \boxed{\mathcal{C} = \dfrac{\epsilon_0 w}{h}.}$

(b) $B = \mu_0 K = \mu_0\dfrac{I}{w}$; $\Phi = Bhl = \dfrac{\mu_0 I}{w}hl = LI \Rightarrow L = \dfrac{\mu_0 h}{w}l \Rightarrow \boxed{\mathcal{L} = \dfrac{\mu_0 h}{w}.}$

(c) $\boxed{\mathcal{C}\mathcal{L} = \mu_0\epsilon_0} = (4\pi\times 10^{-7})(8.85\times 10^{-12}) = \boxed{1.112\times 10^{-17}\,\mathrm{s^2/m^2}.}$
(Propagation speed $1/\sqrt{\mathcal{L}\mathcal{C}} = 1/\sqrt{\mu_0\epsilon_0} = 2.999\times 10^8\,\mathrm{m/s} = c$.)

(d) $D = \sigma$, $E = D/\epsilon = \sigma/\epsilon$, so just replace ϵ_0 by ϵ;
$H = K$, $B = \mu H = \mu K$, so just replace μ_0 by μ. $\Big\}$ $\boxed{\mathcal{L}\mathcal{C} = \epsilon\mu;}$ $\boxed{v = 1/\sqrt{\epsilon\mu}.}$

Problem 7.59

(a) $\mathbf{J} = \sigma(\mathbf{E} + \mathbf{v}\times\mathbf{B})$; $\mathbf{J}$ finite, $\sigma = \infty \Rightarrow \mathbf{E} + (\mathbf{v}\times\mathbf{B}) = 0$. Take the curl: $\nabla\times\mathbf{E} + \nabla\times(\mathbf{v}\times\mathbf{B}) = 0$. But Faraday's law says $\nabla\times\mathbf{E} = -\dfrac{\partial\mathbf{B}}{\partial t}$. So $\dfrac{\partial\mathbf{B}}{\partial t} = \nabla\times(\mathbf{v}\times\mathbf{B})$. qed

(b) $\nabla\cdot\mathbf{B} = 0 \Rightarrow \oint\mathbf{B}\cdot d\mathbf{a} = 0$ for any closed surface. Apply this at time $(t+dt)$ to the surface consisting of $\mathcal{S}$, $\mathcal{S}'$, and $\mathcal{R}$:

$$\int_{\mathcal{S}'}\mathbf{B}(t+dt)\cdot d\mathbf{a} + \int_{\mathcal{R}}\mathbf{B}(t+dt)\cdot d\mathbf{a} - \int_{\mathcal{S}}\mathbf{B}(t+dt)\cdot d\mathbf{a} = 0$$

(the sign change in the third term comes from switching ***outward*** **da** to ***inward*** **da**).

$$d\Phi = \int_{\mathcal{S}'}\mathbf{B}(t+dt)\cdot d\mathbf{a} - \int_{\mathcal{S}}\mathbf{B}(t)\cdot d\mathbf{a} = \int_{\mathcal{S}}\underbrace{[\mathbf{B}(t+dt) - \mathbf{B}(t)]}_{\frac{\partial\mathbf{B}}{\partial t}dt\text{ (for infinitesimal }dt)}\cdot d\mathbf{a} - \int_{\mathcal{R}}\mathbf{B}(t+dt)\cdot d\mathbf{a}$$

$$d\Phi = \left\{\int_{\mathcal{S}}\frac{\partial\mathbf{B}}{\partial t}\cdot d\mathbf{a}\right\}dt - \int_{\mathcal{R}}\mathbf{B}(t+dt)\cdot[(d\mathbf{l}\times\mathbf{v})\,dt] \quad \text{(Figure 7.13).}$$

Since the second term is already first order in dt, we can replace $\mathbf{B}(t+dt)$ by $\mathbf{B}(t)$ (the distinction would be second order):

$$d\Phi = dt\int_{\mathcal{S}}\frac{\partial\mathbf{B}}{\partial t}\cdot d\mathbf{a} - dt\oint_{\mathcal{C}}\underbrace{\mathbf{B}\cdot(d\mathbf{l}\times\mathbf{v})}_{(\mathbf{v}\times\mathbf{B})\cdot d\mathbf{l}} = dt\left\{\int_{\mathcal{S}}\left(\frac{\partial\mathbf{B}}{\partial t}\right)\cdot d\mathbf{a} - \int_{\mathcal{S}}\nabla\times(\mathbf{v}\times\mathbf{B})\cdot d\mathbf{a}\right\}.$$

$$\frac{d\Phi}{dt} = \int_S \left[\frac{\partial \mathbf{B}}{\partial t} - \nabla\times(\mathbf{v}\times\mathbf{B})\right]\cdot d\mathbf{a} = 0. \quad \text{qed}$$

Problem 7.60

(a)

$$\begin{aligned}
\nabla\cdot\mathbf{E}' &= (\nabla\cdot\mathbf{E})\cos\alpha + c(\nabla\cdot\mathbf{B})\sin\alpha = \frac{1}{\epsilon_0}\rho_e\cos\alpha + c\mu_0\rho_m\sin\alpha \\
&= \frac{1}{\epsilon_0}(\rho_e\cos\alpha + c\mu_0\epsilon_0\rho_m\sin\alpha) = \frac{1}{\epsilon_0}(\rho_e\cos\alpha + \frac{1}{c}\rho_m\sin\alpha) = \frac{1}{\epsilon_0}\rho_e'. \checkmark \\
\nabla\cdot\mathbf{B}' &= (\nabla\cdot\mathbf{B})\cos\alpha - \frac{1}{c}(\nabla\cdot\mathbf{E})\sin\alpha = \mu_0\rho_m\cos\alpha - \frac{1}{c\epsilon_0}\rho_e\sin\alpha \\
&= \mu_0(\rho_m\cos\alpha - \frac{1}{c\mu_0\epsilon_0}\rho_e\sin\alpha) = \mu_0(\rho_m\cos\alpha - c\rho_e\sin\alpha) = \mu_0\rho_m'. \checkmark \\
\nabla\times\mathbf{E}' &= (\nabla\times\mathbf{E})\cos\alpha + c(\nabla\times\mathbf{B})\sin\alpha = \left(-\mu_0\mathbf{J}_m - \frac{\partial\mathbf{B}}{\partial t}\right)\cos\alpha + c\left(\mu_0\mathbf{J}_e + \mu_0\epsilon_0\frac{\partial\mathbf{E}}{\partial t}\right)\sin\alpha \\
&= -\mu_0(\mathbf{J}_m\cos\alpha - c\mathbf{J}_e\sin\alpha) - \frac{\partial}{\partial t}\left(\mathbf{B}\cos\alpha - \frac{1}{c}\mathbf{E}\sin\alpha\right) = -\mu_0\mathbf{J}_m' - \frac{\partial\mathbf{B}'}{\partial t}. \checkmark \\
\nabla\times\mathbf{B}' &= (\nabla\times\mathbf{B})\cos\alpha - \frac{1}{c}(\nabla\times\mathbf{E})\sin\alpha = \left(\mu_0\mathbf{J}_e + \mu_0\epsilon_0\frac{\partial\mathbf{E}}{\partial t}\right)\cos\alpha - \frac{1}{c}\left(-\mu_0\mathbf{J}_m - \frac{\partial\mathbf{B}}{\partial t}\right)\sin\alpha \\
&= \mu_0(\mathbf{J}_e\cos\alpha + \frac{1}{c}\mathbf{J}_m\sin\alpha) + \mu_0\epsilon_0\frac{\partial}{\partial t}(\mathbf{E}\cos\alpha + c\mathbf{B}\sin\alpha) = \mu_0\mathbf{J}_e' + \mu_0\epsilon_0\frac{\partial\mathbf{E}'}{\partial t}. \checkmark
\end{aligned}$$

(b)

$$\begin{aligned}
\mathbf{F}' &= q_e'(\mathbf{E}' + \mathbf{v}\times\mathbf{B}') + q_m'(\mathbf{B}' - \frac{1}{c^2}\mathbf{v}\times\mathbf{E}') \\
&= \left(q_e\cos\alpha + \frac{1}{c}q_m\sin\alpha\right)\left[(\mathbf{E}\cos\alpha + c\mathbf{B}\sin\alpha) + \mathbf{v}\times\left(\mathbf{B}\cos\alpha - \frac{1}{c}\mathbf{E}\sin\alpha\right)\right] \\
&\quad + (q_m\cos\alpha - cq_e\sin\alpha)\left[\left(\mathbf{B}\cos\alpha - \frac{1}{c}\mathbf{E}\sin\alpha\right) - \frac{1}{c^2}\mathbf{v}\times(\mathbf{E}\cos\alpha + c\mathbf{B}\sin\alpha)\right] \\
&= q_e\Big[(\mathbf{E}\cos^2\alpha + c\mathbf{B}\sin\alpha\cos\alpha - c\mathbf{B}\sin\alpha\cos\alpha + \mathbf{E}\sin^2\alpha) \\
&\quad +\mathbf{v}\times\left(\mathbf{B}\cos^2\alpha - \frac{1}{c}\mathbf{E}\sin\alpha\cos\alpha + \frac{1}{c}\mathbf{E}\sin\alpha\cos\alpha + \mathbf{B}\sin^2\alpha\right)\Big] \\
&\quad +q_m\Big[\left(\frac{1}{c}\mathbf{E}\sin\alpha\cos\alpha + \mathbf{B}\sin^2\alpha + \mathbf{B}\cos^2\alpha - \frac{1}{c}\mathbf{E}\sin\alpha\cos\alpha\right) \\
&\quad +\mathbf{v}\times\left(\frac{1}{c}\mathbf{B}\sin\alpha\cos\alpha - \frac{1}{c^2}\mathbf{E}\sin^2\alpha - \frac{1}{c^2}\mathbf{E}\cos^2\alpha - \frac{1}{c}\mathbf{B}\sin\alpha\cos\alpha\right)\Big] \\
&= q_e(\mathbf{E} + \mathbf{v}\times\mathbf{B}) + q_m\left(\mathbf{B} - \frac{1}{c^2}\mathbf{v}\times\mathbf{E}\right) = \mathbf{F}. \quad \text{qed}
\end{aligned}$$

Chapter 8

Conservation Laws

Problem 8.1

Example 7.13.

$$\left.\begin{array}{l}\mathbf{E}=\dfrac{\lambda}{2\pi\epsilon_0}\dfrac{1}{s}\,\hat{\mathbf{s}}\\[2ex] \mathbf{B}=\dfrac{\mu_0 I}{2\pi}\dfrac{1}{s}\,\hat{\boldsymbol{\phi}}\end{array}\right\}\mathbf{S}=\frac{1}{\mu_0}(\mathbf{E}\times\mathbf{B})=\frac{\lambda I}{4\pi^2\epsilon_0}\frac{1}{s^2}\,\hat{\mathbf{z}};$$

$$P=\int \mathbf{S}\cdot d\mathbf{a}=\int_a^b S\,2\pi s\,ds=\frac{\lambda I}{2\pi\epsilon_0}\int_a^b \frac{1}{s}\,ds=\frac{\lambda I}{2\pi\epsilon_0}\ln(b/a).$$

$$\text{But } V=\int_a^b \mathbf{E}\cdot d\mathbf{l}=\frac{\lambda}{2\pi\epsilon_0}\int_a^b\frac{1}{s}\,ds=\frac{\lambda}{2\pi\epsilon_0}\ln(b/a),\text{ so }\boxed{P=IV.}$$

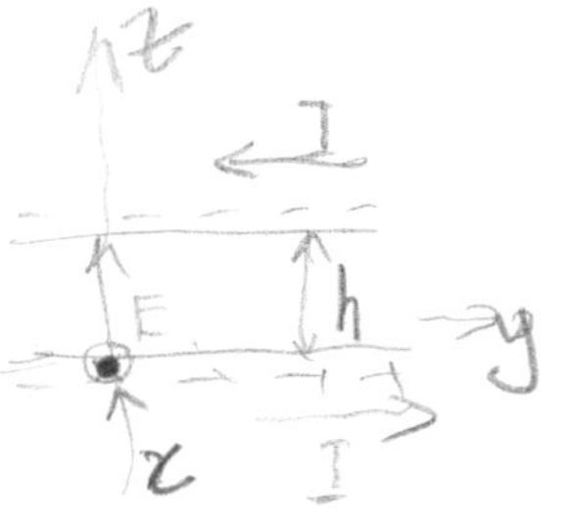

Problem 7.58.

$$\left.\begin{array}{l}\mathbf{E}=\dfrac{\sigma}{\epsilon_0}\,\hat{\mathbf{z}}\\[2ex] \mathbf{B}=\mu_0 K\,\hat{\mathbf{x}}=\dfrac{\mu_0 I}{w}\,\hat{\mathbf{x}}\end{array}\right\}\mathbf{S}=\frac{1}{\mu_0}(\mathbf{E}\times\mathbf{B})=\frac{\sigma I}{\epsilon_0 w}\,\hat{\mathbf{y}};$$

$$P=\int \mathbf{S}\cdot d\mathbf{a}=Swh=\frac{\sigma I h}{\epsilon_0},\text{ but } V=\int \mathbf{E}\cdot d\mathbf{l}=\frac{\sigma}{\epsilon_0}h,\text{ so }\boxed{P=IV.}$$

Problem 8.2

(a) $\mathbf{E}=\dfrac{\sigma}{\epsilon_0}\,\hat{\mathbf{z}};\ \sigma=\dfrac{Q}{\pi a^2};\ Q(t)=It\ \Rightarrow\ \mathbf{E}(t)=\boxed{\dfrac{It}{\pi\epsilon_0 a^2}\,\hat{\mathbf{z}}.}$

$$B\,2\pi s=\mu_0\epsilon_0\frac{\partial E}{\partial t}\pi s^2=\mu_0\epsilon_0\frac{I\pi s^2}{\pi\epsilon_0 a^2}\ \Rightarrow\ \mathbf{B}(s,t)=\boxed{\frac{\mu_0 I s}{2\pi a^2}\,\hat{\boldsymbol{\phi}}.}$$

(b) $u_{\text{em}}=\dfrac{1}{2}\left(\epsilon_0 E^2+\dfrac{1}{\mu_0}B^2\right)=\dfrac{1}{2}\left[\epsilon_0\left(\dfrac{It}{\pi\epsilon_0 a^2}\right)^2+\dfrac{1}{\mu_0}\left(\dfrac{\mu_0 I s}{2\pi a^2}\right)^2\right]=\boxed{\dfrac{\mu_0 I^2}{2\pi^2 a^4}\left[(ct)^2+(s/2)^2\right].}$

$$\mathbf{S}=\frac{1}{\mu_0}(\mathbf{E}\times\mathbf{B})=\frac{1}{\mu_0}\left(\frac{It}{\pi\epsilon_0 a^2}\right)\left(\frac{\mu_0 I s}{2\pi a^2}\right)(-\hat{\mathbf{s}})=\boxed{-\frac{I^2 t}{2\pi^2\epsilon_0 a^4}\,s\,\hat{\mathbf{s}}.}$$

$$\frac{\partial u_{\rm em}}{\partial t} = \frac{\mu_0 I^2}{2\pi^2 a^4} 2c^2 t = \frac{I^2 t}{\pi^2 \epsilon_0 a^4}; \quad -\boldsymbol{\nabla}\cdot\mathbf{S} = \frac{I^2 t}{2\pi^2\epsilon_0 a^4}\boldsymbol{\nabla}\cdot(s\,\hat{\mathbf{s}}) = \frac{I^2 t}{\pi^2 \epsilon_0 a^2} = \frac{\partial u_{\rm em}}{\partial t}. \ \checkmark$$

(c) $$U_{\rm em} = \int u_{\rm em} w 2\pi s\,ds = 2\pi w \frac{\mu_0 I^2}{2\pi^2 a^4}\int_0^b [(ct)^2 + (s/2)^2] s\,ds = \frac{\mu_0 w I^2}{\pi a^4}\left[(ct)^2\frac{s^2}{2} + \frac{1}{4}\frac{s^4}{4}\right]\Bigg|_0^b$$

$$= \boxed{\frac{\mu_0 w I^2 b^2}{2\pi a^4}\left[(ct)^2 + \frac{b^2}{16}\right]}.$$ Over a surface at radius b: $$P_{\rm in} = -\int \mathbf{S}\cdot d\mathbf{a} = \frac{I^2 t}{2\pi^2\epsilon_0 a^4}\left[b\,\hat{\mathbf{s}}\cdot(2\pi b w\,\hat{\mathbf{s}})\right] = \boxed{\frac{I^2 w t b^2}{\pi\epsilon_0 a^4}}.$$

$$\frac{dU_{\rm em}}{dt} = \frac{\mu_0 w I^2 b^2}{2\pi a^4}2c^2 t = \frac{I^2 w t b^2}{\pi \epsilon_0 a^4} = P_{\rm in}.\ \checkmark$$ (Set $b = a$ for *total.*)

Problem 8.3

$$\mathbf{F} = \oint \overleftrightarrow{\mathbf{T}}\cdot d\mathbf{a} - \mu_0\epsilon_0\frac{d}{dt}\int \mathbf{S}\,d\tau.$$

The fields are constant, so the second term is zero. The force is clearly in the z direction, so we need

$$\begin{aligned}(\overleftrightarrow{\mathbf{T}}\cdot d\mathbf{a})_z &= T_{zx}\,da_x + T_{zy}\,da_y + T_{zz}\,da_z = \frac{1}{\mu_0}\left(B_z B_x\,da_x + B_z B_y\,da_y + B_z B_z\,da_z - \frac{1}{2}B^2\,da_z\right)\\ &= \frac{1}{\mu_0}\left[B_z(\mathbf{B}\cdot d\mathbf{a}) - \frac{1}{2}B^2\,da_z\right].\end{aligned}$$

Now $\mathbf{B} = \frac{2}{3}\mu_0\sigma R\omega\,\hat{\mathbf{z}}$ (inside) and $\mathbf{B} = \frac{\mu_0 m}{4\pi r^3}(2\cos\theta\,\hat{\mathbf{r}} + \sin\theta\,\hat{\boldsymbol{\theta}})$ (outside), where $m = \frac{4}{3}\pi R^3(\sigma\omega R)$. (From Eq. 5.68, Prob. 5.36, and Eq. 5.86.) We want a surface that encloses the entire upper hemisphere—say a hemispherical cap just outside $r = R$ plus the equatorial circular disk.

Hemisphere:

$$\begin{aligned}B_z &= \frac{\mu_0 m}{4\pi R^3}\left[2\cos\theta\,(\hat{\mathbf{r}})_z + \sin\theta\,(\hat{\boldsymbol{\theta}})_z\right] = \frac{\mu_0 m}{4\pi R^3}\left[2\cos^2\theta - \sin^2\theta\right] = \frac{\mu_0 m}{4\pi R^3}\left(3\cos^2\theta - 1\right).\\ d\mathbf{a} &= R^2\sin\theta\,d\theta\,d\phi\,\hat{\mathbf{r}};\ \mathbf{B}\cdot d\mathbf{a} = \frac{\mu_0 m}{4\pi R^3}(2\cos\theta)R^2\sin\theta\,d\theta\,d\phi;\ da_z = R^2\sin\theta\,d\theta\,d\phi\,\cos\theta;\\ B^2 &= \left(\frac{\mu_0 m}{4\pi R^3}\right)^2\left(4\cos^2\theta + \sin^2\theta\right) = \left(\frac{\mu_0 m}{4\pi R^3}\right)^2\left(3\cos^2\theta + 1\right).\\ (\overleftrightarrow{\mathbf{T}}\cdot d\mathbf{a})_z &= \frac{1}{\mu_0}\left(\frac{\mu_0 m}{4\pi R^3}\right)^2\left[\left(3\cos^2\theta - 1\right)2\cos\theta R^2\sin\theta\,d\theta\,d\phi - \frac{1}{2}\left(3\cos^2\theta + 1\right)R^2\sin\theta\cos\theta\,d\theta\,d\phi\right]\\ &= \mu_0\left(\frac{\sigma\omega R}{3}\right)^2\left[\frac{1}{2}R^2\sin\theta\cos\theta\,d\theta\,d\phi\right]\left(12\cos^2\theta - 4 - 3\cos^2\theta - 1\right)\\ &= \frac{\mu_0}{2}\left(\frac{\sigma\omega R^2}{3}\right)^2\left(9\cos^2\theta - 5\right)\sin\theta\cos\theta\,d\theta\,d\phi.\\ (F_{\rm hemi})_z &= \frac{\mu_0}{2}\left(\frac{\sigma\omega R^2}{3}\right)^2 2\pi\int_0^{\pi/2}\left(9\cos^3\theta - 5\cos\theta\right)\sin\theta\,d\theta = \mu_0\pi\left(\frac{\sigma\omega R^2}{3}\right)^2\left[-\frac{9}{4}\cos^4\theta + \frac{5}{2}\cos^2\theta\right]\Bigg|_0^{\pi/2}\\ &= \mu_0\pi\left(\frac{\sigma\omega R^2}{3}\right)^2\left(0 + \frac{9}{4} - \frac{5}{2}\right) = -\frac{\mu_0\pi}{4}\left(\frac{\sigma\omega R^2}{3}\right)^2.\end{aligned}$$

Disk:

$$B_z = \frac{2}{3}\mu_0\sigma R\omega; \quad d\mathbf{a} = r\,dr\,d\phi\,\hat{\boldsymbol{\phi}} = -r\,dr\,d\phi\,\hat{\mathbf{z}};$$

$$\mathbf{B}\cdot d\mathbf{a} = -\frac{2}{3}\mu_0\sigma R\omega r\,dr\,d\phi; \quad B^2 = \left(\frac{2}{3}\mu_0\sigma R\omega\right)^2; \quad da_z = -r\,dr\,d\phi.$$

$$(\overleftrightarrow{\mathbf{T}}\cdot d\mathbf{a})_z = \frac{1}{\mu_0}\left(\frac{2}{3}\mu_0\sigma R\omega\right)^2\left[-r\,dr\,d\phi + \frac{1}{2}r\,dr\,d\phi\right] = -\frac{1}{2\mu_0}\left(\frac{2}{3}\mu_0\sigma R\omega\right)^2 r\,dr\,d\phi.$$

$$(F_{\text{disk}})_z = -2\mu_0\left(\frac{\sigma\omega R}{3}\right)^2 2\pi\int_0^R r\,dr = -2\pi\mu_0\left(\frac{\sigma\omega R^2}{3}\right)^2.$$

Total:

$$\mathbf{F} = -\pi\mu_0\left(\frac{\sigma\omega R^2}{3}\right)^2\left(2+\frac{1}{4}\right)\hat{\mathbf{z}} = \boxed{-\pi\mu_0\left(\frac{\sigma\omega R^2}{2}\right)^2\hat{\mathbf{z}}} \text{ (agrees with Prob. 5.42).}$$

Problem 8.4

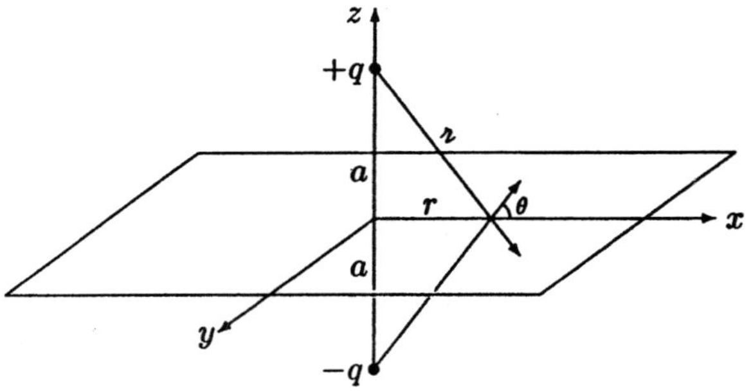

(a) $$(\overleftrightarrow{\mathbf{T}}\cdot d\mathbf{a})_z = T_{zx}\,da_x + T_{zy}\,da_y + T_{zz}\,da_z.$$

But for the xy plane $da_x = da_y = 0$, and $da_z = -r\,dr\,d\phi$ (I'll calculate the force on the *upper* charge).

$$(\overleftrightarrow{\mathbf{T}}\cdot d\mathbf{a})_z = \epsilon_0\left(E_zE_z - \frac{1}{2}E^2\right)(-r\,dr\,d\phi).$$

Now $\mathbf{E} = \frac{1}{4\pi\epsilon_0}2\frac{q}{\imath^2}\cos\theta\,\hat{\mathbf{r}}$, and $\cos\theta = \frac{r}{\imath}$, so $E_z = 0$, $E^2 = \left(\frac{q}{2\pi\epsilon_0}\right)^2\frac{r^2}{(r^2+a^2)^3}$. Therefore

$$F_z = \frac{1}{2}\epsilon_0\left(\frac{q}{2\pi\epsilon_0}\right)^2 2\pi\int_0^{\epsilon fty}\frac{r^3\,dr}{(r^2+a^2)^3} = \frac{q^2}{4\pi\epsilon_0}\frac{1}{2}\int_0^\infty\frac{u\,du}{(u+a^2)^3} \quad (\text{letting } u\equiv r^2)$$

$$= \frac{q^2}{4\pi\epsilon_0}\frac{1}{2}\left[-\frac{1}{(u+a^2)}+\frac{a^2}{2\,(u+a^2)^3}\right]\Bigg|_0^\infty = \frac{q^2}{4\pi\epsilon_0}\frac{1}{2}\left[0+\frac{1}{a^2}-\frac{a^2}{2a^4}\right] = \boxed{\frac{q^2}{4\pi\epsilon_0}\frac{1}{(2a)^2}.} \checkmark$$

(b) In this case $\mathbf{E} = -\frac{1}{4\pi\epsilon_0}2\frac{q}{\imath^2}\sin\theta\,\hat{\mathbf{z}}$, and $\sin\theta = \frac{a}{\imath}$, so

$$E^2 = E_z^2 = \left(\frac{qa}{2\pi\epsilon_0}\right)^2\frac{1}{(r^2+a^2)^3}, \text{ and hence } (\overleftrightarrow{\mathbf{T}}\cdot d\mathbf{a})_z = -\frac{\epsilon_0}{2}\left(\frac{qa}{2\pi\epsilon_0}\right)^2\frac{r\,dr\,d\phi}{(r^2+a^2)^3}. \text{ Therefore}$$

$$F_z = -\frac{\epsilon_0}{2}\left(\frac{qa}{2\pi\epsilon_0}\right)^2 2\pi\int_0^\infty\frac{r\,dr}{(r^2+a^2)^3} = -\frac{q^2a^2}{4\pi\epsilon_0}\left[-\frac{1}{4}\frac{1}{(r^2+a^2)^2}\right]_0^\infty = -\frac{q^2a^2}{4\pi\epsilon_0}\left[0+\frac{1}{4a^4}\right] = \boxed{-\frac{q^2}{4\pi\epsilon_0}\frac{1}{(2a)^2}.} \checkmark$$

Problem 8.5

(a) $E_x = E_y = 0,\ E_z = -\sigma/\epsilon_0$. Therefore

$$T_{xy} = T_{xz} = T_{yz} = \cdots = 0;\quad T_{xx} = T_{yy} = -\frac{\epsilon_0}{2}E^2 = -\frac{\sigma^2}{2\epsilon_0};\quad T_{zz} = \epsilon_0\left(E_z^2 - \frac{1}{2}E^2\right) = \frac{\epsilon_0}{2}E^2 = \frac{\sigma^2}{2\epsilon_0}.$$

$$\boxed{\overleftrightarrow{\mathbf{T}} = \frac{\sigma^2}{2\epsilon_0}\begin{pmatrix} -1 & 0 & 0 \\ 0 & -1 & 0 \\ 0 & 0 & +1 \end{pmatrix}.}$$

(b) $\mathbf{F} = \oint \overleftrightarrow{\mathbf{T}} \cdot d\mathbf{a}$ ($\mathbf{S} = 0$, since $\mathbf{B} = 0$); integrate over the xy plane: $d\mathbf{a} = -dx\,dy\,\hat{\mathbf{z}}$ (negative because *outward* with respect to a surface enclosing the upper plate). Therefore

$$F_z = \int T_{zz}\,da_z = -\frac{\sigma^2}{2\epsilon_0}A, \text{ and the force per unit area is } \mathbf{f} = \frac{\mathbf{F}}{A} = \boxed{-\frac{\sigma^2}{2\epsilon_0}\hat{\mathbf{z}}.}$$

(c) $-T_{zz} = \boxed{\sigma^2/2\epsilon_0}$ is the momentum in the z direction crossing a surface perpendicular to z, per unit area, per unit time (Eq. 8.31).

(d) The recoil force is the momentum delivered per unit time, so the force per unit area on the top plate is

$$\boxed{\mathbf{f} = -\frac{\sigma^2}{2\epsilon_0}\hat{\mathbf{z}}} \quad \text{(same as (b)).}$$

Problem 8.6

(a) $\wp_{em} = \epsilon_0(\mathbf{E}\times\mathbf{B}) = \epsilon_0 EB\,\hat{\mathbf{y}};\quad \mathbf{p}_{em} = \boxed{\epsilon_0 EBAd\,\hat{\mathbf{y}}.}$

(b) $\mathbf{I} = \int_0^\infty \mathbf{F}\,dt = \int_0^\infty I(\mathbf{l}\times\mathbf{B})\,dt = \int_0^\infty IBd(\hat{\mathbf{z}}\times\hat{\mathbf{x}})\,dt = (Bd\,\hat{\mathbf{y}})\int_0^\infty\left(-\frac{dQ}{dt}\right)dt$ $= -(Bd\hat{\mathbf{y}})[Q(\infty) - Q(0)] = BQd\hat{\mathbf{y}}$. But the original field was $E = \sigma/\epsilon_0 = Q/\epsilon_0 A$, so $Q = \epsilon_0 EA$, and hence $\mathbf{I} = \boxed{\epsilon_0 EBAd\,\hat{\mathbf{y}};}$ as expected, the momentum originally stored in the fields (a) is delivered as a kick to the capacitor.

(c) $\oint \mathbf{E}\cdot d\mathbf{l} = -\frac{d\Phi}{dt} = -\frac{dB}{dt}ld$ (for a length l in the y direction). $-lE(d) + lE(0) = -ld\frac{dB}{dt} \Rightarrow$ $E(d) - E(0) = d\frac{dB}{dt}$. $\mathbf{F} = -\sigma AE(d)\,\hat{\mathbf{y}} + \sigma AE(0)\,\hat{\mathbf{y}} = -\sigma A[E(d) - E(0)]\,\hat{\mathbf{y}} = -\sigma Ad\frac{dB}{dt}\,\hat{\mathbf{y}}$. $\mathbf{I} = \int_0^\infty \mathbf{F}\,dt =$ $-(\sigma Ad\,\hat{\mathbf{y}})\int_0^\infty \frac{dB}{dt}\,dt = -(\sigma Ad\,\hat{\mathbf{y}})[B(\infty) - B(0)] = \sigma AdB\,\hat{\mathbf{y}}$. But $E = \frac{\sigma}{\epsilon_0}$, so $\mathbf{I} = \boxed{\epsilon_0 EBAd\,\hat{\mathbf{y}},}$ as before.

Problem 8.7

$\mathbf{B} = \mu_0 nI\,\hat{\mathbf{z}}$ (for $a < r < R$; outside the solenoid $B = 0$). The force on a segment dr of spoke is

$$d\mathbf{F} = I'd\mathbf{l}\times\mathbf{B} = I'\mu_0 nI\,dr(\hat{\mathbf{r}}\times\hat{\mathbf{z}}) = -I'\mu_0 nI\,dr\,\hat{\boldsymbol{\phi}}.$$

The torque on the spoke is

$$\mathbf{N} = \int \mathbf{r}\times d\mathbf{F} = I'\mu_0 nI\int_a^R r\,dr(-\hat{\mathbf{r}}\times\hat{\boldsymbol{\phi}}) = I'\mu_0 nI\frac{1}{2}\left(R^2 - a^2\right)(-\hat{\mathbf{z}}).$$

Therefore the angular momentum of the cylinders is $\mathbf{L} = \int \mathbf{N}\,dt = -\frac{1}{2}\mu_0 n I(R^2 - a^2)\,\hat{\mathbf{z}} \int I'dt$. But $\int I'dt = Q$, so

$$\boxed{\mathbf{L} = -\frac{1}{2}\mu_0 n I Q(R^2 - a^2)\,\hat{\mathbf{z}}} \quad \text{(in agreement with Eq. 8.35).}$$

Problem 8.8

(a)

$$\mathbf{E} = \begin{cases} 0, & (r < R) \\ \dfrac{1}{4\pi\epsilon_0}\dfrac{Q}{r^2}\,\hat{\mathbf{r}}, & (r > R) \end{cases}; \quad \mathbf{B} = \begin{cases} \frac{2}{3}\mu_0 M\,\hat{\mathbf{z}}, & (r < R) \\ \dfrac{\mu_0}{4\pi}\dfrac{m}{r^3}\left[2\cos\theta\,\hat{\mathbf{r}} + \sin\theta\,\hat{\boldsymbol{\theta}}\right], & (r > R) \end{cases} \quad \text{(Ex. 6.1)}$$

(where $m = \frac{4}{3}\pi R^3 M$); $\wp = \epsilon_0(\mathbf{E}\times\mathbf{B}) = \dfrac{\mu_0}{(4\pi)^2}\dfrac{Qm}{r^5}(\hat{\mathbf{r}}\times\hat{\boldsymbol{\theta}})\sin\theta$, and $(\hat{\mathbf{r}}\times\hat{\boldsymbol{\theta}}) = \hat{\boldsymbol{\phi}}$, so

$$\boldsymbol{\ell} = \mathbf{r}\times\wp = \frac{\mu_0}{(4\pi)^2}\frac{mQ}{r^4}\sin\theta(\hat{\mathbf{r}}\times\hat{\boldsymbol{\phi}}).$$

But $(\hat{\mathbf{r}}\times\hat{\boldsymbol{\phi}}) = -\hat{\boldsymbol{\theta}}$, and only the z component will survive integration, so (since $(\hat{\boldsymbol{\theta}})_z = -\sin\theta$):

$$\mathbf{L} = \frac{\mu_0 mQ}{(4\pi)^2}\,\hat{\mathbf{z}}\int\frac{\sin^2\theta}{r^4}\left(r^2\sin\theta\,dr\,d\theta\,d\phi\right). \quad \int_0^{2\pi} d\phi = 2\pi; \quad \int_0^{\pi}\sin^3\theta\,d\theta = \frac{4}{3}; \quad \int_R^{\infty}\frac{1}{r^2}\,dr = \left(-\frac{1}{r}\right)\Big|_R^{\infty} = \frac{1}{R}.$$

$$\mathbf{L} = \frac{\mu_0 mQ}{(4\pi)^2}\,\hat{\mathbf{z}}\,(2\pi)\left(\frac{4}{3}\right)\left(\frac{1}{R}\right) = \boxed{\frac{2}{9}\mu_0 MQR^2\,\hat{\mathbf{z}}.}$$

(b) Apply Faraday's law to the ring shown:

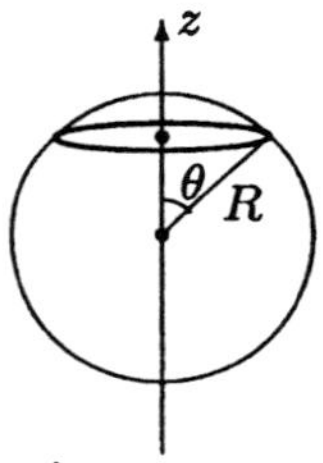

$$\oint\mathbf{E}\cdot d\mathbf{l} = E(2\pi r\sin\theta) = -\frac{d\Phi}{dt} = -\pi(r\sin\theta)^2\left(\frac{2}{3}\mu_0\frac{dM}{dt}\right)$$

$$\Rightarrow \boxed{\mathbf{E} = -\frac{\mu_0}{3}\frac{dM}{dt}(r\sin\theta)\,\hat{\boldsymbol{\phi}}.}$$

The force on a patch of surface (da) is $d\mathbf{F} = \sigma\mathbf{E}\,da = -\dfrac{\mu_0\sigma}{3}\dfrac{dM}{dt}(r\sin\theta)\,da\,\hat{\boldsymbol{\phi}}$ $\left(\sigma = \dfrac{Q}{4\pi R^2}\right)$.

The torque on the patch is $d\mathbf{N} = \mathbf{r}\times d\mathbf{F} = -\dfrac{\mu_0\sigma}{3}\dfrac{dM}{dt}\left(r^2\sin\theta\right)da\,(\hat{\mathbf{r}}\times\hat{\boldsymbol{\phi}})$. But $(\hat{\mathbf{r}}\times\hat{\boldsymbol{\phi}}) = -\hat{\boldsymbol{\theta}}$, and we want only the z component $(\hat{\boldsymbol{\theta}}_z = -\sin\theta)$:

$$\mathbf{N} = -\frac{\mu_0\sigma}{3}\frac{dM}{dt}\,\hat{\mathbf{z}}\int r^2\sin^2\theta\left(r^2\sin\theta\,d\theta\,d\phi\right).$$

Here $r = R$; $\displaystyle\int_0^{\pi}\sin^3\theta\,d\theta = \frac{4}{3}$; $\displaystyle\int_0^{2\pi}d\phi = 2\pi$, so $\mathbf{N} = -\dfrac{\mu_0\sigma}{3}\dfrac{dM}{dt}\,\hat{\mathbf{z}}\,R^4\left(\dfrac{4}{3}\right)(2\pi) = \boxed{-\dfrac{2\mu_0}{9}QR^2\dfrac{dM}{dt}\,\hat{\mathbf{z}}.}$

$$\mathbf{L} = \int\mathbf{N}\,dt = -\frac{2\mu_0}{9}QR^2\,\hat{\mathbf{z}}\int_M^0 dM = \boxed{\frac{2\mu_0}{9}MQR^2\,\hat{\mathbf{z}}} \quad \text{(same as (a)).}$$

(c) Let the charge on the sphere at time t be $q(t)$; the charge *density* is $\sigma = \frac{q(t)}{4\pi R^2}$. The charge below ("south of") the ring in the figure is

$$q_s = \sigma\,(2\pi R^2)\int_0^{\pi} \sin\theta'\,d\theta' = \frac{q}{2}\,(-\cos\theta')\big|_\theta^\pi = \frac{q}{2}(1+\cos\theta).$$

So the total current crossing the ring (flowing "north") is $I(t) = -\frac{1}{2}\frac{dq}{dt}(1+\cos\theta)$, and hence $\mathbf{K}(t) = \frac{I}{2\pi R\sin\theta}(-\hat{\boldsymbol{\theta}}) = \frac{1}{4\pi R}\frac{dq}{dt}\frac{(1+\cos\theta)}{\sin\theta}\,\hat{\boldsymbol{\theta}}$. The force on a patch of area da is $d\mathbf{F} = (\mathbf{K}\times\mathbf{B})\,da$.

$$\mathbf{B}_{\text{ave}} = \left[\frac{2}{3}\mu_0 M\,\hat{\mathbf{z}} + \frac{\mu_0}{4\pi}\frac{\frac{4}{3}\pi R^3 M}{R^3}(2\cos\theta\,\hat{\mathbf{r}} + \sin\theta\,\hat{\boldsymbol{\theta}})\right]\frac{1}{2} = \frac{\mu_0 M}{6}[2\,\hat{\mathbf{z}} + 2\cos\theta\,\hat{\mathbf{r}} + \sin\theta\,\hat{\boldsymbol{\theta}}];$$

$$\mathbf{K}\times\mathbf{B} = \frac{1}{4\pi R}\frac{dq}{dt}\frac{\mu_0 M}{6}\frac{(1+\cos\theta)}{\sin\theta}[2(\hat{\boldsymbol{\theta}}\times\hat{\mathbf{z}}) + 2\cos\theta\,\underbrace{(\hat{\boldsymbol{\theta}}\times\hat{\mathbf{r}})}_{-\hat{\boldsymbol{\phi}}}].$$

$$\begin{aligned}
d\mathbf{N} &= R\,\hat{\mathbf{r}}\times d\mathbf{F} = \frac{\mu_0 M}{24\pi}\left(\frac{dq}{dt}\right)\frac{(1+\cos\theta)}{\sin\theta}2[\underbrace{\hat{\mathbf{r}}\times(\hat{\boldsymbol{\theta}}\times\hat{\mathbf{z}})}_{\hat{\boldsymbol{\theta}}(\hat{\mathbf{r}}\cdot\hat{\mathbf{z}}) - \hat{\mathbf{z}}(\hat{\mathbf{r}}\cdot\hat{\boldsymbol{\theta}})} - \cos\theta\,\underbrace{(\hat{\mathbf{r}}\times\hat{\boldsymbol{\phi}})}_{-\hat{\boldsymbol{\theta}}}]R^2\sin\theta\,d\theta\,d\phi \\
&= \frac{\mu_0 M}{12\pi}\left(\frac{dq}{dt}\right)(1+\cos\theta)R^2[\cos\theta\,\hat{\boldsymbol{\theta}} + \cos\theta\,\hat{\boldsymbol{\theta}}]\,d\theta\,d\phi = \frac{\mu_0 M R^2}{6\pi}\left(\frac{dq}{dt}\right)(1+\cos\theta)\cos\theta\,d\theta\,d\phi\,\hat{\boldsymbol{\theta}}.
\end{aligned}$$

The x and y components integrate to zero; $(\hat{\boldsymbol{\theta}})_z = -\sin\theta$, so (using $\int_0^{2\pi} d\phi = 2\pi$):

$$\begin{aligned}
N_z &= -\frac{\mu_0 M R^2}{6\pi}\left(\frac{dq}{dt}\right)(2\pi)\int_0^{\pi}(1+\cos\theta)\cos\theta\sin\theta\,d\theta = -\frac{\mu_0 M R^2}{3}\left(\frac{dq}{dt}\right)\left(\frac{\sin^2\theta}{2} - \frac{\cos^3\theta}{3}\right)\Bigg|_0^\pi \\
&= -\frac{\mu_0 M R^2}{3}\left(\frac{dq}{dt}\right)\left(\frac{2}{3}\right) = -\frac{2\mu_0}{9}MR^2\frac{dq}{dt}. \quad \boxed{\mathbf{N} = -\frac{2\mu_0}{9}MR^2\frac{dq}{dt}\,\hat{\mathbf{z}}.}
\end{aligned}$$

Therefore

$$\mathbf{L} = \int \mathbf{N}\,dt = -\frac{2\mu_0}{9}MR^2\,\hat{\mathbf{z}}\int_Q^0 dq = \boxed{\frac{2\mu_0}{9}MR^2 Q\,\hat{\mathbf{z}}}\ \text{(same as (a))}.$$

(I used the *average* field at the discontinuity—which is the correct thing to do—but in this case you'd get the same answer using either the inside field or the outside field.)

Problem 8.9

(a) $\mathcal{E} = -\frac{d\Phi}{dt}$; $\Phi = \pi a^2 B$; $B = \mu_0 n I_s$; $\mathcal{E} = I_r R$. So $\boxed{I_r = -\frac{1}{R}\left(\mu_0\pi a^2 n\right)\frac{dI_s}{dt}.}$

(b) $\oint \mathbf{E}\cdot d\mathbf{l} = -\frac{d\Phi}{dt} \Rightarrow E(2\pi a) = -\mu_0\pi a^2 n\frac{dI_s}{dt} \Rightarrow \mathbf{E} = -\frac{1}{2}\mu_0 a n\frac{dI_s}{dt}\,\hat{\boldsymbol{\phi}}$. $\mathbf{B} = \frac{\mu_0 I_r}{2}\frac{b^2}{(b^2+z^2)^{3/2}}\,\hat{\mathbf{z}}$ (Eq. 5.38).

$$\mathbf{S} = \frac{1}{\mu_0}(\mathbf{E}\times\mathbf{B}) = \frac{1}{\mu_0}\left(-\frac{\mu_0 a n}{2}\frac{dI_s}{dt}\right)\left(\frac{\mu_0 I_r}{2}\frac{b^2}{(b^2+z^2)^{3/2}}\right)(\hat{\boldsymbol{\phi}}\times\hat{\mathbf{z}}) = \boxed{-\frac{1}{4}\mu_0 I_r\frac{dI_s}{dt}\frac{ab^2 n}{(b^2+z^2)^{3/2}}\,\hat{\mathbf{r}}.}$$

Power:

$$P = \int \mathbf{S}\cdot d\mathbf{a} = \int_{-\infty}^{\infty} (S)(2\pi a)\,dz = -\frac{1}{2}\pi\mu_0 a^2 b^2 n I_n \frac{dI_s}{dt}\int_{-\infty}^{\infty}\frac{1}{(b^2+z^2)^{3/2}}\,dx$$

The integral is $\frac{z}{b^2\sqrt{z^2+b^2}}\Big|_{-\infty}^{\infty} = \frac{1}{b^2} - \left(-\frac{1}{b^2}\right) = \frac{2}{b^2}$.

$$= -\left(\pi\mu_0 a^2 n\frac{dI_s}{dt}\right) I_r = (RI_r)I_r = I_r^2 R. \quad \text{qed}$$

Problem 8.10

According to Eqs. 3.104, 4.14, 5.87, and 6.16, the fields are

$$\mathbf{E} = \begin{cases} -\dfrac{1}{3\epsilon_0}\mathbf{P}, & (r<R), \\ \dfrac{1}{4\pi\epsilon_0}\dfrac{1}{r^3}[3(\mathbf{p}\cdot\hat{\mathbf{r}})\hat{\mathbf{r}} - \mathbf{p}], & (r>R), \end{cases} \qquad \mathbf{B} = \begin{cases} \dfrac{2}{3}\mu_0\mathbf{M}, & (r<R), \\ \dfrac{\mu_0}{4\pi}\dfrac{m}{r^3}[3(\mathbf{m}\cdot\hat{\mathbf{r}})\hat{\mathbf{r}} - \mathbf{m}], & (r>R), \end{cases}$$

where $\mathbf{p} = (4/3)\pi R^3\mathbf{P}$, and $\mathbf{m} = (4/3)\pi R^3\mathbf{M}$. Now $\mathbf{p} = \epsilon_0\int(\mathbf{E}\times\mathbf{B})\,d\tau$, and there are two contributions, one from inside the sphere and one from outside.

Inside:

$$\mathbf{p}_{\text{in}} = \epsilon_0\int\left(-\frac{1}{3\epsilon_0}\mathbf{P}\right)\times\left(\frac{2}{3}\mu_0\mathbf{M}\right)d\tau = -\frac{2}{9}\mu_0(\mathbf{P}\times\mathbf{M})\int d\tau = -\frac{2}{9}\mu_0(\mathbf{P}\times\mathbf{M})\frac{4}{3}\pi R^3 = \frac{8}{27}\mu_0\pi R^3(\mathbf{M}\times\mathbf{P}).$$

Outside:

$$\mathbf{p}_{\text{out}} = \epsilon_0\frac{1}{4\pi\epsilon_0}\frac{\mu_0}{4\pi}\int\frac{1}{r^6}\{[3(\mathbf{p}\cdot\hat{\mathbf{r}})\hat{\mathbf{r}} - \mathbf{p}]\times[3(\mathbf{m}\cdot\hat{\mathbf{r}})\hat{\mathbf{r}} - \mathbf{m}]\}\,d\tau.$$

Now $\hat{\mathbf{r}}\times(\mathbf{p}\times\mathbf{m}) = \mathbf{p}(\hat{\mathbf{r}}\cdot\mathbf{m}) - \mathbf{m}(\hat{\mathbf{r}}\cdot\mathbf{p})$, so $\hat{\mathbf{r}}\times[\hat{\mathbf{r}}\times(\mathbf{p}\times\mathbf{m})] = (\hat{\mathbf{r}}\cdot\mathbf{m})(\hat{\mathbf{r}}\times\mathbf{p}) - (\hat{\mathbf{r}}\cdot\mathbf{p})(\hat{\mathbf{r}}\times\mathbf{m})$, whereas using the BAC-CAB rule directly gives $\hat{\mathbf{r}}\times[\hat{\mathbf{r}}\times(\mathbf{p}\times\mathbf{m})] = \hat{\mathbf{r}}[\hat{\mathbf{r}}\cdot(\mathbf{p}\times\mathbf{m})] - (\mathbf{p}\times\mathbf{m})(\hat{\mathbf{r}}\cdot\hat{\mathbf{r}})$. So $\{[3(\mathbf{p}\cdot\hat{\mathbf{r}})\hat{\mathbf{r}} - \mathbf{p}]\times[3(\mathbf{m}\cdot\hat{\mathbf{r}})\hat{\mathbf{r}} - \mathbf{m}]\} = -3(\mathbf{p}\cdot\hat{\mathbf{r}})(\hat{\mathbf{r}}\times\mathbf{m}) + 3(\mathbf{m}\cdot\hat{\mathbf{r}})(\hat{\mathbf{r}}\times\mathbf{p}) + (\mathbf{p}\times\mathbf{m}) = 3\{\hat{\mathbf{r}}[\hat{\mathbf{r}}\cdot(\mathbf{p}\times\mathbf{m})] - (\mathbf{p}\times\mathbf{m})\} + (\mathbf{p}\times\mathbf{m}) = -2(\mathbf{p}\times\mathbf{m}) + 3\hat{\mathbf{r}}[\hat{\mathbf{r}}\cdot(\mathbf{p}\times\mathbf{m})]$.

$$\mathbf{p}_{\text{out}} = \frac{\mu_0}{16\pi^2}\int\frac{1}{r^6}\{-2(\mathbf{p}\times\mathbf{m}) + 3\,\hat{\mathbf{r}}[\hat{\mathbf{r}}\cdot(\mathbf{p}\times\mathbf{m})]\}\,r^2\sin\theta\,dr\,d\theta\,d\phi.$$

To evaluate the integral, set the z axis along $(\mathbf{p}\times\mathbf{m})$; then $\hat{\mathbf{r}}\cdot(\mathbf{p}\times\mathbf{m}) = |\mathbf{p}\times\mathbf{m}|\cos\theta$. Meanwhile, $\hat{\mathbf{r}} = \sin\theta\cos\phi\,\hat{\mathbf{x}} + \sin\theta\sin\phi\,\hat{\mathbf{y}} + \cos\theta\,\hat{\mathbf{z}}$. But $\sin\phi$ and $\cos\phi$ integrate to zero, so the $\hat{\mathbf{x}}$ and $\hat{\mathbf{y}}$ terms drop out, leaving

$$\begin{aligned}
\mathbf{p}_{\text{out}} &= \frac{\mu_0}{16\pi^2}\left(\int_0^{\infty}\frac{1}{r^4}\,dr\right)\left\{-2(\mathbf{p}\times\mathbf{m})\int\sin\theta\,d\theta\,d\phi + 3|\mathbf{p}\times\mathbf{m}|\,\hat{\mathbf{z}}\int\cos^2\theta\sin\theta\,d\theta\,d\phi\right\} \\
&= \frac{\mu_0}{16\pi^2}\left(-\frac{1}{3r^3}\right)\Big|_R^{\infty}\left[-2(\mathbf{p}\times\mathbf{m})4\pi + 3(\mathbf{p}\times\mathbf{m})\frac{4\pi}{3}\right] = -\frac{\mu_0}{12\pi R^3}(\mathbf{p}\times\mathbf{m}) \\
&= -\frac{\mu_0}{12\pi R^3}\left(\frac{4}{3}\pi R^3\mathbf{P}\right)\times\left(\frac{4}{3}\pi R^3\mathbf{M}\right) = \frac{4\mu_0}{27}R^3(\mathbf{M}\times\mathbf{P}). \\
\mathbf{p}_{\text{tot}} &= \left(\frac{8}{27}+\frac{4}{27}\right)\mu_0 R^3(\mathbf{M}\times\mathbf{P}) = \boxed{\frac{4}{9}\mu_0 R^3(\mathbf{M}\times\mathbf{P}).}
\end{aligned}$$

Problem 8.11

(a) From Eq. 5.68 and Prob. 5.36,

$$\begin{cases} r < R: \ \mathbf{E} = 0, \ \mathbf{B} = \frac{2}{3}\mu_0 \sigma R \omega \,\hat{\mathbf{z}}, \ \text{ with } \sigma = \frac{e}{4\pi R^2}; \\ r > R: \ \mathbf{E} = \frac{1}{4\pi\epsilon_0}\frac{e}{r^2}\,\hat{\mathbf{r}}, \ \mathbf{B} = \frac{\mu_0}{4\pi}\frac{m}{r^3}(2\cos\theta\,\hat{\mathbf{r}} + \sin\theta\,\hat{\boldsymbol{\theta}}), \ \text{ with } m = \frac{4}{3}\pi\sigma\omega R^4. \end{cases}$$

The energy stored in the electric field is (Ex. 2.8):

$$W_E = \frac{1}{8\pi\epsilon_0}\frac{e^2}{R}.$$

The energy density of the internal magnetic field is:

$$u_B = \frac{1}{2\mu_0}B^2 = \frac{1}{2\mu_0}\left(\frac{2}{3}\mu_0 R\omega\frac{e}{4\pi R^2}\right)^2 = \frac{\mu_0\omega^2 e^2}{72\pi^2 R^2}, \text{ so } W_{B_{\text{in}}} = \frac{\mu_0\omega^2 e^2}{72\pi^2 R^2}\frac{4}{3}\pi R^3 = \frac{\mu_0 e^2\omega^2 R}{54\pi}.$$

The energy density in the external magnetic field is:

$$u_B = \frac{1}{2\mu_0}\frac{\mu_0^2}{16\pi^2}\frac{m^2}{r^6}\left(4\cos^2\theta + \sin^2\theta\right) = \frac{e^2\omega^2 R^4\mu_0}{18(16\pi^2)}\frac{1}{r^6}\left(3\cos^2\theta + 1\right), \text{ so}$$

$$W_{B_{\text{out}}} = \frac{\mu_0 e^2\omega^2 R^4}{(18)(16)\pi^2}\int_R^\infty \frac{1}{r^6}r^2\,dr\int_0^\pi \left(3\cos^2\theta + 1\right)\sin\theta\,d\theta\int_0^{2\pi} d\phi = \frac{\mu_0 e^2\omega^2 R^4}{(18)(16)\pi^2}\left(\frac{1}{3R^3}\right)(4)(2\pi) = \frac{\mu_0 e^2\omega^2 R}{108\pi}.$$

$$W_B = W_{B_{\text{in}}} + W_{b_{\text{out}}} = \frac{\mu_0 e^2\omega^2 R}{108\pi}(2+1) = \frac{\mu_0 e^2\omega^2 R}{36\pi}; \ W = W_E + W_B = \boxed{\frac{1}{8\pi\epsilon_0}\frac{e^2}{R} + \frac{\mu_0 e^2\omega^2 R}{36\pi}.}$$

(b) Same as Prob. 8.8(a), with $Q \to e$ and $m \to \frac{1}{3}e\omega R^2$: $\boxed{\mathbf{L} = \frac{\mu_0 e^2\omega R}{18\pi}\,\hat{\mathbf{z}}.}$

(c) $\frac{\mu_0 e^2}{18\pi}\omega R = \frac{\hbar}{2} \Rightarrow \omega R = \frac{9\pi\hbar}{\mu_0 e^2} = \frac{(9)(\pi)(1.05\times10^{-34})}{(4\pi\times10^{-7})(1.60\times10^{-19})^2} = \boxed{9.23\times10^{10}\,\text{m/s.}}$

$$\frac{1}{8\pi\epsilon_0}\frac{e^2}{R}\left[1 + \frac{2}{9}\left(\frac{\omega R}{c}\right)^2\right] = mc^2; \ \left[1 + \frac{2}{9}\left(\frac{\omega R}{c}\right)^2\right] = 1 + \frac{2}{9}\left(\frac{9.23\times10^{10}}{3\times10^8}\right)^2 = 2.10\times10^4;$$

$$R = \frac{(2.01\times10^4)(1.6\times10^{-19})^2}{8\pi(8.85\times10^{-12})(9.11\times10^{-31})(3\times10^8)^2} = \boxed{2.95\times10^{-11}\,\text{m};} \quad \omega = \frac{9.23\times10^{-10}}{2.95\times10^{-11}} = \boxed{3.13\times10^{21}\,\text{rad/s.}}$$

Since ωR, the speed of a point on the equator, is 300 times the speed of light, this "classical" model is clearly unrealistic.

Problem 8.12

$$\mathbf{E} = \frac{q_e}{4\pi\epsilon_0}\frac{\mathbf{r}}{r^3};$$

$$\mathbf{B} = \frac{\mu_0 q_m}{4\pi}\frac{\mathbf{r}'}{r'^3} = \frac{\mu_0 q_m}{4\pi}\frac{(\mathbf{r} - d\,\hat{\mathbf{z}})}{(r^2 + d^2 - 2rd\cos\theta)^{3/2}}.$$

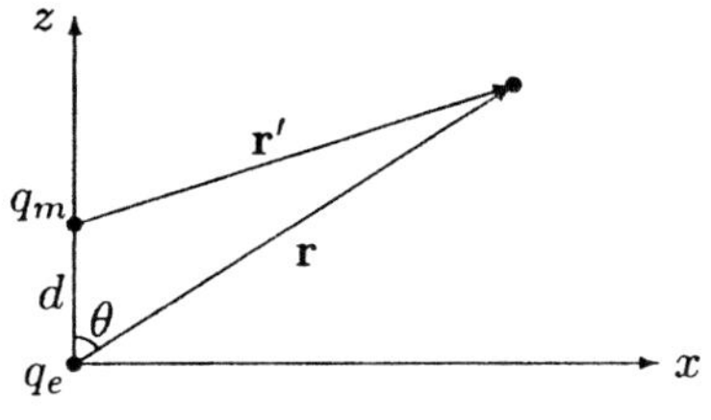

Momentum density (Eq. 8.33):

$$\wp = \epsilon_0(\mathbf{E}\times\mathbf{B}) = \frac{\mu_0 q_e q_m}{(4\pi)^2}\frac{(-d)(\mathbf{r}\times\hat{\mathbf{z}})}{r^3\left(r^2+d^2-2rd\cos\theta\right)^{3/2}}.$$

Angular momentum density (Eq. 8.34):

$$\boldsymbol{\ell} = (\mathbf{r}\times\wp) = -\frac{\mu_0 q_e q_m d}{(4\pi)^2}\frac{\mathbf{r}\times(\mathbf{r}\times\hat{\mathbf{z}})}{r^3\left(r^2+d^2-2rd\cos\theta\right)^{3/2}}.$$ But $\mathbf{r}\times(\mathbf{r}\times\hat{\mathbf{z}}) = \mathbf{r}(\mathbf{r}\cdot\hat{\mathbf{z}}) - r^2\,\hat{\mathbf{z}} = r^2\cos\theta\,\hat{\mathbf{r}} - r^2\,\hat{\mathbf{z}}$.

The x and y components will integrate to zero; using $(\hat{\mathbf{r}})_z = \cos\theta$, we have:

$$\begin{aligned}
\mathbf{L} &= -\frac{\mu_0 q_e q_m d}{(4\pi)^2}\,\hat{\mathbf{z}}\int\frac{r^2(\cos^2\theta-1)}{r^3\left(r^2+d^2-2rd\cos\theta\right)^{3/2}}r^2\sin\theta\,dr\,d\theta\,d\phi. \quad \text{Let } u\equiv\cos\theta: \\
&= \frac{\mu_0 q_e q_m d}{(4\pi)^2}\,\hat{\mathbf{z}}\,(2\pi)\int_{-1}^{1}\int_0^\infty\frac{r\left(1-u^2\right)}{\left(r^2+d^2-2rdu\right)^{3/2}}\,du\,dr.
\end{aligned}$$

Do the r integral first:

$$\int_0^\infty\frac{r\,dr}{\left(r^2+d^2-2rdu\right)^{3/2}} = \left.\frac{(ru-d)}{d(1-u^2)\sqrt{r^2+d^2-2rdu}}\right|_0^\infty = \frac{u}{d\,(1-u^2)} + \frac{d}{d\,(1-u^2)\,d} = \frac{u+1}{d\,(1-u^2)} = \frac{1}{d(1-u)}.$$

Then

$$\mathbf{L} = \frac{\mu_0 q_e q_m d}{8\pi}\,\hat{\mathbf{z}}\,\frac{1}{d}\int_{-1}^{1}\frac{(1-u^2)}{(1-u)}\,du = \frac{\mu_0 q_e q_m}{8\pi}\,\hat{\mathbf{z}}\int_{-1}^{1}(1+u)du = \frac{\mu_0 q_e q_m}{8\pi}\,\hat{\mathbf{z}}\left.\left(u+\frac{u^2}{2}\right)\right|_{-1}^{1} = \boxed{\frac{\mu_0 q_e q_m}{4\pi}\,\hat{\mathbf{z}}.}$$

Problem 8.13

(a) The rotating shell at radius b produces a solenoidal magnetic field:

$$\mathbf{B} = \mu_0 K\,\hat{\mathbf{z}},\ \text{where } K = \sigma_b\omega_b b,\ \text{and } \sigma_b = -\frac{Q}{2\pi bl}.\ \text{So } \mathbf{B} = -\frac{\mu_0\omega_b Q}{2\pi l}\,\hat{\mathbf{z}}\ (a<s<b).$$

The shell at a also produces a magnetic field $(\mu_0\omega_a Q/2\pi l)\,\hat{\mathbf{z}}$, in the region $s<a$, so the total field inside the inner shell is

$$\mathbf{B} = \frac{\mu_0 Q}{2\pi l}\left(\omega_a-\omega_b\right)\hat{\mathbf{z}},\ (s<a).$$

Meanwhile, the electric field is

$$\mathbf{E} = \frac{1}{2\pi\epsilon_0}\frac{\lambda}{s}\,\hat{\mathbf{s}} = \frac{Q}{2\pi\epsilon_0 ls}\,\hat{\mathbf{s}},\quad (a<s<b).$$

$$\wp = \epsilon_0(\mathbf{E}\times\mathbf{B}) = \epsilon_0\left(\frac{Q}{2\pi\epsilon_0 ls}\right)\left(-\frac{\mu_0\omega_b Q}{2\pi l}\right)(\hat{\mathbf{s}}\times\hat{\mathbf{z}}) = \frac{\mu_0\omega_b Q^2}{4\pi^2l^2s}\,\hat{\boldsymbol{\phi}};\quad \boldsymbol{\ell} = \mathbf{r}\times\wp = \frac{\mu_0\omega_b Q^2}{4\pi^2l^2s}(\mathbf{r}\times\hat{\boldsymbol{\phi}}).$$

Now $\mathbf{r}\times\hat{\boldsymbol{\phi}} = (s\,\hat{\mathbf{s}}+z\,\hat{\mathbf{z}})\times\hat{\boldsymbol{\phi}} = s\,\hat{\mathbf{z}} - z\,\hat{\mathbf{s}}$, and the $\hat{\mathbf{s}}$ term integrates to zero, so

$$\mathbf{L} = \frac{\mu_0\omega_b Q^2}{4\pi^2l^2}\,\hat{\mathbf{z}}\int d\tau = \frac{\mu_0\omega_b Q^2}{4\pi^2l^2}\pi(b^2-a^2)l\,\hat{\mathbf{z}} = \boxed{\frac{\mu_0\omega_b Q^2(b^2-a^2)}{4\pi l}\,\hat{\mathbf{z}}.}$$

(b) The extra electric field induced by the changing magnetic field due to the rotating shells is given by $E\,2\pi s = -\frac{d\Phi}{dt} \Rightarrow \mathbf{E} = -\frac{1}{2\pi s}\frac{d\Phi}{dt}\,\hat{\boldsymbol{\phi}}$, and in the region $a < s < b$

$$\Phi = \frac{\mu_0 Q}{2\pi l}(\omega_a - \omega_b)\,\pi a^2 - \frac{\mu_0 Q\omega_b}{2\pi l}\pi\left(s^2 - a^2\right) = \frac{\mu_0 Q}{2l}\left(\omega_a a^2 - \omega_b s^2\right);\ \mathbf{E}(s) = -\frac{1}{2\pi s}\frac{\mu_0 Q}{2l}\left(a^2\frac{d\omega_a}{dt} - s^2\frac{d\omega_b}{dt}\right)\hat{\boldsymbol{\phi}}.$$

In particular,

$$\mathbf{E}(a) = -\frac{\mu_0 Q a}{4\pi l}\left(\frac{d\omega_a}{dt} - \frac{d\omega_b}{dt}\right)\hat{\boldsymbol{\phi}}, \quad \text{and } \mathbf{E}(b) = -\frac{\mu_0 Q}{4\pi l b}\left(a^2\frac{d\omega_a}{dt} - b^2\frac{d\omega_b}{dt}\right)\hat{\boldsymbol{\phi}}.$$

The torque on a shell is $\mathbf{N} = \mathbf{r}\times q\mathbf{E} = qsE\,\hat{\mathbf{z}}$, so

$$\begin{aligned}
\mathbf{N}_a &= Qa\left(-\frac{\mu_0 Q a}{4\pi l}\right)\left(\frac{d\omega_a}{dt} - \frac{d\omega_b}{dt}\right)\hat{\mathbf{z}};\quad \mathbf{L}_a = \int_0^\infty \mathbf{N}_a\,dt = -\frac{\mu_0 Q^2 a^2}{4\pi l}(\omega_a - \omega_b)\,\hat{\mathbf{z}}.\\
\mathbf{N}_b &= -Qb\left(-\frac{\mu_0 Q}{4\pi l b}\right)\left(a^2\frac{d\omega_a}{dt} - b^2\frac{d\omega_b}{dt}\right)\hat{\mathbf{z}};\quad \mathbf{L}_b = \int_0^\infty \mathbf{N}_b\,dt = \frac{\mu_0 Q^2}{4\pi l}\left(a^2\omega_a - b^2\omega_b\right)\hat{\mathbf{z}}.\\
\mathbf{L}_{\text{tot}} &= \mathbf{L}_a + \mathbf{L}_b = \frac{\mu_0 Q^2}{4\pi l}\left(a^2\omega_a - b^2\omega_b - a^2\omega_a + a^2\omega_b\right)\hat{\mathbf{z}} = \boxed{-\frac{\mu_0 Q^2\omega_b}{4\pi l}\left(b^2 - a^2\right)\hat{\mathbf{z}}.}
\end{aligned}$$

Thus the reduction in the final mechanical angular momentum (b) is equal to the residual angular momentum in the fields (a). ✓

Problem 8.14

$\mathbf{B} = \mu_0 nI\,\hat{\mathbf{z}}$, $(s < R)$; $\quad \mathbf{E} = \frac{q}{4\pi\epsilon_0}\frac{\boldsymbol{\mathscr{r}}}{\mathscr{r}^3}$, where $\boldsymbol{\mathscr{r}} = (x - a, y, z)$.

$$\boldsymbol{\wp} = \epsilon_0(\mathbf{E}\times\mathbf{B}) = \epsilon_0(\mu_0 nI)\left(\frac{q}{4\pi\epsilon_0}\right)\frac{1}{\mathscr{r}^3}(\boldsymbol{\mathscr{r}}\times\hat{\mathbf{z}}) = \frac{\mu_0 qnI}{4\pi\mathscr{r}^3}[y\,\hat{\mathbf{x}} - (x - a)\,\hat{\mathbf{y}}].$$

Linear Momentum.

$$\mathbf{p} = \int\boldsymbol{\wp}\,d\tau = \frac{\mu_0 qnI}{4\pi}\int\frac{y\,\hat{\mathbf{x}} - (x - a)\,\hat{\mathbf{y}}}{[(x - a)^2 + y^2 + z^2]^{3/2}}\,dx\,dy\,dz.$$ The $\hat{\mathbf{x}}$ term is odd in y; it integrates to zero.

$$= -\frac{\mu_0 qnI}{4\pi}\,\hat{\mathbf{y}}\int\frac{(x - a)}{[(x - a)^2 + y^2 + z^2]^{3/2}}\,dx\,dy\,dz.$$ Do the z integral first :

$$\left.\frac{z}{[(x - a)^2 + y^2]\sqrt{(x - a)^2 + y^2 + z^2}}\right|_{-\infty}^{\infty} = \frac{2}{[(x - a)^2 + y^2]}.$$

$$= -\frac{\mu_0 qnI}{2\pi}\,\hat{\mathbf{y}}\int\frac{(x - a)}{[(x - a)^2 + y^2]}\,dx\,dy.$$ Switch to polar coordinates :

$x = s\cos\phi,\ y = s\sin\phi,\ dx\,dy \Rightarrow s\,ds\,d\phi;\ [(x - a)^2 + y^2] = s^2 + a^2 - 2sa\cos\phi.$

$$= -\frac{\mu_0 qnI}{2\pi}\,\hat{\mathbf{y}}\int\frac{(s\cos\phi - a)}{(s^2 + a^2 - 2sa\cos\phi)}\,s\,ds\,d\phi$$

Now $\displaystyle\int_0^{2\pi}\frac{\cos\phi\,d\phi}{(A + B\cos\phi)} = \frac{2\pi}{B}\left(1 - \frac{A}{\sqrt{A^2 - B^2}}\right)$; $\displaystyle\int_0^{2\pi}\frac{d\phi}{(A + B\cos\phi)} = \frac{2\pi}{\sqrt{A^2 - B^2}}.$

Here $A^2 - B^2 = (s^2 + a^2)^2 - 4s^2a^2 = s^4 + 2s^2a^2 + a^4 - 4s^2a^2 = (s^2 - a^2)^2$; $\sqrt{A^2 - B^2} = a^2 - s^2$.

$$= \frac{\mu_0 qnI}{2a}\,\hat{\mathbf{y}}\int\left[1 - \left(\frac{a^2 + s^2}{a^2 - s^2}\right) + \frac{2a^2}{(a^2 - s^2)}\right]s\,ds = \frac{\mu_0 qnI}{a}\,\hat{\mathbf{y}}\int_0^R s\,ds = \boxed{\frac{\mu_0 qnIR^2}{2a}\,\hat{\mathbf{y}}.}$$

Angular Momentum.

$$\boldsymbol{\ell} = \mathbf{r}\times\boldsymbol{\wp} = \frac{\mu_0 qnI}{4\pi\mathscr{r}^3}\,\mathbf{r}\times[y\,\hat{\mathbf{x}} - (x-a)\,\hat{\mathbf{y}}] = \frac{\mu_0 qnI}{4\pi\mathscr{r}^3}\left\{z(x-a)\,\hat{\mathbf{x}} + zy\,\hat{\mathbf{y}} - [x(x-a)+y^2]\,\hat{\mathbf{z}}\right\}.$$

The $\hat{\mathbf{x}}$ and $\hat{\mathbf{y}}$ terms are odd in z, and integrate to zero, so

$$\mathbf{L} = -\frac{\mu_0 qnI}{4\pi}\,\hat{\mathbf{z}}\int\frac{x^2+y^2-xa}{[(x-a)^2+y^2+z^2]^{3/2}}\,dx\,dy\,dz.$$ The z integral is the same as before.

$$= -\frac{\mu_0 qnI}{2\pi}\,\hat{\mathbf{z}}\int\frac{x^2+y^2-xa}{[(x-a)^2+y^2]}\,dx\,dy = -\frac{\mu_0 qnI}{2\pi}\,\hat{\mathbf{z}}\int\frac{s-a\cos\phi}{(s^2+a^2-2sa\cos\phi)}s^2\,ds\,d\phi$$

$$= -\mu_0 qnI\,\hat{\mathbf{z}}\int\left[\frac{s^2}{a^2-s^2}+\left(1-\frac{a^2+s^2}{a^2-s^2}\right)\right]s\,ds = -\mu_0 qnI\,\hat{\mathbf{z}}\int_0^R\frac{s^2-s^2}{a^2-s^2}s\,ds = \boxed{\text{zero.}}$$

Problem 8.15

(a) If we're only interested in the work done on *free* charges and currents, Eq. 8.6 becomes $\frac{dW}{dt} = \int_{\mathcal{V}}(\mathbf{E}\cdot\mathbf{J}_f)\,d\tau$. But $\mathbf{J}_f = \boldsymbol{\nabla}\times\mathbf{H} - \frac{\partial\mathbf{D}}{\partial t}$ (Eq. 7.55), so $\mathbf{E}\cdot\mathbf{J}_f = \mathbf{E}\cdot(\boldsymbol{\nabla}\times\mathbf{H}) - \mathbf{E}\cdot\frac{\partial\mathbf{D}}{\partial t}$. From product rule #6, $\boldsymbol{\nabla}\cdot(\mathbf{E}\times\mathbf{H}) = \mathbf{H}(\boldsymbol{\nabla}\times\mathbf{E}) - \mathbf{E}\cdot(\boldsymbol{\nabla}\times\mathbf{H})$, while $\boldsymbol{\nabla}\times\mathbf{E} = -\frac{\partial\mathbf{B}}{\partial t}$, so $\mathbf{E}\cdot(\boldsymbol{\nabla}\times\mathbf{H}) = -\mathbf{H}\cdot\frac{\partial\mathbf{B}}{\partial t} - \boldsymbol{\nabla}\cdot(\mathbf{E}\times\mathbf{H})$. Therefore $\mathbf{E}\cdot\mathbf{J}_f = -\mathbf{H}\cdot\frac{\partial\mathbf{B}}{\partial t} - \mathbf{E}\cdot\frac{\partial\mathbf{D}}{\partial t} - \boldsymbol{\nabla}\cdot(\mathbf{E}\times\mathbf{H})$, and hence

$$\frac{dW}{dt} = -\int_{\mathcal{V}}\left(\mathbf{E}\cdot\frac{\partial\mathbf{D}}{\partial t} + \mathbf{H}\cdot\frac{\partial\mathbf{B}}{\partial t}\right)d\tau - \oint_{\mathcal{S}}(\mathbf{E}\times\mathbf{H})\cdot d\mathbf{a}.$$

This is Poynting's theorem for the fields in matter. Evidently the Poynting vector, representing the power per unit area transported by the fields, is $\mathbf{S} = \mathbf{E}\times\mathbf{H}$, and the rate of change of the electromagnetic energy density is $\frac{\partial u_{\text{em}}}{\partial t} = \mathbf{E}\cdot\frac{\partial\mathbf{D}}{\partial t} + \mathbf{H}\cdot\frac{\partial\mathbf{B}}{\partial t}$.

For *linear* media, $\mathbf{D} = \epsilon\mathbf{E}$ and $\mathbf{H} = \frac{1}{\mu}\mathbf{B}$, with ϵ and μ constant (in time); then

$$\frac{\partial u_{\text{em}}}{\partial t} = \epsilon\mathbf{E}\cdot\frac{\partial\mathbf{E}}{\partial t} + \frac{1}{\mu}\mathbf{B}\cdot\frac{\partial\mathbf{B}}{\partial t} = \frac{1}{2}\epsilon\frac{\partial}{\partial t}(\mathbf{E}\cdot\mathbf{E}) + \frac{1}{2\mu}\frac{\partial}{\partial t}(\mathbf{B}\cdot\mathbf{B}) = \frac{1}{2}\frac{\partial}{\partial t}(\mathbf{E}\cdot\mathbf{D}+\mathbf{B}\cdot\mathbf{H}),$$

so $u_{\text{em}} = \frac{1}{2}(\mathbf{E}\cdot\mathbf{D}+\mathbf{B}\cdot\mathbf{H})$. qed

(b) If we're only interested in the force on *free* charges and currents, Eq. 8.15 becomes $\mathbf{f} = \rho_f\mathbf{E} + \mathbf{J}_f\times\mathbf{B}$. But $\rho_f = \boldsymbol{\nabla}\cdot\mathbf{D}$, and $\mathbf{J}_f = \boldsymbol{\nabla}\times\mathbf{H} - \frac{\partial\mathbf{D}}{\partial t}$, so $\mathbf{f} = \mathbf{E}(\boldsymbol{\nabla}\cdot\mathbf{D}) + (\boldsymbol{\nabla}\times\mathbf{H})\times\mathbf{B} - \left(\frac{\partial\mathbf{D}}{\partial t}\right)\times\mathbf{B}$. Now $\frac{\partial}{\partial t}(\mathbf{D}\times\mathbf{B}) = \frac{\partial\mathbf{D}}{\partial t}\times\mathbf{B} + \mathbf{D}\times\left(\frac{\partial\mathbf{B}}{\partial t}\right)$, and $\frac{\partial\mathbf{B}}{\partial t} = -\boldsymbol{\nabla}\times\mathbf{E}$, so $\frac{\partial\mathbf{D}}{\partial t}\times\mathbf{B} = \frac{\partial}{\partial t}(\mathbf{D}\times\mathbf{B}) + \mathbf{D}\times(\boldsymbol{\nabla}\times\mathbf{E})$, and hence $\mathbf{f} = \mathbf{E}(\boldsymbol{\nabla}\cdot\mathbf{D}) - \mathbf{D}\times(\boldsymbol{\nabla}\times\mathbf{E}) - \mathbf{B}\times(\boldsymbol{\nabla}\times\mathbf{H}) - \frac{\partial}{\partial t}(\mathbf{D}\times\mathbf{B})$. As before, we can with impunity add the term $\mathbf{H}(\boldsymbol{\nabla}\cdot\mathbf{B})$, so

$$\mathbf{f} = \{[\mathbf{E}(\boldsymbol{\nabla}\cdot\mathbf{D}) - \mathbf{D}\times(\boldsymbol{\nabla}\times\mathbf{E})] + [\mathbf{H}(\boldsymbol{\nabla}\cdot\mathbf{B}) - \mathbf{B}\times(\boldsymbol{\nabla}\times\mathbf{H})]\} - \frac{\partial}{\partial t}(\mathbf{D}\times\mathbf{B}).$$

The term in curly brackets can be written as the divergence of a stress tensor (as in Eq. 8.21), and the last term is (minus) the rate of change of the momentum density, $\boldsymbol{\wp} = \mathbf{D}\times\mathbf{B}$.

Chapter 9

Electromagnetic Waves

Problem 9.1

$$\begin{aligned}
\frac{\partial f_1}{\partial z} &= -2Ab(z-vt)e^{-b(z-vt)^2};\ \frac{\partial^2 f_1}{\partial z^2} = -2Ab\left[e^{-b(z-vt)^2} - 2b(z-vt)^2e^{-b(z-vt)^2}\right];\\
\frac{\partial f_1}{\partial t} &= 2Abv(z-vt)e^{-b(z-vt)^2};\ \frac{\partial^2 f_1}{\partial t^2} = 2Abv\left[-ve^{-b(z-vt)^2} + 2bv(z-vt)^2e^{-b(z-vt)^2}\right] = v^2\frac{\partial^2 f_1}{\partial z^2}.\ \checkmark\\
\frac{\partial f_2}{\partial z} &= Ab\cos[b(z-vt)];\ \frac{\partial^2 f_2}{\partial z^2} = -Ab^2\sin[b(z-vt)];\\
\frac{\partial f_2}{\partial t} &= -Abv\cos[b(z-vt)];\ \frac{\partial^2 f_2}{\partial t^2} = -Ab^2v^2\sin[b(z-vt)] = v^2\frac{\partial^2 f_2}{\partial z^2}.\ \checkmark\\
\frac{\partial f_3}{\partial z} &= \frac{-2Ab(z-vt)}{[b(z-vt)^2+1]^2};\ \frac{\partial^2 f_3}{\partial z^2} = \frac{-2Ab}{[b(z-vt)^2+1]^2} + \frac{8Ab^2(z-vt)^2}{[b(z-vt)^2+1]^3};\\
\frac{\partial f_3}{\partial t} &= \frac{2Abv(z-vt)}{[b(z-vt)^2+1]^2};\ \frac{\partial^2 f_3}{\partial t^2} = \frac{-2Abv^2}{[b(z-vt)^2+1]^2} + \frac{8Ab^2v^2(z-vt)^2}{[b(z-vt)^2+1]^3} = v^2\frac{\partial^2 f_3}{\partial z^2}.\ \checkmark\\
\frac{\partial f_4}{\partial z} &= -2Ab^2ze^{-b(bz^2+vt)};\ \frac{\partial^2 f_4}{\partial z^2} = -2Ab^2\left[e^{-b(bz^2+vt)} - 2b^2z^2e^{-b(bz^2+vt)}\right];\\
\frac{\partial f_4}{\partial t} &= -Abve^{-b(bz^2+vt)};\ \frac{\partial^2 f_4}{\partial t^2} = Ab^2v^2e^{-b(bz^2+vt)} \neq v^2\frac{\partial^2 f_4}{\partial z^2}.\\
\frac{\partial f_5}{\partial z} &= Ab\cos(bz)\cos(bvt)^3;\ \frac{\partial^2 f_5}{\partial z^2} = -Ab^2\sin(bz)\cos(bvt)^3;\ \frac{\partial f_5}{\partial t} = -3Ab^3v^3t^2\sin(bz)\sin(bvt)^3;\\
\frac{\partial^2 f_5}{\partial t^2} &= -6Ab^3v^3t\sin(bz)\sin(bvt)^3 - 9Ab^6v^6t^4\sin(bz)\cos(bvt)^3 \neq v^2\frac{\partial^2 f_5}{\partial z^2}.
\end{aligned}$$

Problem 9.2

$$\begin{aligned}
\frac{\partial f}{\partial z} &= Ak\cos(kz)\cos(kvt);\ \frac{\partial^2 f}{\partial z^2} = -Ak^2\sin(kz)\cos(kvt);\\
\frac{\partial f}{\partial t} &= -Akv\sin(kz)\sin(kvt);\ \frac{\partial^2 f}{\partial t^2} = -Ak^2v^2\sin(kz)\cos(kvt) = v^2\frac{\partial^2 f}{\partial z^2}.\ \checkmark
\end{aligned}$$

Use the trig identity $\sin\alpha\cos\beta = \frac{1}{2}[\sin(\alpha+\beta)+\sin(\alpha-\beta)]$ to write

$$f = \boxed{\frac{A}{2}\left\{\sin[k(z+vt)] + \sin[k(z-vt)]\right\},}$$

which is of the form 9.6, with $g = (A/2)\sin[k(z - vt)]$ and $h = (A/2)\sin[k(z + vt)]$.

Problem 9.3

$$\begin{aligned}
(A_3)^2 &= \left(A_3 e^{i\delta_3}\right)\left(A_3 e^{-i\delta_3}\right) = \left(A_1 e^{i\delta_1} + A_2 e^{i\delta_2}\right)\left(A_1 e^{-i\delta_1} + A_2 e^{-i\delta_2}\right) \\
&= (A_1)^2 + (A_2)^2 + A_1A_2\left(e^{i\delta_1}e^{-i\delta_2} + e^{-i\delta_1}e^{i\delta_2}\right) = (A_1)^2 + (A_2)^2 + A_1A_2 2\cos(\delta_1 - \delta_2); \\
A_3 &= \boxed{\sqrt{(A_1)^2 + (A_2)^2 + 2A_1A_2\cos(\delta_1 - \delta_2)}.} \\
A_3 e^{i\delta_3} &= A_3(\cos\delta_3 + i\sin\delta_3) = A_1(\cos\delta_1 + i\sin\delta_1) + A_2(\cos\delta_2 + i\sin\delta_2) \\
&= (A_1\cos\delta_1 + A_2\cos\delta_2) + i(A_1\sin\delta_1 + A_2\sin\delta_2). \quad \tan\delta_3 = \frac{A_3\sin\delta_3}{A_3\cos\delta_3} = \frac{A_1\sin\delta_1 + A_2\sin\delta_2}{A_1\cos\delta_1 + A_2\cos\delta_2}; \\
\delta_3 &= \boxed{\tan^{-1}\left(\frac{A_1\sin\delta_1 + A_2\sin\delta_2}{A_1\cos\delta_1 + A_2\cos\delta_2}\right).}
\end{aligned}$$

Problem 9.4

The wave equation (Eq. 9.2) says $\frac{\partial^2 f}{\partial z^2} = \frac{1}{v^2}\frac{\partial^2 f}{\partial t^2}$. Look for solutions of the form $f(z,t) = Z(z)T(t)$. Plug this in: $T\frac{d^2Z}{dz^2} = \frac{1}{v^2}Z\frac{d^2T}{dt^2}$. Divide by ZT : $\frac{1}{Z}\frac{d^2Z}{dz^2} = \frac{1}{v^2T}\frac{d^2T}{dt^2}$. The left side depends only on z, and the right side only on t, so both must be constant. Call the constant $-k^2$.

$$\left\{\begin{aligned}
\frac{d^2Z}{dz^2} &= -k^2 Z \quad \Rightarrow \quad Z(z) = Ae^{ikz} + Be^{-ikz}, \\
\frac{d^2T}{dt^2} &= -(kv)^2 T \quad \Rightarrow \quad T(t) = Ce^{ikvt} + De^{-ikvt}.
\end{aligned}\right\}$$

(Note that k must be *real*, else Z and T blow up; with no loss of generality we can assume k is *positive*.)
$f(z,t) = \left(Ae^{ikz} + Be^{-ikz}\right)\left(Ce^{ikvt} + De^{-ikvt}\right) = A_1e^{i(kz+kvt)} + A_2e^{i(kz-kvt)} + A_3e^{i(-kz+kvt)} + A_4e^{i(-kz-kvt)}$.
The general linear combination of separable solutions is therefore

$$f(z,t) = \int_0^\infty \left[A_1(k)e^{i(kz+\omega t)} + A_2(k)e^{i(kz-\omega t)} + A_3(k)e^{i(-kz+\omega t)} + A_4(k)e^{i(-kz-\omega t)}\right] dk,$$

where $\omega \equiv kv$. But we can combine the third term with the first, by allowing k to run *negative* ($\omega = |k|v$ remains positive); likewise the second and the fourth:

$$f(z,t) = \int_{-\infty}^\infty \left[A_1(k)e^{i(kz+\omega t)} + A_2(k)e^{i(kz-\omega t)}\right] dk.$$

Because (in the end) we shall only want the the *real part* of f, it suffices to keep only *one* of these terms (since k goes negative, both terms include waves traveling in both directions); the second is traditional (though either would do). Specifically,

$$\text{Re}(f) = \int_{-\infty}^\infty [\text{Re}(A_1)\cos(kz+\omega t) - \text{Im}(A_1)\sin(kz+\omega t) + \text{Re}(A_2)\cos(kz-\omega t) - \text{Im}(A_2)\sin(kz-\omega t)]\, dk.$$

The first term, $\cos(kz+\omega t) = \cos(-kz-\omega t)$, combines with the third, $\cos(kz-\omega t)$, since the negative k is picked up in the other half of the range of integration, and the second, $\sin(kz+\omega t) = -\sin(-kz-\omega t)$, combines with the fourth for the same reason. So the general solution, for our purposes, can be written in the form

$$\tilde{f}(z,t) = \int_{-\infty}^\infty \tilde{A}(k)e^{i(kz-\omega t)}\, dk \quad \text{qed (the tildes remind us that we want the real part).}$$

Problem 9.5

Equation 9.26 $\Rightarrow g_I(-v_1t)+h_R(v_1t)=g_T(-v_2t)$. Now $\frac{\partial g_I}{\partial z}=-\frac{1}{v_1}\frac{\partial g_I}{\partial t}$; $\frac{\partial h_R}{\partial z}=\frac{1}{v_1}\frac{\partial h_R}{\partial t}$; $\frac{\partial g_T}{\partial z}=-\frac{1}{v_2}\frac{\partial g_T}{\partial t}$.
Equation 9.27 $\Rightarrow -\frac{1}{v_1}\frac{\partial g_I(-v_1t)}{\partial t}+\frac{1}{v_1}\frac{\partial h_R(v_1t)}{\partial t}=-\frac{1}{v_2}\frac{\partial g_T(-v_2t)}{\partial t} \Rightarrow g_I(-v_1t)-h_R(v_1t)=\frac{v_1}{v_2}g_T(-v_2t)+\kappa$ (where κ is a constant).

Adding these equations, we get $2g_I(-v_1t)=\left(1+\frac{v_1}{v_2}\right)g_T(-v_2t)+\kappa$, or $g_T(-v_2t)=\left(\frac{2v_2}{v_1+v_2}\right)g_I(-v_1t)+\kappa'$ (where $\kappa'\equiv-\kappa\frac{v_2}{v_1+v_2}$). Now $g_I(z,t)$, $g_T(z,t)$, and $h_R(z,t)$ are each functions of a single variable u (in the first case $u=z-v_1t$, in the second $u=z-v_2t$, and in the third $u=z+v_1t$). Thus

$$\boxed{g_T(u)=\left(\frac{2v_2}{v_1+v_2}\right)g_I(v_1u/v_2)+\kappa'.}$$

Multiplying the first equation by v_1/v_2 and subtracting, $\left(1-\frac{v_1}{v_2}\right)g_I(-v_1t)-\left(1+\frac{v_1}{v_2}\right)h_R(v_1t)=\kappa\Rightarrow$
$h_R(v_1t)=\left(\frac{v_2-v_1}{v_1+v_2}\right)g_I(-v_1t)-\kappa\left(\frac{v_2}{v_1+v_2}\right)$, or $\boxed{h_R(u)=\left(\frac{v_2-v_1}{v_1+v_2}\right)g_I(-u)+\kappa'.}$

[The notation is tricky, so here's an example: for a sinusoidal wave,

$$\begin{cases} g_I = A_I\cos(k_1z-\omega t) = A_I\cos[k_1(z-v_1t)] \Rightarrow g_I(u)=A_I\cos(k_1u). \\ g_T = A_T\cos(k_2z-\omega t) = A_T\cos[k_2(z-v_2t)] \Rightarrow g_T(u)=A_T\cos(k_2u). \\ h_R = A_R\cos(-k_1z-\omega t) = A_R\cos[-k_1(z+v_1t)] \Rightarrow h_R(u)=A_R\cos(-k_1u). \end{cases}$$

Here $\kappa'=0$, and the boundary conditions say $\frac{A_T}{A_I}=\frac{2v_2}{v_1+v_2}$, $\frac{A_R}{A_I}=\frac{v_2-v_1}{v_1+v_2}$ (same as Eq. 9.32), and $\frac{v_1}{v^2}k_1=k_2$ (consistent with Eq. 9.24).]

Problem 9.6

(a) $T\sin\theta_+-T\sin\theta_-=ma\Rightarrow \boxed{T\left(\left.\frac{\partial f}{\partial z}\right|_{0^+}-\left.\frac{\partial f}{\partial z}\right|_{0^-}\right)=m\left.\frac{\partial^2 f}{\partial t^2}\right|_0.}$

(b) $\tilde{A}_I+\tilde{A}_R=\tilde{A}_T$; $T[ik_2\tilde{A}_T-ik_1(\tilde{A}_I-\tilde{A}_R)]=m(-\omega^2\tilde{A}_T)$, or $k_1(\tilde{A}_I-\tilde{A}_R)=\left(k_2-\frac{im\omega^2}{T}\right)\tilde{A}_T$.

Multiply first equation by k_1 and add: $2k_1\tilde{A}_I=\left(k_1+k_2-i\frac{m\omega^2}{T}\right)\tilde{A}_T$, or $\tilde{A}_T=\left(\frac{2k_1}{k_1+k_2-im\omega^2/T}\right)\tilde{A}_I$.

$$\tilde{A}_R=\tilde{A}_T-\tilde{A}_I=\frac{2k_1-(k_1+k_2-im\omega^2/T)}{k_1+k_2-im\omega^2/T}\tilde{A}_I=\left(\frac{k_1-k_2+im\omega^2/T}{k_1+k_2-im\omega^2/T}\right)\tilde{A}_I.$$

If the second string is massless, so $v_2=\sqrt{T/\mu_2}=\infty$, then $k_2/k_1=0$, and we have $\tilde{A}_T=\left(\frac{2}{1-i\beta}\right)\tilde{A}_I$,
$\tilde{A}_R=\left(\frac{1+i\beta}{1-i\beta}\right)\tilde{A}_I$, where $\beta\equiv\frac{m\omega^2}{k_1T}=\frac{m(k_1v_1)^2}{k_1T}=\frac{mk_1}{T}\frac{T}{\mu_1}$, or $\boxed{\beta=m\frac{k_1}{\mu_1}.}$ Now $\left(\frac{1+i\beta}{1-i\beta}\right)=Ae^{i\phi}$, with
$A^2=\left(\frac{1+i\beta}{1-i\beta}\right)\left(\frac{1-i\beta}{1+i\beta}\right)=1\Rightarrow A=1$, and $e^{i\phi}=\frac{(1+i\beta)^2}{(1-i\beta)(1+i\beta)}=\frac{1+2i\beta-\beta^2}{1+\beta^2}\Rightarrow$
$\tan\phi=\frac{2\beta}{1-\beta^2}$. Thus $A_Re^{i\delta_R}=e^{i\phi}A_Ie^{i\delta_I}\Rightarrow \boxed{A_R=A_I,}$ $\boxed{\delta_R=\delta_I+\tan^{-1}\left(\frac{2\beta}{1-\beta^2}\right).}$

Similarly, $\left(\frac{2}{1-i\beta}\right)=Ae^{i\phi}\Rightarrow A^2=\left(\frac{2}{1-i\beta}\right)\left(\frac{2}{1+i\beta}\right)=\frac{4}{1+\beta^2}\Rightarrow A=\frac{2}{\sqrt{1+\beta^2}}.$

$$Ae^{i\phi} = \frac{2(1+i\beta)}{(1-i\beta)(1+i\beta)} = \frac{2(1+i\beta)}{(1+\beta^2)} \Rightarrow \tan\phi = \beta. \text{ So } A_T e^{i\delta_T} = \frac{2}{\sqrt{1+\beta^2}} e^{i\phi} A_I e^{i\delta_I};$$

$$\boxed{A_T = \frac{2}{\sqrt{1+\beta^2}} A_I;} \quad \boxed{\delta_T = \delta_I + \tan^{-1}\beta.}$$

Problem 9.7

(a) $F = T\dfrac{\partial^2 f}{\partial z^2}\Delta z - \gamma \dfrac{\partial f}{\partial t}\Delta z = \mu \Delta z \dfrac{\partial^2 f}{\partial t^2}$, or $\boxed{T\dfrac{\partial^2 f}{\partial z^2} = \mu \dfrac{\partial^2 f}{\partial t^2} + \gamma \dfrac{\partial f}{\partial t}.}$

(b) Let $\tilde{f}(z,t) = \tilde{F}(z)e^{-i\omega t}$; then $Te^{-i\omega t}\dfrac{d^2\tilde{F}}{dz^2} = \mu(-\omega^2)\tilde{F}e^{-i\omega t} + \gamma(-i\omega)\tilde{F}e^{-i\omega t} \Rightarrow$

$T\dfrac{d^2\tilde{F}}{dz^2} = -\omega(\mu\omega + i\gamma)\tilde{F}$, $\dfrac{d^2\tilde{F}}{dz^2} = -\tilde{k}^2\tilde{F}$, where $\tilde{k}^2 \equiv \dfrac{\omega}{T}(\mu\omega + i\gamma)$. Solution : $\tilde{F}(z) = \tilde{A}e^{i\tilde{k}z} + \tilde{B}e^{-i\tilde{k}z}$.

Resolve $\tilde{k}$ into its real and imaginary parts: $\tilde{k} = k + i\kappa \Rightarrow \tilde{k}^2 = k^2 - \kappa^2 + 2ik\kappa = \dfrac{\omega}{T}(\mu\omega + i\gamma)$.

$2k\kappa = \dfrac{\omega\gamma}{T} \Rightarrow \kappa = \dfrac{\omega\gamma}{2kT}$; $k^2 - \kappa^2 = k^2 - \left(\dfrac{\omega\gamma}{2T}\right)^2 \dfrac{1}{k^2} = \dfrac{\mu\omega^2}{T}$; or $k^4 - k^2(\mu\omega^2/T) - (\omega\gamma/2T)^2 = 0 \Rightarrow$

$k^2 = \dfrac{1}{2}\left[(\mu\omega^2/T) \pm \sqrt{(\mu\omega^2/T)^2 + 4(\omega\gamma/2T)^2}\right] = \dfrac{\mu\omega^2}{2T}\left[1 \pm \sqrt{1 + (\gamma/\mu\omega)^2}\right]$. But k is real, so k^2 is positive, so we need the plus sign: $k = \omega\sqrt{\dfrac{\mu}{2T}}\sqrt{1 + \sqrt{1+(\gamma/\mu\omega)^2}}$. $\kappa = \dfrac{\omega\gamma}{2kT} = \dfrac{\gamma}{\sqrt{2T\mu}}\left[1 + \sqrt{1+(\gamma/\mu\omega)^2}\right]^{-1/2}$.

Plugging this in, $\tilde{F} = Ae^{i(k+i\kappa)z} + Be^{-i(k+i\kappa)z} = Ae^{-\kappa z}e^{ikz} + Be^{\kappa z}e^{-ikz}$. But the B term gives an exponentially *increasing* function, which we don't want (I assume the waves are propagating in the $+z$ direction), so $B = 0$, and the solution is $\boxed{\tilde{f}(z,t) = \tilde{A}e^{-\kappa z}e^{i(kz-\omega t)}.}$ (The actual displacement of the string is the real part of this, of course.)

(c) The wave is attenuated by the factor $e^{-\kappa z}$, which becomes $1/e$ when

$z = \dfrac{1}{\kappa} = \boxed{\dfrac{\sqrt{2T\mu}}{\gamma}\sqrt{1 + \sqrt{1+(\gamma/\mu\omega)^2}};}$ this is the characteristic penetration depth.

(d) This is the same as before, except that $k_2 \to k + i\kappa$. From Eq. 9.29, $\tilde{A}_R = \left(\dfrac{k_1 - k - i\kappa}{k_1 + k + i\kappa}\right)\tilde{A}_I$;

$$\left(\frac{A_R}{A_I}\right)^2 = \left(\frac{k_1 - k - i\kappa}{k_1 + k + i\kappa}\right)\left(\frac{k_1 - k + i\kappa}{k_1 + k - i\kappa}\right) = \frac{(k_1-k)^2 + \kappa^2}{(k_1+k)^2 + \kappa^2}. \quad \boxed{A_R = \sqrt{\frac{(k_1-k)^2+\kappa^2}{(k_1+k)^2+\kappa^2}}A_I}$$

(where $k_1 = \omega/v_1 = \omega\sqrt{\mu_1/T}$, while k and κ are defined in part b). Meanwhile

$$\left(\frac{k_1 - k - i\kappa}{k_1 + k + i\kappa}\right) = \frac{(k_1 - k - i\kappa)(k_1 + k + i\kappa)}{(k_1+k)^2 + \kappa^2} = \frac{(k_1)^2 - k^2 - \kappa^2 - 2i\kappa k_1}{(k_1+k)^2 + \kappa^2} \Rightarrow \boxed{\delta_R = \tan^{-1}\left(\frac{-2k_1\kappa}{(k_1)^2 - k^2 - \kappa^2}\right).}$$

Problem 9.8

(a) $\mathbf{f}_v(z,t) = A\cos(kz - \omega t)\,\hat{\mathbf{x}}$; $\mathbf{f}_h(z,t) = A\cos(kz - \omega t + 90°)\,\hat{\mathbf{y}} = -A\sin(kz - \omega t)\,\hat{\mathbf{y}}$. Since $f_v^2 + f_h^2 = A^2$, the vector sum $\mathbf{f} = \mathbf{f}_v + \mathbf{f}_h$ lies on a circle of radius A. At time $t = 0$, $\mathbf{f} = A\cos(kz)\,\hat{\mathbf{x}} - A\sin(kz)\,\hat{\mathbf{y}}$. At time $t = \pi/2\omega$, $\mathbf{f} = A\cos(kz - 90°)\,\hat{\mathbf{x}} - A\sin(kz - 90°)\,\hat{\mathbf{y}} = A\sin(kz)\,\hat{\mathbf{x}} + A\cos(kz)\,\hat{\mathbf{y}}$. Evidently it circles $\boxed{\text{counterclockwise}}$. To make a wave circling the other way, use $\delta_h = -90°$.

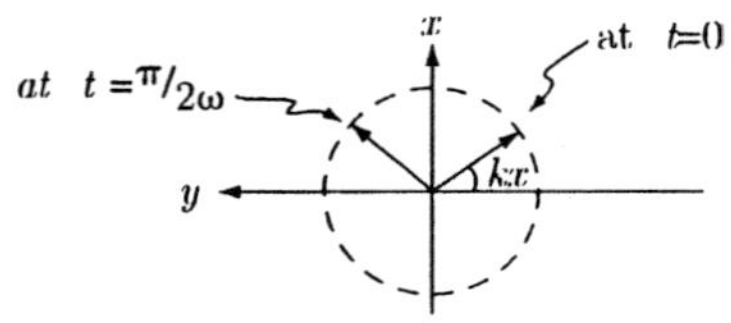

(b)

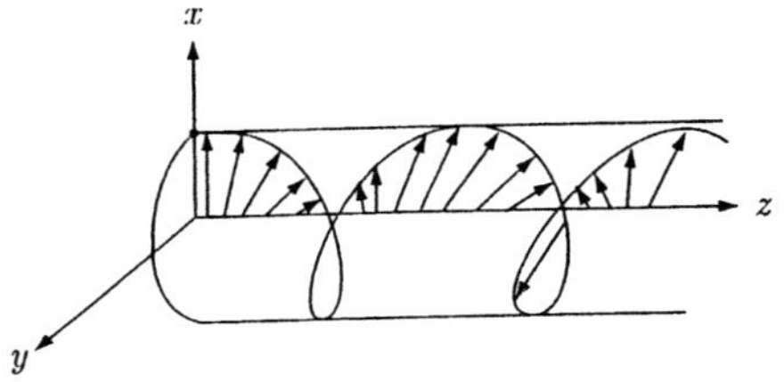

(c) Shake it around in a circle, instead of up and down.

Problem 9.9

(a) $\boxed{\mathbf{k} = -\frac{\omega}{c}\,\hat{\mathbf{x}};\ \hat{\mathbf{n}} = \hat{\mathbf{z}}.}$ $\mathbf{k}\cdot\mathbf{r} = \left(-\frac{\omega}{c}\,\hat{\mathbf{x}}\right)\cdot(x\,\hat{\mathbf{x}} + y\,\hat{\mathbf{y}} + z\,\hat{\mathbf{z}}) = -\frac{\omega}{c}x;\ \mathbf{k}\times\hat{\mathbf{n}} = -\hat{\mathbf{x}}\times\hat{\mathbf{z}} = \hat{\mathbf{y}}.$

$$\boxed{\mathbf{E}(x,t) = E_0\cos\left(\frac{\omega}{c}x + \omega t\right)\hat{\mathbf{z}};\quad \mathbf{B}(x,t) = \frac{E_0}{c}\cos\left(\frac{\omega}{c}x + \omega t\right)\hat{\mathbf{y}}.}$$

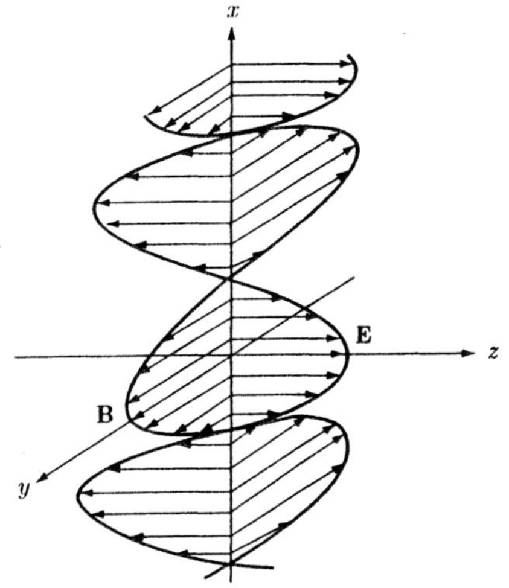

(a)

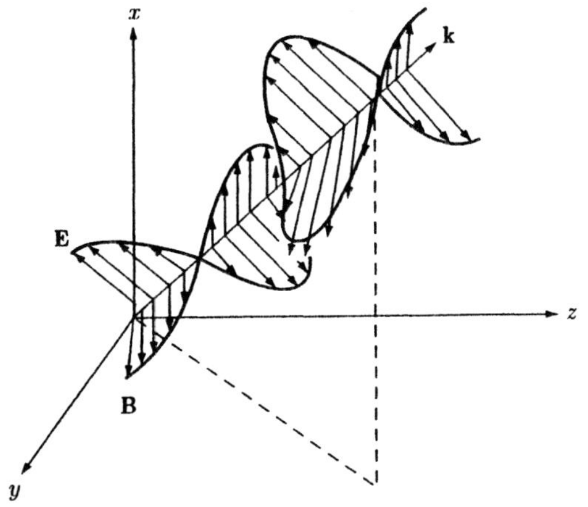

(b)

(b) $\boxed{\mathbf{k} = \frac{\omega}{c}\left(\frac{\hat{\mathbf{x}}+\hat{\mathbf{y}}+\hat{\mathbf{z}}}{\sqrt{3}}\right);\ \hat{\mathbf{n}} = \frac{\hat{\mathbf{x}}-\hat{\mathbf{z}}}{\sqrt{2}}.}$ (Since $\hat{\mathbf{n}}$ is parallel to the xz plane, it must have the form $\alpha\,\hat{\mathbf{x}} + \beta\,\hat{\mathbf{z}}$; since $\hat{\mathbf{n}}\cdot\mathbf{k} = 0, \beta = -\alpha$; and since it is a unit vector, $\alpha = 1/\sqrt{2}$.)

$$\mathbf{k}\cdot\mathbf{r} = \frac{\omega}{\sqrt{3}c}(\hat{\mathbf{x}}+\hat{\mathbf{y}}+\hat{\mathbf{z}})\cdot(x\,\hat{\mathbf{x}}+y\,\hat{\mathbf{y}}+z\,\hat{\mathbf{z}}) = \frac{\omega}{\sqrt{3}c}(x+y+z);\ \hat{\mathbf{k}}\times\hat{\mathbf{n}} = \frac{1}{\sqrt{6}}\begin{vmatrix}\hat{\mathbf{x}} & \hat{\mathbf{y}} & \hat{\mathbf{z}}\\ 1 & 1 & 1\\ 1 & 0 & -1\end{vmatrix} = \frac{1}{\sqrt{6}}(-\hat{\mathbf{x}} + 2\,\hat{\mathbf{y}} - \hat{\mathbf{z}}).$$

$$\boxed{\begin{aligned}\mathbf{E}(x,y,z,t) &= E_0\cos\left[\frac{\omega}{\sqrt{3}c}(x+y+z) - \omega t\right]\left(\frac{\hat{\mathbf{x}}-\hat{\mathbf{z}}}{\sqrt{2}}\right);\\ \mathbf{B}(x,y,z,t) &= \frac{E_0}{c}\cos\left[\frac{\omega}{\sqrt{3}c}(x+y+z) - \omega t\right]\left(\frac{-\hat{\mathbf{x}}+2\hat{\mathbf{y}}-\hat{\mathbf{z}}}{\sqrt{6}}\right).\end{aligned}}$$

Problem 9.10

$P = \frac{I}{c} = \frac{1.3\times10^3}{3.0\times10^8} = \boxed{4.3\times10^{-6}\ \mathrm{N/m^2}.}$ For a perfect reflector the pressure is twice as great: $\boxed{8.6\times10^{-6}\ \mathrm{N/m^2}.}$ Atmospheric pressure is $1.03\times10^5\ \mathrm{N/m^2}$, so the pressure of light on a reflector is $(8.6\times10^{-6})/(1.03\times10^5) = \boxed{8.3\times10^{-11}\ \text{atmospheres.}}$

Problem 9.11

$$\begin{aligned}\langle fg\rangle &= \frac{1}{T}\int_0^T a\cos(\mathbf{k}\cdot\mathbf{r}-\omega t+\delta_a)b\cos(\mathbf{k}\cdot\mathbf{r}-\omega t+\delta_b)\,dt\\ &= \frac{ab}{2T}\int_0^T[\cos(2\mathbf{k}\cdot\mathbf{r}-2\omega t+\delta_a+\delta_b)+\cos(\delta_a-\delta_b)]\,dt=\frac{ab}{2T}\cos(\delta_a-\delta_b)T=\frac{1}{2}ab\cos(\delta_a-\delta_b).\end{aligned}$$

Meanwhile, in the complex notation: $\tilde{f}=\tilde{a}e^{i\mathbf{k}\cdot\mathbf{r}-\omega t)}$, $\tilde{g}=\tilde{b}e^{i\mathbf{k}\cdot\mathbf{r}-\omega t)}$, where $\tilde{a}=ae^{i\delta_a}$, $\tilde{b}=be^{i\delta_b}$. So $\frac{1}{2}\tilde{f}\tilde{g}^*=\frac{1}{2}\tilde{a}e^{i(\mathbf{k}\cdot\mathbf{r}-\omega t)}\tilde{b}^*e^{-i(\mathbf{k}\cdot\mathbf{r}-\omega t)}=\frac{1}{2}\tilde{a}\tilde{b}^*=\frac{1}{2}abe^{i(\delta_a-\delta_b)}$, $\text{Re}\left(\frac{1}{2}\tilde{f}\tilde{g}^*\right)=\frac{1}{2}ab\cos(\delta_a-\delta_b)=\langle fg\rangle$. qed

Problem 9.12

$$T_{ij}=\epsilon_0\left(E_iE_j-\frac{1}{2}\delta_{ij}E^2\right)+\frac{1}{\mu_0}\left(B_iB_j-\frac{1}{2}\delta_{ij}B^2\right).$$

With the fields in Eq. 9.48, **E** has only an x component, and **B** only a y component. So all the "off-diagonal" ($i\neq j$) terms are zero. As for the "diagonal" elements:

$$\begin{aligned}T_{xx} &= \epsilon_0\left(E_xE_x-\frac{1}{2}E^2\right)+\frac{1}{\mu_0}\left(-\frac{1}{2}B^2\right)=\frac{1}{2}\left(\epsilon_0E^2-\frac{1}{\mu_0}B^2\right)=0.\\ T_{yy} &= \epsilon_0\left(-\frac{1}{2}E^2\right)+\frac{1}{\mu_0}\left(B_yB_y-\frac{1}{2}B^2\right)=\frac{1}{2}\left(-\epsilon_0E^2+\frac{1}{\mu_0}B^2\right)=0.\\ T_{zz} &= \epsilon_0\left(-\frac{1}{2}E^2\right)+\frac{1}{\mu_0}\left(-\frac{1}{2}B^2\right)=-u.\end{aligned}$$

So $\boxed{T_{zz}=-\epsilon_0E_0^2\cos^2(kz-\omega t+\delta)}$ (all other elements zero).

The momentum of these fields is in the z direction, and it is being *transported* in the z direction, so *yes*, it does make sense that T_{zz} should be the only nonzero element in T_{ij}. According to Sect. 8.2.3, $-\overleftrightarrow{T}\cdot d\mathbf{a}$ is the rate at which momentum crosses an area $d\mathbf{a}$. Here we have *no* momentum crossing areas oriented in the x or y direction; the momentum per unit time per unit area flowing across a surface oriented in the z direction is $-T_{zz}=u=\wp c$ (Eq. 9.59), so $\Delta p=\wp cA\Delta t$, and hence $\Delta p/\Delta t=\wp cA=$ momentum per unit time crossing area A. Evidently $\boxed{\text{momentum flux density} = \text{energy density.}}$ ✓

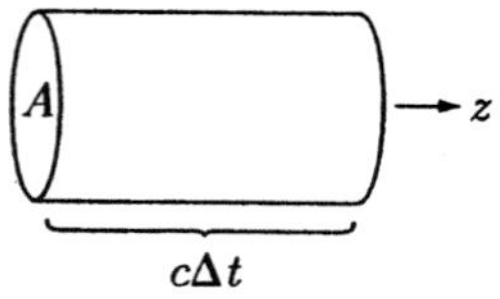

Problem 9.13

$R=\left(\frac{E_{0_R}}{E_{0_I}}\right)^2$ (Eq. 9.86) $\Rightarrow \boxed{R=\left(\frac{1-\beta}{1+\beta}\right)^2}$ (Eq. 9.82), where $\beta\equiv\frac{\mu_1v_1}{\mu_2v_2}$. $T=\frac{\epsilon_2v_2}{\epsilon_1v_1}\left(\frac{E_{0_T}}{E_{0_I}}\right)^2$ (Eq. 9.87)

$\Rightarrow \boxed{T=\beta\left(\frac{2}{1+\beta}\right)^2}$ (Eq. 9.82). [Note that $\frac{\epsilon_2v_2}{\epsilon_1v_1}=\frac{\mu_1}{\mu_2}\frac{\epsilon_2\mu_2}{\epsilon_1\mu_1}\frac{v_2}{v_1}=\frac{\mu_1}{\mu_2}\left(\frac{v_1}{v_2}\right)^2\frac{v_2}{v_1}=\frac{\mu_1v_1}{\mu_2v_2}=\beta$.]

$$T+R=\frac{1}{(1+\beta)^2}\left[4\beta+(1-\beta)^2\right]=\frac{1}{(1+\beta)^2}(4\beta+1-2\beta+\beta^2)=\frac{1}{(1+\beta)^2}(1+2\beta+\beta^2)=1.\ ✓$$

Problem 9.14

Equation 9.78 is replaced by $\tilde{E}_{0_I}\hat{\mathbf{x}} + \tilde{E}_{0_R}\hat{\mathbf{n}}_R = \tilde{E}_{0_T}\hat{\mathbf{n}}_T$, and Eq. 9.80 becomes $\tilde{E}_{0_I}\hat{\mathbf{y}} - \tilde{E}_{0_R}(\hat{\mathbf{z}} \times \hat{\mathbf{n}}_R) = \beta\tilde{E}_{0_T}(\hat{\mathbf{z}} \times \hat{\mathbf{n}}_T)$. The y component of the first equation is $\tilde{E}_{0_R}\sin\theta_R = \tilde{E}_{0_T}\sin\theta_T$; the x component of the second is $\tilde{E}_{0_R}\sin\theta_R = -\beta\tilde{E}_{0_T}\sin\theta_T$. Comparing these two, we conclude that $\sin\theta_R = \sin\theta_T = 0$, and hence $\theta_R = \theta_T = 0$. qed

Problem 9.15

$Ae^{iax} + Be^{ibx} = Ce^{icx}$ for all x, so (using $x = 0$), $A + B = C$.

Differentiate: $iaAe^{iax} + ibBe^{ibx} = icCe^{icx}$, so (using $x = 0$), $aA + bB = cC$.

Differentiate again: $-a^2Ae^{iax} - b^2Be^{ibx} = -c^2Ce^{icx}$, so (using $x = 0$), $a^2A + b^2B = c^2C$.

$a^2A + b^2B = c(cC) = c(aA + bB)$; $(A + B)(a^2A + b^2B) = (A + B)c(aA + bB) = cC(aA + bB)$;

$a^2A^2 + b^2AB + a^2AB + b^2B^2 = (aA + bB)^2 = a^2A^2 + 2abAB + b^2B^2$, or $(a^2 + b^2 - 2ab)AB = 0$, or

$(a - b)^2AB = 0$. But A and B are nonzero, so $a = b$. Therefore $(A + B)e^{iax} = Ce^{icx}$.

$a(A + B) = cC$, or $aC = cC$, so (since $C \neq 0$) $a = c$. *Conclusion:* $a = b = c$. qed

Problem 9.16

$$\left\{\begin{array}{lcl} \tilde{\mathbf{E}}_I &=& \tilde{E}_{0_I}e^{i(\mathbf{k}_I\cdot\mathbf{r}-\omega t)}\hat{\mathbf{y}}, \\ \tilde{\mathbf{B}}_I &=& \dfrac{1}{v_1}\tilde{E}_{0_I}e^{i(\mathbf{k}_I\cdot\mathbf{r}-\omega t)}(-\cos\theta_1\,\hat{\mathbf{x}} + \sin\theta_1\,\hat{\mathbf{z}}); \end{array}\right\}$$

$$\left\{\begin{array}{lcl} \tilde{\mathbf{E}}_R &=& \tilde{E}_{0_R}e^{i(\mathbf{k}_R\cdot\mathbf{r}-\omega t)}\hat{\mathbf{y}}, \\ \tilde{\mathbf{B}}_R &=& \dfrac{1}{v_1}\tilde{E}_{0_R}e^{i(\mathbf{k}_R\cdot\mathbf{r}-\omega t)}(\cos\theta_1\,\hat{\mathbf{x}} + \sin\theta_1\,\hat{\mathbf{z}}); \end{array}\right\}$$

$$\left\{\begin{array}{lcl} \tilde{\mathbf{E}}_T &=& \tilde{E}_{0_T}e^{i(\mathbf{k}_T\cdot\mathbf{r}-\omega t)}\hat{\mathbf{y}}, \\ \tilde{\mathbf{B}}_T &=& \dfrac{1}{v_2}\tilde{E}_{0_T}e^{i(\mathbf{k}_T\cdot\mathbf{r}-\omega t)}(-\cos\theta_2\,\hat{\mathbf{x}} + \sin\theta_1\,\hat{\mathbf{z}}); \end{array}\right\}$$

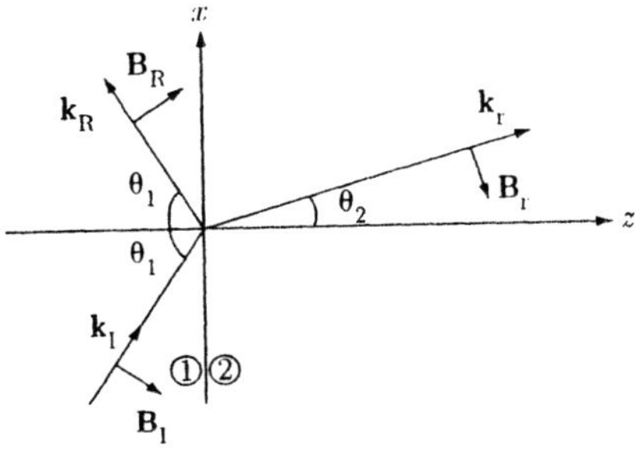

Boundary conditions: $\left\{\begin{array}{ll} \text{(i) } \epsilon_1 E_1^{\perp} = \epsilon_2 E_2^{\perp}, & \text{(iii) } \mathbf{E}_1^{\parallel} = \mathbf{E}_2^{\parallel}, \\ \text{(ii) } B_1^{\perp} = B_2^{\perp}, & \text{(iv) } \frac{1}{\mu_1}\mathbf{B}_1^{\parallel} = \frac{1}{\mu_2}\mathbf{B}_2^{\parallel}. \end{array}\right.$

Law of refraction: $\dfrac{\sin\theta_2}{\sin\theta_1} = \dfrac{v_2}{v_1}$. [Note: $\mathbf{k}_I\cdot\mathbf{r} - \omega t = \mathbf{k}_R\cdot\mathbf{r} - \omega t = \mathbf{k}_T\cdot\mathbf{r} - \omega t$, at $z = 0$, so we can drop all exponential factors in applying the boundary conditions.]

Boundary condition (i): $0 = 0$ (trivial). Boundary condition (iii): $\boxed{\tilde{E}_{0_I} + \tilde{E}_{0_R} = \tilde{E}_{0_T}.}$

Boundary condition (ii): $\dfrac{1}{v_1}\tilde{E}_{0_I}\sin\theta_1 + \dfrac{1}{v_1}\tilde{E}_{0_R}\sin\theta_1 = \dfrac{1}{v_2}\tilde{E}_{0_T}\sin\theta_2 \Rightarrow \tilde{E}_{0_I} + \tilde{E}_{0_R} = \left(\dfrac{v_1\sin\theta_2}{v_2\sin\theta_1}\right)\tilde{E}_{0_T}$.

But the term in parentheses is 1, by the law of refraction, so this is the same as (ii).

Boundary condition (iv): $\dfrac{1}{\mu_1}\left[\dfrac{1}{v_1}\tilde{E}_{0_I}(-\cos\theta_1) + \dfrac{1}{v_1}\tilde{E}_{0_R}\cos\theta_1\right] = \dfrac{1}{\mu_2 v_2}\tilde{E}_{0_T}(-\cos\theta_2) \Rightarrow$

$\tilde{E}_{0_I} - \tilde{E}_{0_R} = \left(\dfrac{\mu_1 v_1\cos\theta_2}{\mu_2 v_2\cos\theta_1}\right)\tilde{E}_{0_T}$. Let $\boxed{\alpha \equiv \dfrac{\cos\theta_2}{\cos\theta_1};\ \beta \equiv \dfrac{\mu_1 v_1}{\mu_2 v_2}.}$ Then $\boxed{\tilde{E}_{0_I} - \tilde{E}_{0_R} = \alpha\beta\tilde{E}_{0_T}.}$

Solving for $\tilde{E}_{0_R}$ and $\tilde{E}_{0_T}$: $2\tilde{E}_{0_I} = (1 + \alpha\beta)\tilde{E}_{0_T} \Rightarrow \tilde{E}_{0_T} = \left(\dfrac{2}{1 + \alpha\beta}\right)\tilde{E}_{0_I}$;

$\tilde{E}_{0_R} = \tilde{E}_{0_T} - \tilde{E}_{0_I} = \left(\dfrac{2}{1 + \alpha\beta} - \dfrac{1 + \alpha\beta}{1 + \alpha\beta}\right)\tilde{E}_{0_I} \Rightarrow \tilde{E}_{0_R} = \left(\dfrac{1 - \alpha\beta}{1 + \alpha\beta}\right)\tilde{E}_{0_I}$.

Since α and β are positive, it follows that $2/(1 + \alpha\beta)$ is positive, and hence the *transmitted* wave is *in phase* with the incident wave, and the (real) amplitudes are related by $\boxed{E_{0_T} = \left(\dfrac{2}{1 + \alpha\beta}\right)E_{0_I}.}$ The *reflected* wave is

in phase if $\alpha\beta < 1$ and 180° out of phase if $\alpha\beta < 1$; the (real) amplitudes are related by $\boxed{E_{0_R} = \left|\frac{1-\alpha\beta}{1+\alpha\beta}\right| E_{0_I}.}$
These are the **Fresnel equations** for polarization perpendicular to the plane of incidence.

To construct the graphs, note that $\alpha\beta = \beta\frac{\sqrt{1-\sin^2\theta/\beta^2}}{\cos\theta} = \frac{\sqrt{\beta^2-\sin^2\theta}}{\cos\theta}$, where θ is the angle of incidence, so, for $\beta = 1.5$, $\alpha\beta = \frac{\sqrt{2.25-\sin^2\theta}}{\cos\theta}$.

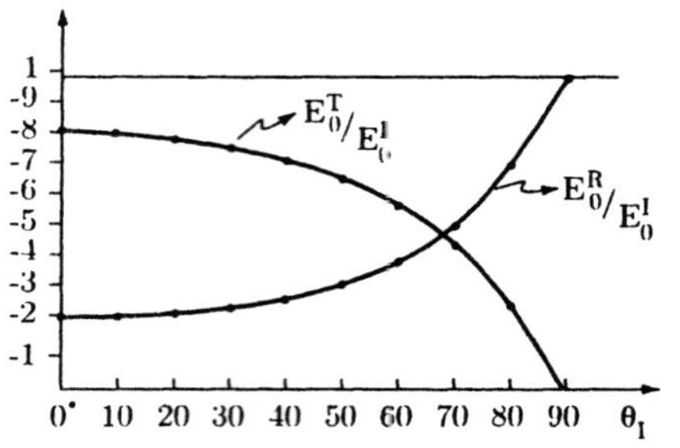

Is there a Brewster's angle? Well, $E_{0_R} = 0$ would mean that $\alpha\beta = 1$, and hence that
$\alpha = \frac{\sqrt{1-(v_2/v_1)^2\sin^2\theta}}{\cos\theta} = \frac{1}{\beta} = \frac{\mu_2 v_2}{\mu_1 v_1}$, or $1 - \left(\frac{v_2}{v_1}\right)^2 \sin^2\theta = \left(\frac{\mu_2 v_2}{\mu_1 v_1}\right)^2 \cos^2\theta$, so
$1 = \left(\frac{v_2}{v_1}\right)^2 \left[\sin^2\theta + (\mu_2/\mu_1)^2\cos^2\theta\right]$. Since $\mu_1 \approx \mu_2$, this means $1 \approx (v_2/v_1)^2$, which is only true for optically indistinguishable media, in which case there is of *course* no reflection—but that would be true at *any* angle, not just at a special "Brewster's angle". [If μ_2 were substantially different from μ_1, and the relative velocities were just right, it *would* be possible to get a Brewster's angle for this case, at

$$\left(\frac{v_1}{v_2}\right)^2 = 1 - \cos^2\theta + \left(\frac{\mu_2}{\mu_1}\right)^2 \cos^2\theta \Rightarrow \cos^2\theta = \frac{(v_1/v_2)^2 - 1}{(\mu_2/\mu_1)^2 - 1} = \frac{(\mu_2\epsilon_2/\mu_1\epsilon_1) - 1}{(\mu_2/\mu_1)^2 - 1} = \frac{(\epsilon_2/\epsilon_1) - (\mu_1/\mu_2)}{(\mu_2/\mu_1) - (\mu_1/\mu_2)}.$$

But the media would be very peculiar.]

By the same token, δ_R is either always 0, or always π, for a given interface—it does not switch over as you change θ, the way it does for polarization in the plane of incidence. In particular, if $\beta = 3/2$, then $\alpha\beta > 1$, for

$$\alpha\beta = \frac{\sqrt{2.25-\sin^2\theta}}{\cos\theta} > 1 \text{ if } 2.25 - \sin^2\theta > \cos^2\theta, \text{ or } 2.25 > \sin^2\theta + \cos^2\theta = 1. \checkmark$$

In general, for $\beta > 1$, $\alpha\beta > 1$, and hence $\delta_R = \pi$. For $\beta < 1$, $\alpha\beta < 1$, and $\delta_R = 0$.

At normal incidence, $\alpha = 1$, so Fresnel's equations reduce to $E_{0_T} = \left(\frac{2}{1+\beta}\right) E_{0_I}$; $E_{0_R} = \left|\frac{1-\beta}{1+\beta}\right| E_{0_I}$, consistent with Eq. 9.82.

Reflection and Transmission coefficients: $\boxed{R = \left(\frac{E_{0_R}}{E_{0_I}}\right)^2 = \left(\frac{1-\alpha\beta}{1+\alpha\beta}\right)^2.}$ Referring to Eq. 9.116,

$$T = \frac{\epsilon_2 v_2}{\epsilon_1 v_1}\alpha\left(\frac{E_{0_T}}{E_{0_I}}\right)^2 = \boxed{\alpha\beta\left(\frac{2}{1+\alpha\beta}\right)^2}.$$

$$R+T = \frac{(1-\alpha\beta)^2+4\alpha\beta}{(1+\alpha\beta)^2} = \frac{1-2\alpha\beta+\alpha^2\beta^2+4\alpha\beta}{(1+\alpha\beta)^2} = \frac{(1+\alpha\beta)^2}{(1+\alpha\beta)^2} = 1. \checkmark$$

Problem 9.17

Equation 9.106 $\Rightarrow \beta = 2.42$; Eq. 9.110 $\Rightarrow$

$$\alpha = \frac{\sqrt{1-(\sin\theta/2.42)^2}}{\cos\theta}.$$

(a) $\theta = 0 \Rightarrow \alpha = 1$. Eq. 9.109 $\Rightarrow \left(\frac{E_{0_R}}{E_{0_I}}\right) = \frac{\alpha-\beta}{\alpha+\beta} = \frac{1-2.42}{1+2.42} = -\frac{1.42}{3.42} = \boxed{-0.415;}$
$\left(\frac{E_{0_T}}{E_{0_I}}\right) = \frac{2}{\alpha+\beta} = \frac{2}{3.42} = \boxed{0.585.}$

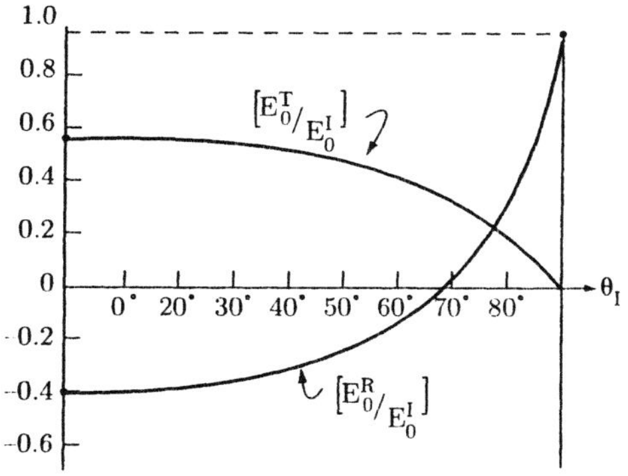

(b) Equation 9.112 $\Rightarrow \theta_B = \tan^{-1}(2.42) = \boxed{67.5°.}$
(c) $E_{0_R} = E_{0_T} \Rightarrow \alpha - \beta = 2; \alpha = \beta + 2 = 4.42$;
$(4.42)^2\cos^2\theta = 1 - \sin^2\theta/(2.42)^2$;
$(4.42)^2(1-\sin^2\theta) = (4.42)^2 - (4.42)^2\sin^2\theta$
$= 1 - 0.171\sin^2\theta$; $19.5 - 1 = (19.5 - 0.17)\sin^2\theta$;
$18.5 = 19.3\sin^2\theta$; $\sin^2\theta = 18.5/19.3 = 0.959$;
$\sin\theta = 0.979$; $\boxed{\theta = 78.3°.}$

Problem 9.18

(a) Equation 9.120 $\Rightarrow \tau = \epsilon/\sigma$. Now $\epsilon = \epsilon_0\epsilon_r$ (Eq. 4.34), $\epsilon_r \cong n^2$ (Eq. 9.70), and for glass the index of refraction is typically around 1.5, so $\epsilon \approx (1.5)^2 \times 8.85 \times 10^{-12} = 2 \times 10^{-11}\ \mathrm{C^2/N\,m^2}$, while $\sigma = 1/\rho \approx 10^{-12}\ \Omega\,\mathrm{m}$ (Table 7.1). Then $\tau = (2\times 10^{-11})/10^{-12} = \boxed{20\ \mathrm{s.}}$ (But the resistivity of glass varies enormously from one type to another, so this answer could be off by a factor of 100 in either direction.)

(b) For silver, $\rho = 1.59 \times 10^{-8}$ (Table 7.1), and $\epsilon \approx \epsilon_0$, so $\omega\epsilon = 2\pi \times 10^{10} \times 8.85 \times 10^{-12} = 0.56$. Since $\sigma = 1/\rho = 6.25 \times 10^7 \gg \omega\epsilon$, the skin depth (Eq. 9.128) is

$$d = \frac{1}{\kappa} \cong \sqrt{\frac{2}{\omega\sigma\mu}} = \sqrt{\frac{2}{2\pi\times 10^{10}\times 6.25\times 10^7 \times 4\pi \times 10^{-7}}} = 6.4\times 10^{-7}\ \mathrm{m} = 6.4 \times 10^{-4}\ \mathrm{mm}.$$

I'd plate silver to a depth of about $\boxed{0.001\,\mathrm{mm};}$ there's no point in making it any thicker, since the fields don't penetrate much beyond this anyway.

(c) For copper, Table 7.1 gives $\sigma = 1/(1.68\times 10^{-8}) = 6\times 10^7$, $\omega\epsilon_0 = (2\pi\times 10^6)\times(8.85\times 10^{-12}) = 6\times 10^{-5}$. Since $\sigma \gg \omega\epsilon$, Eq. 9.126 $\Rightarrow k \approx \sqrt{\frac{\omega\sigma\mu}{2}}$, so (Eq. 9.129)

$$\lambda = 2\pi\sqrt{\frac{2}{\omega\sigma\mu_0}} = 2\pi\sqrt{\frac{2}{2\pi\times 10^6 \times 6\times 10^7 \times 4\pi\times 10^{-7}}} = 4\times 10^{-4}\ \mathrm{m} = \boxed{0.4\,\mathrm{mm.}}$$

¿From Eq. 9.129, the propagation speed is $v = \frac{\omega}{k} = \frac{\omega}{2\pi}\lambda = \lambda\nu = (4\times 10^{-4})\times 10^6 = \boxed{400\,\mathrm{m/s.}}$ In vacuum, $\lambda = \frac{c}{\nu} = \frac{3\times 10^8}{10^6} = \boxed{300\,\mathrm{m};}$ $v = c = \boxed{3\times 10^8\ \mathrm{m/s.}}$ (But really, in a good conductor the skin depth is so small, compared to the wavelength, that the notions of "wavelength" and "propagation speed" lose their meaning.)

Problem 9.19

(a) Use the binomial expansion for the square root in Eq. 9.126:

$$\kappa \cong \omega\sqrt{\frac{\epsilon\mu}{2}}\left[1+\frac{1}{2}\left(\frac{\sigma}{\epsilon\omega}\right)^2-1\right]^{1/2} = \omega\sqrt{\frac{\epsilon\mu}{2}}\frac{1}{\sqrt{2}}\frac{\sigma}{\epsilon\omega} = \frac{\sigma}{2}\sqrt{\frac{\mu}{\epsilon}}.$$

So (Eq. 9.128) $d = \dfrac{1}{\kappa} \cong \dfrac{2}{\sigma}\sqrt{\dfrac{\epsilon}{\mu}}$. qed

For pure water, $\begin{cases} \epsilon = \epsilon_r\epsilon_0 = 80.1\,\epsilon_0 \quad \text{(Table 4.2)}, \\ \mu = \mu_0(1+\chi_m) = \mu_0(1-9.0\times10^{-6}) \cong \mu_0 \quad \text{(Table 6.1)}, \\ \sigma = 1/(2.5\times10^5) \quad \text{(Table 7.1)}. \end{cases}$

So $d = (2)(2.5\times10^5)\sqrt{\dfrac{(80.1)(8.85\times10^{-12})}{4\pi\times10^{-7}}} = \boxed{1.19\times10^4\text{ m.}}$

(b) In this case $(\sigma/\epsilon\omega)^2$ dominates, so (Eq. 9.126) $k \cong \kappa$, and hence (Eqs. 9.128 and 9.129) $\lambda = \dfrac{2\pi}{k} \cong \dfrac{2\pi}{\kappa} = 2\pi d$, or $d = \dfrac{\lambda}{2\pi}$. qed

Meanwhile $\kappa \cong \omega\sqrt{\dfrac{\epsilon\mu}{2}}\sqrt{\dfrac{\sigma}{\epsilon\omega}} = \sqrt{\dfrac{\omega\mu\sigma}{2}} = \sqrt{\dfrac{(10^{15})(4\pi\times10^{-7})(10^7)}{2}} = 8\times10^7$; $d = \dfrac{1}{\kappa} = \dfrac{1}{8\times10^7} = 1.3\times10^{-8} = \boxed{13\,\text{nm.}}$ So the fields do not penetrate far into a metal—which is what accounts for their opacity.

(c) Since $k \cong \kappa$, as we found in (b), Eq. 9.134 says $\phi = \tan^{-1}(1) = 45°$. qed

Meanwhile, Eq. 9.137 says $\dfrac{B_0}{E_0} \cong \sqrt{\epsilon\mu\dfrac{\sigma}{\epsilon\omega}} = \boxed{\sqrt{\dfrac{\sigma\mu}{\omega}}.}$ For a typical metal, then, $\dfrac{B_0}{E_0} = \sqrt{\dfrac{(10^7)(4\pi\times10^{-7})}{10^{15}}} = \boxed{10^{-7}\text{ s/m.}}$ (In vacuum, the ratio is $1/c = 1/(3\times10^8) = 3\times10^{-9}$ s/m, so the magnetic field is comparatively about 100 times larger in a metal.)

Problem 9.20

(a) $u = \dfrac{1}{2}\left(\epsilon E^2 + \dfrac{1}{\mu}B^2\right) = \dfrac{1}{2}e^{-2\kappa z}\left[\epsilon E_0^2\cos^2(kz-\omega t+\delta_E) + \dfrac{1}{\mu}B_0^2\cos^2(kz-\omega t+\delta_E+\phi)\right]$. Averaging over a full cycle, using $\langle\cos^2\rangle = \frac{1}{2}$ and Eq. 9.137:

$$\langle u\rangle = \frac{1}{2}e^{-2\kappa z}\left[\frac{\epsilon}{2}E_0^2 + \frac{1}{2\mu}B_0^2\right] = \frac{1}{4}e^{-2\kappa z}\left[\epsilon E_0^2 + \frac{1}{\mu}E_0^2\epsilon\mu\sqrt{1+\left(\frac{\sigma}{\epsilon\omega}\right)^2}\right] = \frac{1}{4}e^{-2\kappa z}\epsilon E_0^2\left[1+\sqrt{1+\left(\frac{\sigma}{\epsilon\omega}\right)^2}\right].$$

But Eq. 9.126 $\Rightarrow 1+\sqrt{1+\left(\dfrac{\sigma}{\epsilon\omega}\right)^2} = \dfrac{2}{\epsilon\mu}\dfrac{k^2}{\omega^2}$, so $\langle u\rangle = \dfrac{1}{4}e^{-2\kappa z}\epsilon E_0^2\dfrac{2}{\epsilon\mu}\dfrac{k^2}{\omega^2} = \boxed{\dfrac{k^2}{2\mu\omega^2}E_0^2e^{-2\kappa z}.}$ So the ratio of the magnetic contribution to the electric contribution is

$$\frac{\langle u_{\text{mag}}\rangle}{\langle u_{\text{elec}}\rangle} = \frac{B_0^2/\mu}{E_0^2\epsilon} = \frac{1}{\mu\epsilon}\mu\epsilon\sqrt{1+\left(\frac{\sigma}{\epsilon\omega}\right)^2} = \sqrt{1+\left(\frac{\sigma}{\epsilon\omega}\right)^2} > 1. \quad \text{qed}$$

(b) $\mathbf{S} = \dfrac{1}{\mu}(\mathbf{E}\times\mathbf{B}) = \dfrac{1}{\mu}E_0B_0e^{-2\kappa z}\cos(kz-\omega t+\delta_E)\cos(kz-\omega t+\delta_E+\phi)\,\hat{\mathbf{z}}$; $\langle\mathbf{S}\rangle = \dfrac{1}{2\mu}E_0B_0e^{-2\kappa z}\cos\phi\,\hat{\mathbf{z}}$. [The average of the product of the cosines is $(1/2\pi)\int_0^{2\pi}\cos\theta\cos(\theta+\phi)\,d\theta = (1/2)\cos\phi$.] So $I = \dfrac{1}{2\mu}E_0B_0e^{-2\kappa z}\cos\phi = \dfrac{1}{2\mu}E_0^2e^{-2\kappa z}\left(\dfrac{K}{\omega}\cos\phi\right)$, while, from Eqs. 9.133 and 9.134, $K\cos\phi = k$, so $\boxed{I = \dfrac{k}{2\mu\omega}E_0^2e^{-2\kappa z}.}$ qed

Problem 9.21

According to Eq. 9.147, $R = \left|\frac{\tilde{E}_{0_R}}{\tilde{E}_{0_I}}\right|^2 = \left|\frac{1-\tilde{\beta}}{1+\tilde{\beta}}\right|^2 = \left(\frac{1-\tilde{\beta}}{1+\tilde{\beta}}\right)\left(\frac{1-\tilde{\beta}^*}{1+\tilde{\beta}^*}\right)$, where $\tilde{\beta} = \frac{\mu_1 v_1}{\mu_2 \omega}\tilde{k}_2$ $= \frac{\mu_1 v_1}{\mu_2 \omega}(k_2 + i\kappa_2)$ (Eqs. 9.125 and 9.146). Since silver is a good conductor ($\sigma \gg \epsilon\omega$), Eq. 9.126 reduces to

$\kappa_2 \cong k_2 \cong \omega\sqrt{\frac{\epsilon_2\mu_2}{2}}\sqrt{\frac{\sigma}{\epsilon_2\omega}} = \sqrt{\frac{\sigma\omega\mu_2}{2}}$, so $\tilde{\beta} = \frac{\mu_1 v_1}{\mu_2\omega}\sqrt{\frac{\sigma\omega\mu_2}{2}}(1+i) = \mu_1 v_1\sqrt{\frac{\sigma}{2\mu_2\omega}}(1+i)$.

Let $\gamma \equiv \mu_1 v_1\sqrt{\frac{\sigma}{2\mu_2\omega}} = \mu_0 c\sqrt{\frac{\sigma}{2\mu_0\omega}} = c\sqrt{\frac{\sigma\mu_0}{2\omega}} = (3\times10^8)\sqrt{\frac{(6\times10^7)(4\pi\times10^{-7})}{(2)(4\times10^{15})}} = 29.$ Then

$R = \left(\frac{1-\gamma-i\gamma}{1+\gamma+i\gamma}\right)\left(\frac{1-\gamma+i\gamma}{1+\gamma-i\gamma}\right) = \frac{(1-\gamma)^2+\gamma^2}{(1+\gamma)^2+\gamma^2} = \boxed{0.93.}$ Evidently 93% of the light is reflected.

Problem 9.22

(a) We are told that $v = \alpha\sqrt{\lambda}$, where α is a constant. But $\lambda = 2\pi/k$ and $v = \omega/k$, so $\omega = \alpha k\sqrt{2\pi/k} = \alpha\sqrt{2\pi k}$. From Eq. 9.150, $v_g = \frac{d\omega}{dk} = \alpha\sqrt{2\pi}\frac{1}{2\sqrt{k}} = \frac{1}{2}\alpha\sqrt{\frac{2\pi}{k}} = \frac{1}{2}\alpha\sqrt{\lambda} = \frac{1}{2}v$, or $\boxed{v = 2v_g.}$

(b) $\frac{i(px-Et)}{\hbar} = i(kx-\omega t) \Rightarrow k = \frac{p}{\hbar},\ \omega = \frac{E}{\hbar} = \frac{p^2}{2m\hbar} = \frac{\hbar k^2}{2m}$. Therefore $\boxed{v = \frac{\omega}{k} = \frac{E}{p} = \frac{p}{2m} = \frac{\hbar k}{2m};}$ $v_g = \frac{d\omega}{dk} = \frac{2\hbar k}{2m} = \frac{\hbar k}{m} = \boxed{\frac{p}{m}.}$ So $\boxed{v = \frac{1}{2}v_g.}$ Since $p = mv_c$ (where v_c is the classical speed of the particle), it follows that $\boxed{v_g \text{ (not } v\text{) corresponds to the classical veloctity.}}$

Problem 9.23

$E = \frac{1}{4\pi\epsilon_0}\frac{qd}{a^3} \Rightarrow F = -qE = -\left(\frac{1}{4\pi\epsilon_0}\frac{q^2}{a^3}\right)x = -k_{\text{spring}}x = -m\omega_0^2 x$ (Eq. 9.151). So $\boxed{\omega_0 = \sqrt{\frac{q^2}{4\pi\epsilon_0 m a^3}}.}$

$\nu_0 = \frac{\omega_0}{2\pi} = \frac{1}{2\pi}\sqrt{\frac{(1.6\times10^{-19})^2}{4\pi(8.85\times10^{-12})(9.11\times10^{-31})(0.5\times10^{-10})^3}} = \boxed{7.16\times10^{15}\text{ Hz.}}$ This is $\boxed{\text{ultraviolet.}}$

¿From Eqs. 9.173 and 9.174,

$$A = \frac{nq^2}{2m\epsilon_0}\frac{f}{\omega_0^2},\ \begin{cases} N = \#\text{ of molecules per unit volume} = \frac{\text{Avogadro's }\#}{22.4\text{ liters}} = \frac{6.02\times10^{23}}{22.4\times10^{-3}} = 2.69\times10^{25}, \\ f = \#\text{ of electrons per molecule} = 2\text{ (for H}_2\text{)}. \end{cases}$$

$$= \frac{(2.69\times10^{25})(1.6\times10^{-19})^2}{(9.11\times10^{-31})(8.85\times10^{-12})(4.5\times10^{16})^2} = \boxed{4.2\times10^{-5}}\text{ (which is about 1/3 the actual value);}$$

$$B = \left(\frac{2\pi c}{\omega_0}\right)^2 = \left(\frac{2\pi\times3\times10^8}{4.5\times10^{16}}\right)^2 = \boxed{1.8\times10^{-15}\text{ m}^2}\text{ (which is about 1/4 the actual value).}$$

So even this extremely crude model is in the right ball park.

Problem 9.24

Equation 9.170 $\Rightarrow n = 1 + \frac{Nq^2}{2m\epsilon_0}\frac{(\omega_0^2-\omega^2)}{[(\omega_0^2-\omega^2)^2+\gamma^2\omega^2]}$. Let the denominator $\equiv D$. Then

$\frac{dn}{d\omega} = \frac{Nq^2}{2m\epsilon_0}\left\{\frac{-2\omega}{D} - \frac{(\omega_0^2-\omega^2)}{D^2}\left[2(\omega_0^2-\omega^2)(-2\omega)+\gamma^2 2\omega\right]\right\} = 0 \Rightarrow 2\omega D = (\omega_0^2-\omega^2)\left[2(\omega_0^2-\omega^2)-\gamma^2\right]2\omega;$
$(\omega_0^2-\omega^2)^2+\gamma^2\omega^2 = 2(\omega_0^2-\omega^2)^2-\gamma^2(\omega_0^2-\omega^2)$, or $(\omega_0^2-\omega^2)^2 = \gamma^2(\omega^2+\omega_0^2-\omega^2) = \gamma^2\omega_0^2 \Rightarrow (\omega_0^2-\omega^2) = \pm\omega_0\gamma;$

$\omega^2 = \omega_0^2 \mp \omega_0\gamma$, $\omega = \omega_0\sqrt{1 \mp \gamma/\omega_0} \cong \omega_0(1 \mp \gamma/2\omega_0) = \omega_0 \mp \gamma/2$. So $\omega_2 = \omega_0 + \gamma/2$, $\omega_1 = \omega_0 - \gamma/2$, and the width of the anomalous region is $\boxed{\Delta\omega = \omega_2 - \omega_1 = \gamma.}$

From Eq. 9.171, $\alpha = \dfrac{Nq^2\omega^2}{m\epsilon_0 c}\dfrac{\gamma}{(\omega_0^2-\omega^2)^2+\gamma^2\omega^2}$, so at the maximum ($\omega = \omega_0$), $\alpha_{\max} = \dfrac{Nq^2}{m\epsilon_0 c\gamma}$.

At ω_1 and ω_2, $\omega^2 = \omega_0^2 \mp \omega_0\gamma$, so $\alpha = \dfrac{Nq^2\omega^2}{m\epsilon_0 c}\dfrac{\gamma}{\gamma^2\omega_0^2+\gamma^2\omega^2} = \alpha_{\max}\left(\dfrac{\omega^2}{\omega^2+\omega_0^2}\right)$. But

$$\frac{\omega^2}{\omega^2+\omega_0^2} = \frac{\omega_0^2 \mp \omega_0\gamma}{2\omega_0^2 \mp \omega_0\gamma} = \frac{1}{2}\frac{(1 \mp \gamma/\omega_0)}{(1 \mp \gamma/2\omega_0)} \cong \frac{1}{2}\left(1 \mp \frac{\gamma}{\omega_0}\right)\left(1 \pm \frac{\gamma}{2\omega_0}\right) \cong \frac{1}{2}\left(1 \mp \frac{\gamma}{2\omega_0}\right) \cong \frac{1}{2}.$$

So $\alpha \cong \frac{1}{2}\alpha_{\max}$ at ω_1 and ω_2. qed

Problem 9.25

$$k = \frac{\omega}{c}\left[1 + \frac{Nq^2}{2m\epsilon_0}\sum\frac{f_j}{(\omega_j^2-\omega^2)}\right]. \quad v_g = \frac{d\omega}{dk} = \frac{1}{(dk/d\omega)}.$$

$$\frac{dk}{d\omega} = \frac{1}{c}\left[1 + \frac{Nq^2}{2m\epsilon_0}\sum\frac{f_j}{(\omega_j^2-\omega^2)} + \omega\sum f_j\frac{-(-2\omega)}{(\omega_j^2-\omega^2)^2}\right] = \frac{1}{c}\left[1 + \frac{Nq^2}{2m\epsilon_0}\sum f_j\frac{(\omega_j^2+\omega^2)}{(\omega_j^2-\omega^2)^2}\right].$$

$$\boxed{v_g = c\left[1 + \frac{Nq^2}{2m\epsilon_0}\sum f_j\frac{(\omega_j^2+\omega^2)}{(\omega_j^2-\omega^2)^2}\right]^{-1}.}$$ Since the second term in square brackets is ***positive***, it follows that

$\boxed{v_g < c,}$ whereas $v = \dfrac{\omega}{k} = c\left[1 + \dfrac{Nq^2}{2m\epsilon_0}\sum\dfrac{f_j}{(\omega_j^2-\omega^2)}\right]^{-1}$ is greater than c or less than c, depending on ω.

Problem 9.26

(a) From Eqs. 9.176 and 9.177, $\nabla\times\tilde{\mathbf{E}} = -\dfrac{\partial\tilde{\mathbf{B}}}{\partial t} = i\omega\tilde{\mathbf{B}}_0 e^{i(kz-\omega t)}$; $\nabla\times\tilde{\mathbf{B}} = \dfrac{1}{c^2}\dfrac{\partial\tilde{\mathbf{E}}}{\partial t} = -\dfrac{i\omega}{c^2}\tilde{\mathbf{E}}_0 e^{i(kz-\omega t)}$.
In the terminology of Eq. 9.178:

$$(\nabla\times\tilde{\mathbf{E}})_x = \frac{\partial\tilde{E}_z}{\partial y} - \frac{\partial\tilde{E}_y}{\partial z} = \left(\frac{\partial\tilde{E}_{0_z}}{\partial y} - ik\tilde{E}_{0_y}\right)e^{i(kz-\omega t)}. \text{ So (ii) } \frac{\partial E_z}{\partial y} - ikE_y = i\omega B_x.$$

$$(\nabla\times\tilde{\mathbf{E}})_y = \frac{\partial\tilde{E}_x}{\partial z} - \frac{\partial\tilde{E}_z}{\partial x} = \left(ik\tilde{E}_{0_x} - \frac{\partial\tilde{E}_{0_z}}{\partial x}\right)e^{i(kz-\omega t)}. \text{ So (iii) } ikE_x - \frac{\partial E_z}{\partial x} = i\omega B_y.$$

$$(\nabla\times\tilde{\mathbf{E}})_z = \frac{\partial\tilde{E}_y}{\partial x} - \frac{\partial\tilde{E}_x}{\partial y} = \left(\frac{\partial\tilde{E}_{0_y}}{\partial x} - \frac{\partial\tilde{E}_{0_x}}{\partial y}\right)e^{i(kz-\omega t)}. \text{ So (i) } \frac{\partial E_y}{\partial x} - \frac{\partial E_x}{\partial y} = i\omega B_z.$$

$$(\nabla\times\tilde{\mathbf{B}})_x = \frac{\partial\tilde{B}_z}{\partial y} - \frac{\partial\tilde{B}_y}{\partial z} = \left(\frac{\partial\tilde{B}_{0_z}}{\partial y} - ik\tilde{B}_{0_y}\right)e^{i(kz-\omega t)}. \text{ So (v) } \frac{\partial B_z}{\partial y} - ikB_y = -\frac{i\omega}{c^2}E_x.$$

$$(\nabla\times\tilde{\mathbf{B}})_y = \frac{\partial\tilde{B}_x}{\partial z} - \frac{\partial\tilde{B}_z}{\partial x} = \left(ik\tilde{B}_{0_x} - \frac{\partial\tilde{B}_{0_z}}{\partial x}\right)e^{i(kz-\omega t)}. \text{ So (vi) } ikB_x - \frac{\partial B_z}{\partial x} = -\frac{i\omega}{c^2}E_y.$$

$$(\nabla\times\tilde{\mathbf{B}})_z = \frac{\partial\tilde{B}_y}{\partial x} - \frac{\partial\tilde{B}_x}{\partial y} = \left(\frac{\partial\tilde{B}_{0_y}}{\partial x} - \frac{\partial\tilde{B}_{0_x}}{\partial y}\right)e^{i(kz-\omega t)}. \text{ So (iv) } \frac{\partial B_y}{\partial x} - \frac{\partial B_x}{\partial y} = -\frac{i\omega}{c^2}E_z.$$

This confirms Eq. 9.179. Now multiply (iii) by k, (v) by ω, and subtract: $ik^2E_x - k\dfrac{\partial E_z}{\partial x} - \omega\dfrac{\partial B_z}{\partial y} + i\omega kB_y =$
$ik\omega B_y + \dfrac{i\omega^2}{c^2}E_x \Rightarrow i\left(k^2 - \dfrac{\omega^2}{c^2}\right)E_x = k\dfrac{\partial E_z}{\partial x} + \omega\dfrac{\partial B_z}{\partial y}$, or (i) $E_x = \dfrac{i}{(\omega/c)^2 - k^2}\left(k\dfrac{\partial E_z}{\partial x} + \omega\dfrac{\partial B_z}{\partial y}\right)$.

Multiply (ii) by k, (vi) by ω, and add: $k\dfrac{\partial E_z}{\partial y} - ik^2E_y + i\omega kB_x - \omega\dfrac{\partial B_z}{\partial x} = i\omega kB_x - \dfrac{i\omega^2}{c^2}E_y \Rightarrow i\left(\dfrac{\omega^2}{c^2} - k^2\right)E_y =$

$-k\dfrac{\partial E_z}{\partial y} + \omega\dfrac{\partial B_z}{\partial x}$, or (ii) $E_y = \dfrac{i}{(\omega/c)^2 - k^2}\left(k\dfrac{\partial E_z}{\partial y} - \omega\dfrac{\partial B_z}{\partial x}\right)$.

Multiply (ii) by ω/c^2, (vi) by k, and add: $\dfrac{\omega}{c^2}\dfrac{\partial E_z}{\partial y} - i\dfrac{\omega k}{c^2}E_y + ik^2 B_x - k\dfrac{\partial B_z}{\partial x} = i\dfrac{\omega^2}{c^2}B_x - i\dfrac{\omega k}{c^2}E_y \Rightarrow$
$i\left(k^2 - \dfrac{\omega^2}{c^2}\right)B_x = k\dfrac{\partial B_z}{\partial x} - \dfrac{\omega}{c^2}\dfrac{\partial E_z}{\partial y}$, or (iii) $B_x = \dfrac{i}{(\omega/c)^2 - k^2}\left(k\dfrac{\partial B_z}{\partial x} - \dfrac{\omega}{c^2}\dfrac{\partial E_z}{\partial y}\right)$.

Multiply (iii) by ω/c^2, (v) by k, and subtract: $i\dfrac{\omega k}{c^2}E_x - \dfrac{\omega}{c^2}\dfrac{\partial E_z}{\partial x} - k\dfrac{\partial B_z}{\partial y} + ik^2 B_y = i\dfrac{\omega^2}{c^2}B_y + \dfrac{i\omega k}{c^2}E_x \Rightarrow$
$i\left(k^2 - \dfrac{\omega^2}{c^2}\right)B_y = \dfrac{\omega}{c^2}\dfrac{\partial E_z}{\partial x} + k\dfrac{\partial B_z}{\partial y}$, or (iv) $B_y = \dfrac{i}{(\omega/c)^2 - k^2}\left(k\dfrac{\partial B_z}{\partial y} + \dfrac{\omega}{c^2}\dfrac{\partial E_z}{\partial x}\right)$.

This completes the confirmation of Eq. 9.180.

(b) $\nabla \cdot \tilde{\mathbf{E}} = \dfrac{\partial \tilde{E}_x}{\partial x} + \dfrac{\partial \tilde{E}_y}{\partial y} + \dfrac{\partial \tilde{E}_z}{\partial z} = \left(\dfrac{\partial \tilde{E}_{0_x}}{\partial x} + \dfrac{\partial \tilde{E}_{0_y}}{\partial y} + ik\tilde{E}_{0_z}\right)e^{i(kz-\omega t)} = 0 \Rightarrow \dfrac{\partial E_x}{\partial x} + \dfrac{\partial E_y}{\partial y} + ikE_z = 0.$

Using Eq. 9.180, $\dfrac{i}{(\omega/c)^2 - k^2}\left(k\dfrac{\partial^2 E_z}{\partial x^2} + \omega\dfrac{\partial^2 B_z}{\partial x \partial y}\right) + \dfrac{i}{(\omega/c)^2 - k^2}\left(k\dfrac{\partial^2 E_z}{\partial^2 y} - \omega\dfrac{\partial^2 B_z}{\partial x \partial y}\right) + ikE_z = 0,$

or $\dfrac{\partial^2 E_z}{\partial x^2} + \dfrac{\partial^2 E_z}{\partial^2 y} + \left[(\omega/c)^2 - k^2\right]E_z = 0.$

Likewise, $\nabla \cdot \tilde{\mathbf{B}} = 0 \Rightarrow \dfrac{\partial B_x}{\partial x} + \dfrac{\partial B_y}{\partial y} + ikB_z = 0 \Rightarrow$

$\dfrac{i}{(\omega/c)^2 - k^2}\left(k\dfrac{\partial^2 B_z}{\partial x^2} - \dfrac{\omega}{c^2}\dfrac{\partial^2 E_z}{\partial x \partial y}\right) + \dfrac{i}{(\omega/c)^2 - k^2}\left(k\dfrac{\partial^2 B_z}{\partial y^2} + \dfrac{\omega}{c^2}\dfrac{\partial E_z}{\partial x \partial y}\right) + ikB_z = 0 \Rightarrow$
$\dfrac{\partial^2 B_z}{\partial x^2} + \dfrac{\partial^2 B_z}{\partial^2 y} + \left[(\omega/c)^2 - k^2\right]B_z = 0.$

This confirms Eqs. 9.181. [You can also do it by putting Eq. 9.180 into Eq. 9.179 (i) and (iv).]

Problem 9.27

Here $E_z = 0$ (TE) and $\omega/c = k$ ($n = m = 0$), so Eq. 9.179(ii) $\Rightarrow E_y = -cB_x$, Eq. 9.179(iii) $\Rightarrow E_x = cB_y$, Eq. 9.179(v) $\Rightarrow \dfrac{\partial B_z}{\partial y} = i\left(kB_y - \dfrac{\omega}{c^2}E_x\right) = i\left(kB_y - \dfrac{\omega}{c}B_y\right) = 0$, Eq. 9.179(vi) $\Rightarrow \dfrac{\partial B_z}{\partial x} = i\left(kB_x + \dfrac{\omega}{c^2}E_y\right) = i\left(kB_x - \dfrac{\omega}{c}B_x\right) = 0$. So $\dfrac{\partial B_z}{\partial x} = \dfrac{\partial B_z}{\partial y} = 0$, and since B_z is a function only of x and y, this says B_z is in fact a *constant* (as Eq. 9.186 also suggests). Now Faraday's law (in integral form) says $\oint \mathbf{E}\cdot d\mathbf{l} = -\int \dfrac{\partial \mathbf{B}}{\partial t}\cdot d\mathbf{a}$, and Eq. 9.176 $\Rightarrow \dfrac{\partial \mathbf{B}}{\partial t} = -i\omega\mathbf{B}$, so $\oint \mathbf{E}\cdot d\mathbf{l} = i\omega\int \mathbf{B}\cdot d\mathbf{a}$. Applied to a cross-section of the waveguide this gives $\oint \mathbf{E}\cdot d\mathbf{l} = i\omega e^{i(kz-\omega t)}\int B_z\, da = i\omega B_z e^{i(kz-\omega t)}(ab)$ (since B_z is constant, it comes outside the integral). But if the boundary is just inside the metal, where $\mathbf{E} = 0$, it follows that $\boxed{B_z = 0.}$ So this would be a TEM mode, which we already know cannot exist for this guide.

Problem 9.28

Here $a = 2.28\,\text{cm}$ and $b = 1.01\,\text{cm}$, so $\nu_{10} = \dfrac{1}{2\pi}\omega_{10} = \dfrac{c}{2a} = 0.66\times10^{10}\,\text{Hz}$; $\nu_{20} = 2\dfrac{c}{2a} = 1.32\times10^{10}\,\text{Hz}$; $\nu_{30} = 3\dfrac{c}{2a} = 1.97\times10^{10}\,\text{Hz}$; $\nu_{01} = \dfrac{c}{2b} = 1.49\times10^{10}\,\text{Hz}$; $\nu_{02} = 2\dfrac{c}{2b} = 2.97\times10^{10}\,\text{Hz}$; $\nu_{11} = \dfrac{c}{2}\sqrt{\dfrac{1}{a^2}+\dfrac{1}{b^2}} = 1.62\times10^{10}\,\text{Hz}$. Evidently just four modes occur: $\boxed{\text{10, 20, 01, and 11.}}$

To get only *one* mode you must drive the waveguide at a frequency between ν_{10} and ν_{20}:
$\boxed{0.66\times10^{10} < \nu < 1.32\times10^{10}\,\text{Hz.}}$ $\lambda = \dfrac{c}{\nu}$, so $\lambda_{10} = 2a$; $\lambda_{20} = a$. $\boxed{2.28\,\text{cm} < \lambda < 4.56\,\text{cm.}}$

Problem 9.29

From Prob. 9.11, $\langle \mathbf{S} \rangle = \frac{1}{2\mu_0}(\tilde{\mathbf{E}} \times \tilde{\mathbf{B}}^*)$. Here (Eq. 9.176) $\tilde{\mathbf{E}} = \tilde{\mathbf{E}}_0 e^{i(kz-\omega t)}$, $\tilde{\mathbf{B}}^* = \tilde{\mathbf{B}}_0^* e^{-i(kz-\omega t)}$, and, for the TE$_{mn}$ mode (Eqs. 9.180 and 9.186)

$$\begin{aligned}
B_x^* &= \frac{-ik}{(\omega/c)^2 - k^2}\left(\frac{-m\pi}{a}\right) B_0 \sin\left(\frac{m\pi x}{a}\right)\cos\left(\frac{n\pi y}{b}\right); \\
B_y^* &= \frac{-ik}{(\omega/c)^2 - k^2}\left(\frac{-n\pi}{b}\right) B_0 \cos\left(\frac{m\pi x}{a}\right)\sin\left(\frac{n\pi y}{b}\right); \\
B_z^* &= B_0 \cos\left(\frac{m\pi x}{a}\right)\cos\left(\frac{n\pi y}{b}\right); \\
E_x &= \frac{i\omega}{(\omega/c)^2 - k^2}\left(\frac{-n\pi}{b}\right) B_0 \cos\left(\frac{m\pi x}{a}\right)\sin\left(\frac{n\pi y}{b}\right); \\
E_y &= \frac{-i\omega}{(\omega/c)^2 - k^2}\left(\frac{-m\pi}{a}\right) B_0 \sin\left(\frac{m\pi x}{a}\right)\cos\left(\frac{n\pi y}{b}\right); \\
E_z &= 0.
\end{aligned}$$

So

$$\begin{aligned}
\langle \mathbf{S} \rangle &= \frac{1}{2\mu_0}\left\{\frac{i\pi\omega B_0^2}{(\omega/c)^2 - k^2}\left(\frac{m}{a}\right)\sin\left(\frac{m\pi x}{a}\right)\cos\left(\frac{m\pi x}{a}\right)\cos^2\left(\frac{n\pi y}{b}\right)\hat{\mathbf{x}}\right. \\
&+ \frac{i\pi\omega B_0^2}{(\omega/c)^2 - k^2}\left(\frac{n}{b}\right)\cos^2\left(\frac{m\pi x}{a}\right)\sin\left(\frac{n\pi y}{b}\right)\cos\left(\frac{n\pi y}{b}\right)\hat{\mathbf{y}} \\
&+ \left.\frac{\omega k\pi^2 B_0^2}{[(\omega/c)^2 - k^2]^2}\left[\left(\frac{n}{b}\right)^2\cos^2\left(\frac{m\pi x}{a}\right)\sin^2\left(\frac{n\pi y}{b}\right) + \left(\frac{m}{a}\right)^2\sin^2\left(\frac{m\pi x}{a}\right)\cos^2\left(\frac{n\pi y}{b}\right)\right]\hat{\mathbf{z}}\right\}.
\end{aligned}$$

$$\int \langle \mathbf{S} \rangle \cdot d\mathbf{a} = \boxed{\frac{1}{8\mu_0}\frac{\omega k\pi^2 B_0^2}{[(\omega/c)^2 - k^2]^2} ab\left[\left(\frac{m}{a}\right)^2 + \left(\frac{n}{b}\right)^2\right].}$$

[In the last step I used $\int_0^a \sin^2(m\pi x/a)\,dx = \int_0^a \cos^2(m\pi x/a)\,dx = a/2$; $\int_0^b \sin^2(n\pi y/b)\,dy = \int_0^b \cos^2(n\pi y/b)\,dy = b/2$.]

Similarly,

$$\begin{aligned}
\langle u \rangle &= \frac{1}{4}\left(\epsilon_0 \tilde{\mathbf{E}}\cdot\tilde{\mathbf{E}}^* + \frac{1}{\mu_0}\tilde{\mathbf{B}}\cdot\tilde{\mathbf{B}}^*\right) \\
&= \frac{\epsilon_0}{4}\frac{\omega^2\pi^2 B_0^2}{[(\omega/c)^2 - k^2]^2}\left[\left(\frac{n}{b}\right)^2\cos^2\left(\frac{m\pi x}{a}\right)\sin^2\left(\frac{n\pi y}{b}\right) + \left(\frac{m}{a}\right)^2\sin^2\left(\frac{m\pi x}{a}\right)\cos^2\left(\frac{n\pi y}{b}\right)\right] \\
&+ \frac{1}{4\mu_0}\left\{B_0^2\cos^2\left(\frac{m\pi x}{a}\right)\cos^2\left(\frac{n\pi y}{b}\right)\right. \\
&+ \left.\frac{k^2\pi^2 B_0^2}{[(\omega/c)^2 - k^2]^2}\left[\left(\frac{n}{b}\right)^2\cos^2\left(\frac{m\pi x}{a}\right)\sin^2\left(\frac{n\pi y}{b}\right) + \left(\frac{m}{a}\right)^2\sin^2\left(\frac{m\pi x}{a}\right)\cos^2\left(\frac{n\pi y}{b}\right)\right]\right\}.
\end{aligned}$$

$$\int \langle u \rangle\, da = \boxed{\frac{ab}{4}\left\{\frac{\epsilon_0}{4}\frac{\omega^2\pi^2 B_0^2}{[(\omega/c)^2 - k^2]^2}\left[\left(\frac{n}{b}\right)^2 + \left(\frac{m}{a}\right)^2\right] + \frac{B_0^2}{4\mu_0} + \frac{1}{4\mu_0}\frac{k^2\pi^2 B_0^2}{[(\omega/c)^2 - k^2]^2}\left[\left(\frac{n}{b}\right)^2 + \left(\frac{m}{a}\right)^2\right]\right\}.}$$

These results can be simplified, using Eq. 9.190 to write $\left[(\omega/c)^2 - k^2\right] = (\omega_{mn}/c)^2$, $\epsilon_0\mu_0 = 1/c^2$ to eliminate ϵ_0, and Eq. 9.188 to write $\left[(m/a)^2 + (n/b)^2\right] = (\omega_{mn}/\pi c)^2$:

$$\int \langle \mathbf{S}\rangle \cdot d\mathbf{a} = \frac{\omega k a b c^2}{8\mu_0\omega_{mn}^2}B_0^2; \quad \int \langle u\rangle\, da = \frac{\omega^2 ab}{8\mu_0\omega_{mn}^2}B_0^2.$$

Evidently

$$\frac{\text{energy per unit time}}{\text{energy per unit length}} = \frac{\int \langle \mathbf{S}\rangle\cdot d\mathbf{a}}{\int\langle u\rangle\, da} = \frac{kc^2}{\omega} = \frac{c}{\omega}\sqrt{\omega^2-\omega_{mn}^2} = v_g \text{ (Eq. 9.192).} \quad \text{qed}$$

Problem 9.30

Following Sect. 9.5.2, the problem is to solve Eq. 9.181 with $E_z \neq 0, B_z = 0$, subject to the boundary conditions 9.175. Let $E_z(x,y) = X(x)Y(y)$; as before, we obtain $X(x) = A\sin(k_x x) + B\cos(k_x x)$. But the boundary condition requires $E_z = 0$ (and hence $X = 0$) when $x = 0$ and $x = a$, so $B = 0$ and $k_x = m\pi/a$. But this time $m = 1, 2, 3, \ldots$, but *not* zero, since $m = 0$ would kill X entirely. The same goes for $Y(y)$. Thus $\boxed{E_z = E_0 \sin\left(\frac{m\pi x}{a}\right)\sin\left(\frac{n\pi y}{b}\right) \text{ with } n, m = 1, 2, 3, \ldots}$.

The rest is the same as for TE waves: $\boxed{\omega_{mn} = c\pi\sqrt{(m/a)^2+(n/b)^2}}$ is the cutoff frequency, the wave velocity is $v = c/\sqrt{1-(\omega_{mn}/\omega)^2}$, and the group velocity is $v_g = c\sqrt{1-(\omega_{mn}/\omega)^2}$. The lowest TM mode is 11, with cutoff frequency $\omega_{11} = c\pi\sqrt{(1/a)^2+(1/b)^2}$. So the ratio of the lowest TM frequency to the lowest TE frequency is $\dfrac{c\pi\sqrt{(1/a)^2+(1/b)^2}}{(c\pi/a)} = \boxed{\sqrt{1+(a/b)^2}.}$

Problem 9.31

(a) $\nabla\cdot\mathbf{E} = \frac{1}{s}\frac{\partial}{\partial s}(sE_s) = 0\ \checkmark$; $\nabla\cdot\mathbf{B} = \frac{1}{s}\frac{\partial}{\partial\phi}(B_\phi) = 0\ \checkmark$; $\nabla\times\mathbf{E} = \frac{\partial E_s}{\partial z}\hat{\boldsymbol{\phi}} - \frac{1}{s}\frac{\partial E_s}{\partial\phi}\hat{\mathbf{z}} = -\frac{E_0 k\sin(kz-\omega t)}{s}\hat{\boldsymbol{\phi}} \stackrel{?}{=} -\frac{\partial\mathbf{B}}{\partial t} = -\frac{E_0\omega}{c}\frac{\sin(kz-\omega t)}{s}\hat{\boldsymbol{\phi}}\ \checkmark$ (since $k=\omega/c$); $\nabla\times\mathbf{B} = -\frac{\partial B_\phi}{\partial z}\hat{\mathbf{s}} + \frac{1}{s}\frac{\partial}{\partial s}(sB_\phi)\hat{\mathbf{z}} = \frac{E_0 k}{c}\frac{\sin(kz-\omega t)}{s}\hat{\mathbf{s}} \stackrel{?}{=} \frac{1}{c^2}\frac{\partial\mathbf{E}}{\partial t} = \frac{E_0\omega}{c^2}\frac{\sin(kz-\omega t)}{s}\hat{\mathbf{s}}\ \checkmark$. Boundary conditions: $E^{\parallel} = E_z = 0\ \checkmark$; $B^{\perp} = B_s = 0\ \checkmark$.

(b) To determine λ, use Gauss's law for a cylinder of radius s and length dz:

$$\oint \mathbf{E}\cdot d\mathbf{a} = E_0\frac{\cos(kz-\omega t)}{s}(2\pi s)\,dz = \frac{1}{\epsilon_0}Q_{\text{enc}} = \frac{1}{\epsilon_0}\lambda\,dz \Rightarrow \boxed{\lambda = 2\pi\epsilon_0 E_0\cos(kz-\omega t).}$$

To determine I, use Ampére's law for a circle of radius s (note that the displacement current through this loop is zero, since $\mathbf{E}$ is in the $\hat{\mathbf{s}}$ direction): $\oint \mathbf{B}\cdot d\mathbf{l} = \frac{E_0}{c}\frac{\cos(kz-\omega t)}{s}(2\pi s) = \mu_0 I_{\text{enc}} \Rightarrow \boxed{I = \frac{2\pi E_0}{\mu_0 c}\cos(kz-\omega t).}$

The charge and current on the outer conductor are precisely the $\boxed{\text{opposite}}$ of these, since $\mathbf{E} = \mathbf{B} = 0$ *inside* the metal, and hence the ***total*** enclosed charge and current must be zero.

Problem 9.32

$\tilde{f}(z,0) = \int_{-\infty}^{\infty}\tilde{A}(k)e^{ikz}\,dk \Rightarrow \tilde{f}(z,0)^* = \int_{-\infty}^{\infty}\tilde{A}(k)^*e^{-ikz}\,dk$. Let $l \equiv -k$; then $\tilde{f}(z,0)^* = \int_{\infty}^{-\infty}\tilde{A}(-l)^*e^{ilz}(-dl) = \int_{-\infty}^{\infty}\tilde{A}(-l)^*e^{ilz}\,dl = \int_{-\infty}^{\infty}\tilde{A}(-k)^*e^{ikz}\,dk$ (renaming the dummy variable $l\to k$).

$f(z,0) = \text{Re}\left[\tilde{f}(z,0)\right] = \frac{1}{2}\left[\tilde{f}(z,0)+\tilde{f}(z,0)^*\right] = \int_{-\infty}^{\infty}\frac{1}{2}\left[\tilde{A}(k)+\tilde{A}(-k)^*\right]e^{ikz}\,dk$. Therefore

$\frac{1}{2}\left[\tilde{A}(k)+\tilde{A}(-k)^*\right] = \frac{1}{2\pi}\int_{-\infty}^{\infty} f(z,0)e^{-ikz}\,dz.$

Meanwhile, $\dot{\tilde{f}}(z,t) = \int_{-\infty}^{\infty} \tilde{A}(k)(-i\omega)e^{i(kz-\omega t)}\,dk \Rightarrow \dot{\tilde{f}}(z,0) = \int_{-\infty}^{\infty} [-i\omega\tilde{A}(k)]e^{ikz}\,dk.$

(Note that $\omega = |k|v$, here, so it does *not* come outside the integral.)

$$\begin{aligned}
\dot{\tilde{f}}(z,0)^* &= \int_{-\infty}^{\infty}[i\omega\tilde{A}(k)^*]e^{-ikz}\,dk = \int_{-\infty}^{\infty}[i|k|v\tilde{A}(k)^*]e^{-ikz}\,dk = \int_{\infty}^{-\infty}[i|l|v\tilde{A}(-l)^*]e^{ilz}(-dl)\\
&= \int_{-\infty}^{\infty}[i|k|v\tilde{A}(-k)^*]e^{ikz}\,dk = \int_{-\infty}^{\infty}[i\omega\tilde{A}(-k)^*]e^{ikz}\,dk.\\
\dot{f}(z,0) &= \mathrm{Re}\left[\dot{\tilde{f}}(z,0)\right] = \frac{1}{2}\left[\dot{\tilde{f}}(z,0)+\dot{\tilde{f}}(z,0)^*\right] = \int_{-\infty}^{\infty}\frac{1}{2}[-i\omega\tilde{A}(k)+i\omega\tilde{A}(-k)^*]e^{ikz}\,dk.
\end{aligned}$$

$\frac{-i\omega}{2}\left[\tilde{A}(k)-\tilde{A}(-k)^*\right] = \frac{1}{2\pi}\int_{-\infty}^{\infty}\dot{f}(z,0)e^{-ikz}\,dz$, or $\frac{1}{2}\left[\tilde{A}(k)-\tilde{A}(-k)^*\right] = \frac{1}{2\pi}\int_{-\infty}^{\infty}\left[\frac{i}{\omega}\dot{f}(z,0)\right]e^{-ikz}\,dz.$

Adding these two results, we get $\boxed{\tilde{A}(k) = \frac{1}{2\pi}\int_{-\infty}^{\infty}\left[f(z,0)+\frac{i}{\omega}\dot{f}(z,0)\right]e^{-ikz}\,dz.}$ qed

---

Problem 9.33

(a) (i) *Gauss's law:* $\nabla\cdot\mathbf{E} = \frac{1}{r\sin\theta}\frac{\partial E_\phi}{\partial\phi} = 0.$ ✓

(ii) *Faraday's law:*

$$\begin{aligned}
-\frac{\partial\mathbf{B}}{\partial t} &= \nabla\times\mathbf{E} = \frac{1}{r\sin\theta}\frac{\partial}{\partial\theta}(\sin\theta E_\phi)\,\hat{\mathbf{r}} - \frac{1}{r}\frac{\partial}{\partial r}(rE_\phi)\,\hat{\boldsymbol{\theta}}\\
&= \frac{1}{r\sin\theta}\frac{\partial}{\partial\theta}\left[E_0\frac{\sin^2\theta}{r}\left(\cos u - \frac{1}{kr}\sin u\right)\right]\hat{\mathbf{r}} - \frac{1}{r}\frac{\partial}{\partial r}\left[E_0\sin\theta\left(\cos u - \frac{1}{kr}\sin u\right)\right]\hat{\boldsymbol{\theta}}.
\end{aligned}$$

But $\frac{\partial}{\partial r}\cos u = -k\sin u$; $\frac{\partial}{\partial r}\sin u = k\cos u.$

$$= \frac{1}{r\sin\theta}\frac{E_0}{r}2\sin\theta\cos\theta\left(\cos u - \frac{1}{kr}\sin u\right)\hat{\mathbf{r}} - \frac{1}{r}E_0\sin\theta\left(-k\sin u + \frac{1}{kr^2}\sin u - \frac{1}{r}\cos u\right)\hat{\boldsymbol{\theta}}.$$

Integrating with respect to t, and noting that $\int\cos u\,dt = -\frac{1}{\omega}\sin u$ and $\int\sin u\,dt = \frac{1}{\omega}\cos u$, we obtain

$$\boxed{\mathbf{B} = \frac{2E_0\cos\theta}{\omega r^2}\left(\sin u + \frac{1}{kr}\cos u\right)\hat{\mathbf{r}} + \frac{E_0\sin\theta}{\omega r}\left(-k\cos u + \frac{1}{kr^2}\cos u + \frac{1}{r}\sin u\right)\hat{\boldsymbol{\theta}}.}$$

(iii) *Divergence of* $\mathbf{B}$:

$$\begin{aligned}
\nabla\cdot\mathbf{B} &= \frac{1}{r^2}\frac{\partial}{\partial r}(r^2B_r) + \frac{1}{r\sin\theta}\frac{\partial}{\partial\theta}(\sin\theta B_\theta)\\
&= \frac{1}{r^2}\frac{\partial}{\partial r}\left[\frac{2E_0\cos\theta}{\omega}\left(\sin u + \frac{1}{kr}\cos u\right)\right] + \frac{1}{r\sin\theta}\frac{\partial}{\partial\theta}\left[\frac{E_0\sin^2\theta}{\omega r}\left(-k\cos u + \frac{1}{kr^2}\cos u + \frac{1}{r}\sin u\right)\right]\\
&= \frac{1}{r^2}\frac{2E_0\cos\theta}{\omega}\left(k\cos u - \frac{1}{kr^2}\cos u - \frac{1}{r}\sin u\right)\\
&\quad + \frac{1}{r\sin\theta}\frac{2E_0\sin\theta\cos\theta}{\omega r}\left(-k\cos u + \frac{1}{kr^2}\cos u + \frac{1}{r}\sin u\right)
\end{aligned}$$

$$= \frac{2E_0\cos\theta}{\omega r^2}\left(k\cos u - \frac{1}{kr^2}\cos u - \frac{1}{r}\sin u - k\cos u + \frac{1}{kr^2}\cos u + \frac{1}{r}\sin u\right) = 0. \checkmark$$

(iv) *Ampére/Maxwell:*

$$\begin{aligned}
\nabla\times\mathbf{B} &= \frac{1}{r}\left[\frac{\partial}{\partial r}(rB_\theta) - \frac{\partial B_r}{\partial\theta}\right]\hat{\boldsymbol{\phi}} \\
&= \frac{1}{r}\left\{\frac{\partial}{\partial r}\left[\frac{E_0\sin\theta}{\omega}\left(-k\cos u + \frac{1}{kr^2}\cos u + \frac{1}{r}\sin u\right)\right] - \frac{\partial}{\partial\theta}\left[\frac{2E_0\cos\theta}{\omega r^2}\left(\sin u + \frac{1}{kr}\cos u\right)\right]\right\}\hat{\boldsymbol{\phi}} \\
&= \frac{E_0\sin\theta}{\omega r}\left(k^2\sin u - \frac{2}{kr^3}\cos u - \frac{1}{r^2}\sin u - \frac{1}{r^2}\sin u + \frac{k}{r}\cos u + \frac{2}{r^2}\sin u + \frac{2}{kr^3}\cos u\right)\hat{\boldsymbol{\phi}} \\
&= \frac{k}{\omega}\frac{E_0\sin\theta}{r}\left(k\sin u + \frac{1}{r}\cos u\right)\hat{\boldsymbol{\phi}} = \frac{1}{c}\frac{E_0\sin\theta}{r}\left(k\sin u + \frac{1}{r}\cos u\right)\hat{\boldsymbol{\phi}}. \\
\frac{1}{c^2}\frac{\partial\mathbf{E}}{\partial t} &= \frac{1}{c^2}\frac{E_0\sin\theta}{r}\left(\omega\sin u + \frac{\omega}{kr}\cos u\right)\hat{\boldsymbol{\phi}} = \frac{1}{c^2}\frac{\omega}{k}\frac{E_0\sin\theta}{r}\left(k\sin u + \frac{1}{r}\cos u\right)\hat{\boldsymbol{\phi}} \\
&= \frac{1}{c}\frac{E_0\sin\theta}{r}\left(k\sin u + \frac{1}{r}\cos u\right)\hat{\boldsymbol{\phi}} = \nabla\times\mathbf{B}. \checkmark
\end{aligned}$$

(b) *Poynting Vector:*

$$\begin{aligned}
\mathbf{S} &= \frac{1}{\mu_0}(\mathbf{E}\times\mathbf{B}) = \frac{E_0\sin\theta}{\mu_0 r}\left(\cos u - \frac{1}{kr}\sin u\right)\left[\frac{2E_0\cos\theta}{\omega r^2}\left(\sin u + \frac{1}{kr}\cos u\right)\hat{\boldsymbol{\theta}}\right. \\
&\quad \left. + \frac{E_0\sin\theta}{\omega r}\left(-k\cos u + \frac{1}{kr^2}\cos u + \frac{1}{r}\sin u\right)(-\hat{\mathbf{r}})\right] \\
&= \frac{E_0^2\sin\theta}{\mu_0\omega r^2}\left\{\frac{2\cos\theta}{r}\left[\sin u\cos u + \frac{1}{kr}(\cos^2 u - \sin^2 u) - \frac{1}{k^2r^2}\sin u\cos u\right]\hat{\boldsymbol{\theta}}\right. \\
&\quad \left. - \sin\theta\left(-k\cos^2 u + \frac{1}{kr^2}\cos^2 u + \frac{1}{r}\sin u\cos u + \frac{1}{r}\sin u\cos u - \frac{1}{k^2r^3}\sin u\cos u - \frac{1}{kr^2}\sin^2 u\right)\hat{\mathbf{r}}\right\} \\
&= \boxed{\begin{aligned}&\frac{E_0^2\sin\theta}{\mu_0\omega r^2}\left\{\frac{2\cos\theta}{r}\left[\left(1 - \frac{1}{k^2r^2}\right)\sin u\cos u + \frac{1}{kr}(\cos^2 u - \sin^2 u)\right]\hat{\boldsymbol{\theta}}\right. \\ &\quad \left. + \sin\theta\left[\left(-\frac{2}{r} + \frac{1}{k^2r^3}\right)\sin u\cos u + k\cos^2 u + \frac{1}{kr^2}(\sin^2 u - \cos^2 u)\right]\hat{\mathbf{r}}\right\}.\end{aligned}}
\end{aligned}$$

Averaging over a full cycle, using $\langle\sin u\cos u\rangle = 0$, $\langle\sin^2 u\rangle = \langle\cos^2 u\rangle = \frac{1}{2}$, we get the intensity:

$$\mathbf{I} = \langle\mathbf{S}\rangle = \frac{E_0^2\sin\theta}{\mu_0\omega r^2}\left(\frac{k}{2}\sin\theta\right)\hat{\mathbf{r}} = \boxed{\frac{E_0^2\sin^2\theta}{2\mu_0 c r^2}\hat{\mathbf{r}}.}$$

It points in the $\hat{\mathbf{r}}$ direction, and falls off as $1/r^2$, as we would expect for a spherical wave.

(c) $P = \int \mathbf{I}\cdot d\mathbf{a} = \frac{E_0^2}{2\mu_0 c}\int\frac{\sin^2\theta}{r^2}r^2\sin\theta\,d\theta\,d\phi = \frac{E_0^2}{2\mu_0 c}2\pi\int_0^\pi \sin^3\theta\,d\theta = \boxed{\frac{4\pi}{3}\frac{E_0^2}{\mu_0 c}.}$

Problem 9.34

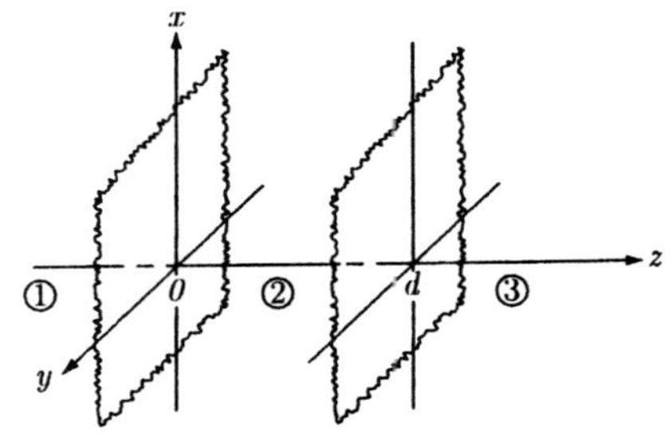

$$z<0: \quad \begin{cases} \tilde{\mathbf{E}}_I(z,t) = \tilde{E}_I e^{i(k_1 z-\omega t)}\,\hat{\mathbf{x}}, & \tilde{\mathbf{B}}_I(z,t) = \frac{1}{v_1}\tilde{E}_I e^{i(k_1 z-\omega t)}\,\hat{\mathbf{y}} \\ \tilde{\mathbf{E}}_R(z,t) = \tilde{E}_R e^{i(-k_1 z-\omega t)}\,\hat{\mathbf{x}}, & \tilde{\mathbf{B}}_R(z,t) = -\frac{1}{v_1}\tilde{E}_R e^{i(-k_1 z-\omega t)}\,\hat{\mathbf{y}}. \end{cases}$$

$$0<z<d: \quad \begin{cases} \tilde{\mathbf{E}}_r(z,t) = \tilde{E}_r e^{i(k_2 z-\omega t)}\,\hat{\mathbf{x}}, & \tilde{\mathbf{B}}_r(z,t) = \frac{1}{v_2}\tilde{E}_r e^{i(k_2 z-\omega t)}\,\hat{\mathbf{y}} \\ \tilde{\mathbf{E}}_l(z,t) = \tilde{E}_l e^{i(-k_2 z-\omega t)}\,\hat{\mathbf{x}}, & \tilde{\mathbf{B}}_l(z,t) = -\frac{1}{v_2}\tilde{E}_l e^{i(-k_2 z-\omega t)}\,\hat{\mathbf{y}}. \end{cases}$$

$$z>d: \quad \left\{ \tilde{\mathbf{E}}_T(z,t) = \tilde{E}_T e^{i(k_3 z-\omega t)}\,\hat{\mathbf{x}}, \quad \tilde{\mathbf{B}}_T(z,t) = \frac{1}{v_3}\tilde{E}_T e^{i(k_3 z-\omega t)}\,\hat{\mathbf{y}}. \right.$$

Boundary conditions: $\mathbf{E}_1^{\|} = \mathbf{E}_2^{\|}$, $\mathbf{B}_1^{\|} = \mathbf{B}_2^{\|}$, at each boundary (assuming $\mu_1 = \mu_2 = \mu_3 = \mu_0$):

$$z=0: \quad \begin{cases} \tilde{E}_I + \tilde{E}_R = \tilde{E}_r + \tilde{E}_l; \\ \frac{1}{v_1}\tilde{E}_I - \frac{1}{v_1}\tilde{E}_R = \frac{1}{v_2}\tilde{E}_r - \frac{1}{v_2}\tilde{E}_l \Rightarrow \tilde{E}_I - \tilde{E}_R = \beta(\tilde{E}_r - \tilde{E}_l), \text{ where } \beta \equiv v_1/v_2. \end{cases}$$

$$z=d: \quad \begin{cases} \tilde{E}_r e^{ik_2 d} + \tilde{E}_l e^{-ik_2 d} = \tilde{E}_T e^{ik_3 d}; \\ \frac{1}{v_2}\tilde{E}_r e^{ik_2 d} - \frac{1}{v_2}\tilde{E}_l e^{-ik_2 d} = \frac{1}{v_3}\tilde{E}_T e^{ik_3 d} \Rightarrow \tilde{E}_r e^{ik_2 d} - \tilde{E}_l e^{-ik_2 d} = \alpha \tilde{E}_T e^{ik_3 d}, \text{ where } \alpha \equiv v_2/v_3. \end{cases}$$

We have here four equations; the problem is to eliminate $\tilde{E}_R$, $\tilde{E}_r$, and $\tilde{E}_l$, to obtain a single equation for $\tilde{E}_T$ in terms of $\tilde{E}_I$.

Add the first two to eliminate $\tilde{E}_R$: $\quad 2\tilde{E}_I = (1+\beta)\tilde{E}_r + (1-\beta)\tilde{E}_l$;
Add the last two to eliminate $\tilde{E}_l$: $\quad 2\tilde{E}_r e^{ik_2 d} = (1+\alpha)\tilde{E}_T e^{ik_3 d}$;
Subtract the last two to eliminate $\tilde{E}_r$: $\quad 2\tilde{E}_l e^{-ik_2 d} = (1-\alpha)\tilde{E}^T e^{ik_3 d}$.

Plug the last two of these into the first:

$$\begin{aligned} 2\tilde{E}_I &= (1+\beta)\frac{1}{2}e^{-ik_2 d}(1+\alpha)\tilde{E}_T e^{ik_3 d} + (1-\beta)\frac{1}{2}e^{ik_2 d}(1-\alpha)\tilde{E}_T e^{ik_3 d} \\ 4\tilde{E}_I &= \left[(1+\alpha)(1+\beta)e^{-ik_2 d} + (1-\alpha)(1-\beta)e^{ik_2 d}\right]\tilde{E}_T e^{ik_3 d} \\ &= \left[(1+\alpha\beta)\left(e^{-ik_2 d} + e^{ik_2 d}\right) + (\alpha+\beta)\left(e^{-ik_2 d} - e^{ik_2 d}\right)\right]\tilde{E}_T e^{ik_3 d} \\ &= 2\left[(1+\alpha\beta)\cos(k_2 d) - i(\alpha+\beta)\sin(k_2 d)\right]\tilde{E}_T e^{ik_3 d}. \end{aligned}$$

Now the transmission coefficient is $T = \dfrac{v_3\epsilon_3 E_{T_0}^2}{v_1\epsilon_1 E_{I_0}^2} = \dfrac{v_3}{v_1}\left(\dfrac{\mu_0\epsilon_3}{\mu_0\epsilon_1}\right)\dfrac{|\tilde{E}_T|^2}{|\tilde{E}_I|^2} = \dfrac{v_1}{v_3}\dfrac{|\tilde{E}_T|^2}{|\tilde{E}_I|^2} = \alpha\beta\dfrac{|\tilde{E}_T|^2}{|\tilde{E}_I|^2}$, so

$$\begin{aligned}
T^{-1} &= \frac{1}{\alpha\beta}\frac{|\tilde{E}_I|^2}{|\tilde{E}_T|^2} = \frac{1}{\alpha\beta}\left|\frac{1}{2}\left[(1+\alpha\beta)\cos(k_2d) - i(\alpha+\beta)\sin(k_2d)\right]e^{ik_3d}\right|^2 \\
&= \frac{1}{4\alpha\beta}\left[(1+\alpha\beta)^2\cos^2(k_2d) + (\alpha+\beta)^2\sin^2(k_2d)\right]. \quad \text{But } \cos^2(k_2d) = 1-\sin^2(k_2d). \\
&= \frac{1}{4\alpha\beta}\left[(1+\alpha\beta)^2 + (\alpha^2+2\alpha\beta+\beta^2-1-2\alpha\beta-\alpha^2\beta^2)\sin^2(k_2d)\right] \\
&= \frac{1}{4\alpha\beta}\left[(1+\alpha\beta)^2 - (1-\alpha^2)(1-\beta^2)\sin^2(k_2d)\right].
\end{aligned}$$

But $n_1 = \dfrac{c}{v_1}$, $n_2 = \dfrac{c}{v_2}$, $n_3 = \dfrac{c}{v_3}$, so $\alpha = \dfrac{n_3}{n_3}$, $\beta = \dfrac{n_2}{n_1}$.

$$= \boxed{\frac{1}{4n_1n_3}\left[(n_1+n_3)^2 + \frac{(n_1^2-n_2^2)(n_3^2-n_2^2)}{n_2^2}\sin^2(k_2d)\right]}.$$

Problem 9.35

$T = 1 \Rightarrow \sin kd = 0 \Rightarrow kd = 0, \pi, 2\pi \ldots$. The *minimum* (nonzero) thickness is $d = \pi/k$. But $k = \omega/v = 2\pi\nu/v = 2\pi\nu n/c$, and $n = \sqrt{\epsilon\mu/\epsilon_0\mu_0}$ (Eq. 9.69), where (presumably) $\mu \approx \mu_0$. So $n = \sqrt{\epsilon/\epsilon_0} = \sqrt{\epsilon_r}$, and hence $d = \dfrac{\pi c}{2\pi\nu\sqrt{\epsilon_r}} = \dfrac{c}{2\nu\sqrt{\epsilon_r}} = \dfrac{3\times 10^8}{2(10\times 10^9)\sqrt{2.5}} = 9.49\times 10^{-3}\,\text{m}$, or $\boxed{9.5\,\text{mm.}}$

Problem 9.36

From Eq. 9.199,

$$\begin{aligned}
T^{-1} &= \frac{1}{4(4/3)(1)}\left\{[(4/3)+1]^2 + \frac{[(16/9)-(9/4)][1-(9/4)]}{(9/4)}\sin^2(3\omega d/2c)\right\} \\
&= \frac{3}{16}\left[\frac{49}{9} + \frac{(-17/36)(-5/4)}{(9/4)}\sin^2(3\omega d/2c)\right] = \frac{49}{48} + \frac{85}{(48)(36)}\sin^2(3\omega d/2c). \\
T &= \frac{48}{49 + (85/36)\sin^2(3\omega d/2c)}.
\end{aligned}$$

Since $\sin^2(3\omega d/2c)$ ranges from 0 to 1, $T_{\min} = \dfrac{48}{49+(85/36)} = \boxed{0.935;}$ $T_{\max} = \dfrac{48}{49} = \boxed{0.980.}$ Not much variation, and the transmission is good (over 90%) for *all* frequencies. Since Eq. 9.199 is unchanged when you switch 1 and 3, the transmission is the same either direction, and the $\boxed{\text{fish sees you just as well as you see it.}}$

Problem 9.37

(a) Equation 9.91 $\Rightarrow$ $\tilde{\mathbf{E}}_T(\mathbf{r},t) = \tilde{\mathbf{E}}_{0_T}e^{i(\mathbf{k}_T\cdot\mathbf{r}-\omega t)}$; $\mathbf{k}_T\cdot\mathbf{r} = k_T(\sin\theta_T\,\hat{\mathbf{x}} + \cos\theta_T\,\hat{\mathbf{z}})\cdot(x\,\hat{\mathbf{x}} + y\,\hat{\mathbf{y}} + z\,\hat{\mathbf{z}}) = k_T(x\sin\theta_T + z\cos\theta_T) = xk_T\sin\theta_T + izk_T\sqrt{\sin^2\theta_T - 1} = kx + i\kappa z$, where

$$\begin{aligned}
k &\equiv k_T\sin\theta_T = \left(\frac{\omega n_2}{c}\right)\frac{n_1}{n_2}\sin\theta_I = \frac{\omega n_1}{c}\sin\theta_I, \\
\kappa &\equiv k_T\sqrt{\sin^2\theta_T - 1} = \frac{\omega n_2}{c}\sqrt{(n_1/n_2)^2\sin^2\theta_I - 1} = \frac{\omega}{c}\sqrt{n_1^2\sin^2\theta_I - n_2^2}. \quad \text{So} \\
\tilde{\mathbf{E}}_T(\mathbf{r},t) &= \tilde{\mathbf{E}}_{0_T}e^{-\kappa z}e^{i(kx-\omega t)}. \quad \text{qed}
\end{aligned}$$

(b) $R = \left|\dfrac{\tilde{E}_{0_R}}{\tilde{E}_{0_I}}\right|^2 = \left|\dfrac{\alpha-\beta}{\alpha+\beta}\right|^2$. Here β is real (Eq. 9.106) and α is purely imaginary (Eq. 9.108); write $\alpha = ia$, with a real: $R = \left(\dfrac{ia-\beta}{ia+\beta}\right)\left(\dfrac{-ia-\beta}{-ia+\beta}\right) = \dfrac{a^2+\beta^2}{a^2+\beta^2} = \boxed{1.}$

(c) From Prob. 9.16, $E_{0_R} = \left|\dfrac{1-\alpha\beta}{1+\alpha\beta}\right| E_{0_I}$, so $R = \left|\dfrac{1-\alpha\beta}{1+\alpha\beta}\right|^2 = \left|\dfrac{1-ia\beta}{1+ia\beta}\right|^2 = \dfrac{(1-ia\beta)(1+ia\beta)}{(1+ia\beta)(1-ia\beta)} = \boxed{1.}$

(d) From the solution to Prob. 9.16, the transmitted wave is

$$\tilde{\mathbf{E}}(\mathbf{r},t) = \tilde{E}_{0_T} e^{i(\mathbf{k}_T\cdot\mathbf{r}-\omega t)}\,\hat{\mathbf{y}}, \quad \tilde{\mathbf{B}}(\mathbf{r},t) = \frac{1}{v_2}\tilde{E}_{0_T} e^{i(\mathbf{k}_T\cdot\mathbf{r}-\omega t)}(-\cos\theta_T\,\hat{\mathbf{x}} + \sin\theta_T\,\hat{\mathbf{z}}).$$

Using the results in (a): $\mathbf{k}_T\cdot\mathbf{r} = kx + i\kappa z - \omega t$, $\sin\theta_T = \dfrac{ck}{\omega n_2}$, $\cos\theta_T = i\dfrac{c\kappa}{\omega n_2}$:

$$\tilde{\mathbf{E}}(\mathbf{r},t) = \tilde{E}_{0_T} e^{-\kappa z} e^{i(kx-\omega t)}\,\hat{\mathbf{y}}, \quad \tilde{\mathbf{B}}(\mathbf{r},t) = \frac{1}{v_2}\tilde{E}_{0_T} e^{-\kappa z} e^{i(kx-\omega t)}\left(-i\frac{c\kappa}{\omega n_2}\,\hat{\mathbf{x}} + \frac{ck}{\omega n_2}\,\hat{\mathbf{z}}\right).$$

We may as well choose the phase constant so that $\tilde{E}_{0_T}$ is *real.* Then

$$\begin{aligned}
\mathbf{E}(\mathbf{r},t) &= E_0 e^{-\kappa z}\cos(kx-\omega t)\,\hat{\mathbf{y}};\\
\mathbf{B}(\mathbf{r},t) &= \frac{1}{v_2}E_0 e^{-\kappa z}\frac{c}{\omega n_2}\mathrm{Re}\left\{[\cos(kx-\omega t) + i\sin(kx-\omega t)]\,[-i\kappa\,\hat{\mathbf{x}} + k\,\hat{\mathbf{z}}]\right\}\\
&= \frac{1}{\omega}E_0 e^{-\kappa z}\left[\kappa\sin(kx-\omega t)\,\hat{\mathbf{x}} + k\cos(kx-\omega t)\,\hat{\mathbf{z}}\right]. \quad \text{qed}
\end{aligned}$$

(I used $v_2 = c/n_2$ to simplfy $\mathbf{B}$.)

(e) (i)
$$\nabla\cdot\mathbf{E} = \frac{\partial}{\partial y}\left[E_0 e^{-\kappa z}\cos(kx-\omega t)\right] = 0.\ \checkmark$$

(ii)
$$\begin{aligned}
\nabla\cdot\mathbf{B} &= \frac{\partial}{\partial x}\left[\frac{E_0}{\omega}e^{-\kappa z}\kappa\sin(kx-\omega t)\right] + \frac{\partial}{\partial z}\left[\frac{E_0}{\omega}e^{-\kappa z}k\cos(kx-\omega t)\right]\\
&= \frac{E_0}{\omega}\left[e^{-\kappa z}\kappa k\cos(kx-\omega t) - \kappa e^{-\kappa z}k\cos(kx-\omega t)\right] = 0.\ \checkmark
\end{aligned}$$

(iii)
$$\begin{aligned}
\nabla\times\mathbf{E} &= \begin{vmatrix}\hat{\mathbf{x}} & \hat{\mathbf{y}} & \hat{\mathbf{z}}\\ \partial/\partial x & \partial/\partial y & \partial/\partial z\\ 0 & E_y & 0\end{vmatrix} = -\frac{\partial E_y}{\partial z}\,\hat{\mathbf{x}} + \frac{\partial E_y}{\partial x}\,\hat{\mathbf{z}}\\
&= \kappa E_0 e^{-\kappa z}\cos(kx-\omega t)\,\hat{\mathbf{x}} - E_0 e^{-\kappa z}k\sin(kx-\omega t)\,\hat{\mathbf{z}}.\\
-\frac{\partial\mathbf{B}}{\partial t} &= -\frac{E_0}{\omega}e^{-\kappa z}\left[-\kappa\omega\cos(kx-\omega t)\,\hat{\mathbf{x}} + k\omega\sin(kx-\omega t)\,\hat{\mathbf{z}}\right]\\
&= \kappa E_0 e^{-\kappa z}\cos(kx-\omega t)\,\hat{\mathbf{x}} - kE_0 e^{-\kappa z}\sin(kx-\omega t)\,\hat{\mathbf{z}} = \nabla\times\mathbf{E}.\ \checkmark
\end{aligned}$$

(iv)
$$\begin{aligned}
\nabla\times\mathbf{B} &= \begin{vmatrix}\hat{\mathbf{x}} & \hat{\mathbf{y}} & \hat{\mathbf{z}}\\ \partial/\partial x & \partial/\partial y & \partial/\partial z\\ B_x & 0 & B_z\end{vmatrix} = \left(\frac{\partial B_x}{\partial z} - \frac{\partial B_z}{\partial x}\right)\hat{\mathbf{y}}\\
&= \left[-\frac{E_0}{\omega}\kappa^2 e^{-\kappa z}\sin(kx-\omega t) + \frac{E_0}{\omega}e^{-\kappa z}k^2\sin(kx-\omega t)\right]\hat{\mathbf{y}} = (k^2-\kappa^2)\frac{E_0}{\omega}e^{-\kappa z}\sin(kx-\omega t)\,\hat{\mathbf{y}}.
\end{aligned}$$

Eq. 9.202 $\Rightarrow k^2 - \kappa^2 = \left(\dfrac{\omega}{c}\right)^2\left[n_1^2\sin^2\theta_I - (n_1\sin\theta_I)^2 + (n_2)^2\right] = \left(\dfrac{n_2\omega}{c}\right)^2 = \omega^2\epsilon_2\mu_2.$

$$
\begin{aligned}
&= \epsilon_2\mu_2\omega E_0 e^{-\kappa z}\sin(kx-\omega t)\,\hat{\mathbf{y}}.\\
\mu_2\epsilon_2\frac{\partial \mathbf{E}}{\partial t} &= \mu_2\epsilon_2 E_0 e^{-\kappa z}\omega\sin(kx-\omega t)\,\hat{\mathbf{y}} = \boldsymbol{\nabla}\times\mathbf{B}\ \checkmark.
\end{aligned}
$$

(f)

$$
\mathbf{S} = \frac{1}{\mu_2}(\mathbf{E}\times\mathbf{B}) = \frac{1}{\mu_2}\frac{E_0^2}{\omega}e^{-2\kappa z}\begin{vmatrix}\hat{\mathbf{x}} & \hat{\mathbf{y}} & \hat{\mathbf{z}}\\ 0 & \cos(kx-\omega t) & 0\\ \kappa\sin(kx-\omega t) & 0 & k\cos(kx-\omega t)\end{vmatrix}
$$

$$
= \boxed{\frac{E_0^2}{\mu_2\omega}e^{-2\kappa z}\left[k\cos^2(kx-\omega t)\,\hat{\mathbf{x}} - \kappa\sin(kx-\omega t)\cos(kx-\omega t)\,\hat{\mathbf{z}}\right].}
$$

Averaging over a complete cycle, using $\langle\cos^2\rangle = 1/2$ and $\langle\sin\cos\rangle = 0$, $\langle\mathbf{S}\rangle = \dfrac{E_0^2 k}{2\mu_2\omega}e^{-2\kappa z}\,\hat{\mathbf{x}}$. On average, then, no energy is transmitted in the z direction, only in the x direction (parallel to the interface). qed

Problem 9.38

Look for solutions of the form $\mathbf{E} = \mathbf{E}_0(x,y,z)e^{-i\omega t}$, $\mathbf{B} = \mathbf{B}_0(x,y,z)e^{-i\omega t}$, subject to the boundary conditions $\mathbf{E}^{\parallel} = 0$, $B^{\perp} = 0$ at all surfaces. Maxwell's equations, in the form of Eq. 9.177, give

$$
\left\{\begin{array}{llll}\boldsymbol{\nabla}\cdot\mathbf{E}=0 & \Rightarrow \boldsymbol{\nabla}\cdot\mathbf{E}_0 = 0; & \boldsymbol{\nabla}\times\mathbf{E} = -\frac{\partial \mathbf{B}}{\partial t} & \Rightarrow \boldsymbol{\nabla}\times\mathbf{E}_0 = i\omega\mathbf{B}_0;\\ \boldsymbol{\nabla}\cdot\mathbf{B}=0 & \Rightarrow \boldsymbol{\nabla}\cdot\mathbf{B}_0 = 0; & \boldsymbol{\nabla}\times\mathbf{B} = \frac{1}{c^2}\frac{\partial \mathbf{E}}{\partial t} & \Rightarrow \boldsymbol{\nabla}\times\mathbf{B}_0 = -\frac{i\omega}{c^2}\mathbf{E}_0.\end{array}\right\}
$$

From now on I'll leave off the subscript (0). The problem is to solve the (time independent) equations

$$
\left\{\begin{array}{ll}\boldsymbol{\nabla}\cdot\mathbf{E}=0; & \boldsymbol{\nabla}\times\mathbf{E} = i\omega\mathbf{B};\\ \boldsymbol{\nabla}\cdot\mathbf{B}=0; & \boldsymbol{\nabla}\times\mathbf{B} = -\frac{i\omega}{c^2}\mathbf{E}.\end{array}\right\}
$$

From $\boldsymbol{\nabla}\times\mathbf{E} = i\omega\mathbf{B}$ it follows that I can get $\mathbf{B}$ once I know $\mathbf{E}$, so I'll concentrate on the latter for the moment.

$\boldsymbol{\nabla}\times(\boldsymbol{\nabla}\times\mathbf{E}) = \boldsymbol{\nabla}(\boldsymbol{\nabla}\cdot\mathbf{E}) - \nabla^2\mathbf{E} = -\nabla^2\mathbf{E} = \boldsymbol{\nabla}\times(i\omega\mathbf{B}) = i\omega\left(-\dfrac{i\omega}{c^2}\mathbf{E}\right) = \dfrac{\omega^2}{c^2}\mathbf{E}$. So

$\nabla^2 E_x = -\left(\dfrac{\omega}{c}\right)^2 E_x$; $\nabla^2 E_y = -\left(\dfrac{\omega}{c}\right)^2 E_y$; $\nabla^2 E_z = -\left(\dfrac{\omega}{c}\right)^2 E_z$. Solve each of these by separation of variables:

$E_x(x,y,z) = X(x)Y(y)Z(z) \Rightarrow YZ\dfrac{d^2X}{dx^2} + ZX\dfrac{d^2Y}{dy^2} + XY\dfrac{d^2Z}{dz^2} = -\left(\dfrac{\omega}{c}\right)^2 XYZ$, or $\dfrac{1}{X}\dfrac{d^2X}{dx^2} + \dfrac{1}{Y}\dfrac{d^2Y}{dy^2} + \dfrac{1}{Z}\dfrac{d^2Z}{dz^2} = -(\omega/c)^2$. Each term must be a constant, so $\dfrac{d^2X}{dx^2} = -k_x^2 X$, $\dfrac{d^2Y}{dy^2} = -k_y^2 Y$, $\dfrac{d^2Z}{dz^2} = -k_z^2 Z$, with $k_x^2 + k_y^2 + k_z^2 = -(\omega/c)^2$. The solution is

$$
E_x(x,y,z) = [A\sin(k_x x) + B\cos(k_x x)][C\sin(k_y y) + D\cos(k_y y)][E\sin(k_z z) + F\cos(k_z z)].
$$

But $\mathbf{E}^{\parallel} = 0$ at the boundaries $\Rightarrow E_x = 0$ at $y = 0$ and $z = 0$, so $D = F = 0$, and $E_x = 0$ at $y = b$ and $z = d$, so $k_y = n\pi/b$ and $k_z = l\pi/d$, where n and l are integers. A similar argument applies to E_y and E_z. *Conclusion:*

$$
\begin{aligned}
E_x(x,y,z) &= [A\sin(k_x x) + B\cos(k_x x)]\sin(k_y y)\sin(k_z z),\\
E_y(x,y,z) &= \sin(k_x x)[C\sin(k_y y) + D\cos(k_y y)]\sin(k_z z),\\
E_z(x,y,z) &= \sin(k_x x)\sin(k_y y)[E\sin(k_z z) + F\cos(k_z z)],
\end{aligned}
$$

where $k_x = m\pi/a$. (Actually, there is no reason at this stage to assume that k_x, k_y, and k_z are the *same* for all three components, and I should really affix a second subscript (x for E_x, y for E_y, and z for E_z), but in a moment we shall see that *in fact* they *do* have to be the same, so to avoid cumbersome notation I'll *assume* they are from the start.)

Now $\boldsymbol{\nabla}\cdot\mathbf{E} = 0 \Rightarrow k_x[A\cos(k_x x) - B\sin(k_x x)]\sin(k_y y)\sin(k_z z) + k_y\sin(k_x x)[C\cos(k_y y) - D\sin(k_y y)]\sin(k_z z) + k_z\sin(k_x x)\sin(k_y y)[E\cos(k_z z) - F\sin(k_z z)] = 0$. In particular, putting in $x = 0$, $k_x A\sin(k_y y)\sin(k_z z) = 0$, and hence $A = 0$. Likewise $y = 0 \Rightarrow C = 0$ and $z = 0 \Rightarrow E = 0$. (Moreover, if the k's were *not* equal for different components, then by Fourier analysis this equation could not be satisfied (for all x, y, and z) unless the other three constants were *also* zero, and we'd be left with no field at all.) It follows that $-(Bk_x + Dk_y + Fk_z) = 0$ (in order that $\nabla \cdot \mathbf{E} = 0$), and we are left with

$$\boxed{\begin{array}{l}\mathbf{E} = B\cos(k_x x)\sin(k_y y)\sin(k_z z)\,\hat{\mathbf{x}} + D\sin(k_x x)\cos(k_y y)\sin(k_z z)\,\hat{\mathbf{y}} + F\sin(k_x x)\sin(k_y y)\cos(k_z z)\,\hat{\mathbf{z}},\\ \text{with } k_x = (m\pi/a),\ k_y = (n\pi/b),\ k_z = (l\pi/d)\ (l,\ m,\ n \text{ all integers}), \text{ and } Bk_x + Dk_y + Fk_z = 0.\end{array}}$$

The corresponding magnetic field is given by $\mathbf{B} = -(i/\omega)\nabla \times \mathbf{E}$:

$$\begin{aligned}
B_x &= -\frac{i}{\omega}\left(\frac{\partial E_z}{\partial y} - \frac{\partial E_y}{\partial z}\right) = -\frac{i}{\omega}\left[Fk_y\sin(k_x x)\cos(k_y y)\cos(k_z z) - Dk_z\sin(k_x x)\cos(k_y y)\cos(k_z z)\right],\\
B_y &= -\frac{i}{\omega}\left(\frac{\partial E_x}{\partial z} - \frac{\partial E_z}{\partial x}\right) = -\frac{i}{\omega}\left[Bk_z\cos(k_x x)\sin(k_y y)\cos(k_z z) - Fk_x\cos(k_x x)\sin(k_y y)\cos(k_z z)\right],\\
B_z &= -\frac{i}{\omega}\left(\frac{\partial E_y}{\partial x} - \frac{\partial E_x}{\partial y}\right) = -\frac{i}{\omega}\left[Dk_x\cos(k_x x)\cos(k_y y)\sin(k_z z) - Bk_y\cos(k_x x)\cos(k_y y)\sin(k_z z)\right].
\end{aligned}$$

Or:

$$\boxed{\begin{aligned}\mathbf{B} = & -\frac{i}{\omega}(Fk_y - Dk_z)\sin(k_x x)\cos(k_y y)\cos(k_z z)\,\hat{\mathbf{x}} - \frac{i}{\omega}(Bk_z - Fk_x)\cos(k_x x)\sin(k_y y)\cos(k_z z)\,\hat{\mathbf{y}}\\ & - \frac{i}{\omega}(Dk_x - Bk_y)\cos(k_x x)\cos(k_y y)\sin(k_z z)\,\hat{\mathbf{z}}.\end{aligned}}$$

These *automatically* satisfy the boundary condition $B^{\perp} = 0$ ($B_x = 0$ at $x = 0$ and $x = a$, $B_y = 0$ at $y = 0$ and $y = b$, and $B_z = 0$ at $z = 0$ and $z = d$).

As a check, let's see if $\nabla \cdot \mathbf{B} = 0$:

$$\begin{aligned}
\nabla \cdot \mathbf{B} &= -\frac{i}{\omega}(Fk_y - Dk_z)k_x\cos(k_x x)\cos(k_y y)\cos(k_z z) - \frac{i}{\omega}(Bk_z - Fk_x)k_y\cos(k_x x)\cos(k_y y)\cos(k_z z)\\
&\quad - \frac{i}{\omega}(Dk_x - Bk_y)k_z\cos(k_x x)\cos(k_y y)\cos(k_z z)\\
&= -\frac{i}{\omega}(Fk_xk_y - Dk_xk_z + Bk_zk_y - Fk_xk_y + Dk_xk_z - Bk_yk_z)\cos(k_x x)\cos(k_y y)\cos(k_z z) = 0. \checkmark
\end{aligned}$$

The boxed equations satisfy all of Maxwell's equations, and they meet the boundary conditions. For TE modes, we pick $E_z = 0$, so $F = 0$ (and hence $Bk_x + Dk_y = 0$, leaving only the overall amplitude undetermined, for given l, m, and n); for TM modes we want $B_z = 0$ (so $Dk_x - Bk_y = 0$, again leaving only one amplitude undetermined, since $Bk_x + Dk_y + Fk_z = 0$). In either case (TE_{lmn} or TM_{lmn}), the frequency is given by $\omega^2 = c^2(k_x^2 + k_y^2 + k_z^2) = c^2\left[(m\pi/a)^2 + (n\pi/b)^2 + (l\pi/d)^2\right]$, or $\boxed{\omega = c\pi\sqrt{(m/a)^2 + (n/b)^2 + (l/d)^2}.}$

Chapter 10

Potentials and Fields

Problem 10.1

$$\Box^2 V + \frac{\partial L}{\partial t} = \nabla^2 V - \mu_0\epsilon_0\frac{\partial^2 V}{\partial t^2} + \frac{\partial}{\partial t}(\boldsymbol{\nabla}\cdot\mathbf{A}) + \mu_0\epsilon_0\frac{\partial^2 V}{\partial t^2} = \nabla^2 V + \frac{\partial}{\partial t}(\boldsymbol{\nabla}\cdot\mathbf{A}) = -\frac{1}{\epsilon_0}\rho. \checkmark$$

$$\Box^2\mathbf{A} - \boldsymbol{\nabla}L = \nabla^2\mathbf{A} - \mu_0\epsilon_0\frac{\partial^2\mathbf{A}}{\partial t^2} - \boldsymbol{\nabla}\left(\boldsymbol{\nabla}\cdot\mathbf{A} + \mu_0\epsilon_0\frac{\partial V}{\partial t}\right) = -\mu_0\mathbf{J}. \checkmark$$

Problem 10.2

(a) $W = \frac{1}{2}\int\left(\epsilon_0 E^2 + \frac{1}{\mu_0}B^2\right)d\tau$. At $t_1 = d/c$, $x \geq d = ct_1$, so $\mathbf{E} = 0$, $\mathbf{B} = 0$, and hence $\boxed{W(t_1) = 0.}$

At $T_2 = (d+h)/c$, $ct_2 = d+h$:

$$\mathbf{E} = -\frac{\mu_0\alpha}{2}(d+h-x)\,\hat{\mathbf{z}}, \quad \mathbf{B} = \frac{1}{c}\frac{\mu_0\alpha}{2}(d+h-x)\,\hat{\mathbf{y}},$$

so $B^2 = \frac{1}{c^2}E^2$, and

$$\left(\epsilon_0 E^2 + \frac{1}{\mu_0}B^2\right) = \epsilon_0\left(E^2 + \frac{1}{\mu_0\epsilon_0}\frac{1}{c^2}E^2\right) = 2\epsilon_0 E^2.$$

Therefore

$$W(t_2) = \frac{1}{2}(2\epsilon_0)\frac{\mu_0^2\alpha^2}{4}\int_d^{(d+h)}(d+h-x)^2\,dx\,(lw) = \frac{\epsilon_0\mu_0^2\alpha^2 lw}{4}\left[-\frac{(d+h-x)^3}{3}\right]_d^{d+h} = \boxed{\frac{\epsilon_0\mu_0^2\alpha^2 lwh^3}{12}}.$$

(b) $\mathbf{S}(x) = \frac{1}{\mu_0}(\mathbf{B}\times\mathbf{E}) = \frac{1}{\mu_0 c}E^2[-\hat{\mathbf{z}}\times(\pm\hat{\mathbf{y}})] = \pm\frac{1}{\mu_0 c}E^2\,\hat{\mathbf{x}} = \boxed{\pm\frac{\mu_0\alpha^2}{4c}(ct-|x|)^2\,\hat{\mathbf{x}}}$

(plus sign for $x > 0$, as here). For $|x| > ct$, $\mathbf{S} = 0$.

So the energy per unit time entering the box in this time interval is

$$\frac{dW}{dt} = P = \int\mathbf{S}(d)\cdot d\mathbf{a} = \boxed{\frac{\mu_0\alpha^2 lw}{4c}(ct-d)^2.}$$

Note that no energy flows out the top, since $\mathbf{S}(d+h) = 0$.

(c) $$W = \int_{t_1}^{t_2} P\,dt = \frac{\mu_0 \alpha^2 l w}{4c} \int_{d/x}^{(d+h)/c} (ct-d)^2\,dt = \frac{\mu_0 \alpha^2 l w}{4c}\left[\frac{(ct-d)^3}{3c}\right]_{d/c}^{(d+h)/c} = \boxed{\frac{\mu_0 \alpha^2 l w h^3}{12c^2}.}$$

Since $1/c^2 = \mu_0\epsilon_0$, this agrees with the answer to (a).

Problem 10.3

$$\mathbf{E} = -\nabla V - \frac{\partial \mathbf{A}}{\partial t} = \boxed{\frac{1}{4\pi\epsilon_0}\frac{q}{r^2}\,\hat{\mathbf{r}}.} \qquad \mathbf{B} = \nabla\times\mathbf{A} = \boxed{0.}$$

This is a funny set of potentials for a $\boxed{\text{stationary point charge}}$ q at the origin. ($V = \frac{1}{4\pi\epsilon_0}\frac{q}{r}$, $\mathbf{A} = 0$ would, of course, be the customary choice.) Evidently $\boxed{\rho = q\delta^3(\mathbf{r});\ \mathbf{J} = 0.}$

Problem 10.4

$$\begin{aligned}
\mathbf{E} &= -\nabla V - \frac{\partial \mathbf{A}}{\partial t} = -A_0\cos(kx-\omega t)\,\hat{\mathbf{y}}(-\omega) = \boxed{A_0\omega\cos(kx-\omega t)\,\hat{\mathbf{y}},}\\
\mathbf{B} &= \nabla\times\mathbf{A} = \hat{\mathbf{z}}\,\frac{\partial}{\partial x}\left[A_0\sin(kx-\omega t)\right] = \boxed{A_0 k\cos(kx-\omega t)\,\hat{\mathbf{z}}.}
\end{aligned}$$

Hence $\nabla\cdot\mathbf{E} = 0$ ✓, $\nabla\cdot\mathbf{B} = 0$ ✓.

$$\nabla\times\mathbf{E} = \hat{\mathbf{z}}\,\frac{\partial}{\partial x}\left[A_0\omega\cos(kx-\omega t)\right] = -A_0\omega k\sin(kx-\omega t)\,\hat{\mathbf{z}}, \qquad -\frac{\partial \mathbf{B}}{\partial t} = -A_0\omega k\sin(kx-\omega t)\,\hat{\mathbf{z}},$$

so $\nabla\times\mathbf{E} = -\frac{\partial \mathbf{B}}{\partial t}$ ✓.

$$\nabla\times\mathbf{B} = -\hat{\mathbf{y}}\,\frac{\partial}{\partial x}\left[A_0 k\cos(kx-\omega t)\right] = A_0 k^2\sin(kx-\omega t)\,\hat{\mathbf{y}}, \qquad \frac{\partial \mathbf{E}}{\partial t} = A_0\omega^2\sin(kx-\omega t)\,\hat{\mathbf{y}}.$$

So $\nabla\times\mathbf{B} = \mu_0\epsilon_0\frac{\partial \mathbf{E}}{\partial t}$ *provided* $\boxed{k^2 = \mu_0\epsilon_0\omega^2,}$ or, since $c^2 = 1/\mu_0\epsilon_0$, $\boxed{\omega = ck.}$

Problem 10.5

$$V' = V - \frac{\partial\lambda}{\partial t} = 0 - \left(-\frac{1}{4\pi\epsilon_0}\frac{q}{r}\right) = \boxed{\frac{1}{4\pi\epsilon_0}\frac{q}{r};} \ \mathbf{A}' = \mathbf{A} + \nabla\lambda = -\frac{1}{4\pi\epsilon_0}\frac{qt}{r^2}\,\hat{\mathbf{r}} + \left(-\frac{1}{4\pi\epsilon_0}qt\right)\left(-\frac{1}{r^2}\,\hat{\mathbf{r}}\right) = \boxed{0.}$$

This gauge function transforms the "funny" potentials of Prob. 10.3 into the "ordinary" potentials of a stationary point charge.

Problem 10.6

Ex. 10.1: $\nabla\cdot\mathbf{A} = 0$; $\frac{\partial V}{\partial t} = 0$. $\boxed{\text{Both Coulomb and Lorentz.}}$

Prob. 10.3: $\nabla\cdot\mathbf{A} = -\frac{qt}{4\pi\epsilon_0}\,\nabla\cdot\left(\frac{\hat{\mathbf{r}}}{r^2}\right) = -\frac{qt}{\epsilon_0}\delta^3(\mathbf{r})$; $\frac{\partial V}{\partial t} = 0$. $\boxed{\text{Neither.}}$

Prob. 10.4: $\nabla\cdot\mathbf{A} = 0$; $\frac{\partial V}{\partial t} = 0$. $\boxed{\text{Both.}}$

Problem 10.7

Suppose $\nabla\cdot\mathbf{A} \neq -\mu_0\epsilon_0\frac{\partial V}{\partial t}$. (Let $\nabla\cdot\mathbf{A} + \mu_0\epsilon_0\frac{\partial V}{\partial t} = \Phi$—some known function.) We want to pick λ such that $\mathbf{A}'$ and V' (Eq. 10.7) *do* obey $\nabla\cdot\mathbf{A}' = -\mu_0\epsilon_0\frac{\partial V'}{\partial t}$.

$$\nabla\cdot\mathbf{A}' + \mu_0\epsilon_0\frac{\partial V'}{\partial t} = \nabla\cdot\mathbf{A} + \nabla^2\lambda + \mu_0\epsilon_0\frac{\partial V}{\partial t} - \mu_0\epsilon_0\frac{\partial^2\lambda}{\partial t^2} = \Phi + \Box^2\lambda.$$

This will be zero provided we pick for λ the solution to $\Box^2\lambda = -\Phi$, which by hypothesis (and in fact) we know how to solve.

We *could* always find a gauge in which $V' = 0$, simply by picking $\lambda = \int_0^t V\,dt'$. We *cannot* in general pick $\mathbf{A} = 0$—this would make $\mathbf{B} = 0$. [Finding such a gauge function would amount to expressing $\mathbf{A}$ as $-\nabla\lambda$, and we know that vector functions *cannot* in general be written as gradients—only if they happen to have curl zero, which $\mathbf{A}$ (ordinarily) does *not*.]

Problem 10.8

¿From the product rule:

$$\nabla\cdot\left(\frac{\mathbf{J}}{\eta}\right) = \frac{1}{\eta}(\nabla\cdot\mathbf{J}) + \mathbf{J}\cdot\left(\nabla\frac{1}{\eta}\right), \quad \nabla'\cdot\left(\frac{\mathbf{J}}{\eta}\right) = \frac{1}{\eta}(\nabla'\cdot\mathbf{J}) + \mathbf{J}\cdot\left(\nabla'\frac{1}{\eta}\right).$$

But $\nabla\frac{1}{\eta} = -\nabla'\frac{1}{\eta}$, since $\boldsymbol{\eta} = \mathbf{r} - \mathbf{r}'$. So

$$\nabla\cdot\left(\frac{\mathbf{J}}{\eta}\right) = \frac{1}{\eta}(\nabla\cdot\mathbf{J}) - \mathbf{J}\cdot\left(\nabla'\frac{1}{\eta}\right) = \frac{1}{\eta}(\nabla\cdot\mathbf{J}) + \frac{1}{\eta}(\nabla'\cdot\mathbf{J}) - \nabla'\cdot\left(\frac{\mathbf{J}}{\eta}\right).$$

But

$$\nabla\cdot\mathbf{J} = \frac{\partial J_x}{\partial x} + \frac{\partial J_y}{\partial y} + \frac{\partial J_z}{\partial z} = \frac{\partial J_x}{\partial t_r}\frac{\partial t_r}{\partial x} + \frac{\partial J_y}{\partial t_r}\frac{\partial t_r}{\partial y} + \frac{\partial J_z}{\partial t_r}\frac{\partial t_r}{\partial z},$$

and

$$\frac{\partial t_r}{\partial x} = -\frac{1}{c}\frac{\partial \eta}{\partial x}, \quad \frac{\partial t_r}{\partial y} = -\frac{1}{c}\frac{\partial \eta}{\partial y}, \quad \frac{\partial t_r}{\partial z} = -\frac{1}{c}\frac{\partial \eta}{\partial z},$$

so

$$\nabla\cdot\mathbf{J} = -\frac{1}{c}\left[\frac{\partial J_x}{\partial t_r}\frac{\partial \eta}{\partial x} + \frac{\partial J_y}{\partial t_r}\frac{\partial \eta}{\partial y} + \frac{\partial J_z}{\partial t_r}\frac{\partial \eta}{\partial z}\right] = -\frac{1}{c}\frac{\partial \mathbf{J}}{\partial t_r}\cdot(\nabla\eta).$$

Similarly,

$$\nabla'\cdot\mathbf{J} = -\frac{\partial\rho}{\partial t} - \frac{1}{c}\frac{\partial\mathbf{J}}{\partial t_r}\cdot(\nabla'\eta).$$

[The first term arises when we differentiate with respect to the *explicit* $\mathbf{r}'$, and use the continuity equation.] thus

$$\nabla\cdot\left(\frac{\mathbf{J}}{\eta}\right) = \frac{1}{\eta}\left[-\frac{1}{c}\frac{\partial\mathbf{J}}{\partial t_r}\cdot(\nabla'\eta)\right] + \frac{1}{\eta}\left[-\frac{\partial\rho}{\partial t} - \frac{1}{c}\frac{\partial\mathbf{J}}{\partial t_r}\cdot(\nabla'\eta)\right] - \nabla\cdot\left(\frac{\mathbf{J}}{\eta}\right) = -\frac{1}{\eta}\frac{\partial\rho}{\partial t} - \nabla'\cdot\left(\frac{\mathbf{J}}{\eta}\right)$$

(the other two terms cancel, since $\nabla\eta = -\nabla'\eta$). Therefore:

$$\nabla\cdot\mathbf{A} = \frac{\mu_0}{4\pi}\left[-\frac{\partial}{\partial t}\int\frac{\rho}{\eta}\,d\tau - \int\nabla'\cdot\left(\frac{\mathbf{J}}{\eta}\right)d\tau\right] = -\mu_0\epsilon_0\frac{\partial}{\partial t}\left[\frac{1}{4\pi\epsilon_0}\int\frac{\rho}{\eta}\,d\tau\right] - \frac{\mu_0}{4\pi}\oint\frac{\mathbf{J}}{\eta}\cdot d\mathbf{a}.$$

The last term is over the suface at "infinity", where $\mathbf{J} = 0$, so it's zero. Therefore $\nabla\cdot\mathbf{A} = -\mu_0\epsilon_0\frac{\partial V}{\partial t}$. ✓

Problem 10.9

(a) As in Ex. 10.2, for $t < r/c$, $\mathbf{A} = 0$; for $t > r/c$,

$$\mathbf{A}(r,t) = \left(\frac{\mu_0}{4\pi}\hat{\mathbf{z}}\right)2\int_0^{\sqrt{(ct)^2-r^2}} \frac{k(t-\sqrt{r^2+z^2}/c)}{\sqrt{r^2+z^2}}\,dz = \frac{\mu_0 k}{2\pi}\hat{\mathbf{z}}\left\{t\int_0^{\sqrt{(ct)^2-r^2}}\frac{dz}{\sqrt{r^2+z^2}} - \frac{1}{c}\int_0^{\sqrt{(ct)^2-r^2}} dz\right\}$$

$$= \left(\frac{\mu_0 k}{2\pi}\hat{\mathbf{z}}\right)\left[t\ln\left(\frac{ct+\sqrt{(ct)^2-r^2}}{r}\right) - \frac{1}{c}\sqrt{(ct)^2-r^2}\right].$$ Accordingly,

$$\mathbf{E}(r,t) = -\frac{\partial\mathbf{A}}{\partial t} = -\frac{\mu_0 k}{2\pi}\hat{\mathbf{z}}\left\{\ln\left(\frac{ct+\sqrt{(ct)^2-r^2}}{r}\right) + t\left(\frac{r}{ct+\sqrt{(ct)^2-r^2}}\right)\left(\frac{1}{r}\right)\left(c+\frac{1}{2}\frac{2c^2t}{\sqrt{(ct)^2-r^2}}\right) - \frac{1}{2c}\frac{2c^2t}{\sqrt{(ct)^2-r^2}}\right\}$$

$$= -\frac{\mu_0 k}{2\pi}\hat{\mathbf{z}}\left\{\ln\left(\frac{ct+\sqrt{(ct)^2-r^2}}{r}\right) + \frac{ct}{\sqrt{(ct)^2-r^2}} - \frac{ct}{\sqrt{(ct)^2-r^2}}\right\}$$

$$= \boxed{-\frac{\mu_0 k}{2\pi}\ln\left(\frac{ct+\sqrt{(ct)^2-r^2}}{r}\right)\hat{\mathbf{z}}}$$ (or zero, for $t < r/c$).

$$\mathbf{B}(r,t) = -\frac{\partial A_z}{\partial r}\hat{\boldsymbol{\phi}}$$

$$= -\frac{\mu_0 k}{2\pi}\left\{t\left(\frac{r}{ct+\sqrt{(ct)^2-r^2}}\right)\frac{\left[r\frac{1}{2}\frac{(-2r)}{\sqrt{(ct)^2-r^2}} - ct - \sqrt{(ct)^2-r^2}\right]}{r^2} - \frac{1}{2c}\frac{(-2r)}{\sqrt{(ct)^2-r^2}}\right\}\hat{\boldsymbol{\phi}}$$

$$= -\frac{\mu_0 k}{2\pi}\left\{\frac{-ct^2}{r\sqrt{(ct)^2-r^2}} + \frac{r}{c\sqrt{(ct)^2-r^2}}\right\}\hat{\boldsymbol{\phi}} = -\frac{\mu_0 k}{2\pi}\frac{(-c^2t^2+r^2)}{rc\sqrt{(ct)^2-r^2}}\hat{\boldsymbol{\phi}} = \boxed{\frac{\mu_0 k}{2\pi rc}\sqrt{(ct)^2-r^2}\,\hat{\boldsymbol{\phi}}.}$$

(b) $\mathbf{A}(r,t) = \frac{\mu_0}{4\pi}\hat{\mathbf{z}}\int_{-\infty}^{\infty}\frac{q_0\delta(t-\mathscr{r}/c)}{\mathscr{r}}\,dz$. But $\mathscr{r} = \sqrt{r^2+z^2}$, so the integrand is even in z:

$$\mathbf{A}(r,t) = \left(\frac{\mu_0 q_0}{4\pi}\hat{\mathbf{z}}\right)2\int_0^{\infty}\frac{\delta(t-\mathscr{r}/c)}{\mathscr{r}}\,dz.$$

Now $z = \sqrt{\mathscr{r}^2-r^2} \Rightarrow dz = \frac{1}{2}\frac{2\mathscr{r}\,d\mathscr{r}}{\sqrt{\mathscr{r}^2-r^2}} = \frac{\mathscr{r}\,d\mathscr{r}}{\sqrt{\mathscr{r}^2-r^2}}$, and $z = 0 \Rightarrow \mathscr{r} = r$, $z = \infty \Rightarrow \mathscr{r} = \infty$. So:

$$\mathbf{A}(r,t) = \frac{\mu_0 q_0}{2\pi}\hat{\mathbf{z}}\int_r^{\infty}\frac{1}{\mathscr{r}}\delta\left(t-\frac{\mathscr{r}}{c}\right)\frac{\mathscr{r}\,d\mathscr{r}}{\sqrt{\mathscr{r}^2-r^2}}.$$

Now $\delta(t - \imath/c) = c\delta(\imath - ct)$ (Ex. 1.15); therefore $\mathbf{A} = \frac{\mu_0 q_0}{2\pi}\,\hat{\mathbf{z}}\,c\int_r^\infty \frac{\delta(\imath - ct)}{\sqrt{\imath^2 - r^2}}\,d\imath$, so

$$\mathbf{A}(r,t) = \frac{\mu_0 q_0 c}{2\pi}\frac{1}{\sqrt{(ct)^2 - r^2}}\,\hat{\mathbf{z}} \quad \text{(or zero, if } ct < r);$$

$$\mathbf{E}(r,t) = -\frac{\partial \mathbf{A}}{\partial t} = -\frac{\mu_0 q_0 c}{2\pi}\left(-\frac{1}{2}\right)\frac{2c^2 t}{[(ct)^2 - r^2]^{3/2}}\,\hat{\mathbf{z}} = \boxed{\frac{\mu_0 q_0 c^3 t}{2\pi[(ct)^2 - r^2]^{3/2}}\,\hat{\mathbf{z}}} \quad \text{(or zero, for } t < r/c);$$

$$\mathbf{B}(r,t) = -\frac{\partial \mathbf{A}_z}{\partial t}\,\hat{\boldsymbol{\phi}} = -\frac{\mu_0 q_0 c}{2\pi}\left(-\frac{1}{2}\right)\frac{-2r}{[(ct)^2 - r^2]^{3/2}}\,\hat{\boldsymbol{\phi}} = \boxed{\frac{-\mu_0 q_0 c r}{2\pi[(ct)^2 - r^2]^{3/2}}\,\hat{\boldsymbol{\phi}}} \quad \text{(or zero, for } t < r/c).$$

Problem 10.10

$$\mathbf{A} = \frac{\mu_0}{4\pi}\int \frac{\mathbf{I}(t_r)}{\imath}\,dl = \frac{\mu_0 k}{4\pi}\int \frac{(t - \imath/c)}{\imath}\,d\mathbf{l} = \frac{\mu_0 k}{4\pi}\left\{t\int \frac{d\mathbf{l}}{\imath} - \frac{1}{c}\int d\mathbf{l}\right\}.$$

But for the complete loop, $\int d\mathbf{l} = 0$, so $\mathbf{A} = \frac{\mu_0 k t}{4\pi}\left\{\frac{1}{a}\int_1 d\mathbf{l} + \frac{1}{b}\int_2 d\mathbf{l} + 2\,\hat{\mathbf{x}}\int_a^b \frac{dx}{x}\right\}$. Here $\int_1 d\mathbf{l} = 2a\,\hat{\mathbf{x}}$ (inner circle), $\int_2 d\mathbf{l} = -2b\,\hat{\mathbf{x}}$ (outer circle), so

$$\mathbf{A} = \frac{\mu_0 k t}{4\pi}\left[\frac{1}{a}(2a) + \frac{1}{b}(-2b) + 2\ln(b/a)\right]\hat{\mathbf{x}} \Rightarrow \boxed{\mathbf{A} = \frac{\mu_0 k t}{2\pi}\ln(b/a)\,\hat{\mathbf{x}},} \quad \mathbf{E} = -\frac{\partial \mathbf{A}}{\partial t} = \boxed{-\frac{\mu_0 k}{2\pi}\ln(b/a)\,\hat{\mathbf{x}}.}$$

The changing magnetic field induces the electric field. Since we only know $\mathbf{A}$ at *one point* (the center), we can't compute $\boldsymbol{\nabla}\times\mathbf{A}$ to get $\mathbf{B}$.

Problem 10.11

In this case $\dot{\rho}(\mathbf{r},t) = \dot{\rho}(\mathbf{r},0)$ and $\dot{\mathbf{J}}(\mathbf{r},t) = 0$, so Eq. 10.29 $\Rightarrow$

$$\mathbf{E}(\mathbf{r},t) = \frac{1}{4\pi\epsilon_0}\int\left[\frac{\rho(\mathbf{r}',0) + \dot{\rho}(\mathbf{r}',0)t_r}{\imath^2} + \frac{\dot{\rho}(\mathbf{r}',0)}{c\imath}\right]\hat{\boldsymbol{\imath}}\,d\tau', \text{ but } t_r = t - \frac{\imath}{c} \text{ (Eq. 10.18), so}$$

$$= \frac{1}{4\pi\epsilon_0}\int\left[\frac{\rho(\mathbf{r}',0) + \dot{\rho}(\mathbf{r}',0)t}{\imath^2} - \frac{\dot{\rho}(\mathbf{r}',0)(\imath/c)}{\imath^2} + \frac{\dot{\rho}(\mathbf{r}',0)}{c\imath}\right]\hat{\boldsymbol{\imath}}\,d\tau' = \frac{1}{4\pi\epsilon_0}\int \frac{\rho(\mathbf{r}',t)}{\imath^2}\hat{\boldsymbol{\imath}}\,d\tau'. \quad \text{qed}$$

Problem 10.12

In this approximation we're dropping the higher derivatives of $\mathbf{J}$, so $\dot{\mathbf{J}}(t_r) = \dot{\mathbf{J}}(t)$, and Eq. 10.31 $\Rightarrow$

$$\mathbf{B}(\mathbf{r},t) = \frac{\mu_0}{4\pi}\int \frac{1}{\imath^2}\left[\mathbf{J}(\mathbf{r}',t) + (t_r - t)\dot{\mathbf{J}}(\mathbf{r}',t) + \frac{\imath}{c}\dot{\mathbf{J}}(\mathbf{r}',t)\right]\times\hat{\boldsymbol{\imath}}\,d\tau', \text{ but } t_r - t = -\frac{\imath}{c} \text{ (Eq. 10.18), so}$$

$$= \frac{\mu_0}{4\pi}\int \frac{\mathbf{J}(\mathbf{r}',t)\times\hat{\boldsymbol{\imath}}}{\imath^2}\,d\tau'. \quad \text{qed}$$

Problem 10.13

At time t the charge is at $\mathbf{r}(t) = a[\cos(\omega t)\,\hat{\mathbf{x}} + \sin(\omega t)\,\hat{\mathbf{y}}]$, so $\mathbf{v}(t) = \omega a[-\sin(\omega t)\,\hat{\mathbf{x}} + \cos(\omega t)\,\hat{\mathbf{y}}]$. Therefore $\boldsymbol{\imath} = z\,\hat{\mathbf{z}} - a[\cos(\omega t_r)\,\hat{\mathbf{x}} + \sin(\omega t_r)\,\hat{\mathbf{y}}]$, and hence $\imath^2 = z^2 + a^2$ (of course), and $\imath = \sqrt{z^2 + a^2}$.

$$\hat{\boldsymbol{\imath}}\cdot\mathbf{v} = \frac{1}{\imath}(\boldsymbol{\imath}\cdot\mathbf{v}) = \frac{1}{\imath}\left\{-\omega a^2[-\sin(\omega t_r)\cos(\omega t_r) + \sin(\omega t_r)\cos(\omega t_r)]\right\} = 0, \text{ so } \left(1 - \frac{\hat{\boldsymbol{\imath}}\cdot\mathbf{v}}{c}\right) = 1.$$

Therefore

$$V(z,t) = \boxed{\frac{1}{4\pi\epsilon_0}\frac{q}{\sqrt{z^2+a^2}}};\ \mathbf{A}(z,t) = \boxed{\frac{q\omega a}{4\pi\epsilon_0 c^2\sqrt{z^2+a^2}}[-\sin(\omega t_r)\,\hat{\mathbf{x}} + \cos(\omega t_r)\,\hat{\mathbf{y}}),}\ \text{where}\ \boxed{t_r = t - \frac{\sqrt{z^2+a^2}}{c}.}$$

Problem 10.14

Term under square root in (Eq. 9.98) is:

$$\begin{aligned} I &= c^4t^2 - 2c^2t(\mathbf{r}\cdot\mathbf{v}) + (\mathbf{r}\cdot\mathbf{v})^2 + c^2r^2 - c^4t^2 - v^2r^2 + v^2c^2t^2 \\ &= (\mathbf{r}\cdot\mathbf{v})^2 + (c^2-v^2)r^2 + c^2(vt)^2 - 2c^2(\mathbf{r}\cdot\mathbf{v}t). \quad \text{put in } \mathbf{v}t = \mathbf{r} - \mathbf{R}^2. \\ &= (\mathbf{r}\cdot\mathbf{v})^2 + (c^2-v^2)r^2 + c^2(r^2+R^2-2\mathbf{r}\cdot\mathbf{R}) - 2c^2(r^2 - \mathbf{r}\cdot\mathbf{R}) = (\mathbf{r}\cdot\mathbf{v})^2 - r^2v^2 + c^2R^2. \end{aligned}$$

but

$$\begin{aligned} (\mathbf{r}\cdot\mathbf{v})^2 - r^2v^2 &= ((\mathbf{R}+\mathbf{v}t)\cdot\mathbf{v})^2 - (\mathbf{R}+\mathbf{v}t)^2v^2 \\ &= (\mathbf{R}\cdot\mathbf{v})^2 + v^4t^2 + 2(\mathbf{R}\cdot\mathbf{v})v^2t - R^2v^2 - 2(\mathbf{R}\cdot\mathbf{v})tv^2 - v^2t^2v^2 \\ &= (\mathbf{R}\cdot\mathbf{v})^2 - R^2v^2 = R^2v^2\cos^2\theta - R^2v^2 = -R^2v^2\left(1-\cos^2\theta\right) \\ &= -R^2v^2\sin^2\theta. \end{aligned}$$

Therefore

$$I = -R^2v^2\sin^2\theta + c^2R^2 = c^2R^2\left(1 - \frac{v^2}{c^2}\sin^2\theta\right).$$

Hence

$$V(\mathbf{r},t) = \frac{1}{4\pi\epsilon_0}\frac{q}{R\sqrt{1-\frac{v^2}{c^2}\sin^2\theta}}. \quad \text{qed}$$

Problem 10.15

Once seen, from a given point x, the particle will forever remain in view—to disappear it would have to travel faster than light.

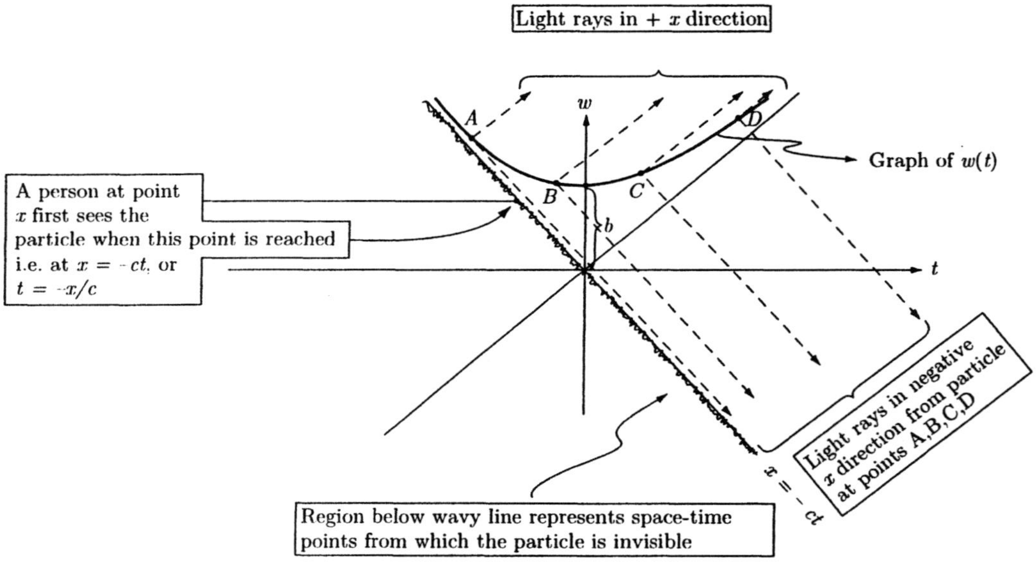

Problem 10.16

First calculate t_r: $t_r = t - |\mathbf{r} - \mathbf{w}(t_r)|/c \Rightarrow$

$-c(t_r - t) = x - \sqrt{b^2 + c^2 t_r^2} \Rightarrow c(t_r - t) + x = \sqrt{b^2 + c^2 t_r^2}$;

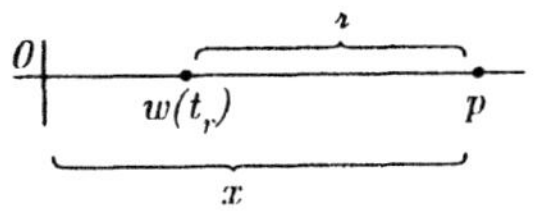

$c^2 t_r^2 - 2c^2 t_r t + c^2 t^2 + 2xct_r - 2xct + x^2 = b^2 + c^2 t_r^2$;

$2ct_r(x - ct) + (x^2 - 2xct + c^2t^2) = b^2$;

$$2ct_r(x - ct) = b^2 - (x - ct)^2, \text{ or } t_r = \frac{b^2 - (x - ct)^2}{2c(x - ct)}.$$

Now $V(x, t) = \dfrac{1}{4\pi\epsilon_0} \dfrac{qc}{(\mathscr{r}c - \boldsymbol{\mathscr{r}} \cdot \mathbf{v})}$, and $\mathscr{r}c - \boldsymbol{\mathscr{r}} \cdot \mathbf{v} = \mathscr{r}(c - v)$; $\mathscr{r} = c(t - t_r)$.

$$v = \frac{1}{2} \frac{1}{\sqrt{b^2 + c^2 t_r^2}} 2c^2 t_r = \frac{c^2 t_r}{c(t_r - t) + x} = \frac{c^2 t_r}{ct_r + (x - ct)}; \quad (c - v) = \frac{c^2 t_r + c(x - ct) - c^2 t_r}{ct_r + (x - ct)} = \frac{c(x - ct)}{ct_r + (x - ct)};$$

$$\mathscr{r}c - \boldsymbol{\mathscr{r}} \cdot \mathbf{v} = \frac{c(t - t_r)c(x - ct)}{ct_r + (x - ct)} = \frac{c^2(t - t_r)(x - ct)}{ct_r + (x - ct)}; \quad ct_r + (x - ct) = \frac{b^2 - (x - ct)^2}{2(x - ct)} + (x - ct) = \frac{b^2 + (x - ct)^2}{2(x - ct)};$$

$$t - t_r = \frac{2ct(x - ct) - b^2 + (x - ct)^2}{2c(x - ct)} = \frac{(x - ct)(x + ct) - b^2}{2c(x - ct)} = \frac{(x^2 - c^2t^2 - b^2)}{2c(x - ct)}. \quad \text{Therefore}$$

$$\frac{1}{\mathscr{r}c - \boldsymbol{\mathscr{r}} \cdot \mathbf{v}} = \left[\frac{b^2 + (x - ct)^2}{2(x - ct)}\right] \frac{1}{c^2(x - ct)} \frac{2c(x - ct)}{[2ct(x - ct) - b^2 + (x - ct)^2]} = \frac{b^2 + (x - ct)^2}{c(x - ct)\,[2ct(x - ct) - b^2 + (x - ct)^2]}.$$

The term in square brackets simplifies to $(2ct + x - ct)(x - ct) - b^2 = (x + ct)(x - ct) - b^2 = x^2 - c^2t^2 - b^2$.

So $\boxed{V(x, t) = \dfrac{q}{4\pi\epsilon_0} \dfrac{b^2 + (x - ct)^2}{(x - ct)(x^2 - c^2t^2 - b^2)}.}$

Meanwhile

$$\begin{aligned} \mathbf{A} &= \frac{V}{c^2}\mathbf{v} = \frac{c^2 t_r}{ct_r + (x - ct)} \frac{V}{c^2} \hat{\mathbf{x}} = \left[\frac{b^2 - (x - ct)^2}{2c(x - ct)}\right] \frac{2(x - ct)}{b^2 + (x - ct)^2} \frac{q}{4\pi\epsilon_0} \frac{b^2 + (x - ct)^2}{(x - ct)(x^2 - c^2t^2 - b^2)} \hat{\mathbf{x}} \\ &= \boxed{\frac{q}{4\pi\epsilon_0 c} \frac{b^2 - (x - ct)^2}{(x - ct)(x^2 - c^2t^2 - b^2)} \hat{\mathbf{x}}.} \end{aligned}$$

Problem 10.17

From Eq. 10.33, $c(t - t_r) = \mathscr{r} \Rightarrow c^2(t - t_r)^2 = \mathscr{r}^2 = \boldsymbol{\mathscr{r}} \cdot \boldsymbol{\mathscr{r}}$. Differentiate with respect to t:

$$2c^2(t - t_r)\left(1 - \frac{\partial t_r}{\partial t}\right) = 2\boldsymbol{\mathscr{r}} \cdot \frac{\partial \boldsymbol{\mathscr{r}}}{\partial t}, \text{ or } c\mathscr{r}\left(1 - \frac{\partial t_r}{\partial t}\right) = \boldsymbol{\mathscr{r}} \cdot \frac{\partial \boldsymbol{\mathscr{r}}}{\partial t}. \text{ Now } \boldsymbol{\mathscr{r}} = \mathbf{r} - \mathbf{w}(t_r), \text{ so}$$

$$\frac{\partial \boldsymbol{\mathscr{r}}}{\partial t} = -\frac{\partial \mathbf{w}}{\partial t} = -\frac{\partial \mathbf{w}}{\partial t_r}\frac{\partial t_r}{\partial t} = -\mathbf{v}\frac{\partial t_r}{\partial t}; \quad c\mathscr{r}\left(1 - \frac{\partial t_r}{\partial t}\right) = -\boldsymbol{\mathscr{r}} \cdot \mathbf{v}\frac{\partial t_r}{\partial t}; \quad c\mathscr{r} = \frac{\partial t_r}{\partial t}(c\mathscr{r} - \boldsymbol{\mathscr{r}} \cdot \mathbf{v}) = \frac{\partial t_r}{\partial t}(\boldsymbol{\mathscr{r}} \cdot \mathbf{u}) \text{ (Eq. 10.64)},$$

and hence $\dfrac{\partial t_r}{\partial t} = \dfrac{c\mathscr{r}}{\boldsymbol{\mathscr{r}} \cdot \mathbf{u}}$. qed

Now Eq. 10.40 says $\mathbf{A}(\mathbf{r},t) = \dfrac{\mathbf{v}}{c^2}V(\mathbf{r},t)$, so

$$\begin{aligned}
\frac{\partial \mathbf{A}}{\partial t} &= \frac{1}{c^2}\left(\frac{\partial \mathbf{v}}{\partial t}V + \mathbf{v}\frac{\partial V}{\partial t}\right) = \frac{1}{c^2}\left(\frac{\partial \mathbf{v}}{\partial t_r}\frac{\partial t_r}{\partial t}V + \mathbf{v}\frac{\partial V}{\partial t}\right)\\
&= \frac{1}{c^2}\left[\mathbf{a}\frac{\partial t_r}{\partial t}\frac{1}{4\pi\epsilon_0}\frac{qc}{\boldsymbol{\mathscr{r}}\cdot\mathbf{u}} + \mathbf{v}\frac{1}{4\pi\epsilon_0}\frac{-qc}{(\boldsymbol{\mathscr{r}}\cdot\mathbf{u})^2}\frac{\partial}{\partial t}(\mathscr{r}c - \boldsymbol{\mathscr{r}}\cdot\mathbf{v})\right]\\
&= \frac{1}{c^2}\frac{qc}{4\pi\epsilon_0}\left[\frac{\mathbf{a}}{\boldsymbol{\mathscr{r}}\cdot\mathbf{u}}\frac{\partial t_r}{\partial t} - \frac{\mathbf{v}}{(\boldsymbol{\mathscr{r}}\cdot\mathbf{u})^2}\left(c\frac{\partial \mathscr{r}}{\partial t} - \frac{\partial \boldsymbol{\mathscr{r}}}{\partial t}\cdot\mathbf{v} - \boldsymbol{\mathscr{r}}\cdot\frac{\partial \mathbf{v}}{\partial t}\right)\right].
\end{aligned}$$

But $\mathscr{r} = c(t-t_r) \Rightarrow \dfrac{\partial \mathscr{r}}{\partial t} = c\left(1 - \dfrac{\partial t_r}{\partial t}\right)$, $\boldsymbol{\mathscr{r}} = \mathbf{r} - \mathbf{w}(t_r) \Rightarrow \dfrac{\partial \boldsymbol{\mathscr{r}}}{\partial t} = -\mathbf{v}\dfrac{\partial t_r}{\partial t}$ (as above), and

$$\frac{\partial \mathbf{v}}{\partial t} = \frac{\partial \mathbf{v}}{\partial t_r}\frac{\partial t_r}{\partial t} = \mathbf{a}\frac{\partial t_r}{\partial t}.$$

$$\begin{aligned}
&= \frac{q}{4\pi\epsilon_0 c(\boldsymbol{\mathscr{r}}\cdot\mathbf{u})^2}\left\{\mathbf{a}(\boldsymbol{\mathscr{r}}\cdot\mathbf{u})\frac{\partial t_r}{\partial t} - \mathbf{v}\left[c^2\left(1-\frac{\partial t_r}{\partial t}\right) + v^2\frac{\partial t_r}{\partial t} - \boldsymbol{\mathscr{r}}\cdot\mathbf{a}\frac{\partial t_r}{\partial t}\right]\right\}\\
&= \frac{q}{4\pi\epsilon_0 c(\boldsymbol{\mathscr{r}}\cdot\mathbf{u})^2}\left\{-c^2\mathbf{v} + \left[(\boldsymbol{\mathscr{r}}\cdot\mathbf{u})\mathbf{a} + (c^2 - v^2 + \boldsymbol{\mathscr{r}}\cdot\mathbf{a})\mathbf{v}\right]\frac{\partial t_r}{\partial t}\right\}\\
&= \frac{q}{4\pi\epsilon_0 c(\boldsymbol{\mathscr{r}}\cdot\mathbf{u})^2}\left\{-c^2\mathbf{v} + \left[(\boldsymbol{\mathscr{r}}\cdot\mathbf{u})\mathbf{a} + (c^2 - v^2 + \boldsymbol{\mathscr{r}}\cdot\mathbf{a})\mathbf{v}\right]\frac{c\mathscr{r}}{\boldsymbol{\mathscr{r}}\cdot\mathbf{u}}\right\}\\
&= \frac{q}{4\pi\epsilon_0 c(\boldsymbol{\mathscr{r}}\cdot\mathbf{u})^3}\left[-c^2\mathbf{v}(\boldsymbol{\mathscr{r}}\cdot\mathbf{u}) + c\mathscr{r}(\boldsymbol{\mathscr{r}}\cdot\mathbf{u})\mathbf{a} + c\mathscr{r}(c^2 - v^2 + \boldsymbol{\mathscr{r}}\cdot\mathbf{a})\mathbf{v}\right]\\
&= \frac{qc}{4\pi\epsilon_0}\frac{1}{(\mathscr{r}c - \boldsymbol{\mathscr{r}}\cdot\mathbf{v})^3}\left[(\mathscr{r}c - \boldsymbol{\mathscr{r}}\cdot\mathbf{v})\left(-\mathbf{v} + \frac{\mathscr{r}}{c}\mathbf{a}\right) + \frac{\mathscr{r}}{c}(c^2 - v^2 + \boldsymbol{\mathscr{r}}\cdot\mathbf{a})\mathbf{v}\right]. \quad \text{qed}
\end{aligned}$$

Problem 10.18

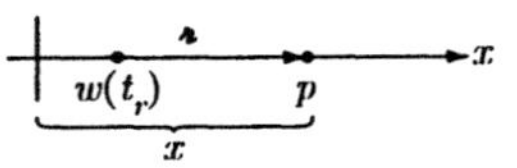

$\mathbf{E} = \dfrac{q}{4\pi\epsilon_0}\dfrac{\mathscr{r}}{(\boldsymbol{\mathscr{r}}\cdot\mathbf{u})^3}\left[(c^2 - v^2)\mathbf{u} + \boldsymbol{\mathscr{r}}\times(\mathbf{u}\times\mathbf{a})\right]$. Here $\mathbf{v} = v\,\hat{\mathbf{x}}$, $\mathbf{a} = a\,\hat{\mathbf{x}}$, and, for points to the *right*, $\hat{\boldsymbol{\mathscr{r}}} = \hat{\mathbf{x}}$. So $\mathbf{u} = (c-v)\,\hat{\mathbf{x}}$, $\mathbf{u}\times\mathbf{a} = 0$, and $\boldsymbol{\mathscr{r}}\cdot\mathbf{u} = \mathscr{r}(c-v)$.

$$\begin{aligned}
\mathbf{E} &= \frac{q}{4\pi\epsilon_0}\frac{\mathscr{r}}{\mathscr{r}^3(c-v)^3}(c^2 - v^2)(c-v)\,\hat{\mathbf{x}} = \frac{q}{4\pi\epsilon_0}\frac{1}{\mathscr{r}^2}\frac{(c+v)(c-v)^2}{(c-v)^3}\,\hat{\mathbf{x}} = \frac{q}{4\pi\epsilon_0}\frac{1}{\mathscr{r}^2}\left(\frac{c+v}{c-v}\right)\hat{\mathbf{x}};\\
\mathbf{B} &= \frac{1}{c}\hat{\boldsymbol{\mathscr{r}}}\times\mathbf{E} = 0. \quad \text{qed}
\end{aligned}$$

For field points to the *left*, $\hat{\boldsymbol{\mathscr{r}}} = -\hat{\mathbf{x}}$ and $\mathbf{u} = -(c+v)\,\hat{\mathbf{x}}$, so $\boldsymbol{\mathscr{r}}\cdot\mathbf{u} = \mathscr{r}(c+v)$, and

$$\mathbf{E} = -\frac{q}{4\pi\epsilon_0}\frac{\mathscr{r}}{\mathscr{r}^3(c+v)^3}(c^2 - v^2)(c+v)\,\hat{\mathbf{x}} = \boxed{\frac{-q}{4\pi\epsilon_0}\frac{1}{\mathscr{r}^2}\left(\frac{c-v}{c+v}\right)\hat{\mathbf{x}};\ \mathbf{B} = 0.}$$

Problem 10.19

(a) $\mathbf{E} = \dfrac{\lambda}{4\pi\epsilon_0}(1 - v^2/c^2)\displaystyle\int \frac{\hat{\mathbf{R}}}{R^2}\frac{dx}{\left[1 - (v/c)^2\sin^2\theta\right]^{3/2}}$.

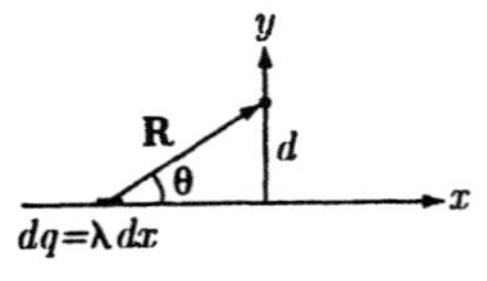

The horizontal components cancel; the vertical component of $\hat{\mathbf{R}}$ is $\sin\theta$ (see diagram). Here $d = R\sin\theta$, so $\dfrac{1}{R^2} = \dfrac{\sin^2\theta}{d^2}$; $-\dfrac{x}{d} = \cot\theta$, so $dx = -d(-\csc^2\theta)\,d\theta = \dfrac{d}{\sin^2\theta}\,d\theta$;

$\frac{1}{R^2}\,dx = \frac{d}{\sin^2\theta}\frac{\sin^2\theta}{d^2}\,d\theta = \frac{d\theta}{d}$. Thus

$$\begin{aligned}
\mathbf{E} &= \frac{\lambda}{4\pi\epsilon_0}(1-v^2/c^2)\left(\frac{\hat{\mathbf{y}}}{d}\right)\int_0^{\pi}\frac{\sin\theta}{\left[1-(v/c)^2\sin^2\theta\right]^{3/2}}\,d\theta. \quad \text{Let } z\equiv\cos\theta, \text{ so } \sin^2\theta = 1-z^2.\\
&= \frac{\lambda(1-v^2/c^2)\,\hat{\mathbf{y}}}{4\pi\epsilon_0 d}\int_{-1}^{1}\frac{1}{[1-(v/c)^2+(v/c)^2z^2]^{3/2}}\,dz\\
&= \frac{\lambda(1-v^2/c^2)\,\hat{\mathbf{y}}}{4\pi\epsilon_0 d}\left[\frac{1}{(v/c)^3}\frac{z}{(c^2/v^2-1)\sqrt{(c/v)^2-1+z^2}}\right]\Bigg|_{-1}^{+1}\\
&= \frac{\lambda(1-v^2/c^2)}{4\pi\epsilon_0 d}\frac{c}{v}\frac{1}{(1-c^2/v^2)}\frac{2}{\sqrt{(c/v)^2-1+1}}\,\hat{\mathbf{y}} = \boxed{\frac{1}{4\pi\epsilon_0}\frac{2\lambda}{d}\,\hat{\mathbf{y}}} \text{ (same as for a line charge at rest).}
\end{aligned}$$

(b) $\mathbf{B} = \frac{1}{c^2}(\mathbf{v}\times\mathbf{E})$ for each segment $dq = \lambda\,dx$. Since $\mathbf{v}$ is constant, it comes outside the integral, and the same formula holds for the *total* field:

$$\mathbf{B} = \frac{1}{c^2}(\mathbf{v}\times\mathbf{E}) = \frac{1}{c^2}v\frac{1}{4\pi\epsilon_0}\frac{2\lambda}{d}(\hat{\mathbf{x}}\times\hat{\mathbf{y}}) = \mu_0\epsilon_0 v\frac{1}{4\pi\epsilon_0}\frac{2\lambda}{d}\,\hat{\mathbf{z}} = \frac{\mu_0}{4\pi}\frac{2\lambda v}{d}\,\hat{\mathbf{z}}.$$

But $\lambda v = I$, so $\boxed{\mathbf{B} = \frac{\mu_0}{4\pi}\frac{2I}{d}\,\hat{\boldsymbol{\phi}}}$ (the same as we got in magnetostatics, Eq. 5.36 and Ex. 5.7).

Problem 10.20

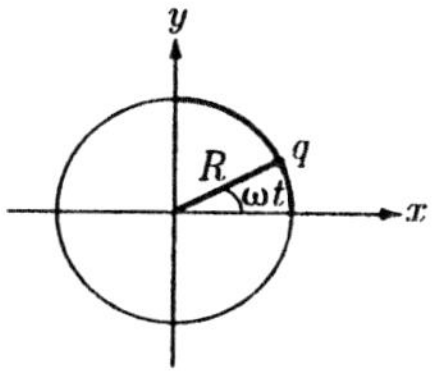

$\mathbf{w}(t) = R[\cos(\omega t)\,\hat{\mathbf{x}} + \sin(\omega t)\,\hat{\mathbf{y}}]$;
$\mathbf{v}(t) = R\omega[-\sin(\omega t)\,\hat{\mathbf{x}} + \cos(\omega t)\,\hat{\mathbf{y}}]$;
$\mathbf{a}(t) = -R\omega^2[\cos(\omega t)\,\hat{\mathbf{x}} + \sin(\omega t)\,\hat{\mathbf{y}}] = -\omega^2\mathbf{w}(t)$;
$\boldsymbol{\mathscr{r}} = -\mathbf{w}(t_r)$;
$\mathscr{r} = R$;
$t_r = t - R/c$;
$\hat{\boldsymbol{\mathscr{r}}} = -[\cos(\omega t_r)\,\hat{\mathbf{x}} + \sin(\omega t_r)\,\hat{\mathbf{y}}]$;

$$\begin{aligned}
\mathbf{u} &= c\hat{\boldsymbol{\mathscr{r}}} - \mathbf{v}(t_r) = -c[\cos(\omega t_r)\,\hat{\mathbf{x}} + \sin(\omega t_r)\,\hat{\mathbf{y}}] - \omega R[-\sin(\omega t_r)\,\hat{\mathbf{x}} + \cos(\omega t_r)\,\hat{\mathbf{y}}]\\
&= -\left\{[c\cos(\omega t_r) - \omega R\sin(\omega t_r)]\,\hat{\mathbf{x}} + [c\sin(\omega t_r) + \omega R\cos(\omega t_r)]\,\hat{\mathbf{y}}\right\};\\
\boldsymbol{\mathscr{r}}\times(\mathbf{u}\times\mathbf{a}) &= (\boldsymbol{\mathscr{r}}\cdot\mathbf{a})\mathbf{u} - (\boldsymbol{\mathscr{r}}\cdot\mathbf{u})\mathbf{a};\quad \boldsymbol{\mathscr{r}}\cdot\mathbf{a} = -\mathbf{w}\cdot(-\omega^2\mathbf{w}) = \omega^2R^2;\\
\boldsymbol{\mathscr{r}}\cdot\mathbf{u} &= R\left[c\cos^2(\omega t_r) - \omega R\sin(\omega t_r)\cos(\omega t_r) + c\sin^2(\omega t_r) + \omega R\sin(\omega t_r)\cos(\omega t_r)\right] = Rc;
\end{aligned}$$

$v^2 = (\omega R)^2$. So (Eq. 10.65):

$$\begin{aligned}
\mathbf{E} &= \frac{q}{4\pi\epsilon_0}\frac{R}{(Rc)^3}\left[\mathbf{u}\left(c^2-\omega^2R^2\right) + \mathbf{u}(\omega R)^2 - \mathbf{a}(Rc)\right] = \frac{q}{4\pi\epsilon_0}\frac{c\mathbf{u} - R\mathbf{a}}{(Rc)^2}\\
&= \frac{q}{4\pi\epsilon_0}\frac{1}{(Rc)^2}\left\{-[c^2\cos(\omega t_r) - \omega Rc\sin(\omega t_r)]\,\hat{\mathbf{x}} - [c^2\sin(\omega t_r) + \omega Rc\cos(\omega t_r)]\,\hat{\mathbf{y}}\right.\\
&\quad \left. + R^2\omega^2\cos(\omega t_r)\,\hat{\mathbf{x}} + R^2\omega^2\sin(\omega t_r)\,\hat{\mathbf{y}}\right\}\\
&= \boxed{\frac{q}{4\pi\epsilon_0}\frac{1}{(Rc)^2}\left\{\left[\left(\omega^2R^2 - c^2\right)\cos(\omega t_r) + \omega Rc\sin(\omega t_r)\right]\hat{\mathbf{x}} + \left[\left(\omega^2R^2-c^2\right)\sin(\omega t_r) - \omega Rc\cos(\omega t_r)\right]\hat{\mathbf{y}}\right\}.}
\end{aligned}$$

$$\begin{aligned}
\mathbf{B} &= \frac{1}{c}\hat{\boldsymbol{\imath}} \times \mathbf{E} = \frac{1}{c}\left(\hat{\imath}_x E_y - \hat{\imath}_y E_x\right)\hat{\mathbf{z}} \\
&= -\frac{1}{c}\frac{q}{4\pi\epsilon_0}\frac{1}{(Rc)^2}\left\{\cos(\omega t_r)\left[\left(\omega^2R^2 - c^2\right)\sin(\omega t_r) - \omega Rc\cos(\omega t_r)\right]\right. \\
&\quad \left. - \sin(\omega t_r)\left[\left(\omega^2R^2 - c^2\right)\cos(\omega t_r) + \omega Rc\sin(\omega t_r)\right]\right\}\hat{\mathbf{z}} \\
&= -\frac{q}{4\pi\epsilon_0}\frac{1}{R^2c^3}\left[-\omega Rc\cos^2(\omega t_r) - \omega Rc\sin^2(\omega t_r)\right]\hat{\mathbf{z}} = \frac{q}{4\pi\epsilon_0}\frac{1}{R^2c^3}\omega Rc\,\hat{\mathbf{z}} = \boxed{\frac{q}{4\pi\epsilon_0}\frac{\omega}{Rc^2}\hat{\mathbf{z}}.}
\end{aligned}$$

Notice that $\mathbf{B}$ is constant in time.

To obtain the field at the center of a circular *ring* of charge, let $q \to \lambda(2\pi R)$; for this ring to carry current I, we need $I = \lambda v = \lambda\omega R$, so $\lambda = I/\omega R$, and hence $q \to (I/\omega R)(2\pi R) = 2\pi I/\omega$. Thus $\mathbf{B} = \frac{2\pi I}{4\pi\epsilon_0}\frac{1}{Rc^2}\hat{\mathbf{z}}$, or, since $1/c^2 = \epsilon_0\mu_0$, $\boxed{\mathbf{B} = \frac{\mu_0 I}{2R}\hat{\mathbf{z}},}$ the same as Eq. 5.38, in the case $z = 0$.

Problem 10.21

$\lambda(\phi, t) = \lambda_0|\sin(\theta/2)|$, where $\theta = \phi - \omega t$. So the (retarded) scalar potential at the center is (Eq. 10.19)

$$\begin{aligned}
V(t) &= \frac{1}{4\pi\epsilon_0}\int \frac{\lambda}{\imath}\,dl' = \frac{1}{4\pi\epsilon_0}\int_0^{2\pi}\frac{\lambda_0\left|\sin[(\phi - \omega t_r)/2]\right|}{a}a\,d\phi \\
&= \frac{\lambda_0}{4\pi\epsilon_0}\int_0^{2\pi}\sin(\theta/2)\,d\theta = \frac{\lambda_0}{4\pi\epsilon_0}\left[-2\cos(\theta/2)\right]\Big|_0^{2\pi} \\
&= \frac{\lambda_0}{4\pi\epsilon_0}\left[2 - (-2)\right] = \boxed{\frac{\lambda_0}{\pi\epsilon_0}.}
\end{aligned}$$

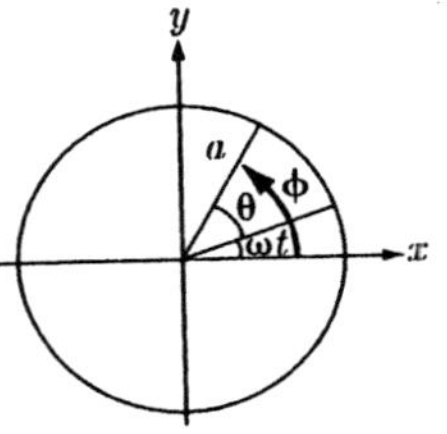

(Note: at fixed t_r, $d\phi = d\theta$, and it goes through one full cycle of ϕ *or* θ.)

Meanwhile $\mathbf{I}(\phi, t) = \lambda\mathbf{v} = \lambda_0\omega a\left|\sin[(\phi - \omega t)/2]\right|\hat{\boldsymbol{\phi}}$. From Eq. 10.19 (again)

$$\mathbf{A}(t) = \frac{\mu_0}{4\pi}\int\frac{\mathbf{I}}{\imath}\,dl' = \frac{\mu_0}{4\pi}\int_0^{2\pi}\frac{\lambda_0\omega a\left|\sin[(\phi - \omega t_r)/2]\right|\hat{\boldsymbol{\phi}}}{a}a\,d\phi.$$

But $t_r = t - a/c$ is again constant, for the ϕ integration, and $\hat{\boldsymbol{\phi}} = -\sin\phi\,\hat{\mathbf{x}} + \cos\phi\,\hat{\mathbf{y}}$.

$$= \frac{\mu_0\lambda_0\omega a}{4\pi}\int_0^{2\pi}\left|\sin[(\phi - \omega t_r)/2]\right|\left(-\sin\phi\,\hat{\mathbf{x}} + \cos\phi\,\hat{\mathbf{y}}\right)d\phi.$$

Again, switch variables to $\theta = \phi - \omega t_r$, and integrate from $\theta = 0$ to $\theta = 2\pi$ (so we don't have to worry about the absolute value).

$$= \frac{\mu_0\lambda_0\omega a}{4\pi}\int_0^{2\pi}\sin(\theta/2)\left[-\sin(\theta + \omega t_r)\,\hat{\mathbf{x}} + \cos(\theta + \omega t_r)\,\hat{\mathbf{y}}\right]d\theta.$$

Now

$$
\begin{aligned}
\int_0^{2\pi} \sin(\theta/2)\sin(\theta+\omega t_r)\,d\theta &= \frac{1}{2}\int_0^{2\pi}\left[\cos(\theta/2+\omega t_r)-\cos(3\theta/2+\omega t_r)\right]d\theta \\
&= \frac{1}{2}\left[2\sin(\theta/2+\omega t_r)-\frac{2}{3}\sin(3\theta/2+\omega t_r)\right]\Bigg|_0^{2\pi} \\
&= \sin(\pi+\omega t_r)-\sin(\omega t_r)-\frac{1}{3}\sin(3\pi+\omega t_r)+\frac{1}{3}\sin(\omega t_r) \\
&= -2\sin(\omega t_r)+\frac{2}{3}\sin(\omega t_r) = -\frac{4}{3}\sin(\omega t_r). \\
\int_0^{2\pi} \sin(\theta/2)\cos(\theta+\omega t_r)\,d\theta &= \frac{1}{2}\int_0^{2\pi}\left[-\sin(\theta/2+\omega t_r)+\sin(3\theta/2+\omega t_r)\right]d\theta \\
&= \frac{1}{2}\left[2\cos(\theta/2+\omega t_r)-\frac{2}{3}\cos(3\theta/2+\omega t_r)\right]\Bigg|_0^{2\pi} \\
&= \cos(\pi+\omega t_r)-\cos(\omega t_r)-\frac{1}{3}\cos(3\pi+\omega t_r)+\frac{1}{3}\cos(\omega t_r) \\
&= -2\cos(\omega t_r)+\frac{2}{3}\cos(\omega t_r) = -\frac{4}{3}\cos(\omega t_r).
\end{aligned}
$$

So

$$
\mathbf{A}(t) = \frac{\mu_0\lambda_0\omega a}{4\pi}\left(\frac{4}{3}\right)\left[\sin(\omega t_r)\,\hat{\mathbf{x}}-\cos(\omega t_r)\,\hat{\mathbf{y}}\right] = \boxed{\frac{\mu_0\lambda_0\omega a}{3\pi}\left\{\sin[\omega(t-a/c)]\,\hat{\mathbf{x}}-\cos[\omega(t-a/c)]\,\hat{\mathbf{y}}\right\}.}
$$

Problem 10.22

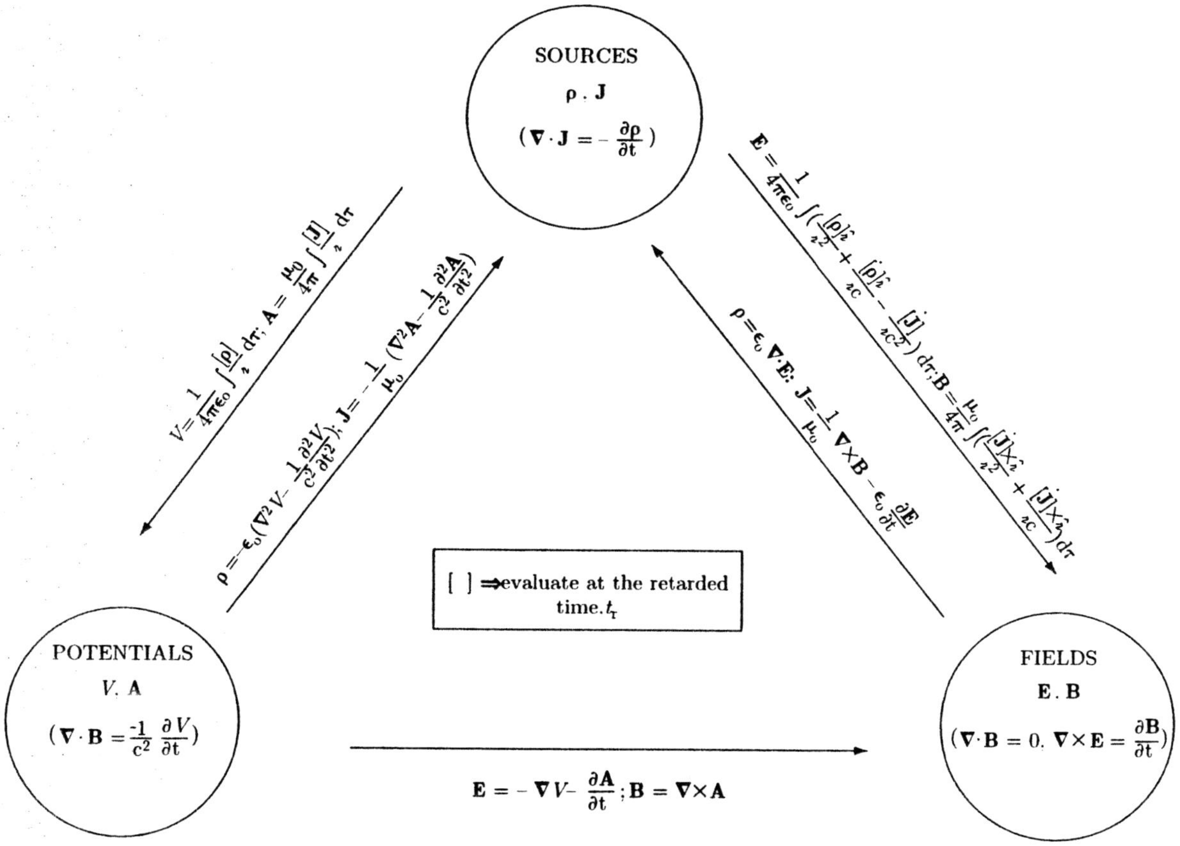

Problem 10.23

Using Product Rule #5, Eq. 10.43 $\Rightarrow$

$$\begin{aligned}
\nabla\cdot\mathbf{A} &= \frac{\mu_0}{4\pi}qc\mathbf{v}\cdot\nabla\left[(c^2t-\mathbf{r}\cdot\mathbf{v})^2+(c^2-v^2)(r^2-c^2t^2)\right]^{-1/2}\\
&= \frac{\mu_0 qc}{4\pi}\mathbf{v}\cdot\left\{-\frac{1}{2}\left[(c^2t-\mathbf{r}\cdot\mathbf{v})^2+(c^2-v^2)(r^2-c^2t^2)\right]^{-3/2}\nabla\left[(c^2t-\mathbf{r}\cdot\mathbf{v})^2+(c^2-v^2)(r^2-c^2t^2)\right]\right\}\\
&= -\frac{\mu_0 qc}{8\pi}\left[(c^2t-\mathbf{r}\cdot\mathbf{v})^2+(c^2-v^2)(r^2-c^2t^2)\right]^{-3/2}\mathbf{v}\cdot\left\{-2(c^2t-\mathbf{r}\cdot\mathbf{v})\nabla(\mathbf{r}\cdot\mathbf{v})+(c^2-v^2)\nabla(r^2)\right\}.
\end{aligned}$$

Product Rule #4 $\Rightarrow$

$$\begin{aligned}
\nabla(\mathbf{r}\cdot\mathbf{v}) &= \mathbf{v}\times(\nabla\times\mathbf{r})+(\mathbf{v}\cdot\nabla)\mathbf{r},\ \text{but }\nabla\times\mathbf{r}=0,\\
(\mathbf{v}\cdot\nabla)\mathbf{r} &= \left(v_x\frac{\partial}{\partial x}+v_y\frac{\partial}{\partial y}+v_z\frac{\partial}{\partial z}\right)(x\,\hat{\mathbf{x}}+y\,\hat{\mathbf{y}}+z\,\hat{\mathbf{z}})=v_x\,\hat{\mathbf{x}}+v_y\,\hat{\mathbf{y}}+v_z\,\hat{\mathbf{z}}=\mathbf{v},\ \text{and}\\
\nabla(r^2) &= \nabla(\mathbf{r}\cdot\mathbf{r})=2\mathbf{r}\times(\nabla\times\mathbf{r})+2(\mathbf{r}\cdot\nabla)\mathbf{r}=2\mathbf{r}.\ \text{So}
\end{aligned}$$

$$\begin{aligned}
\nabla\cdot\mathbf{A} &= -\frac{\mu_0 qc}{8\pi}\left[(c^2t-\mathbf{r}\cdot\mathbf{v})^2+(c^2-v^2)(r^2-c^2t^2)\right]^{-3/2}\mathbf{v}\cdot\left[-2(c^2t-\mathbf{r}\cdot\mathbf{v})\mathbf{v}+(c^2-v^2)2\mathbf{r}\right]\\
&= \frac{\mu_0 qc}{4\pi}\left[(c^2t-\mathbf{r}\cdot\mathbf{v})^2+(c^2-v^2)(r^2-c^2t^2)\right]^{-3/2}\left\{(c^2t-\mathbf{r}\cdot\mathbf{v})v^2-(c^2-v^2)(\mathbf{r}\cdot\mathbf{v})\right\}.
\end{aligned}$$

But the term in curly brackets is : $c^2tv^2-v^2(\mathbf{r}\cdot\mathbf{v})-c^2(\mathbf{r}\cdot\mathbf{v})+v^2(\mathbf{r}\cdot\mathbf{v})=c^2(v^2t-\mathbf{r}\cdot\mathbf{v})$.

$$= \frac{\mu_0 qc^3}{4\pi}\frac{(v^2t-\mathbf{r}\cdot\mathbf{v})}{\left[(c^2t-\mathbf{r}\cdot\mathbf{v})^2+(c^2-v^2)(r^2-c^2t^2)\right]^{3/2}}.$$

Meanwhile, from Eq. 10.42,

$$\begin{aligned}
-\mu_0\epsilon_0\frac{\partial V}{\partial t} &= -\mu_0\epsilon_0\frac{1}{4\pi\epsilon_0}qc\left(-\frac{1}{2}\right)\left[(c^2t-\mathbf{r}\cdot\mathbf{v})^2+(c^2-v^2)(r^2-c^2t^2)\right]^{-3/2}\times\\
&\quad\frac{\partial}{\partial t}\left[(c^2t-\mathbf{r}\cdot\mathbf{v})^2+(c^2-v^2)(r^2-c^2t^2)\right]\\
&= -\frac{\mu_0 qc}{8\pi}\left[(c^2t-\mathbf{r}\cdot\mathbf{v})^2+(c^2-v^2)(r^2-c^2t^2)\right]^{-3/2}\left[2(c^2t-\mathbf{r}\cdot\mathbf{v})c^2+(c^2-v^2)(-2c^2t)\right]\\
&= -\frac{\mu_0 qc^3}{4\pi}\frac{(c^2t-\mathbf{r}\cdot\mathbf{v}-c^2t+v^2t)}{\left[(c^2t-\mathbf{r}\cdot\mathbf{v})^2+(c^2-v^2)(r^2-c^2t^2)\right]^{3/2}}=\nabla\cdot\mathbf{A}.\ \checkmark
\end{aligned}$$

Problem 10.24

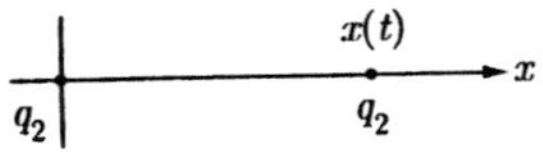

(a) $\boxed{\mathbf{F}_2=\frac{q_1q_2}{4\pi\epsilon_0}\frac{1}{(b^2+c^2t^2)}\hat{\mathbf{x}}.}$

(This is just Coulomb's law, since q_1 is at rest.)

(b) $I_2=\frac{q_1q_2}{4\pi\epsilon_0}\int_{-\infty}^{\infty}\frac{1}{(b^2+c^2t^2)}dt=\frac{q_1q_2}{4\pi\epsilon_0}\left[\frac{1}{bc}\tan^{-1}(ct/b)\right]\Big|_{-\infty}^{\infty}=\frac{q_1q_2}{4\pi\epsilon_0 bc}\left[\tan^{-1}(\infty)-\tan^{-1}(-\infty)\right]$

$=\frac{q_1q_2}{4\pi\epsilon_0 bc}\left[\frac{\pi}{2}-\left(-\frac{\pi}{2}\right)\right]=\boxed{\frac{q_1q_2}{4\pi\epsilon_0}\frac{\pi}{bc}.}$

(c) From Prob. 10.18, $\mathbf{E} = -\frac{q_2}{4\pi\epsilon_0}\frac{1}{x^2}\left(\frac{c-v}{c+v}\right)\hat{\mathbf{x}}$. Here x and v are to be evaluated at the retarded time t_r, which is given by $c(t-t_r) = x(t_r) = \sqrt{b^2+c^2t_r^2} \Rightarrow c^2t^2 - 2ctt_r + c^2t_r^2 = b^2 + c^2t_r^2 \Rightarrow t_r = \frac{c^2t^2-b^2}{2c^2t}$. *Note:* As we found in Prob. 10.15, q_2 first "comes into view" (for q_1) at time $t=0$. Before that it can exert no force on q_1, and there *is* no retarded time. From the graph of t_r versus t we see that t_r ranges all the way from $-\infty$ to ∞ while $t > 0$.

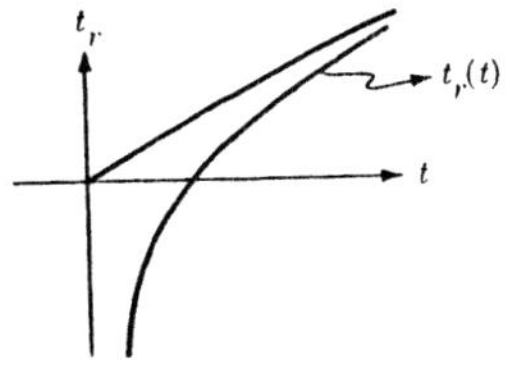

$$x(t_r) = c(t-t_r) = \frac{2c^2t^2 - c^2t^2 + b^2}{2ct} = \frac{b^2+c^2t^2}{2ct} \text{ (for } t>0). \quad v(t) = \frac{1}{2}\frac{2c^2t}{\sqrt{b^2+c^2t^2}} = \frac{c^2t}{x}, \text{ so}$$

$$v(t_r) = \left(\frac{c^2t^2-b^2}{2t}\right)\left(\frac{2ct}{b^2+c^2t^2}\right) = c\left(\frac{c^2t^2-b^2}{c^2t^2+b^2}\right) \text{ (for } t>0). \quad \text{Therefore}$$

$$\frac{c-v}{c+v} = \frac{(c^2t^2+b^2)-(c^2t^2-b^2)}{(c^2t^2+b^2)+(c^2t^2-b^2)} = \frac{2b^2}{2c^2t^2} = \frac{b^2}{c^2t^2} \text{ (for } t>0). \quad \mathbf{E} = -\frac{q_2}{4\pi\epsilon_0}\frac{4c^2t^2}{(b^2+c^2t^2)^2}\frac{b^2}{c^2t^2}\hat{\mathbf{x}} \Rightarrow$$

$$\boxed{\mathbf{F}_1 = \begin{cases} 0, & t<0; \\ -\frac{q_1q_2}{4\pi\epsilon_0}\frac{4b^2}{(b^2+c^2t^2)^2}\hat{\mathbf{x}}, & t>0. \end{cases}}$$

(d) $I_1 = -\frac{q_1q_2}{4\pi\epsilon_0}4b^2\int_0^\infty \frac{1}{(b^2+c^2t^2)^2}\,dt$. The integral is

$$\frac{1}{c^4}\int_0^\infty \frac{1}{[(b/c)^2+t^2]^2}\,dt = \frac{1}{c^4}\left(\frac{c^2}{2b^2}\right)\left[\frac{t}{(b/c)^2+t^2}\Big|_0^\infty + \int_0^\infty \frac{1}{[(b/c)^2+t^2)]}\,dt\right] = \frac{1}{2c^2b^2}\left(\frac{\pi c}{2b}\right) = \frac{\pi}{4cb^3}.$$

So $\boxed{I_1 = -\frac{q_1q_2}{4\pi\epsilon_0}\frac{\pi}{bc}.}$

(e) $\mathbf{F}_1 \neq -\mathbf{F}_2$, so Newton's third law is *not* obeyed. On the other hand, $I_1 = -I_2$ in this instance, which suggests that the *net* momentum delivered from (1) to (2) is equal and opposite to the net momentum delivered from (2) to (1), and hence that the total mechanical momentum is conserved. (In general, the fields might carry off some momentum, leaving the mechanical momentum altered; but that doesn't happen in the present case.)

Problem 10.25

$\mathbf{S} = \frac{1}{\mu_0}(\mathbf{E}\times\mathbf{B})$; $\mathbf{B} = \frac{1}{c^2}(\mathbf{v}\times\mathbf{E})$ (Eq. 10.69).

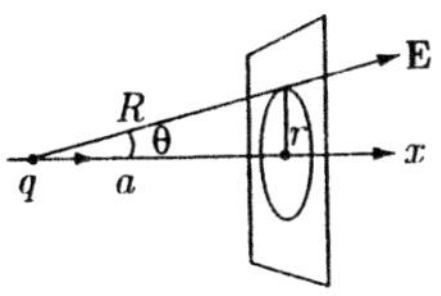

So $\mathbf{S} = \frac{1}{\mu_0 c^2}[\mathbf{E}\times(\mathbf{v}\times\mathbf{E})] = \epsilon_0\left[E^2\mathbf{v} - (\mathbf{v}\cdot\mathbf{E})\mathbf{E}\right]$.

The power crossing the plane is $P = \int \mathbf{S}\cdot d\mathbf{a}$, and $d\mathbf{a} = 2\pi r\, dr\, \hat{\mathbf{x}}$ (see diagram). So

$$\begin{aligned}
P &= \epsilon_0 \int (E^2 v - E_x^2 v) 2\pi r\, dr; \; E_x = E\cos\theta, \text{ so } E^2 - E_x^2 = E^2 \sin^2\theta. \\
&= 2\pi\epsilon_0 v \int E^2 \sin^2\theta\, r\, dr. \text{ From Eq. 10.68, } \mathbf{E} = \frac{q}{4\pi\epsilon_0}\frac{1}{\gamma^2}\frac{\hat{\mathbf{R}}}{R^2\left[1-(v/c)^2\sin^2\theta\right]^{3/2}} \text{ where } \gamma \equiv \frac{1}{\sqrt{1-v^2/c^2}}. \\
&= 2\pi\epsilon_0 v \left(\frac{q}{4\pi\epsilon_0}\right)^2 \frac{1}{\gamma^2}\int_0^\infty \frac{r\sin^2\theta}{R^4\left[1-(v/c)^2\sin^2\theta\right]^3}\, dr. \quad \text{Now } r = a\tan\theta \Rightarrow dr = a\frac{1}{\cos^2\theta}\, d\theta; \; \frac{1}{R} = \frac{\cos\theta}{a}. \\
&= \frac{v}{2\gamma^4}\frac{q^2}{4\pi\epsilon_0}\frac{1}{a^2}\int_0^{\pi/2}\frac{\sin^3\theta\cos\theta}{\left[1-(v/c)^2\sin^2\theta\right]^3}\, d\theta. \quad \text{Let } u \equiv \sin^2\theta, \text{ so } du = 2\sin\theta\cos\theta\, d\theta. \\
&= \frac{vq^2}{16\pi\epsilon_0 a^2\gamma^4}\int_0^1 \frac{u}{\left[1-(v/c)^2 u\right]^3}\, du = \frac{vq^2}{16\pi\epsilon_0 a^2\gamma^4}\left(\frac{\gamma^4}{2}\right) = \boxed{\frac{vq^2}{32\pi\epsilon_0 a^2}}.
\end{aligned}$$

Problem 10.26

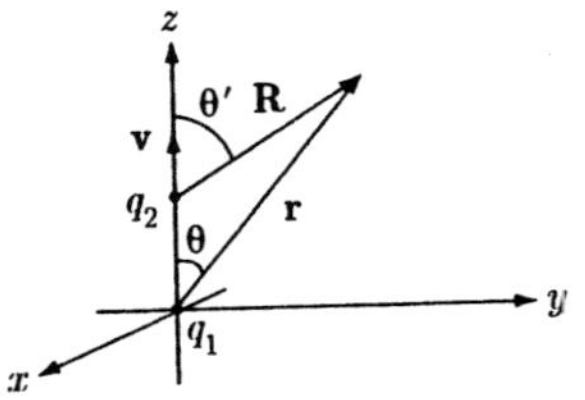

(a) $\boxed{\mathbf{F}_{12}(t) = \frac{1}{4\pi\epsilon_0}\frac{q_1 q_2}{(vt)^2}\hat{\mathbf{z}}.}$

(b) From Eq. 10.68, with $\theta = 180°$, $R = vt$, and $\hat{\mathbf{R}} = -\hat{\mathbf{z}}$:

$$\boxed{\mathbf{F}_{21}(t) = -\frac{1}{4\pi\epsilon_0}\frac{q_1 q_2(1-v^2/c^2)}{(vt)^2}\hat{\mathbf{z}}.}$$

$\boxed{\text{No,}}$ Newton's third law does *not* hold: $\mathbf{F}_{12} \neq \mathbf{F}_{21}$, because of the extra factor $(1-v^2/c^2)$.

(c) From Eq. 8.29, $\mathbf{p} = \epsilon_0\int(\mathbf{E}\times\mathbf{B})\, d\tau$. Here $\mathbf{E} = \mathbf{E}_1 + \mathbf{E}_2$, whereas $\mathbf{B} = \mathbf{B}_2$, so $\mathbf{E}\times\mathbf{B} = (\mathbf{E}_1\times\mathbf{B}_2) + (\mathbf{E}_2\times\mathbf{B}_2)$. But the latter, when integrated over all space, is independent of time. We want only the time-dependent part: $\mathbf{p}(t) = \epsilon_0\int(\mathbf{E}_1\times\mathbf{B}_2)\, d\tau$. Now $\mathbf{E}_1 = \frac{1}{4\pi\epsilon_0}\frac{q_1}{r^2}\hat{\mathbf{r}}$, while, from Eq. 10.69, $\mathbf{B}_2 = \frac{1}{c^2}(\mathbf{v}\times\mathbf{E}_2)$, and (Eq. 10.68)

$\mathbf{E}_2 = \frac{q_2}{4\pi\epsilon_0}\frac{(1-v^2/c^2)}{(1-v^2\sin^2\theta'/c^2)^{3/2}}\frac{\hat{\mathbf{R}}}{R^2}$. But $\mathbf{R} = \mathbf{r} - \mathbf{v}t$; $R^2 = r^2 + v^2t^2 - 2rvt\cos\theta$; $\sin\theta' = \frac{r\sin\theta}{R}$. So

$\mathbf{E}_2 = \frac{q_2}{4\pi\epsilon_0}\frac{(1-v^2/c^2)}{\left[1-(vr\sin\theta/Rc)^2\right]^{3/2}}\frac{(\mathbf{r}-\mathbf{v}t)}{R^3}$. Finally, noting that $\mathbf{v}\times(\mathbf{r}-\mathbf{v}t) = \mathbf{v}\times\mathbf{r} = vr\sin\theta\,\hat{\boldsymbol{\phi}}$, we get

$\mathbf{B}_2 = \frac{q_2(1-v^2/c^2)}{4\pi\epsilon_0 c^2}\frac{vr\sin\theta}{\left[R^2-(vr\sin\theta/c)^2\right]^{3/2}}\hat{\boldsymbol{\phi}}$. So $\mathbf{p}(t) = \epsilon_0\frac{q_1}{4\pi\epsilon_0}\frac{q_2(1-v^2/c^2)v}{4\pi\epsilon_0 c^2}\int\frac{1}{r^2}\frac{r\sin\theta\,(\hat{\mathbf{r}}\times\hat{\boldsymbol{\phi}})}{\left[R^2-(vr\sin\theta/c)^2\right]^{3/2}}$.

But $\hat{\mathbf{r}}\times\hat{\boldsymbol{\phi}} = -\hat{\boldsymbol{\theta}} = -(\cos\theta\cos\phi\,\hat{\mathbf{x}} + \cos\theta\sin\phi\,\hat{\mathbf{y}} - \sin\theta\,\hat{\mathbf{z}})$, and the x and y components integrate to zero, so:

$$\begin{aligned}
\mathbf{p}(t) &= \frac{q_1 q_2 v(1-v^2/c^2)\,\hat{\mathbf{z}}}{(4\pi c)^2\epsilon_0}\int\frac{\sin^2\theta}{r\left[r^2+(vt)^2-2rvt\cos\theta-(vr\sin\theta/c)^2\right]^{3/2}}r^2\sin\theta\, dr\, d\theta\, d\phi \\
&= \frac{q_1 q_2 v(1-v^2/c^2)\,\hat{\mathbf{z}}}{8\pi c^2\epsilon_0}\int\frac{r\sin^3\theta}{\left[r^2+(vt)^2-2rvt\cos\theta-(vr\sin\theta/c)^2\right]^{3/2}}\, dr\, d\theta.
\end{aligned}$$

I'll do the r integral first. According to the CRC Tables,

$$\int_0^\infty \frac{x}{(a+bx+cx^2)^{3/2}}\,dx = -\frac{2(bx+2a)}{(4ac-b^2)\sqrt{a+bx+cx^2}}\bigg|_0^\infty = -\frac{2}{4ac-b^2}\left[\frac{b}{\sqrt{c}}-\frac{2a}{\sqrt{a}}\right]$$
$$= -\frac{2}{\sqrt{c}(4ac-b^2)}\left(b-2\sqrt{ac}\right) = \frac{2}{\sqrt{c}}\frac{(2\sqrt{ac}-b)}{(2\sqrt{ac}-b)(2\sqrt{ac}+b)} = \frac{2}{\sqrt{c}}\left(2\sqrt{ac}+b\right).$$

In this case $x=r$, $a=(vt)^2$, $b=-2vt\cos\theta$, and $c=1-(v/c)^2\sin^2\theta$. So the r integral is

$$\frac{2}{\sqrt{1-(v/c)^2\sin^2\theta}\left[2vt\sqrt{1-(v/c)^2\sin^2\theta}-2vt\cos\theta\right]} = \frac{1}{vt\sqrt{1-(v/c)^2\sin^2\theta}\left[\sqrt{1-(v/c)^2\sin^2\theta}-\cos\theta\right]}$$
$$= \frac{\left[\sqrt{1-(v/c)^2\sin^2\theta}+\cos\theta\right]}{vt\sqrt{1-(v/c)^2\sin^2\theta}\left[1-(v/c)^2\sin^2\theta-\cos^2\theta\right]} = \frac{1}{vt\sin^2\theta(1-v^2/c^2)}\left[1+\frac{\cos\theta}{\sqrt{1-(v/c)^2\sin^2\theta}}\right].$$

So

$$\mathbf{p}(t) = \frac{q_1q_2v(1-v^2/c^2)\,\hat{\mathbf{z}}}{8\pi c^2\epsilon_0}\frac{1}{vt(1-v^2/c^2)}\int_0^\pi \frac{1}{\sin^2\theta}\left[1+\frac{\cos\theta}{\sqrt{1-(v/c)^2\sin^2\theta}}\right]\sin^3\theta\,d\theta$$
$$= \frac{q_1q_2\,\hat{\mathbf{z}}}{8\pi c^2\epsilon_0 t}\left\{\int_0^\pi \sin\theta\,d\theta + \frac{c}{v}\int_0^\pi \frac{\cos\theta\sin\theta}{\sqrt{(c/v)^2-\sin^2\theta}}\,d\theta\right\}.$$

But $\int_0^\pi \sin\theta\,d\theta = 2$. In the second integral let $u\equiv\cos\theta$, so $du=-\sin\theta\,d\theta$:

$$\int_0^\pi \frac{\cos\theta\sin\theta}{\sqrt{(c/v)^2-\sin^2\theta}}\,d\theta = \int_{-1}^1 \frac{u}{\sqrt{(c/v)^2-1+u^2}}\,du = 0$$ (the integrand is odd, and the interval is even).

Conclusion: $\boxed{\mathbf{p}(t) = \frac{\mu_0 q_1 q_2}{4\pi t}\,\hat{\mathbf{z}}}$ (plus a term constant in time).

(d)

$$\mathbf{F}_{12}+\mathbf{F}_{21} = \frac{1}{4\pi\epsilon_0}\frac{q_1q_2}{v^2t^2}\,\hat{\mathbf{z}} - \frac{1}{4\pi\epsilon_0}\frac{q_1q_2(1-v^2/c^2)}{v^2t^2}\,\hat{\mathbf{z}} = \frac{q_1q_2}{4\pi\epsilon_0 v^2t^2}\left(1-1+\frac{v^2}{c^2}\right)\hat{\mathbf{z}} = \frac{q_1q_2}{4\pi\epsilon_0 c^2t^2}\,\hat{\mathbf{z}} = \frac{\mu_0 q_1q_2}{4\pi t^2}\,\hat{\mathbf{z}}.$$
$$-\frac{d\mathbf{p}}{dt} = \frac{\mu_0 q_1q_2}{4\pi t^2}\,\hat{\mathbf{z}} = \mathbf{F}_{12}+\mathbf{F}_{21}. \quad \text{qed}$$

Since q_1 is at rest, and q_2 is moving at constant velocity, there must be another force ($\mathbf{F}_{\text{mech}}$) acting on them, to balance $\mathbf{F}_{12}+\mathbf{F}_{21}$; what we have found is that $\mathbf{F}_{\text{mech}} = d\mathbf{p}_{\text{em}}/dt$, which means that the impulse imparted to the system by the external force ends up as momentum in the fields. [For further discussion of this problem see J. J. G. Scanio, *Am. J. Phys.* **43**, 258 (1975).]

Chapter 11

Radiation

Problem 11.1

From Eq. 11.17, $\mathbf{A} = -\frac{\mu_0 p_0 \omega}{4\pi}\frac{1}{r}\sin[\omega(t-r/c)](\cos\theta\,\hat{\mathbf{r}} - \sin\theta\,\hat{\boldsymbol{\theta}})$, so

$$\begin{aligned}
\nabla\cdot\mathbf{A} &= -\frac{\mu_0 p_0\omega}{4\pi}\left\{\frac{1}{r^2}\frac{\partial}{\partial r}\left[r^2\frac{1}{r}\sin[\omega(t-r/c)]\cos\theta\right] + \frac{1}{r\sin\theta}\frac{\partial}{\partial\theta}\left[-\sin^2\theta\frac{1}{r}\sin[\omega(t-r/c)]\right]\right\}\\
&= -\frac{\mu_0 p_0\omega}{4\pi}\left\{\frac{1}{r^2}\left(\sin[\omega(t-r/c)] - \frac{\omega r}{c}\cos[\omega(t-r/c)]\right)\cos\theta - \frac{2\sin\theta\cos\theta}{r^2\sin\theta}\sin[\omega(t-r/c)]\right\}\\
&= \mu_0\epsilon_0\left\{\frac{p_0\omega}{4\pi\epsilon_0}\left(\frac{1}{r^2}\sin[\omega(t-r/c)] + \frac{\omega}{rc}\cos[\omega(t-r/c)]\right)\cos\theta\right\}.
\end{aligned}$$

Meanwhile, from Eq. 11.12,

$$\begin{aligned}
\frac{\partial V}{\partial t} &= \frac{p_0\cos\theta}{4\pi\epsilon_0 r}\left\{-\frac{\omega^2}{c}\cos[\omega(t-r/c)] - \frac{\omega}{r}\sin[\omega(t-r/c)]\right\}\\
&= -\frac{p_0\omega}{4\pi\epsilon_0}\left\{\frac{1}{r^2}\sin[\omega(t-r/c)] + \frac{\omega}{rc}\cos[\omega(t-r/c)]\right\}\cos\theta.\quad \text{So } \nabla\cdot\mathbf{A} = -\mu_0\epsilon_0\frac{\partial V}{\partial t}.\quad \text{qed}
\end{aligned}$$

Problem 11.2

Eq. 11.14: $\boxed{V(\mathbf{r},t) = -\frac{\omega}{4\pi\epsilon_0 c}\frac{\mathbf{p}_0\cdot\hat{\mathbf{r}}}{r}\sin[\omega(t-r/c)].}$ Eq. 11.17: $\boxed{\mathbf{A}(\mathbf{r},t) = -\frac{\mu_0\omega}{4\pi}\frac{\mathbf{p}_0}{r}\sin[\omega(t-r/c)].}$

Now $\mathbf{p}_0\times\hat{\mathbf{r}} = p_0\sin\theta\,\hat{\boldsymbol{\phi}}$ and $\hat{\mathbf{r}}\times(\mathbf{p}_0\times\hat{\mathbf{r}}) = p_0\sin\theta(\hat{\mathbf{r}}\times\hat{\boldsymbol{\phi}}) = -p_0\sin\theta\,\hat{\boldsymbol{\theta}}$, so

Eq. 11.18: $\boxed{\mathbf{E}(\mathbf{r},t) = \frac{\mu_0\omega^2}{4\pi}\frac{\hat{\mathbf{r}}\times(\mathbf{p}_0\times\hat{\mathbf{r}})}{r}\cos[\omega(t-r/c)].}$ Eq. 11.19: $\boxed{\mathbf{B}(\mathbf{r},t) = -\frac{\mu_0\omega^2}{4\pi c}\frac{(\mathbf{p}_0\times\hat{\mathbf{r}})}{r}\cos[\omega(t-r/c)].}$

Eq. 11.21: $\boxed{\langle\mathbf{S}\rangle = \frac{\mu_0\omega^4}{32\pi^2 c}\frac{(\mathbf{p}_0\times\hat{\mathbf{r}})^2}{r^2}\hat{\mathbf{r}}.}$

Problem 11.3

$P = I^2R = q_0^2\omega^2\sin^2(\omega t)R$ (Eq. 11.15) $\Rightarrow \langle P\rangle = \frac{1}{2}q_0^2\omega^2 R$. Equate this to Eq. 11.22:

$$\frac{1}{2}q_0^2\omega^2 R = \frac{\mu_0 q_0^2 d^2\omega^4}{12\pi c} \Rightarrow \boxed{R = \frac{\mu_0 d^2\omega^2}{6\pi c};}\ \text{or, since } \omega = \frac{2\pi c}{\lambda},$$

$$R = \frac{\mu_0 d^2}{6\pi c}\frac{4\pi^2c^2}{\lambda^2} = \frac{2}{3}\pi\mu_0 c\left(\frac{d}{\lambda}\right)^2 = \frac{2}{3}\pi(4\pi\times10^{-7})(3\times10^8)\left(\frac{d}{\lambda}\right)^2 = 80\pi^2\left(\frac{d}{\lambda}\right)^2\Omega = \boxed{789.6(d/\lambda)^2\,\Omega.}$$

For the wires in an ordinary radio, with $d = 5 \times 10^{-2}$ m and (say) $\lambda = 10^3$ m, $R = 790(5 \times 10^{-5})^2 = 2 \times 10^{-6}\,\Omega$, which is negligible compared to the Ohmic resistance.

Problem 11.4

By the superposition principle, we can *add* the potentials of the two dipoles. Let's first express V (Eq. 11.14) in Cartesian coordinates: $V(x, y, z, t) = -\frac{p_0\omega}{4\pi\epsilon_0 c}\left(\frac{z}{x^2+y^2+z^2}\right)\sin[\omega(t-r/c)]$. That's for an oscillating dipole along the z axis. For one along x or y, we just change z to x or y. In the present case, $\mathbf{p} = p_0[\cos(\omega t)\,\hat{\mathbf{x}} + \cos(\omega t - \pi/2)\,\hat{\mathbf{y}}]$, so the one along y is delayed by a phase angle $\pi/2$: $\sin[\omega(t - r/c)] \to \sin[\omega(t - r/c) - \pi/2] = -\cos[\omega(t - r/c)]$ (just let $\omega t \to \omega t - \pi/2$). Thus

$$V = -\frac{p_0\omega}{4\pi\epsilon_0 c}\left\{\frac{x}{x^2+y^2+z^2}\sin[\omega(t-r/c)] - \frac{y}{x^2+y^2+z^2}\cos[\omega(t-r/c)]\right\}$$
$$= \boxed{-\frac{p_0\omega}{4\pi\epsilon_0 c}\frac{\sin\theta}{r}\{\cos\phi\sin[\omega(t-r/c)] - \sin\phi\cos[\omega(t-r/c)]\}.}\quad \text{Similarly,}$$
$$\mathbf{A} = \boxed{-\frac{\mu_0 p_0\omega}{4\pi r}\{\sin[\omega(t-r/c)]\,\hat{\mathbf{x}} - \cos[\omega(t-r/c)]\,\hat{\mathbf{y}}\}.}$$

We *could* get the fields by differentiating these potentials, but I prefer to work with Eqs. 11.18 and 11.19, using superposition. Since $\hat{\mathbf{z}} = \cos\theta\,\hat{\mathbf{r}} - \sin\theta\,\hat{\boldsymbol{\theta}}$, and $\cos\theta = z/r$, Eq. 11.18 can be written

$\mathbf{E} = \frac{\mu_0 p_0\omega^2}{4\pi r}\cos[\omega(t-r/c)]\left(\hat{\mathbf{z}} - \frac{z}{r}\hat{\mathbf{r}}\right)$. In the case of the rotating dipole, therefore,

$$\mathbf{E} = \boxed{\frac{\mu_0 p_0\omega^2}{4\pi r}\left\{\cos[\omega(t-r/c)]\left(\hat{\mathbf{x}} - \frac{x}{r}\hat{\mathbf{r}}\right) + \sin[\omega(t-r/c)]\left(\hat{\mathbf{y}} - \frac{y}{r}\hat{\mathbf{r}}\right)\right\},}$$
$$\mathbf{B} = \boxed{\frac{1}{c}(\hat{\mathbf{r}} \times \mathbf{E}).}$$

$\mathbf{S} = \frac{1}{\mu_0}(\mathbf{E}\times\mathbf{B}) = \frac{1}{\mu_0 c}[\mathbf{E}\times(\hat{\mathbf{r}}\times\mathbf{E})] = \frac{1}{\mu_0 c}[E^2\,\hat{\mathbf{r}} - (\mathbf{E}\cdot\hat{\mathbf{r}})\mathbf{E}] = \frac{E^2}{\mu_0 c}\hat{\mathbf{r}}$ (notice that $\mathbf{E}\cdot\hat{\mathbf{r}} = 0$). Now

$$E^2 = \left(\frac{\mu_0 p_0\omega^2}{4\pi r}\right)^2\left\{a^2\cos^2[\omega(t-r/c)] + b^2\sin^2[\omega(t-r/c)] + 2(\mathbf{a}\cdot\mathbf{b})\sin[\omega(t-r/c)]\cos[\omega(t-r/c)]\right\},$$

where $\mathbf{a} \equiv \hat{\mathbf{x}} - (x/r)\hat{\mathbf{r}}$ and $\mathbf{b} \equiv \hat{\mathbf{y}} - (y/r)\hat{\mathbf{r}}$. Noting that $\hat{\mathbf{x}}\cdot\mathbf{r} = x$ and $\hat{\mathbf{y}}\cdot\mathbf{r} = y$, we have

$$a^2 = 1 + \frac{x^2}{r^2} - 2\frac{x^2}{r^2} = 1 - \frac{x^2}{r^2};\ b^2 = 1 - \frac{y^2}{r^2};\ \mathbf{a}\cdot\mathbf{b} = -\frac{y}{r}\frac{x}{r} - \frac{x}{r}\frac{y}{r} + \frac{xy}{r^2} = -\frac{xy}{r^2}.$$

$$\begin{aligned}
E^2 &= \left(\frac{\mu_0 p_0 \omega^2}{4\pi r}\right)^2 \left\{\left(1 - \frac{x^2}{r^2}\right)\cos^2[\omega(t - r/c)] + \left(1 - \frac{y^2}{r^2}\right)\sin^2[\omega(t - r/c)] \right. \\
&\quad \left. - 2\frac{xy}{r^2}\sin[\omega(t - r/c)]\cos[\omega(t - r/c)]\right\} \\
&= \left(\frac{\mu_0 p_0 \omega^2}{4\pi r}\right)^2 \left\{1 - \frac{1}{r^2}\left(x^2\cos^2[\omega(t - r/c)] + 2xy\sin[\omega(t - r/c)]\cos[\omega(t - r/c)] + y^2\sin^2[\omega(t - r/c)]\right)\right\} \\
&= \left(\frac{\mu_0 p_0 \omega^2}{4\pi r}\right)^2 \left\{1 - \frac{1}{r^2}\left(x\cos[\omega(t - r/c)] + y\sin[\omega(t - r/c)]\right)^2\right\}
\end{aligned}$$

But $x = r\sin\theta\cos\phi$ and $y = r\sin\theta\sin\phi$.

$$\begin{aligned}
&= \left(\frac{\mu_0 p_0 \omega^2}{4\pi r}\right)^2 \left\{1 - \sin^2\theta\left(\cos\phi\cos[\omega(t - r/c)] + \sin\phi\sin[\omega(t - r/c)]\right)^2\right\} \\
&= \left(\frac{\mu_0 p_0 \omega^2}{4\pi r}\right)^2 \left\{1 - \left(\sin\theta\cos[\omega(t - r/c) - \phi]\right)^2\right\}.
\end{aligned}$$

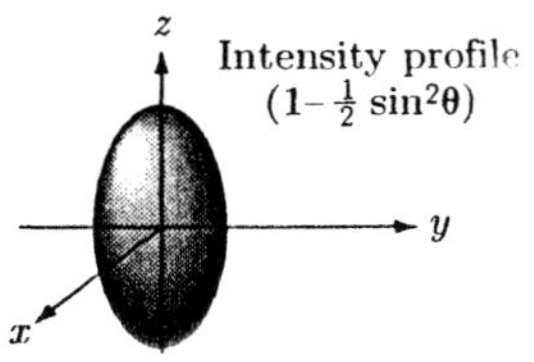

$$\mathbf{S} = \boxed{\frac{\mu_0}{c}\left(\frac{p_0\omega^2}{4\pi r}\right)^2 \left\{1 - \left(\sin\theta\cos[\omega(t - r/c) - \phi]\right)^2\right\}\hat{\mathbf{r}}.}$$

$$\langle\mathbf{S}\rangle = \frac{\mu_0}{c}\left(\frac{p_0\omega^2}{4\pi r}\right)^2 \left[1 - \frac{1}{2}\sin^2\theta\right]\hat{\mathbf{r}}.$$

$$\begin{aligned}
P &= \int \langle\mathbf{S}\rangle\cdot d\mathbf{a} = \frac{\mu_0}{c}\left(\frac{p_0\omega^2}{4\pi}\right)^2 \int \frac{1}{r^2}\left(1 - \frac{1}{2}\sin^2\theta\right) r^2\sin\theta\, d\theta\, d\phi \\
&= \frac{\mu_0 p_0^2\omega^4}{16\pi^2 c} 2\pi \left[\int_0^\pi \sin\theta\, d\theta - \frac{1}{2}\int_0^\pi \sin^3\theta\, d\theta\right] = \frac{\mu_0 p_0^2\omega^4}{8\pi c}\left(2 - \frac{1}{2}\cdot\frac{4}{3}\right) = \boxed{\frac{\mu_0 p_0^2\omega^4}{6\pi c}.}
\end{aligned}$$

This is *twice* the power radiated by either oscillating dipole alone (Eq. 11.22). In general, $\mathbf{S} = \frac{1}{\mu_0}(\mathbf{E}\times\mathbf{B}) = \frac{1}{\mu_0}[(\mathbf{E}_1 + \mathbf{E}_2)\times(\mathbf{B}_1 + \mathbf{B}_2)] = \frac{1}{\mu_0}[(\mathbf{E}_1\times\mathbf{B}_1) + (\mathbf{E}_2\times\mathbf{B}_2) + (\mathbf{E}_1\times\mathbf{B}_2) + (\mathbf{E}_2\times\mathbf{B}_1)] = \mathbf{S}_1 + \mathbf{S}_2 +$ cross terms. In this particular case, the fields of 1 and 2 are 90° out of phase, so the cross terms go to zero in the time averaging, and the total power radiated is just the sum of the two individual powers.

Problem 11.5

Go back to Eq. 11.33:

$$\mathbf{A} = \frac{\mu_0 m_0}{4\pi}\left(\frac{\sin\theta}{r}\right)\left\{\frac{1}{r}\cos[\omega(t - r/c)] - \frac{\omega}{c}\sin[\omega(t - r/c)]\right\}\hat{\boldsymbol{\phi}}.$$

Since $V = 0$ here,

$$\begin{aligned}
\mathbf{E} &= -\frac{\partial \mathbf{A}}{\partial t} = -\frac{\mu_0 m_0}{4\pi}\left(\frac{\sin\theta}{r}\right)\left\{\frac{1}{r}(-\omega)\sin[\omega(t-r/c)] - \frac{\omega}{c}\omega\cos[\omega(t-r/c)]\right\}\hat{\boldsymbol{\phi}} \\
&= \boxed{\frac{\mu_0 m_0 \omega}{4\pi}\left(\frac{\sin\theta}{r}\right)\left\{\frac{1}{r}\sin[\omega(t-r/c)] + \frac{\omega}{c}\cos[\omega(t-r/c)]\right\}\hat{\boldsymbol{\phi}}.} \\
\mathbf{B} &= \boldsymbol{\nabla}\times\mathbf{A} = \frac{1}{r\sin\theta}\frac{\partial}{\partial\theta}(A_\phi\sin\theta)\,\hat{\mathbf{r}} - \frac{1}{r}\frac{\partial}{\partial r}(rA_\phi)\,\hat{\boldsymbol{\theta}} \\
&= \frac{\mu_0 m_0}{4\pi}\left\{\frac{1}{r\sin\theta}\frac{2\sin\theta\cos\theta}{r}\left[\frac{1}{r}\cos[\omega(t-r/c)] - \frac{\omega}{c}\sin[\omega(t-r/c)]\right]\hat{\mathbf{r}}\right. \\
&\quad \left. - \frac{\sin\theta}{r}\left[-\frac{1}{r^2}\cos[\omega(t-r/c)] + \frac{\omega}{rc}\sin[\omega(t-r/c)] - \frac{\omega}{c}\left(-\frac{\omega}{c}\right)\cos[\omega(t-r/c)]\right]\hat{\boldsymbol{\theta}}\right\} \\
&= \boxed{\begin{aligned}\frac{\mu_0 m_0}{4\pi}&\left\{\frac{2\cos\theta}{r^2}\left[\frac{1}{r}\cos[\omega(t-r/c)] - \frac{\omega}{c}\sin[\omega(t-r/c)]\right]\hat{\mathbf{r}}\right. \\ &\left. - \frac{\sin\theta}{r}\left[-\frac{1}{r^2}\cos[\omega(t-r/c)] + \frac{\omega}{rc}\sin[\omega(t-r/c)] + \left(\frac{\omega}{c}\right)^2\cos[\omega(t-r/c)]\right]\hat{\boldsymbol{\theta}}\right\}.\end{aligned}}
\end{aligned}$$

These are precisely the fields we studied in Prob. 9.33, with $A \to \dfrac{\mu_0 m_0 \omega^2}{4\pi c}$. The Poynting vector (quoting the solution to that problem) is

$$\boxed{\begin{aligned}\mathbf{S} = \frac{\mu_0 m_0^2\omega^3}{16\pi^2 c^2}\left(\frac{\sin\theta}{r^2}\right)&\left\{\frac{2\cos\theta}{r}\left[\left(1-\frac{c^2}{\omega^2 r^2}\right)\sin u\cos u + \frac{c}{\omega r}\left(\cos^2 u - \sin^2 u\right)\right]\hat{\boldsymbol{\theta}}\right. \\ &\left. \sin\theta\left[\left(-\frac{2}{r}+\frac{c^2}{\omega^2 r^3}\right)\sin u\cos u + \frac{\omega}{c}\cos^2 u + \frac{c}{\omega r^2}\left(\sin^2 u - \cos^2 u\right)\right]\hat{\mathbf{r}}\right\},\end{aligned}}$$

where $u \equiv -\omega(t - r/c)$. The intensity is $\boxed{\langle\mathbf{S}\rangle = \dfrac{\mu_0 m_0^2\omega^4}{32\pi^2 c^3}\dfrac{\sin^2\theta}{r^2}\,\hat{\mathbf{r}},}$ the same as Eq. 11.39.

Problem 11.6

$$I^2 R = I_0^2 R\cos^2(\omega t) \Rightarrow \langle P\rangle = \frac{1}{2}I_0^2 R = \frac{\mu_0 m_0^2\omega^4}{12\pi c^3} = \frac{\mu_0\pi^2 b^4 I_0^2\omega^4}{12\pi c^3},\ \text{so}\ \boxed{R = \frac{\mu_0\pi b^4\omega^4}{6c^3};}\ \text{or, since}\ \omega = \frac{2\pi c}{\lambda},$$

$$R = \frac{\mu_0\pi b^4}{6c^3}\frac{16\pi^4 c^4}{\lambda^4} = \boxed{\frac{8}{3}\pi^5\mu_0 c\left(\frac{b}{\lambda}\right)^4} = \frac{8}{3}(\pi^5)(4\pi\times10^{-7})(3\times10^8)(b/\lambda)^4 = \boxed{3.08\times10^5(b/\lambda)^4\,\Omega.}$$

Because $b \ll \lambda$, and R goes like the *fourth* power of this small number, R is typically much smaller than the electric radiative resistance (Prob. 11.3). For the dimensions we used in Prob. 11.3 ($b = 5\,\text{cm}$ and $\lambda = 10^3\,\text{m}$), $R = 3\times10^5(5\times10^{-5})^4 = 2\times10^{-12}\,\Omega$, which is a millionth of the comparable electrical radiative resistance.

Problem 11.7

With $\alpha = 90°$, Eq. 7.68 $\Rightarrow \mathbf{E}' = c\mathbf{B}$, $\mathbf{B}' = -\mathbf{E}/c$, $q_m' = -cq_e \Rightarrow m_0 \equiv q_m' d = -cq_e d = -cp_0$. So

$$\mathbf{E}' = c\left\{-\frac{\mu_0(-m_0/c)\omega^2}{4\pi c}\left(\frac{\sin\theta}{r}\right)\cos[\omega(t-r/c)]\,\hat{\boldsymbol{\phi}}\right\} = \boxed{\frac{\mu_0 m_0\omega^2}{4\pi c}\left(\frac{\sin\theta}{r}\right)\cos[\omega(t-r/c)]\,\hat{\boldsymbol{\phi}}.}$$

$$\mathbf{B}' = -\frac{1}{c}\left\{-\frac{\mu_0(-m_0/c)\omega^2}{4\pi}\left(\frac{\sin\theta}{r}\right)\cos[\omega(t-r/c)]\,\hat{\boldsymbol{\theta}}\right\} = \boxed{-\frac{\mu_0 m_0\omega^2}{4\pi c^2}\left(\frac{\sin\theta}{r}\right)\cos[\omega(t-r/c)]\,\hat{\boldsymbol{\theta}}.}$$

These are *identical* to the fields of an Ampére dipole (Eqs. 11.36 and 11.37), which is consistent with our general experience that the two models generate identical fields *except right at* the dipole (not relevant here, since we're in the radiation zone).

Problem 11.8

$\mathbf{p}(t) = p_0[\cos(\omega t)\,\hat{\mathbf{x}} + \sin(\omega t)\,\hat{\mathbf{y}}] \Rightarrow \ddot{\mathbf{p}}(t) = -\omega^2 p_0[\cos(\omega t)\,\hat{\mathbf{x}} + \sin(\omega t)\,\hat{\mathbf{y}}] \Rightarrow$

$[\ddot{\mathbf{p}}(t)]^2 = \omega^4 p_0^2[\cos^2(\omega t) + \sin^2(\omega t)] = p_0^2\omega^4$. So Eq. 11.59 says $\boxed{\mathbf{S} = \frac{\mu_0 p_0^2 \omega^4}{16\pi^2 c}\frac{\sin^2\theta}{r^2}\,\hat{\mathbf{r}}.}$ (This appears to disagree with the answer to Prob. 11.4. The reason is that in Eq. 11.59 the polar axis is along the direction of $\ddot{\mathbf{p}}(t_0)$; as the dipole rotates, so do the axes. Thus the angle θ here is not the same as in Prob. 11.4.) Meanwhile, Eq. 11.60 says $\boxed{P = \frac{\mu_0 p_0^2\omega^4}{6\pi c}.}$ (This *does* agree with Prob. 11.4, because we have now integrated over all angles, and the orientation of the polar axis irrelevant.)

Problem 11.9

At $t = 0$ the dipole moment of the ring is

$$\begin{aligned} \mathbf{p}_0 &= \int \lambda \mathbf{r}\, dl = \int (\lambda_0 \sin\phi)(b\sin\phi\,\hat{\mathbf{y}} + b\cos\phi\,\hat{\mathbf{x}})b\,d\phi = \lambda_0 b^2\left(\hat{\mathbf{y}}\int_0^{2\pi}\sin^2\phi\,d\phi + \hat{\mathbf{x}}\int_0^{2\pi}\sin\phi\cos\phi\,d\phi\right) \\ &= \lambda b^2(\pi\,\hat{\mathbf{y}} + 0\,\hat{\mathbf{x}}) = \pi b^2\lambda_0\,\hat{\mathbf{y}}. \end{aligned}$$

As it rotates (counterclockwise, say) $\mathbf{p}(t) = p_0[\cos(\omega t)\,\hat{\mathbf{y}} - \sin(\omega t)\,\hat{\mathbf{x}}]$, so $\ddot{\mathbf{p}} = -\omega^2\mathbf{p}$, and hence $(\ddot{\mathbf{p}})^2 = \omega^4 p_0^2$. Therefore (Eq. 11.60) $P = \frac{\mu_0}{6\pi c}\omega^4(\pi b^2\lambda_0)^2 = \boxed{\frac{\pi\mu_0\omega^4 b^4\lambda_0^2}{6c}.}$

Problem 11.10

$\mathbf{p} = -ey\,\hat{\mathbf{y}}$, $y = \frac{1}{2}gt^2$, so $\mathbf{p} = -\frac{1}{2}get^2\,\hat{\mathbf{y}}$; $\ddot{\mathbf{p}} = -ge\,\hat{\mathbf{y}}$. Therefore (Eq. 11.60) : $P = \frac{\mu_0}{6\pi c}(ge)^2$. Now, the time it takes to fall a distance h is given by $h = \frac{1}{2}gt^2 \Rightarrow t = \sqrt{2h/g}$, so the energy radiated in falling a distance h is $U_{\text{rad}} = Pt = \frac{\mu_0(ge)^2}{6\pi c}\sqrt{2h/g}$. Meanwhile, the potential energy lost is $U_{\text{pot}} = mgh$. So the *fraction* is

$$f = \frac{U_{\text{rad}}}{U_{\text{pot}}} = \frac{\mu_0 g^2 e^2}{6\pi c}\sqrt{\frac{2h}{g}}\frac{1}{mgh} = \boxed{\frac{\mu_0 e^2}{6\pi mc}\sqrt{\frac{2g}{h}}.} = \frac{(4\pi\times10^{-7})(1.6\times10^{-19})^2}{6\pi(9.11\times10^{-31})(3\times10^8)}\sqrt{\frac{(2)(9.8)}{(0.02)}} = \boxed{2.76\times10^{-22}.}$$

Evidently *almost* all the energy goes into kinetic form (as indeed I *assumed* in saying $y = \frac{1}{2}gt^2$).

Problem 11.11

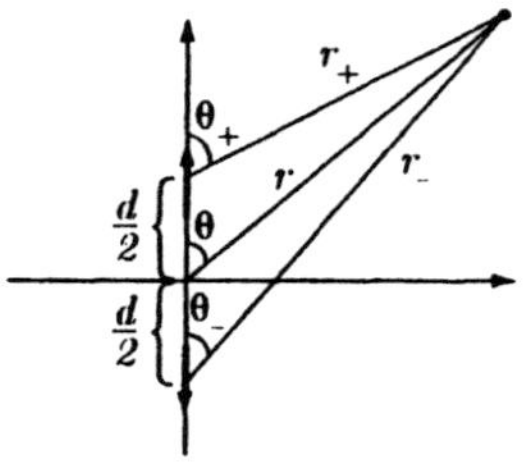

(a)
$$V_\pm = \mp\frac{p_0\omega}{4\pi\epsilon_0 c}\left(\frac{\cos\theta_\pm}{r_\pm}\right)\sin[\omega(t-r_\pm/c)].\quad V_{\text{tot}} = V_+ + V_-.$$

$$r_\pm = \sqrt{r^2+(d/2)^2\mp 2r(d/2)\cos\theta} \cong r\sqrt{1\mp(d/r)\cos\theta} \cong r\left(1\mp\frac{d}{2r}\cos\theta\right).$$

$$\frac{1}{r_\pm} \cong \frac{1}{r}\left(1\pm\frac{d}{2r}\cos\theta\right).$$

$$\cos\theta_\pm = \frac{r\cos\theta\mp(d/2)}{r_\pm} = r\left(\cos\theta\mp\frac{d}{2r}\right)\frac{1}{r}\left(1\pm\frac{d}{2r}\cos\theta\right) = \cos\theta\pm\frac{d}{2r}\cos^2\theta\mp\frac{d}{2r}$$

$$= \cos\theta\mp\frac{d}{2r}(1-\cos^2\theta) = \cos\theta\mp\frac{d}{2r}\sin^2\theta.$$

$$\sin[\omega(t-r_\pm/c)] = \sin\left\{\omega\left[t-\frac{r}{c}\left(1\mp\frac{d}{2r}\cos\theta\right)\right]\right\} = \sin\left(\omega t_0\pm\frac{\omega d}{2c}\cos\theta\right),\ \text{where } t_0\equiv t-r/c.$$

$$= \sin(\omega t_0)\cos\left(\frac{\omega d}{2c}\cos\theta\right)\pm\cos(\omega t_0)\sin\left(\frac{\omega d}{2c}\cos\theta\right) \cong \sin(\omega t_0)\pm\frac{\omega d}{2c}\cos\theta\cos(\omega t_0).$$

$$V_\pm = \mp\frac{p_0\omega}{4\pi\epsilon_0 cr}\left\{\left(1\pm\frac{d}{2r}\cos\theta\right)\left(\cos\theta\mp\frac{d}{2r}\sin^2\theta\right)\left[\sin(\omega t_0)\pm\frac{\omega d}{2c}\cos\theta\cos(\omega t_0)\right]\right\}$$

$$= \mp\frac{p_0\omega}{4\pi\epsilon_0 cr}\left\{\left(\cos\theta\mp\frac{d}{2r}\sin^2\theta\pm\frac{d}{2r}\cos^2\theta\right)\left[\sin(\omega t_0)\pm\frac{\omega d}{2c}\cos\theta\cos(\omega t_0)\right]\right\}$$

$$= \mp\frac{p_0\omega}{4\pi\epsilon_0 cr}\left[\cos\theta\sin(\omega t_0)\pm\frac{\omega d}{2c}\cos^2\theta\cos(\omega t_0)\pm\frac{d}{2r}(\cos^2\theta-\sin^2\theta)\sin(\omega t_0\right].$$

$$V_{\text{tot}} = -\frac{p_0\omega}{4\pi\epsilon_0 cr}\left[\frac{\omega d}{c}\cos^2\theta\cos(\omega t_0)+\frac{d}{r}(\cos^2\theta-\sin^2\theta)\sin(\omega t_0)\right]$$

$$= \boxed{-\frac{p_0\omega^2 d}{4\pi\epsilon_0 c^2 r}\left[\cos^2\theta\cos(\omega t_0)+\frac{c}{\omega r}(\cos^2\theta-\sin^2\theta)\sin(\omega t_0\right].}$$

In the radiation zone ($r \gg \omega/c$) the second term is negligible, so $\boxed{V = -\frac{p_0\omega^2 d}{4\pi\epsilon_0 c^2 r}\cos^2\theta\cos[\omega(t-r/c)].}$

Meanwhile

$$\mathbf{A}_\pm = \mp\frac{\mu_0 p_0\omega}{4\pi r_\pm}\sin[\omega(t-r_\pm/c)]\,\hat{\mathbf{z}}$$

$$= \mp\frac{\mu_0 p_0\omega}{4\pi r}\left\{\left(1\pm\frac{d}{2r}\cos\theta\right)\left[\sin(\omega t_0)\pm\frac{\omega d}{2c}\cos\theta\cos(\omega t_0)\right]\right\}\hat{\mathbf{z}}$$

$$= \mp\frac{\mu_0 p_0\omega}{4\pi r}\left[\sin(\omega t_0)\pm\frac{\omega d}{2c}\cos\theta\cos(\omega t_0)\pm\frac{d}{2r}\cos\theta\sin(\omega t_0)\right]\hat{\mathbf{z}}.$$

$$\mathbf{A}_{\text{tot}} = \mathbf{A}_+ + \mathbf{A}_= -\frac{\mu_0 p_0\omega}{4\pi r}\left[\frac{\omega d}{c}\cos\theta\cos(\omega t_0)+\frac{d}{r}\cos\theta\sin(\omega t_0)\right]\hat{\mathbf{z}}$$

$$= \boxed{-\frac{\mu_0 p_0\omega^2 d}{4\pi cr}\cos\theta\left[\cos(\omega t_0)+\frac{c}{\omega r}\sin(\omega t_0)\right]\hat{\mathbf{z}}.}$$

In the radiation zone, $\boxed{\mathbf{A} = -\frac{\mu_0 p_0 \omega^2 d}{4\pi c r}\cos\theta\cos[\omega(t-r/c)]\,\hat{\mathbf{z}}.}$

(b) To simplify the notation, let $\alpha \equiv -\frac{\mu_0 p_0 \omega^2 d}{4\pi}$. Then

$$\begin{aligned}
V &= \alpha\frac{\cos^2\theta}{r}\cos[\omega(t-r/c)];\\
\nabla V &= \frac{\partial V}{\partial r}\hat{\mathbf{r}} + \frac{1}{r}\frac{\partial V}{\partial\theta}\hat{\boldsymbol{\theta}} = \alpha\cos^2\theta\left\{-\frac{1}{r^2}\cos[\omega(t-r/c)] + \frac{\omega}{rc}\sin[\omega(t-r/c)]\right\}\hat{\mathbf{r}}\\
&\quad + \alpha\frac{-2\cos\theta\sin\theta}{r^2}\cos[\omega(t-r/c)]\,\hat{\boldsymbol{\theta}} = \alpha\frac{\omega}{c}\frac{\cos^2\theta}{r}\sin[\omega(t-r/c)]\,\hat{\mathbf{r}}\quad\text{(in the radiation zone).}\\
\mathbf{A} &= \frac{\alpha}{c}\frac{\cos\theta}{r}\cos[\omega(t-r/c)]\left(\cos\theta\,\hat{\mathbf{r}} - \sin\theta\,\hat{\boldsymbol{\theta}}\right).\quad \frac{\partial\mathbf{A}}{\partial t} = -\frac{\alpha\omega}{c}\frac{\cos\theta}{r}\sin[\omega(t-r/c)]\left(\cos\theta\,\hat{\mathbf{r}} - \sin\theta\,\hat{\boldsymbol{\theta}}\right).
\end{aligned}$$

$$\begin{aligned}
\mathbf{E} &= -\nabla V - \frac{\partial\mathbf{A}}{\partial t} = -\frac{\alpha\omega}{cr}\sin[\omega(t-r/c)]\left(\cos^2\theta\,\hat{\mathbf{r}} - \cos^2\theta\,\hat{\mathbf{r}} + \sin\theta\cos\theta\,\hat{\boldsymbol{\theta}}\right)\\
&= \boxed{-\frac{\alpha\omega}{cr}\sin\theta\cos\theta\sin[\omega(t-r/c)]\,\hat{\boldsymbol{\theta}}.}\\
\mathbf{B} &= \nabla\times\mathbf{A} = \frac{1}{r}\left[\frac{\partial}{\partial r}(rA_\theta) - \frac{\partial A_r}{\partial\theta}\right]\hat{\boldsymbol{\phi}}\\
&= \frac{\alpha}{cr}\left\{\frac{\partial}{\partial r}\left(\cos\theta\cos[\omega(t-r/c)](-\sin\theta)\right) - \frac{\partial}{\partial\theta}\left[\frac{\cos^2\theta}{r}\cos[\omega(t-r/c)]\right]\right\}\hat{\boldsymbol{\phi}}\\
&= \frac{\alpha}{cr}(-\sin\theta\cos\theta)\frac{\omega}{c}\sin[\omega(t-r/c)]\,\hat{\boldsymbol{\phi}}\ \text{(in the radiation zone)} = \boxed{-\frac{\alpha\omega}{c^2 r}\sin\theta\cos\theta\sin[\omega(t-r/c)]\,\hat{\boldsymbol{\phi}}.}
\end{aligned}$$

Notice that $\mathbf{B} = \frac{1}{c}(\hat{\mathbf{r}}\times\mathbf{E})$ and $\mathbf{E}\cdot\hat{\mathbf{r}} = 0$.

$$\begin{aligned}
\mathbf{S} &= \frac{1}{\mu_0}(\mathbf{E}\times\mathbf{B}) = \frac{1}{\mu_0 c}\mathbf{E}\times(\hat{\mathbf{r}}\times\mathbf{E}) = \frac{1}{\mu_0 c}\left[E^2\,\hat{\mathbf{r}} - (\mathbf{E}\cdot\hat{\mathbf{r}})\mathbf{E}\right] = \frac{E^2}{\mu_0 c}\hat{\mathbf{r}}\\
&= \boxed{\frac{1}{\mu_0 c}\left\{\frac{\alpha\omega}{rc}\sin\theta\cos\theta\sin[\omega(t-r/c)]\right\}^2\hat{\mathbf{r}}.}\quad \boxed{I = \frac{1}{2\mu_0 c}\left(\frac{\alpha\omega}{rc}\sin\theta\cos\theta\right)^2.}\\
P &= \int\langle\mathbf{S}\rangle\cdot d\mathbf{a} = \frac{1}{\mu_0 c}\left(\frac{\alpha\omega}{c}\right)^2\int\sin^2\theta\cos^2\theta\sin\theta\,d\theta\,d\phi = \frac{1}{2\mu_0 c}\left(\frac{\alpha\omega}{c}\right)^2 2\pi\int_0^\pi(1-\cos^2\theta)\cos^2\theta\sin\theta\,d\theta.\\
&\quad\text{The integral is : } -\left.\frac{\cos^3\theta}{3}\right|_0^\pi + \left.\frac{\cos^5\theta}{5}\right|_0^\pi = \frac{2}{3} - \frac{2}{5} = \frac{4}{15}.\\
&= \frac{1}{2\mu_0 c}\frac{\omega^2}{c^2}\frac{\mu_0^2}{16\pi^2}(p_0 d)^2\omega^4 2\pi\frac{4}{15} = \boxed{\frac{\mu_0}{60\pi c^3}(p_0 d)^2\omega^6.}
\end{aligned}$$

Notice that it goes like ω^6, whereas *dipole* radiation goes like ω^4.

Problem 11.12

Here $V = 0$ (since the ring is neutral), and the current depends only on t (not on position), so the retarded vector potential (Eq. 11.52) is $\mathbf{A}(\mathbf{r},t) = \frac{\mu_0}{4\pi}\oint\frac{I(t-\imath/c)}{\imath}\,d\mathbf{l}'$. But in this case it does *not* suffice to replace $\imath$ by

r in the denominator—that would lead to Eq. 11.54, and hence to $\mathbf{A} = 0$ (since $\mathbf{p} = 0$). Instead, use Eq. 11.30: $\frac{1}{\imath} \cong \frac{1}{r}\left(1 + \frac{b}{r}\sin\theta\cos\phi'\right)$. Meanwhile, $d\mathbf{l}' = b\,d\phi'\hat{\boldsymbol{\phi}} = b(-\sin\phi'\,\hat{\mathbf{x}} + \cos\phi'\,\hat{\mathbf{y}})\,d\phi'$, and

$$I(t - \imath/c) \cong I(t - r/c + (b/c)\sin\theta\cos\phi') = I(t_0 + (b/c)\sin\theta\cos\phi') \cong I(t_0) + \dot{I}(t_0)\frac{b}{c}\sin\theta\cos\phi'$$

(carrying all terms to first order in b). As always, $t_0 = t - r/c$. (From now on I'll suppress the argument: I, $\dot{I}$, etc. are all to be evaluated at t_0.) Then

$$\begin{aligned}
\mathbf{A}(\mathbf{r}, t) &= \frac{\mu_0}{4\pi}\oint \frac{1}{r}\left(1 + \frac{b}{r}\sin\theta\cos\phi'\right)\left(I + \dot{I}\frac{b}{c}\sin\theta\cos\phi'\right) b(-\sin\phi'\,\hat{\mathbf{x}} + \cos\phi'\,\hat{\mathbf{y}})\,d\phi' \\
&\cong \frac{\mu_0 b}{4\pi r}\int_0^{2\pi}\left[I + \dot{I}\frac{b}{c}\sin\theta\cos\phi' + I\frac{b}{r}\sin\theta\cos\phi'\right](-\sin\phi'\,\hat{\mathbf{x}} + \cos\phi'\,\hat{\mathbf{y}})\,d\phi'.
\end{aligned}$$

But $\int_0^{2\pi}\sin\phi'\,d\phi' = \int_0^{2\pi}\cos\phi'\,d\phi' = \int_0^{2\pi}\sin\phi'\cos\phi'\,d\phi' = 0$, while $\int_0^{2\pi}\cos^2\phi'\,d\phi' = \pi$.

$$= \frac{\mu_0 b}{4\pi r}(\pi\,\hat{\mathbf{y}})\left[\dot{I}\frac{b}{c}\sin\theta + I\frac{b}{r}\sin\theta\right] = \frac{\mu_0 b^2}{4r^2}\sin\theta\left(I + \frac{r}{c}\dot{I}\right)\hat{\mathbf{y}}.$$

In general (i.e. for points *not* on the xz plane) $\hat{\mathbf{y}} \to \hat{\boldsymbol{\phi}}$; moreover, in the radiation zone we are not interested in terms that go like $1/r^2$, so $\boxed{\mathbf{A}(\mathbf{r}, t) = \frac{\mu_0 b^2}{4c}\left[\dot{I}(t - r/c)\right]\frac{\sin\theta}{r}\hat{\boldsymbol{\phi}}.}$

$$\begin{aligned}
\mathbf{E}(\mathbf{r}, t) &= -\frac{\partial\mathbf{A}}{\partial t} = \boxed{-\frac{\mu_0 b^2}{4c}\left[\ddot{I}(t - r/c)\right]\frac{\sin\theta}{r}\hat{\boldsymbol{\phi}}.} \\
\mathbf{B}(\mathbf{r}, t) &= \boldsymbol{\nabla}\times\mathbf{A} = \frac{1}{r\sin\theta}\frac{\partial}{\partial\theta}(A_\phi\sin\theta)\,\hat{\mathbf{r}} - \frac{1}{r}\frac{\partial}{\partial r}(rA_\phi)\,\hat{\boldsymbol{\theta}} \\
&= \frac{\mu_0 b^2}{4c}\left[\frac{\dot{I}}{r\sin\theta}\frac{1}{r}2\sin\theta\cos\theta\,\hat{\mathbf{r}} - \frac{1}{r}\ddot{I}\left(-\frac{1}{c}\right)\sin\theta\,\hat{\boldsymbol{\theta}}\right] = \boxed{\frac{\mu_0 b^2}{4c^2}\ddot{I}\frac{\sin\theta}{r}\hat{\boldsymbol{\theta}}.} \\
\mathbf{S} &= \frac{1}{\mu_0}(\mathbf{E}\times\mathbf{B}) = \frac{1}{\mu_0 c}\left(\frac{\mu_0 b^2}{4c}\ddot{I}\frac{\sin\theta}{r}\right)^2(-\hat{\boldsymbol{\phi}}\times\hat{\boldsymbol{\theta}}) = \boxed{\frac{\mu_0}{16c^3}\left(b^2\ddot{I}\right)^2\frac{\sin^2\theta}{r^2}\hat{\mathbf{r}}.} \\
P &= \int\mathbf{S}\cdot d\mathbf{a} = \frac{\mu_0}{16c^3}\left(b^2\ddot{I}\right)^2\int\frac{\sin^2\theta}{r^2}r^2\sin\theta\,d\theta\,d\phi = \frac{\mu_0}{16c^3}\left(b^2\ddot{I}\right)^2(2\pi)\left(\frac{4}{3}\right) = \frac{\mu_0\pi}{6c^3}\left(b^2\ddot{I}\right)^2 \\
&= \boxed{\frac{\mu_0\ddot{m}^2}{6\pi c^3}.} \quad \text{(Note that } m = I\pi b^2\text{, so } \ddot{m} = \ddot{I}\pi b^2\text{.)}
\end{aligned}$$

Problem 11.13

(a) $P = \frac{\mu_0 q^2 a^2}{6\pi c}$, and the time it takes to come to rest is $t = v_0/a$, so the energy radiated is $U_{\text{rad}} = Pt = \frac{\mu_0 q^2 a^2}{6\pi c}\frac{v_0}{a}$. The initial kinetic energy was $U_{\text{kin}} = \frac{1}{2}mv_0^2$, so the fraction radiated is $f = \frac{U_{\text{rad}}}{U_{\text{kin}}} = \boxed{\frac{\mu_0 q^2 a}{3\pi m v_0 c}.}$

(b) $d = \frac{1}{2}at^2 = \frac{1}{2}a\frac{v_0^2}{a^2} = \frac{v_0^2}{2a}$, so $a = \frac{v_0^2}{2d}$. Then

$$f = \frac{\mu_0 q^2}{3\pi m v_0 c}\frac{v_0^2}{2d} = \frac{\mu_0 q^2 v_0}{6\pi m c d} = \frac{(4\pi\times10^{-7})(1.6\times10^{-19})^2(10^5)}{6\pi(9.11\times10^{-31})(3\times10^8)(3\times10^{-9})} = \boxed{2\times10^{-10}.}$$

So radiative losses due to collisions in an ordinary wire are negligible.

Problem 11.14

$F = \frac{1}{4\pi\epsilon_0}\frac{q^2}{r^2} = ma = m\frac{v^2}{r} \Rightarrow v = \sqrt{\frac{1}{4\pi\epsilon_0}\frac{q^2}{mr}}$. At the beginning ($r_0 = 0.5$ Å),

$$\frac{v}{c} = \left[\frac{(1.6\times 10^{-19})^2}{4\pi(8.85\times 10^{-12})(9.11\times 10^{-31})(5\times 10^{-11})}\right]^{-1/2}\frac{1}{3\times 10^8} = 0.0075,$$

and when the radius is one hundredth of this v/c is only 10 times greater (0.075), so for *most* of the trip the velocity is safely nonrelativistic.

¿From the Larmor formula, $P = \frac{\mu_0 q^2}{6\pi c}\left(\frac{v^2}{r}\right)^2 = \frac{\mu_0 q^2}{6\pi c}\left(\frac{1}{4\pi\epsilon_0}\frac{q^2}{mr^2}\right)^2$ (since $a = v^2/r$), and $P = -dU/dt$, where U is the (total) energy of the electron:

$$U = U_{\text{kin}} + U_{\text{pot}} = \frac{1}{2}mv^2 - \frac{1}{4\pi\epsilon_0}\frac{q^2}{r} = \frac{1}{2}\left(\frac{1}{4\pi\epsilon_0}\frac{q^2}{r}\right) - \frac{1}{4\pi\epsilon_0}\frac{q^2}{r} = -\frac{1}{8\pi\epsilon_0}\frac{q^2}{r}.$$

So $-\frac{dU}{dt} = -\frac{1}{8\pi\epsilon_0}\frac{q^2}{r^2}\frac{dr}{dt} = P = \frac{q^2}{6\pi\epsilon_0 c^3}\left(\frac{1}{4\pi\epsilon_0}\frac{q^2}{mr^2}\right)^2$, and hence $\frac{dr}{dt} = -\frac{1}{3c}\left(\frac{q^2}{2\pi\epsilon_0 mc}\right)^2\frac{1}{r^2}$, or

$dt = -3c\left(\frac{2\pi\epsilon_0 mc}{q^2}\right)^2 r^2\,dr \Rightarrow t = -3c\left(\frac{2\pi\epsilon_0 mc}{q^2}\right)^2\int_{r_0}^{0} r^2\,dr = \boxed{c\left(\frac{2\pi\epsilon_0 mc}{q^2}\right)^2 r_0^3}$

$= (3\times 10^8)\left[\frac{2\pi(8.85\times 10^{-12})(9.11\times 10^{-31})(3\times 10^8)}{(1.6\times 10^{-19})^2}\right]^2(5\times 10^{-11})^3 = \boxed{1.3\times 10^{-11}\text{ s.}}$ (Not very long!)

Problem 11.15

According to Eq. 11.74, the maximum occurs at $\frac{d}{d\theta}\left[\frac{\sin^2\theta}{(1-\beta\cos\theta)^5}\right] = 0$. Thus

$\frac{2\sin\theta\cos\theta}{(1-\beta\cos\theta)5} - \frac{5\sin^2\theta(\beta\sin\theta)}{(1-\beta\cos\theta)^6} = 0 \Rightarrow 2\cos\theta(1-\beta\cos\theta) = 5\beta\sin^2\theta = 5\beta(1-\cos^2\theta);$

$2\cos\theta - 2\beta\cos^2\theta = 5\beta - 5\beta\cos^2\theta$, or $3\beta\cos^2\theta + 2\cos\theta - 5\beta = 0$. So

$\cos\theta = \frac{-2\pm\sqrt{4+60\beta^2}}{6\beta} = \frac{1}{3\beta}\left(\pm\sqrt{1+15\beta^2}-1\right)$. We want the plus sign, since $\theta_m \to 90°(\cos\theta_m = 0)$ when $\beta \to 0$ (Fig. 11.12): $\boxed{\theta_{\text{max}} = \cos^{-1}\left(\frac{\sqrt{1+15\beta^2}-1}{3\beta}\right)}$.

For $v \approx c$, $\beta \approx 1$; write $\beta = 1-\epsilon$ (where $\epsilon \ll 1$), and expand to first order in ϵ:

$$\begin{aligned}\left(\frac{\sqrt{1+15\beta^2}-1}{3\beta}\right) &= \frac{1}{3(1-\epsilon)}\left[\sqrt{1+15(1-\epsilon)^2}-1\right] \cong \frac{1}{3}(1+\epsilon)\left[\sqrt{1+15(1-2\epsilon)}-1\right] \\ &= \frac{1}{3}(1+\epsilon)\left[\sqrt{16-30\epsilon}-1\right] = \frac{1}{3}(1+\epsilon)\left[4\sqrt{1-(15\epsilon/8)}-1\right] = \frac{1}{3}(1+\epsilon)\left[4\left(1-\frac{15}{16}\epsilon\right)-1\right] \\ &= \frac{1}{3}(1+\epsilon)\left(3-\frac{15}{4}\epsilon\right) = (1+\epsilon)(1-\frac{5}{4}\epsilon) \cong 1+\epsilon-\frac{5}{4}\epsilon = 1-\frac{1}{4}\epsilon.\end{aligned}$$

Evidently $\theta_{\text{max}} \approx 0$, so $\cos\theta_{\text{max}} \cong 1 - \frac{1}{2}\theta_{\text{max}}^2 = 1 - \frac{1}{4}\epsilon \Rightarrow \theta_{\text{max}}^2 = \frac{1}{2}\epsilon$, or $\theta_{\text{max}} \cong \sqrt{\epsilon/2} = \boxed{\sqrt{(1-\beta)/2}.}$

Let $f \equiv \dfrac{(dP/d\Omega|_{\theta_m})_{\rm ur}}{(dP/d\Omega|_{\theta_m})_{\rm rest}} = \left[\dfrac{\sin^2\theta_{\rm max}}{(1-\beta\cos\theta_{\rm max})^5}\right]_{\rm ur}$. Now $\sin^2\theta_{\rm max} \cong \epsilon/2$, and

$(1-\beta\cos\theta_{\rm max}) \cong 1-(1-\epsilon)(1-\frac{1}{4}\epsilon) \cong 1-(1-\epsilon-\frac{1}{4}\epsilon) = \frac{5}{4}\epsilon$. So $f = \dfrac{\epsilon/2}{(5\epsilon/4)^5} = \left(\dfrac{4}{5}\right)^5 \dfrac{1}{2\epsilon^4}$. But

$\gamma = \dfrac{1}{\sqrt{1-\beta^2}} = \dfrac{1}{\sqrt{1-(1-\epsilon)^2}} \cong \dfrac{1}{\sqrt{1-(1-2\epsilon)}} = \dfrac{1}{\sqrt{2\epsilon}} \Rightarrow \epsilon = \dfrac{1}{2\gamma^2}$. Therefore

$$f = \left(\frac{4}{5}\right)^5 \frac{1}{2}(2\gamma^2)^4 = \boxed{\frac{1}{4}\left(\frac{8}{5}\right)^5 \gamma^8 = 2.62\gamma^8.}$$

Problem 11.16

Equation 11.72 says $\dfrac{dP}{d\Omega} = \dfrac{q^2}{16\pi^2\epsilon_0}\dfrac{|\hat{\boldsymbol{\imath}}\times(\mathbf{u}\times\mathbf{a})|^2}{(\hat{\boldsymbol{\imath}}\cdot\mathbf{u})^5}$. Let $\beta \equiv v/c$.

$\mathbf{u} = c\hat{\boldsymbol{\imath}} - \mathbf{v} = c\hat{\boldsymbol{\imath}} - v\hat{\mathbf{z}} \Rightarrow \hat{\boldsymbol{\imath}}\cdot\mathbf{u} = c - v(\hat{\boldsymbol{\imath}}\cdot\hat{\mathbf{z}}) = c - v\cos\theta = c\left(1-\dfrac{v}{c}\cos\theta\right) = c(1-\beta\cos\theta)$;

$\mathbf{a}\cdot\mathbf{u} = ac(\hat{\mathbf{x}}\cdot\hat{\boldsymbol{\imath}}) - av(\hat{\mathbf{x}}\cdot\hat{\mathbf{z}}) = ac\sin\theta\cos\phi$; $\quad u^2 = \mathbf{u}\cdot\mathbf{u} = c^2 - 2cv(\hat{\boldsymbol{\imath}}\cdot\hat{\mathbf{z}}) + v^2 = c^2+v^2-2cv\cos\theta$.

$$\begin{aligned}
\hat{\boldsymbol{\imath}}\times(\mathbf{u}\times\mathbf{a}) &= (\hat{\boldsymbol{\imath}}\cdot\mathbf{a})\mathbf{u} - (\hat{\boldsymbol{\imath}}\cdot\mathbf{u})\mathbf{a};\\
|\hat{\boldsymbol{\imath}}\times(\mathbf{u}\times\mathbf{a})|^2 &= (\hat{\boldsymbol{\imath}}\cdot\mathbf{a})^2u^2 - 2(\mathbf{u}\cdot\mathbf{a})(\hat{\boldsymbol{\imath}}\cdot\mathbf{a})(\hat{\boldsymbol{\imath}}\cdot\mathbf{u}) + (\hat{\boldsymbol{\imath}}\cdot\mathbf{u})^2a^2\\
&= (c^2+v^2-2cv\cos\theta)(a\sin\theta\cos\phi)^2 - 2(ac\sin\theta\cos\phi)(a\sin\theta\cos\phi)(c-v\cos\theta) + a^2c^2(1-\beta\cos\theta)^2\\
&= a^2\left[c^2(1-\beta\cos\theta)^2 + (\sin^2\theta\cos^2\phi)(c^2+v^2-2cv\cos\theta-2c^2+2cv\cos\theta\right]\\
&= a^2c^2\left[(1-\beta\cos\theta)^2 - (1-\beta^2)(\sin\theta\cos\phi)^2\right].\\
\frac{dP}{d\Omega} &= \boxed{\frac{\mu_0q^2a^2}{16\pi^2c}\frac{\left[(1-\beta\cos\theta)^2-(1-\beta^2)\sin^2\theta\cos^2\phi\right]}{(1-\beta\cos\theta)^5}.}
\end{aligned}$$

The total power radiated (in all directions) is:

$$\begin{aligned}
P &= \int\frac{dP}{d\Omega}d\Omega = \int\frac{dP}{d\Omega}\sin\theta\,d\theta\,d\phi = \frac{\mu_0q^2a^2}{16\pi^2c}\int\int\frac{\left[(1-\beta\cos\theta)^2-(1-\beta^2)\sin^2\theta\cos^2\phi\right]}{(1-\beta\cos\theta)^5}\sin\theta\,d\theta\,d\phi.\\
&\text{But } \int_0^{2\pi}d\phi = 2\pi \text{ and } \int_0^{2\pi}\cos^2\phi\,d\phi = \pi.\\
&= \frac{\mu_0q^2a^2}{16\pi^2c}\pi\int_0^{\pi}\frac{\left[2(1-\beta\cos\theta)^2-(1-\beta^2)\sin^2\theta\right]}{(1-\beta\cos\theta)^5}\sin\theta\,d\theta.
\end{aligned}$$

Let $w \equiv (1-\beta\cos\theta)$. Then $(1-w)/\beta = \cos\theta$; $\sin^2\theta = \left[\beta^2-(1-w)^2\right]/\beta^2$, and the numerator becomes

$$\begin{aligned}
2w^2 - \frac{(1-\beta^2)}{\beta^2}(\beta^2-1+2w-w^2) &= \frac{1}{\beta^2}\left[2w^2\beta^2+(1-\beta^2)^2-2(1-\beta^2)w+w^2(1-\beta^2)\right]\\
&= \frac{1}{\beta^2}\left[(1-\beta^2)^2-2(1-\beta^2)w+(1+\beta^2)w^2\right];
\end{aligned}$$

$dw = \beta \sin\theta\, d\theta \Rightarrow \sin\theta\, d\theta = \frac{1}{\beta}\, dw$. When $\theta = 0$, $w = (1-\beta)$; when $\theta = \pi$, $w = (1+\beta)$.

$$P = \frac{\mu_0 q^2 a^2}{16\pi c}\frac{1}{\beta^3}\int_{(1-\beta)}^{(1+\beta)} \frac{1}{w^5}\left[(1-\beta^2)^2 - 2(1-\beta^2)w + (1+\beta^2)w^2\right] dw. \quad \text{The integral is}$$

$$\begin{aligned}
\text{Int} &= (1-\beta^2)^2 \int \frac{1}{w^5}\, dw - 2(1-\beta^2)\int \frac{1}{w^4}\, dw + (1+\beta^2)\int \frac{1}{w^3}\, dw \\
&= \left[(1-\beta^2)^2\left(-\frac{1}{4w^4}\right) - 2(1-\beta^2)\left(-\frac{1}{3w^3}\right) + (1+\beta^2)\left(-\frac{1}{2w^2}\right)\right]\Bigg|_{1-\beta}^{1+\beta}.
\end{aligned}$$

$$\begin{aligned}
\frac{1}{w^2}\Bigg|_{1-\beta}^{1+\beta} &= \frac{1}{(1+\beta)^2} - \frac{1}{(1-\beta)^2} = \frac{(1-2\beta+\beta^2) - (1+2\beta+\beta^2)}{(1+\beta)^2(1-\beta)^2} = -\frac{4\beta}{(1-\beta^2)^2}. \\
\frac{1}{w^3}\Bigg|_{1-\beta}^{1+\beta} &= \frac{1}{(1+\beta)^3} - \frac{1}{(1-\beta)^3} = \frac{(1-3\beta+3\beta^2-\beta^3) - (1+3\beta+3\beta^2+\beta^3)}{(1+\beta)^3(1-\beta)^3} = -\frac{2\beta(3+\beta^2)}{(1-\beta^2)^3}. \\
\frac{1}{w^4}\Bigg|_{1-\beta}^{1+\beta} &= \frac{1}{(1+\beta)^4} - \frac{1}{(1-\beta)^4} = \frac{(1-4\beta+6\beta^2-4\beta^3+\beta^4) - (1+4\beta+6\beta^2+4\beta^3+\beta^4)}{(1+\beta)^4(1-\beta)^4} = -\frac{8\beta(1+\beta^2)}{(1-\beta^2)^4}.
\end{aligned}$$

$$\begin{aligned}
\text{Int} &= (1-\beta^2)^2\left(-\frac{1}{4}\right)\frac{-8\beta(1+\beta^2)}{(1-\beta^2)^4} - 2(1-\beta^2)\left(-\frac{1}{3}\right)\frac{-2\beta(3+\beta^2)}{(1-\beta^2)^3} + (1+\beta^2)\left(-\frac{1}{2}\right)\frac{-4\beta}{(1-\beta^2)^2} \\
&= \frac{2\beta}{(1-\beta^2)^2}\left[(1+\beta^2) - \frac{2}{3}(3+\beta^2) + (1+\beta^2)\right] = \frac{8}{3}\frac{\beta^3}{(1-\beta^2)^2}. \\
P &= \frac{\mu_0 q^2 a^2}{16\pi c}\frac{1}{\beta^3}\frac{8}{3}\frac{\beta^3}{(1-\beta^2)^2} = \boxed{\frac{\mu_0 q^2 a^2 \gamma^4}{6\pi c},} \quad \text{where } \gamma = \frac{1}{\sqrt{1-\beta^2}}.
\end{aligned}$$

Is this consistent with the Liénard formula (Eq. 11.73)? Here $\mathbf{v}\times\mathbf{a} = va(\hat{\mathbf{z}}\times\hat{\mathbf{x}}) = va\,\hat{\mathbf{y}}$, so $a^2 - \left(\frac{\mathbf{v}}{c}\times\mathbf{a}\right)^2 = a^2\left(1 - \frac{v^2}{c^2}\right) = (1-\beta^2)a^2 = \frac{1}{\gamma^2}a^2$, so the Liénard formula says $P = \frac{\mu_0 q^2 \gamma^6}{6\pi c}\frac{a^2}{\gamma^2}$. ✓

Problem 11.17

(a) To counteract the radiation reaction (Eq. 11.80), you must exert a force $\mathbf{F}_e = -\frac{\mu_0 q^2}{6\pi c}\dot{\mathbf{a}}$. *For circular motion,* $\mathbf{r}(t) = R\left[\cos(\omega t)\,\hat{\mathbf{x}} + \sin(\omega t)\,\hat{\mathbf{y}}\right]$, $\mathbf{v}(t) = \dot{\mathbf{r}} = R\omega\left[-\sin(\omega t)\,\hat{\mathbf{x}} + \cos(\omega t)\,\hat{\mathbf{y}}\right]$;
$\mathbf{a}(t) = \dot{\mathbf{v}} = -R\omega^2\left[\cos(\omega t)\,\hat{\mathbf{x}} + \sin(\omega t)\,\hat{\mathbf{y}}\right] = -\omega^2\mathbf{r}$; $\dot{\mathbf{a}} = -\omega^2\dot{\mathbf{r}} = -\omega^2\mathbf{v}$. So $\boxed{\mathbf{F}_e = \frac{\mu_0 q^2}{6\pi c}\omega^2\mathbf{v}.}$

$\boxed{P_e = \mathbf{F}_e\cdot\mathbf{v} = \frac{\mu_0 q^2}{6\pi c}\omega^2 v^2.}$ This is the power *you* must supply.

Meanwhile, the power *radiated* is (Eq. 11.70) $P_{\text{rad}} = \frac{\mu_0 q^2 a^2}{6\pi c}$, and $a^2 = \omega^4 r^2 = \omega^4 R^2 = \omega^2 v^2$, so $P_{\text{rad}} = \frac{\mu_0 q^2}{6\pi c}\omega^2 v^2$, and the two expressions agree.

(b) *For simple harmonic motion,* $\mathbf{r}(t) = A\cos(\omega t)\,\hat{\mathbf{z}}$; $\mathbf{v} = \dot{\mathbf{r}} = -A\omega\sin(\omega t)\,\hat{\mathbf{z}}$; $\mathbf{a} = \dot{\mathbf{v}} = -A\omega^2\cos(\omega t)\,\hat{\mathbf{z}} = -\omega^2\mathbf{r}$; $\dot{\mathbf{a}} = -\omega^2\dot{\mathbf{r}} = -\omega^2\mathbf{v}$. So $\boxed{\mathbf{F}_e = \frac{\mu_0 q^2}{6\pi c}\omega^2\mathbf{v};\ P_e = \frac{\mu_0 q^2}{6\pi c}\omega^2 v^2.}$ But this time $a^2 = \omega^4 r^2 = \omega^4 A^2\cos^2(\omega t)$, whereas $\omega^2 v^2 = \omega^4 A^2 \sin^2(\omega t)$, so

$$P_{\text{rad}} = \frac{\mu_0 q^2}{6\pi c}\omega^4 A^2 \cos^2(\omega t) \neq P_e = \frac{\mu_0 q^2}{6\pi c}\omega^4 A^2 \sin^2(\omega t);$$

the power you deliver is *not* equal to the power radiated. However, since the time *averages* of $\sin^2(\omega t)$ and $\cos^2(\omega t)$ are equal (to wit: 1/2), *over a full cycle* the energy radiated is the same as the energy input. (In the mean time energy is evidently being stored temporarily in the nearby fields.)

(c) *In free fall*, $\mathbf{v}(t) = \frac{1}{2}gt^2\,\hat{\mathbf{y}}$; $\mathbf{v} = gt\,\hat{\mathbf{y}}$; $\mathbf{a} = g\,\hat{\mathbf{y}}$; $\dot{\mathbf{a}} = 0$. So $\boxed{\mathbf{F}_e = 0;}$ the radiation reaction is zero, and hence $\boxed{P_e = 0.}$ But there *is* radiation: $\boxed{P_{\text{rad}} = \frac{\mu_0 q^2}{6\pi c}g^2.}$ Evidently energy is being continuously extracted from the nearby fields. This paradox persists even in the *exact* solution (where we do *not* assume $v \ll c$, as in the Larmor formula and the Abraham-Lorentz formula)—see Prob. 11.31.

Problem 11.18

(a) $\gamma = \omega^2\tau$, and $\tau = 6\times10^{-24}\,\text{s}$ (for electrons). Is $\gamma \ll \omega$ (i.e. is $\tau \ll 1/\omega$)? If ω is in the optical region, $\omega = 2\pi\nu = 2\pi(5\times10^{14}) = 3\times10^{15}$; $1/\omega = (1/3)\times10^{-15} = 3\times10^{-16}$, which is much greater than τ, so the damping is indeed "small". ✓

(b) Problem 9.24 gave $\Delta\omega \cong \gamma = \omega_0^2\tau = [2\pi(7\times10^{15})]^2(6\times10^{-24}) = \boxed{1\times10^{10}\,\text{rad/s.}}$ Since we're in the region of $\omega_0 \approx 4\times10^{16}$ rad/s, the width of the anomalous dispersion zone is *very* narrow.

Problem 11.19

(a) $a = \tau\dot{a} + \frac{F}{m} \Rightarrow \frac{dv}{dt} = \tau\frac{da}{dt} + \frac{F}{m} \Rightarrow \int \frac{dv}{dt}\,dt = \tau\int\frac{da}{dt}\,dt + \frac{1}{m}\int F\,dt.$

$[v(t_0+\epsilon) - v(t_0-\epsilon)] = \tau\,[a(t_0+\epsilon) - a(t_0-\epsilon)] + \frac{2\epsilon}{m}F_{\text{ave}}$, where F_{ave} is the average force during the interval. But v is continuous, so as long as F is not a delta function, we are left (in the limit $\epsilon \to 0$) with $[a(t_0+\epsilon) - a(t_0-\epsilon)] = 0$. Thus a, too, is continuous. qed

(b) (i) $a = \tau\dot{a} = \tau\frac{da}{dt} \Rightarrow \frac{da}{a} = \frac{1}{\tau}dt \Rightarrow \int\frac{da}{a} = \frac{1}{\tau}\int dt \Rightarrow \ln a = \frac{t}{\tau} + \text{constant} \Rightarrow \boxed{a(t) = Ae^{t/\tau},}$ where A is a constant.

(ii) $a = \tau\dot{a} + \frac{F}{m} \Rightarrow \tau\frac{da}{dt} = a - \frac{F}{m} \Rightarrow \frac{da}{a - F/m} = \frac{1}{\tau}dt \Rightarrow \ln(a - F/m) = \frac{t}{\tau} + \text{constant} \Rightarrow a - \frac{F}{m} = Be^{t/\tau} \Rightarrow \boxed{a(t) = \frac{F}{m} + Be^{t/\tau},}$ where B is some other constant.

(iii) Same as (i): $\boxed{a(t) = Ce^{t/\tau},}$ where C is a third constant.

(c) At $t = 0$, $A = F/m + B$; at $t = T$, $F/m + Be^{T/\tau} = Ce^{T/\tau} \Rightarrow C = (F/m)e^{-T/\tau} + B$. So

$$\boxed{a(t) = \begin{cases} [(F/m) + B]\,e^{t/\tau}, & t \le 0; \\ \left[(F/m) + Be^{t/\tau}\right], & 0 \le t \le T; \\ \left[(F/m)e^{-T/\tau} + B\right]e^{t/\tau}, & t \ge T. \end{cases}}$$

To eliminate the runaway in region (iii), we'd need $B = -(F/m)e^{-T/\tau}$; to avoid preacceleration in region (i), we'd need $B = -(F/m)$. Obviously, we cannot do both at once.

(d) If we choose to eliminate the runaway, then

$$a(t) = \begin{cases} (F/m)\left[1 - e^{-T/\tau}\right]e^{t/\tau}, & t \le 0; \\ (F/m)\left[1 - e^{(t-T)/\tau}\right], & 0 \le t \le T; \\ 0, & t \ge T. \end{cases}$$

(i) $v = (F/m)\left[1 - e^{-T/\tau}\right]\int e^{t/\tau}dt = (F\tau/m)\left[1 - e^{-T/\tau}\right]e^{t/\tau} + D$, where D is a constant determined by the condition $v(-\infty) = 0 \Rightarrow D = 0$.

(ii) $v = (F/m)\left[t - \tau e^{(t-T)/\tau}\right] + E$, where E is a constant determined by the continuity of v at $t = 0$: $(F\tau/m)\left[1 - e^{-T/\tau}\right] = (F/m)\left[-\tau e^{-T/\tau}\right] + E \Rightarrow E = (F\tau/m)$.

(iii) v is a constant determined by the continuity of v at $t = T$: $v = (F/m)[T + \tau - \tau] = (F/m)T$.

$$v(t) = \begin{cases} (F\tau/m)\left[1 - e^{-T/\tau}\right]e^{t/\tau}, & t \le 0; \\ (F/m)\left[t + \tau - \tau e^{(t-T)/\tau}\right], & 0 \le t \le T; \\ (F/m)T, & t \ge T. \end{cases}$$

(e)

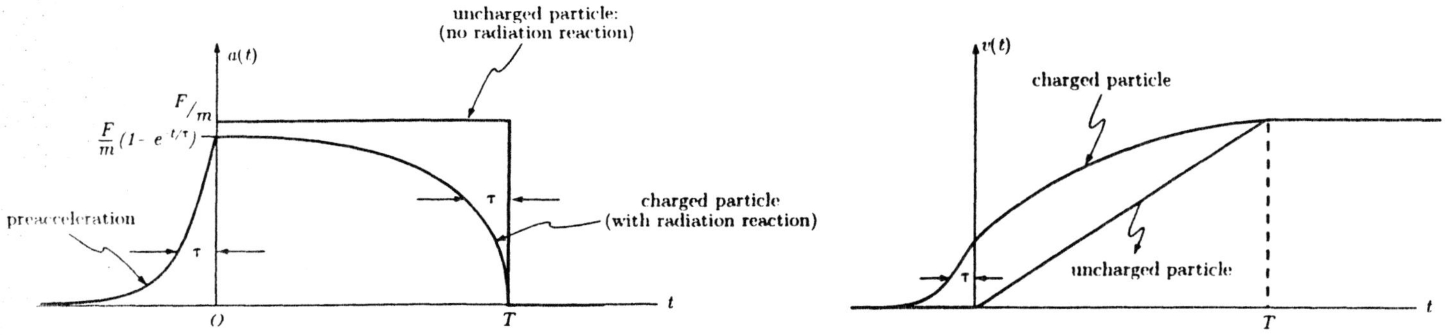

Problem 11.20

(a) From Eq.11.80, $F_{\text{rad}}^{\text{end}} = \dfrac{\mu_0(q/2)^2}{6\pi c}\dot{a}$, so $F_{\text{rad}} = F_{\text{rad}}^{\text{int}} + 2F_{\text{rad}}^{\text{end}} = \dfrac{\mu_0 q^2}{6\pi c}\dot{a}\left[\dfrac{1}{2} + 2\left(\dfrac{1}{4}\right)\right] = \dfrac{\mu_0 q^2}{6\pi c}\dot{a}$. ✓

(b) $F_{\text{rad}} = \dfrac{\mu_0}{12\pi c}\dot{a}\displaystyle\int_0^L \left\{\int_0^{y_1} 2\lambda\, dy_2\right\} 2\lambda\, dy_1$. (Running the y_2 integral up to y_1 insures that $y_1 \ge y_2$, so we don't count the same pair twice. Alternatively, run *both* integrals from 0 to L—*intentionally* double-counting—and divide the result by 2.)

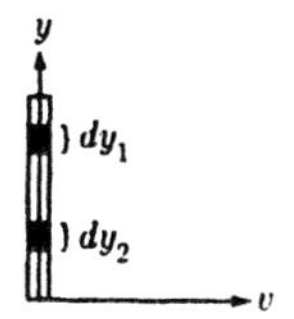

$$F_{\text{rad}} = \frac{\mu_0 \dot{a}}{12\pi c}(4\lambda^2)\int_0^L y_1\,dy_1 = \frac{\mu_0 \dot{a}}{12\pi c}(4\lambda^2)\frac{L^2}{2} = \frac{\mu_0}{6\pi c}(\lambda L)^2 \dot{a} = \frac{\mu_0 q^2}{6\pi c}\dot{a}. \checkmark$$

Problem 11.21

(a) This is an oscillating electric dipole, with amplitude $p_0 = qd$ and frequency $\omega = \sqrt{k/m}$. The (averaged) Poynting vector is given by Eq. 11.21: $\langle \mathbf{S}\rangle = \left(\frac{\mu_0 p_0^2 \omega^4}{32\pi^2 c}\right)\frac{\sin^2\theta}{r^2}\hat{\mathbf{r}}$, so the power per unit area of *floor* is

$$I_f = \langle \mathbf{S}\rangle \cdot \hat{\mathbf{z}} = \left(\frac{\mu_0 p_0^2 \omega^4}{32\pi^2 c}\right)\frac{\sin^2\theta\cos\theta}{r^2}. \quad \text{But } \sin\theta = \frac{R}{r},\ \cos\theta = \frac{h}{r},\ \text{and } r^2 = R^2 + h^2.$$

$$= \boxed{\left(\frac{\mu_0 q^2 d^2 \omega^4}{32\pi^2 c}\right)\frac{R^2 h}{(R^2+h^2)^{5/2}}.}$$

$$\frac{dI_f}{dR} = 0 \Rightarrow \frac{d}{dR}\left[\frac{R^2}{(R^2+h^2)^{5/2}}\right] = 0 \Rightarrow \frac{2R}{(R^2+h^2)^{5/2}} - \frac{5}{2}\frac{R^2}{(R^2+h^2)^{7/2}}2R = 0 \Rightarrow$$

$(R^2+h^2) - \frac{5}{2}R^2 = 0 \Rightarrow h^2 = \frac{3}{2}R^2 \Rightarrow \boxed{R = \sqrt{2/3}h,}$ for maximum intensity.

(b)

$$P = \int I_f(R)\,da = \int I_f(R)\,2\pi R\,dR = 2\pi\left(\frac{\mu_0 (qd)^2\omega^4}{32\pi^2 c}\right) h \int_0^\infty \frac{R^3}{(R^2+h^2)^{5/2}}\,dR. \quad \text{Let } x \equiv R^2:$$

$$\int_0^\infty \frac{R^3}{(R^2+h^2)^{5/2}}\,dR = \frac{1}{2}\int_0^\infty \frac{x}{(x+h^2)^{5/2}}\,dx = \frac{1}{2h}\frac{\Gamma(2)\Gamma(1/2)}{\Gamma(5/2)} = \frac{2}{3h}.$$

$$= 2\pi\left(\frac{\mu_0 q^2 d^2\omega^4}{32\pi^2 c}\right) h \frac{2}{3h} = \boxed{\frac{\mu_0 q^2 d^2 \omega^4}{24\pi c},}$$

which should be (and *is*) half the *total* radiated power (Eq. 11.22)—the rest hits the ceiling, of course.

(c) The amplitude is $x_0(t)$, so $U = \frac{1}{2}kx_0^2$ is the energy, at time t, and $dU/dt = -2P$ is the power radiated:

$$\frac{1}{2}k\frac{d}{dt}(x_0^2) = -\frac{\mu_0\omega^4}{12\pi c}q^2 x_0^2 \Rightarrow \frac{d}{dt}(x_0^2) = -\frac{\mu_0\omega^4 q^2}{6\pi k c}(x_0^2) = -\kappa x_0^2 \Rightarrow x_0^2 = d^2 e^{-\kappa t} \text{ or } x_0(t) = de^{-\kappa t/2}.$$

$$\tau = \frac{2}{\kappa} = \frac{12\pi k c}{\mu_0 q^2 k^2}m^2 = \boxed{\frac{12\pi c m^2}{\mu_0 q^2 k}.}$$

Problem 11.22

(a) From Eq. 11.39, $\langle \mathbf{S}\rangle = \left(\frac{\mu_0 m_0^2\omega^4}{32\pi^2 c^3}\right)\frac{\sin^2\theta}{r^2}\hat{\mathbf{r}}$. Here $\sin\theta = R/r$, $r = \sqrt{R^2+h^2}$, and the total radiated power (Eq. 11.40) is $P = \frac{\mu_0 m_0^2 \omega^4}{12\pi c^3}$. So the intensity is $I(R) = \left(\frac{12P}{32\pi}\right)\frac{R^2}{(R^2+h^2)^2} = \boxed{\frac{3P}{8\pi}\frac{R^2}{(R^2+h^2)^2}.}$

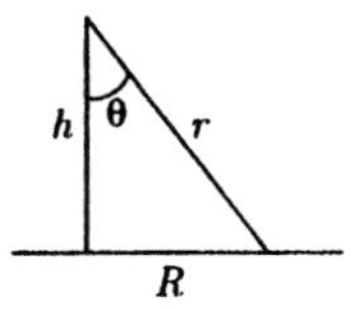

(b) The intensity *directly* below the antenna ($R = 0$) would (ideally) have been *zero*. The engineer *should* have measured it at the position of *maximum* intensity:

$$\frac{dI}{dR} = \frac{3P}{8\pi}\left[\frac{2R}{(R^2+h^2)^2} - \frac{2R^2}{(R^2+h^2)^3}2R\right] = \frac{3P}{8\pi}\frac{2R}{(R^2+h^2)^3}(R^2+h^2-2R^2) = 0 \Rightarrow \boxed{R = h.}$$

At this location the intensity is $I(h) = \frac{3P}{8\pi}\frac{h^2}{(2h^2)^2} = \boxed{\frac{3P}{32\pi h^2}}.$

(c) $I_{\text{max}} = \frac{3(35 \times 10^3)}{32\pi(200)^2} = 0.026\,\text{W/m}^2 = \boxed{2.6\,\mu\text{W/cm}^2.}$ $\boxed{\text{Yes, KRUD is in compliance.}}$

Problem 11.23

(a) $\mathbf{m}(t) = M\cos\psi\,\hat{\mathbf{z}} + M\sin\psi[\cos(\omega t)\,\hat{\mathbf{x}} + \sin(\omega t)\,\hat{\mathbf{y}}]$. As in Prob. 11.4, the power radiated will be twice that of an oscillating magnetic dipole with dipole moment of amplitude $m_0 = M\sin\psi$. Therefore (quoting Eq. 11.40): $\boxed{P = \frac{\mu_0 M^2\omega^4\sin^2\psi}{6\pi c^3}.}$ (Alternatively, you can get this from the answer to Prob. 11.12.)

(b) From Eq. 5.86, with $r \to R$, $m \to M$, and $\theta = \pi/2$: $B = \frac{\mu_0}{4\pi}\frac{M}{R^3}$, so

$$\boxed{M = \frac{4\pi R^3}{\mu_0}B} = \frac{4\pi(6.4 \times 10^6)^3(5 \times 10^{-5})}{4\pi \times 10^{-7}} = \boxed{1.3 \times 10^{23}\,\text{A m}^2.}$$

(c) $P = \frac{(4\pi \times 10^{-7})(1.3 \times 10^{23})^2\sin^2(11°)}{6\pi(3 \times 10^8)^3}\left(\frac{2\pi}{24 \times 60 \times 60}\right)^4 = \boxed{4 \times 10^{-5}\,\text{W}}$ (not much).

(d) $P = \frac{\mu_0(4\pi R^3 B/\mu_0)^2\omega^4\sin^2\psi}{6\pi c^3} = \frac{8\pi}{3\mu_0 c^3}\left(\omega^2 R^3 B\sin\psi\right)^2$. Using the average value (1/2) for $\sin^2\psi$,

$$P = \frac{8\pi}{3(4\pi \times 10^{-7})(3 \times 10^8)^3}\left[\left(\frac{2\pi}{10^{-3}}\right)^2(10^4)^3(10^8)\right]^2\frac{1}{2} = \boxed{2 \times 10^{36}\,\text{W}}\text{ (a lot).}$$

Problem 11.24

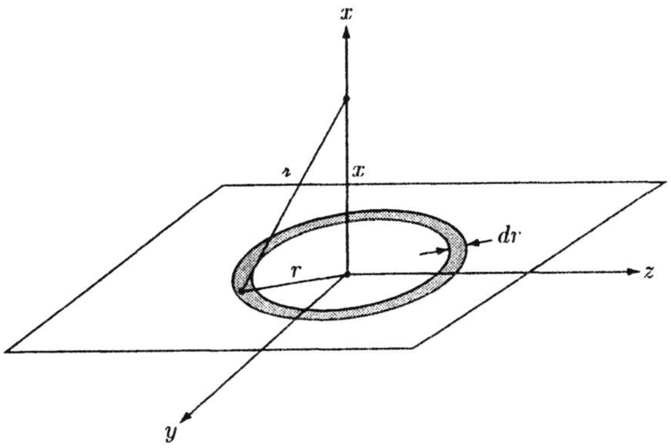

(a) $\mathbf{A}(x,t) = \frac{\mu_0}{4\pi}\int\frac{\mathbf{K}(t_r)}{\mathscr{r}}\,da$

$= \frac{\mu_0\hat{\mathbf{z}}}{4\pi}\int\frac{K(t_r)}{\sqrt{r^2+x^2}}2\pi r\,dr$

$= \frac{\mu_0\hat{\mathbf{z}}}{2}\int\frac{K(t-\sqrt{r^2+x^2}/c)}{\sqrt{r^2+x^2}}r\,dr.$

The maximum r is given by $t - \sqrt{r^2+x^2}/c = 0$; $r_{\text{max}} = \sqrt{c^2t^2 - x^2}$ (since $K(t) = 0$ for $t < 0$).

(i)

$$\mathbf{A}(x,t) = \frac{\mu_0 K_0\hat{\mathbf{z}}}{2}\int_0^{r_m}\frac{r}{\sqrt{r^2+x^2}}dr = \frac{\mu_0 K_0\hat{\mathbf{z}}}{2}\sqrt{r^2+x^2}\Big|_0^{r_m} = \frac{\mu_0 K_0\hat{\mathbf{z}}}{2}\left(\sqrt{r_m^2 - x^2} - x\right) = \frac{\mu_0 K_0(ct-x)}{2}\hat{\mathbf{z}}.$$

$$\mathbf{E}(x,t) = -\frac{\partial\mathbf{A}}{\partial t} = \boxed{-\frac{\mu_0 K_0 c}{2}\hat{\mathbf{z}},}\text{ for } ct > x, \text{ and } 0, \text{ for } ct < x.$$

$$\mathbf{B}(x,t) = \nabla \times \mathbf{A} = -\frac{\partial A_z}{\partial x}\hat{\mathbf{y}} = \boxed{\frac{\mu_0 K_0}{2}\hat{\mathbf{y}},}\text{ for } ct > x, \text{ and } 0, \text{ for } ct < x.$$

(ii)

$$\mathbf{A}(x,t) = \frac{\mu_0\alpha\,\hat{\mathbf{z}}}{2}\int_0^{r_m}\frac{(t-\sqrt{r^2+x^2}/c)}{\sqrt{r^2+x^2}}r\,dr = \frac{\mu_0\alpha\hat{\mathbf{z}}}{2}\left[t\int_0^{r_m}\frac{r}{\sqrt{r^2+x^2}}\,dr - \frac{1}{c}\int_0^{r_m}r\,dr\right]$$

$$= \frac{\mu_0\alpha\,\hat{\mathbf{z}}}{2}\left[t(ct-x)-\frac{1}{2c}(c^2t^2-x^2)\right] = \frac{\mu_0\alpha\,\hat{\mathbf{z}}}{4c}(x^2-2ctx+c^2t^2) = \frac{\mu_0\alpha(x-ct)^2}{4c}\,\hat{\mathbf{z}}.$$

$$\mathbf{E}(x,t) = -\frac{\partial\mathbf{A}}{\partial t} = \boxed{\frac{\mu_0\alpha(x-ct)}{2}\,\hat{\mathbf{z}},}\ \text{for } ct>x, \text{ and } 0, \text{ for } ct<x.$$

$$\mathbf{B}(x,t) = \nabla\times\mathbf{A} = -\frac{\partial A_z}{\partial x}\,\hat{\mathbf{y}} = \boxed{-\frac{\mu_0\alpha}{2c}(x-ct)\,\hat{\mathbf{y}},}\ \text{for } ct>x, \text{ and } 0, \text{ for } ct<x.$$

(b) Let $u \equiv \frac{1}{c}\left(\sqrt{r^2+x^2}-x\right)$, so $du = \frac{1}{c}\left[\frac{1}{2}\frac{1}{\sqrt{r^2+x^2}}2r\,dr\right] = \frac{1}{c}\frac{r}{\sqrt{r^2+x^2}}\,dr$, and $t-\frac{\sqrt{r^2+x^2}}{c} = t-\frac{x}{c}-u$, and as $r:0\to\infty$, $u:0\to\infty$. Then $\mathbf{A}(x,t) = \frac{\mu_0 c\,\hat{\mathbf{z}}}{2}\int_0^\infty K\left(t-\frac{x}{c}-u\right)du.$ qed

$$\mathbf{E}(x,t) = -\frac{\partial\mathbf{A}}{\partial t} = -\frac{\mu_0 c\,\hat{\mathbf{z}}}{2}\int_0^\infty\frac{\partial}{\partial t}K\left(t-\frac{x}{c}-u\right)du. \quad \text{But } \frac{\partial}{\partial t}K\left(t-\frac{x}{c}-u\right) = -\frac{\partial}{\partial u}K\left(t-\frac{x}{c}-u\right).$$

$$= \frac{\mu_0 c}{2}\hat{\mathbf{z}}\int_0^\infty\frac{\partial}{\partial u}K\left(t-\frac{x}{c}-u\right)du = \frac{\mu_0 c}{2}\hat{\mathbf{z}}\left[K\left(t-\frac{x}{c}-u\right)\right]\Big|_0^\infty = -\frac{\mu_0 c}{2}\left[K(t-x/c)-K(-\infty)\right]\hat{\mathbf{z}}$$

$$= \boxed{-\frac{\mu_0 c}{2}K(t-x/c)\,\hat{\mathbf{z}},}\ [\text{if } K(-\infty)=0].$$

Note that (i) and (ii) are consistent with this result. Meanwhile

$$\mathbf{B}(x,t) = -\frac{\partial A_z}{\partial x}\,\hat{\mathbf{y}} = -\frac{\mu_0 c}{c}\hat{\mathbf{y}}\int_0^\infty\frac{\partial}{\partial x}K\left(t-\frac{x}{c}-u\right)du. \quad \text{But } \frac{\partial}{\partial x}K\left(t-\frac{x}{c}-u\right) = \frac{1}{c}\frac{\partial}{\partial u}K\left(t-\frac{x}{c}-u\right).$$

$$= -\frac{\mu_0}{2}\hat{\mathbf{y}}\int_0^\infty\frac{\partial}{\partial u}K\left(t-\frac{x}{c}-u\right)du = -\frac{\mu_0}{2}\hat{\mathbf{y}}\left[K\left(t-\frac{x}{c}-u\right)\right]\Big|_0^\infty = \frac{\mu_0}{2}\left[K(t-x/c)-K(-\infty)\right]\hat{\mathbf{y}}$$

$$= \boxed{\frac{\mu_0}{2}K(t-x/c)\,\hat{\mathbf{y}},}\ [\text{if } K(-\infty)=0].$$

$$\mathbf{S} = \frac{1}{\mu_0}(\mathbf{E}\times\mathbf{B}) = \frac{1}{\mu_0}\left(\frac{\mu_0 c}{2}\right)\left(\frac{\mu_0}{2}\right)K(t-x/c)\left[-\hat{\mathbf{z}}\times\hat{\mathbf{y}}\right] = \frac{\mu_0 c}{4}\left[K(t-x/c)\right]^2\hat{\mathbf{x}}.$$

This is the power per unit area that reaches x at time t; it left the surface at time $(t-x/c)$. Moreover, an equal amount of energy is radiated *downward*, so the total power leaving the surface at time t is $\frac{\mu_0 c}{2}\left[K(t)\right]^2$.

Problem 11.25

$$p(t) = 2qz(t);\ \ddot{p} = 2q\ddot{z};\ F = m\ddot{z} = -\frac{1}{4\pi\epsilon_0}\frac{q^2}{(2z)^2};\ \ddot{z} = -\frac{1}{4\pi\epsilon_0}\frac{q^2}{4mz^2} = -\frac{\mu_0 c^2 q^2}{16\pi m z^2};\ \ddot{p} = -\frac{\mu_0 c^2 q^3}{8\pi m z^2}.$$

Using Eq. 11.60, the power radiated is $P = \frac{\mu_0\ddot{p}}{6\pi c} = \frac{\mu_0}{6\pi c}\left(-\frac{\mu_0 c^2 q^3}{8\pi m z^2}\right)^2 = \frac{\mu_0^3 c^3 q^6}{6(4\pi)^3 m^2 z^4} = \boxed{\left(\frac{\mu_0 c q^2}{4\pi}\right)^3\frac{1}{6m^2 z^4}.}$

Problem 11.26

With $\alpha = 90°$, Eq. 7.68 gives $\mathbf{E}' = c\mathbf{B}$, $\mathbf{B}' = -\frac{1}{c}\mathbf{E}$, $q_m' = -cq_e$. Use this to "translate" Eqs. 10.65, 10.66, and 11.70:

$$\begin{aligned}
\mathbf{E}' &= c\left(\frac{1}{c}\hat{\boldsymbol{\imath}}\times\mathbf{E}\right) = \hat{\boldsymbol{\imath}}\times(-c\mathbf{B}') = -c(\hat{\boldsymbol{\imath}}\times\mathbf{B}').\\
\mathbf{B}' &= -\frac{1}{c}\mathbf{E} = -\frac{1}{c}\frac{q_e}{4\pi\epsilon_0}\frac{\boldsymbol{\imath}}{(\boldsymbol{\imath}\cdot\mathbf{u})^3}\left[(c^2-v^2)\mathbf{u}+\boldsymbol{\imath}\times(\mathbf{u}\times\mathbf{a})\right]\\
&= -\frac{1}{c}\frac{(-q_m'/c)}{4\pi\epsilon_0}\frac{\boldsymbol{\imath}}{(\boldsymbol{\imath}\cdot\mathbf{u})^3}\left[(c^2-v^2)\mathbf{u}+\boldsymbol{\imath}\times(\mathbf{u}\times\mathbf{a})\right] = \frac{\mu_0 q_m'}{4\pi}\frac{\boldsymbol{\imath}}{(\boldsymbol{\imath}\cdot\mathbf{u})^3}\left[(c^2-v^2)\mathbf{u}+\boldsymbol{\imath}\times(\mathbf{u}\times\mathbf{a})\right].\\
P &= \frac{\mu_0 a^2}{6\pi c}q_e^2 = \frac{\mu_0 a^2}{6\pi c}\left(-\frac{1}{c}q_m'\right)^2 = \frac{\mu_0 a^2}{6\pi c^3}(q_m')^2.
\end{aligned}$$

Or, dropping the primes,

$$\boxed{\begin{aligned}
\mathbf{B}(\mathbf{r},t) &= \frac{\mu_0 q_m}{4\pi}\frac{\boldsymbol{\imath}}{(\boldsymbol{\imath}\cdot\mathbf{u})^3}\left[(c^2-v^2)\mathbf{u}+\boldsymbol{\imath}\times(\mathbf{u}\times\mathbf{a})\right].\\
\mathbf{E}(\mathbf{r},t) &= -c(\hat{\boldsymbol{\imath}}\times\mathbf{B}).\\
P &= \frac{\mu_0 q_m^2 a^2}{6\pi c^3}.
\end{aligned}}$$

Problem 11.27

(a) $W_{\text{ext}} = \int F\,dx = F\int_0^T v(t)\,dt$. From Prob. 11.19, $v(t) = \frac{F}{m}\left[t+\tau-\tau e^{(t-T)/\tau}\right]$. So

$$\begin{aligned}
W_{\text{ext}} &= \frac{F^2}{m}\left[\int_0^T t\,dt + \tau\int_0^T dt - \tau e^{-T/\tau}\int_0^T e^{t/\tau}\,dt\right] = \frac{F^2}{m}\left[\frac{t^2}{2}+\tau t-\tau e^{-T/\tau}\tau e^{t/\tau}\right]\Bigg|_0^T\\
&= \frac{F^2}{m}\left[\frac{1}{2}T^2+\tau T-\tau^2 e^{-T/\tau}\left(e^{T/\tau}-1\right)\right] = \boxed{\frac{F^2}{m}\left(\frac{1}{2}T^2+\tau T-\tau^2+\tau^2 e^{-T/\tau}\right).}
\end{aligned}$$

(b) From Prob. 11.19, the final velocity is $v_f = (F/m)T$, so $W_{\text{kin}} = \frac{1}{2}mv_f^2 = \frac{1}{2}m\frac{F^2}{m^2}T^2 = \boxed{\frac{F^2T^2}{2m}.}$

(c) $W_{\text{rad}} = \int P\,dt$. According to the Larmor formula, $P = \frac{\mu_0 q^2 a^2}{6\pi c}$, and (again from Prob. 11.19)

$$a(t) = \begin{cases} (F/m)\left[1-e^{-T/\tau}\right]e^{t/\tau}, & (t\le 0);\\ (F/m)\left[1-e^{(t-T)/\tau}\right], & (0\le t\le T).\end{cases}$$

$$\begin{aligned}
W_{\text{rad}} &= \frac{\mu_0 q^2}{6\pi c}\frac{F^2}{m^2}\left\{\left(1-e^{-T/\tau}\right)^2\int_{-\infty}^{0} e^{2t/\tau}\,dt + \int_0^T\left[1-e^{(t-T)/\tau}\right]^2 dt\right\}\\
&= \tau\frac{F^2}{m}\left\{\left(1-e^{-T/\tau}\right)^2\left(\frac{\tau}{2}e^{2t/\tau}\right)\Big|_{-\infty}^{0} + \int_0^T dt - 2e^{-T/\tau}\int_0^T e^{t/\tau}\,dt + e^{-2T/\tau}\int_0^T e^{2t/\tau}\,dt\right\}\\
&= \frac{\tau F^2}{m}\left[\frac{\tau}{2}\left(1-e^{-T/\tau}\right)^2 + T - 2e^{-T/\tau}\left(\tau e^{t/\tau}\right)\Big|_0^T + e^{-2T/\tau}\left(\frac{\tau}{2}e^{2t/\tau}\right)\Big|_0^T\right\}\\
&= \frac{\tau F^2}{m}\left[\frac{\tau}{2}\left(1-2e^{-T/\tau}+e^{-2T/\tau}\right) + T - 2\tau e^{-T/\tau}\left(e^{T/\tau}-1\right) + \frac{\tau}{2}e^{-2T/\tau}\left(e^{2T/\tau}-1\right)\right]\\
&= \frac{\tau F^2}{m}\left[\frac{\tau}{2} - \tau e^{-T/\tau} + \frac{\tau}{2}e^{-2T/\tau} + T - 2\tau + 2\tau e^{-T/\tau} + \frac{\tau}{2} - \frac{\tau}{2}e^{-2T/\tau}\right] = \boxed{\frac{\tau F^2}{m}\left(T-\tau+\tau e^{-T/\tau}\right).}
\end{aligned}$$

Energy conservation requires that the work done by the external force equal the final kinetic energy plus the energy radiated:

$$W_{\text{kin}} + W_{\text{rad}} = \frac{F^2T^2}{2m} + \frac{\tau F^2}{m}\left(T-\tau+\tau e^{-T/\tau}\right) = \frac{F^2}{m}\left(\frac{1}{2}T^2 + \tau T - \tau^2 + \tau^2 e^{-T/\tau}\right) = W_{\text{ext}}.\ \checkmark$$

Problem 11.28

(a) $a = \tau\dot{a} + \frac{k}{m}\delta(t) \Rightarrow \int_{-\epsilon}^{\epsilon} a(t)\,dt = v(\epsilon) - v(-\epsilon) = \tau\int_{-\epsilon}^{\epsilon}\frac{da}{dt}\,dt + \frac{k}{m}\int_{-\epsilon}^{\epsilon}\delta(t)\,dt = \tau[a(\epsilon)-a(-\epsilon)] + \frac{k}{m}.$

If the velocity is continuous, so $v(\epsilon) = v(-\epsilon)$, then $\boxed{a(\epsilon) - a(-\epsilon) = -\frac{k}{m\tau}.}$

When $t<0$, $a = \tau\dot{a} \Rightarrow a(t) = Ae^{t/\tau}$; when $t>0$, $a = \tau\dot{a} \Rightarrow a(t) = Be^{t/\tau}$; $\Delta a = B - A = -\frac{k}{m\tau}$

$\Rightarrow B = A - \frac{k}{m\tau}$, so the general solution is $\boxed{a(t) = \begin{cases} Ae^{t/\tau}, & (t<0);\\ [A-(k/m\tau)]\,e^{t/\tau}, & (t>0).\end{cases}}$

To eliminate the runaway we'd need $A = k/m\tau$; to eliminate preacceleration we'd need $A = 0$. Obviously, you can't do both. If you choose to eliminate the runaway, then $\boxed{a(t) = \begin{cases}(k/m\tau)e^{t/\tau}, & (t<0);\\ 0, & (t>0).\end{cases}}$

$$v(t) = \int_{-\infty}^{t} a(t)\,dt = \frac{k}{m\tau}\int_{-\infty}^{t} e^{t/\tau}\,dt = \frac{k}{m\tau}\left(\tau e^{t/\tau}\right)\Big|_{-\infty}^{t} = \frac{k}{m}e^{t/\tau}\ (\text{for } t<0);$$

for $t>0$, $v(t) = v(0) + \int_0^t a(t)\,dt = v(0) = \frac{k}{m}$. So $\boxed{v(t) = \begin{cases}(k/m)e^{t/\tau}, & (t<0);\\ (k/m), & (t>0).\end{cases}}$

For an *un*charged particle we would have $a(t) = \frac{k}{m}\delta(t)$, $v(t) = \int_{-\infty}^{t} a(t)\,dt = \begin{cases} 0, & (t<0);\\ (k/m), & (t>0).\end{cases}$

The graphs:

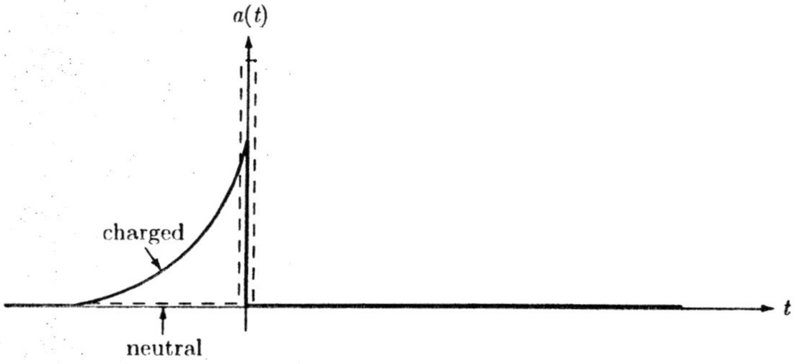

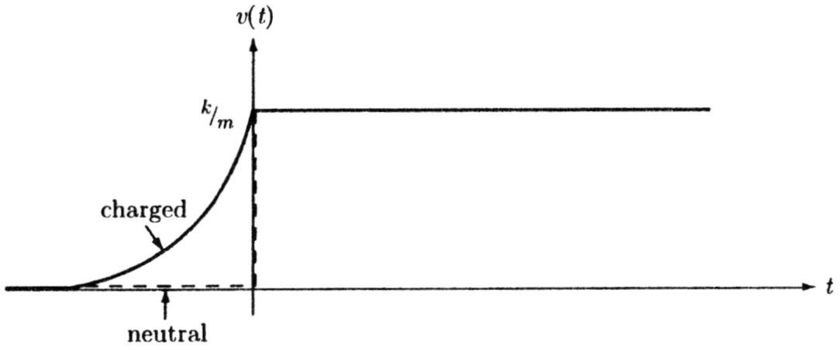

(b)

$$W_{\text{ext}} = \int F\,dx = \int Fv\,dt = k\int \delta(t)v(t)\,dt = kv(0) = \frac{k^2}{m}.$$

$$W_{\text{kin}} = \frac{1}{2}mv_f^2 = \frac{1}{2}m\left(\frac{k}{m}\right)^2 = \frac{k^2}{2m}.$$

$$W_{\text{ext}} = \int P_{\text{rad}}\,dt = \frac{\mu_0 q^2}{6\pi c}\int [a(t)]^2\,dt = \tau m\left(\frac{k}{m\tau}\right)^2\int_{-\infty}^{0} e^{2t/\tau}\,dt = \frac{k^2}{m\tau}\left(\frac{\tau}{2}e^{2t/\tau}\right)\Big|_{-\infty}^{0} = \frac{k^2}{m\tau}\frac{\tau}{2} = \frac{k^2}{2m}.$$

Clearly, $W_{\text{ext}} = W_{\text{kin}} + W_{\text{rad}}$. ✓

Problem 11.29

Our task is to solve the equation $a = \tau\dot{a} + \dfrac{U_0}{m}[-\delta(x) + \delta(x-L)]$, subject to the boundary conditions
(1) x continuous at $x = 0$ and $x = L$;
(2) v continuous at $x = 0$ and $x = L$;
(3) $\Delta a = \pm U_0/m\tau v$ (plus at $x = 0$, minus at $x = L$).

The third of these follows from integrating the equation of motion:

$$\begin{aligned}\int \frac{dv}{dt}\,dt &= \tau\int\frac{da}{dt}\,dt + \frac{U_0}{m}\int[-\delta(x) + \delta(x-L)]\,dt,\\ \Delta v &= \tau\Delta a + \frac{U_0}{m}\int[-\delta(x)+\delta(x-L)]\frac{dt}{dx}\,dx = 0,\\ \Delta a &= -\frac{U_0}{m\tau}\int\frac{1}{v}[-\delta(x)+\delta(x-L)]\,dx = \pm\frac{U_0}{m\tau v}.\end{aligned}$$

In each of the three regions the force is zero (it acts only at $x = 0$ and $x = L$), and the general solution is

$$a(t) = Ae^{t/\tau};\quad v(t) = A\tau e^{t/\tau} + B;\quad x(t) = A\tau^2 e^{t/\tau} + Bt + C.$$

(I'll put subscripts on the constants A, B, and C, to distinguish the three regions.)

Region iii ($x > L$): To avoid the runaway we pick $A_3 = 0$; then $a(t) = 0$, $v(t) = B_3$, $x(t) = B_3 t + C_3$. Let the final velocity be v_f ($= B_3$), set the clock so that $t = 0$ when the particle is at $x = 0$, and let T be the time it takes to traverse the barrier, so $x(T) = L = v_f T + C_3$, and hence $C_3 = L - v_f T$. Then

$$\boxed{a(t) = 0;\quad v(t) = v_f,\quad x(t) = L + v_f(t-T),}\quad (t < T).$$

Region ii ($0 < x < L$): $a = A_2 e^{t/\tau}$, $v = A_2\tau e^{t/\tau} + B_2$, $x = A_2\tau^2 e^{t/\tau} + B_2 t + C_2$.

$$(3) \quad \Rightarrow \quad 0 - A_2 e^{T/\tau} = -\frac{U_0}{m\tau v_f} \Rightarrow A_2 = \frac{U_0}{m\tau v_f} e^{-T/\tau}.$$

$$(2) \quad \Rightarrow \quad v_f = A_2 \tau e^{T/\tau} + B_2 = \frac{U_0}{mv_f} + B_2 \Rightarrow B_2 = v_f - \frac{U_0}{mv_f}.$$

$$(1) \quad \Rightarrow \quad L = A_2\tau^2 e^{T/\tau} + B_2 T + C_2 = \frac{U_0\tau}{mv_f} + v_f T - \frac{U_0 T}{mv_f} + C_2 = v_f T + \frac{U_0}{mv_f}(\tau - T) + C_2 \Rightarrow$$

$$C_2 = L - v_f T + \frac{U_0}{mv_f}(T - \tau).$$

$$\boxed{\begin{aligned} a(t) &= \frac{U_0}{m\tau v_f} e^{(t-T)/\tau}; \\ v(t) &= v_f + \frac{U_0}{mv_f}\left[e^{(t-T)/\tau} - 1\right]; \\ x(t) &= L + v_f(t-T) + \frac{U_0}{mv_f}\left[\tau e^{(t-T)/\tau} - t + T - \tau\right]; \end{aligned}} \qquad (0 < t < T).$$

[*Note:* if the barrier is sufficiently wide (or high) the particle may turn around before reaching L, but we're interested here in the régime where it *does* tunnel through.]

In particular, for $t = 0$ (when $x = 0$):

$$0 = L - v_f T + \frac{U_0}{mv_f}\left[\tau e^{-T/\tau} + T - \tau\right] \Rightarrow L = v_f T - \frac{U_0}{mv_f}\left[\tau e^{-T/\tau} + T - \tau\right]. \quad \text{qed}$$

Region i ($x < 0$): $a = A_1 e^{t/\tau}$, $v = A_1 \tau e^{t/\tau} + B_1$, $x = A_1\tau^2 e^{t/\tau} + B_1 t + C_1$. Let v_i be the incident velocity (at $t \to -\infty$); then $B_1 = v_i$. Condition (3) says

$$\frac{U_0}{m\tau v_f} e^{-T/\tau} - A_1 = \frac{U_0}{m\tau v_0},$$

where v_0 is the speed of the particle as it passes $x = 0$. From the solution in region (ii) it follows that $v_0 = v_f + \frac{U_0}{mv_f}\left(e^{-T/\tau} - 1\right)$. But we can also express it in terms of the solution in region (i): $v_0 = A_1\tau + v_i$. Therefore

$$\begin{aligned} v_i &= v_f + \frac{U_0}{mv_f}\left(e^{-T/\tau} - 1\right) - A_1\tau = v_f + \frac{U_0}{mv_f}\left(e^{-T/\tau} - 1\right) + \frac{U_0}{mv_0} - \frac{U_0}{mv_f} e^{-T/\tau} \\ &= v_f - \frac{U_0}{mv_f} + \frac{U_0}{mv_0} = v_f - \frac{U_0}{mv_f}\left(1 - \frac{v_f}{v_0}\right) = v_f - \frac{U_0}{mv_f}\left\{1 - \frac{v_f}{v_f + (U_0/mv_f)\left[e^{-T/\tau} - 1\right]}\right\} \\ &= v_f - \frac{U_0}{mv_f}\left\{1 - \frac{1}{1 + (U_0/mv_f^2)\left[e^{-T/\tau} - 1\right]}\right\}. \quad \text{qed} \end{aligned}$$

If $\frac{1}{2}mv_f^2 = \frac{1}{2}U_0$, then

$$L = v_f T - v_f\left[\tau e^{-T/\tau} + T - \tau\right] = v_f\left[T - \tau e^{-T/\tau} - T + \tau\right] = \tau v_f\left(1 - e^{-T/\tau}\right);$$

$$v_i = v_f - v_f\left[1 - \frac{1}{1+e^{-T/\tau}-1}\right] = v_f\left(1 - 1 + e^{T/\tau}\right) = v_f e^{T/\tau}.$$

Putting these together,

$$\frac{L}{\tau v_f} = 1 - e^{-T/\tau} \Rightarrow e^{-T/\tau} = 1 - \frac{L}{\tau v_f} \Rightarrow e^{T/\tau} = \frac{1}{1-(L/\tau v_f)} \Rightarrow v_i = \frac{v_f}{1-(L/v_f\tau)}. \quad \text{qed}$$

In particular, for $L = v_f\tau/4$, $v_i = \dfrac{v_f}{1-1/4} = \dfrac{4}{3}v_f$, so $\dfrac{KE_i}{KE_f} = \dfrac{\frac{1}{2}mv_i^2}{\frac{1}{2}mv_f^2} = \left(\dfrac{v_i}{v_f}\right)^2 = \dfrac{16}{9} \Rightarrow$ $KE_i = \dfrac{16}{9}KE_f = \dfrac{16}{9}\dfrac{1}{2}U_0 = \dfrac{8}{9}U_0$.

Problem 11.30

(a) From Eq. 10.65, $\mathbf{E}_1 = \dfrac{(q/2)}{4\pi\epsilon_0}\dfrac{\mathscr{r}}{(\boldsymbol{\mathscr{r}}\cdot\mathbf{u})^3}\left[(c^2-v^2)\mathbf{u} + (\boldsymbol{\mathscr{r}}\cdot\mathbf{a})\mathbf{u} - (\boldsymbol{\mathscr{r}}\cdot\mathbf{u})\mathbf{a}\right]$. Here $\mathbf{u} = c\hat{\boldsymbol{\mathscr{r}}} - \mathbf{v}$, $\boldsymbol{\mathscr{r}} = l\,\hat{\mathbf{x}} + d\,\hat{\mathbf{y}}$, $\mathbf{v} = v\,\hat{\mathbf{x}}$, $\mathbf{a} = a\,\hat{\mathbf{x}}$, so $\boldsymbol{\mathscr{r}}\cdot\mathbf{v} = lv$, $\boldsymbol{\mathscr{r}}\cdot\mathbf{a} = la$, $\boldsymbol{\mathscr{r}}\cdot\mathbf{u} = c\mathscr{r} - \boldsymbol{\mathscr{r}}\cdot\mathbf{v} = c\mathscr{r} - lv$. We want only the x component. Noting that $u_x = (c/\mathscr{r})l - v = (cl - v\mathscr{r})/\mathscr{r}$, we have:

$$\begin{aligned}
E_{1_x} &= \frac{q}{8\pi\epsilon_0}\frac{\mathscr{r}}{(c\mathscr{r}-lv)^3}\left[\frac{1}{\mathscr{r}}(cl-v\mathscr{r})(c^2-v^2+la) - a(c\mathscr{r}-lv)\right]\\
&= \frac{q}{8\pi\epsilon_0}\frac{1}{(c\mathscr{r}-lv)^3}\left[(cl-v\mathscr{r})(c^2-v^2) + cl^2a - v\mathscr{r}la - ac\mathscr{r}^2 + alv\mathscr{r}\right]. \text{ But } \mathscr{r}^2 = l^2+d^2.\\
&= \frac{q}{8\pi\epsilon_0}\frac{1}{(c\mathscr{r}-lv)^3}\left[(cl-v\mathscr{r})(c^2-v^2) - acd^2\right].\\
\mathbf{F}_{\text{self}} &= \frac{q^2}{8\pi\epsilon_0}\frac{1}{(c\mathscr{r}-lv)^3}\left[(cl-v\mathscr{r})(c^2-v^2) - acd^2\right]\hat{\mathbf{x}}. \quad \text{(This generalizes Eq. 11.90.)}
\end{aligned}$$

Now $x(t) - x(t_r) = l = vT + \frac{1}{2}aT^2 + \frac{1}{6}\dot{a}T^3 + \cdots$, where $T = t - t_r$, and v, a, and $\dot{a}$ are all evaluated at the retarded time t_r.

$$(cT)^2 = \mathscr{r}^2 = l^2 + d^2 = d^2 + (vT + \frac{1}{2}aT^2 + \frac{1}{6}\dot{a}T^3)^2 = d^2 + v^2T^2 + vaT^3 + \frac{1}{3}v\dot{a}T^4 + \frac{1}{4}a^2T^4;$$

$$c^2T^2(1 - v^2/c^2) = c^2T^2/\gamma^2 = d^2 + vaT^3 + \left(\frac{1}{3}v\dot{a} + \frac{1}{4}a^2\right)T^4. \text{ Solve for } T \text{ as a power series in } d:$$

$$T = \frac{\gamma d}{c}\left(1 + Ad + Bd^2 + \cdots\right) \Rightarrow \frac{c^2}{\gamma^2}\frac{\gamma^2d^2}{c^2}\left(1 + 2Ad + 2Bd^2 + A^2d^2\right) = d^2 + va\frac{\gamma^3d^3}{c^3}(1+3Ad) + \left(\frac{v\dot{a}}{3} + \frac{a^2}{4}\right)\frac{\gamma^4}{c^4}d^4.$$

Comparing like powers of d: $A = \dfrac{1}{2}va\dfrac{\gamma^3}{c^3}$; $2B + A^2 = \dfrac{3va\gamma^3}{c^3}A + \left(\dfrac{v\dot{a}}{3} + \dfrac{a^2}{4}\right)\dfrac{\gamma^4}{c^4}$.

$$\begin{aligned}
2B &= \frac{3va\gamma^3}{c^3}\frac{1}{2}va\frac{\gamma^3}{c^3} - \frac{1}{4}v^2a^2\frac{\gamma^6}{c^6} + \frac{v\dot{a}}{3}\frac{\gamma^4}{c^4} + \frac{a^2\gamma^4}{4c^4} = \frac{v\dot{a}}{3}\frac{\gamma^4}{c^4} + \frac{\gamma^6a^2}{4c^4}\left(\frac{1}{\gamma^2} - \frac{v^2}{c^2}\right) + \frac{3}{2}\frac{v^2a^2\gamma^6}{c^6}\\
&= \frac{\gamma^4}{c^4}\left[\frac{v\dot{a}}{3} + \frac{a^2\gamma^2}{4}\left(1 - \frac{v^2}{c^2} - \frac{v^2}{c^2} + 6\frac{v^2}{c^2}\right)\right] \Rightarrow B = \frac{\gamma^4}{2c^4}\left[\frac{v\dot{a}}{3} + \frac{\gamma^2a^2}{4}\left(1 + 4\frac{v^2}{c^2}\right)\right].\\
T &= \frac{\gamma d}{c}\left\{1 + \frac{va}{2}\frac{\gamma^3}{c^3}d + \frac{\gamma^4}{2c^4}\left[\frac{v\dot{a}}{3} + \frac{\gamma^2a^2}{4}\left(1 + 4\frac{v^2}{c^2}\right)\right]d^2\right\} + (\ )\,d^4 + \cdots \text{ (generalizing Eq. 11.93).}
\end{aligned}$$

$$
\begin{aligned}
l &= vT + \frac{1}{2}aT^2 + \frac{1}{6}\dot{a}T^3 + \cdots \\
&= \frac{v\gamma d}{c}\left\{1 + \frac{va}{2}\frac{\gamma^3}{c^3}d + \frac{\gamma^4}{2c^4}\left[\frac{v\dot{a}}{3} + \frac{\gamma^2 a^2}{4}\left(1 + 4\frac{v^2}{c^2}\right)\right]d^2\right\} + \frac{1}{2}a\frac{\gamma^2 d^2}{c^2}\left[1 + va\frac{\gamma^3}{c^3}d\right] + \frac{1}{6}\dot{a}\frac{\gamma^3}{c^3}d^3 \\
&= \left(\frac{v\gamma}{c}\right)d + \frac{a}{2}\frac{\gamma^4}{c^2}\left(1 - \frac{v^2}{c^2} + \frac{v^2}{c^2}\right)d^2 + \left\{\frac{v\gamma}{2c}\frac{\gamma^4}{c^4}\left[\frac{v\dot{a}}{3} + \frac{\gamma^2 a^2}{4}\left(1 + 4\frac{v^2}{c^2}\right)\right] + \frac{1}{2}a\frac{\gamma^2}{c^2}va\frac{\gamma^3}{c^3} + \frac{1}{6}\dot{a}\frac{\gamma^3}{c^3}\right\}d^3 \\
&= \left(\frac{v\gamma}{c}\right)d + \left(\frac{a\gamma^4}{2c^2}\right)d^2 + \frac{\gamma^3}{2c^3}\left[\frac{\dot{a}}{3}\left(1 + \gamma^2\frac{v^2}{c^2}\right) + \frac{v\gamma^4 a^2}{c^2}\left(\frac{1}{4} + \frac{v^2}{c^2} + 1 - \frac{v^2}{c^2}\right)\right]d^3 \\
&= \left(\frac{v\gamma}{c}\right)d + \left(\frac{a\gamma^4}{2c^2}\right)d^2 + \frac{\gamma^5}{2c^3}\left[\frac{\dot{a}}{3} + \frac{5}{4}\frac{v\gamma^2 a^2}{c^2}\right]d^3 + (\;)\,d^4 + \cdots
\end{aligned}
$$

$$
\begin{aligned}
\mathscr{r} &= cT = \gamma d\left\{1 + \frac{va}{2}\frac{\gamma^3}{c^3}d + \frac{\gamma^4}{2c^4}\left[\frac{v\dot{a}}{3} + \gamma^2 a^2\left(\frac{1}{4} + \frac{v^2}{c^2}\right)\right]d^2\right\} + (\;)\,d^4 + \cdots \\
c\mathscr{r} - lv &= c\gamma d + \frac{va\gamma^4}{2c^2}d^2 + \frac{\gamma^5}{2c^3}\left[\frac{v\dot{a}}{3} + \gamma^2 a^2\left(\frac{1}{4} + \frac{v^2}{c^2}\right)\right]d^3 - \frac{v^2\gamma}{c}d - \frac{av\gamma^4}{2c^2}d^2 - \frac{\gamma^5 v}{2c^3}\left[\frac{\dot{a}}{3} + \frac{5}{4}\frac{v\gamma^2 a^2}{c^2}\right]d^3 + \cdots \\
&= c\gamma d\left(1 - \frac{v^2}{c^2}\right) + \frac{\gamma^5}{2c^3}\left[\frac{v\dot{a}}{3} + \gamma^2 a^2\left(\frac{1}{4} + \frac{v^2}{c^2}\right) - \frac{v\dot{a}}{3} - \frac{5}{4}\frac{v^2\gamma^2 a^2}{c^2}\right]d^3 + \cdots \\
&= \frac{c}{\gamma}d + \frac{\gamma^5 a^2}{8c^3}d^3 + (\;)\,d^4 + \cdots
\end{aligned}
$$

$$
\begin{aligned}
cl - v\mathscr{r} &= v\gamma d + \frac{a\gamma^4}{2c}d^2 + \frac{\gamma^5}{2c^2}\left(\frac{\dot{a}}{3} + \frac{5}{4}\frac{v\gamma^2 a^2}{c^2}\right)d^3 - v\gamma d - \frac{v^2 a}{2}\frac{\gamma^4}{c^3}d^2 - \frac{v\gamma^5}{2c^4}\left[\frac{v\dot{a}}{3} + \gamma^2 a^2\left(\frac{1}{4} + \frac{v^2}{c^2}\right)\right]d^3 \\
&= \frac{a\gamma^4}{2c}\left(1 - \frac{v^2}{c^2}\right)d^2 + \frac{\gamma^5}{2c^2}\left[\frac{\dot{a}}{3} + \frac{5}{4}\frac{v\gamma^2 a^2}{c^2} - \frac{v^2}{c^2}\frac{\dot{a}}{3} - \frac{v\gamma^2 a^2}{c^2}\left(\frac{1}{4} + \frac{v^2}{c^2}\right)\right]d^3 + (\;)\,d^4 + \cdots \\
&= \left(\frac{a\gamma^2}{2c}\right)d^2 + \frac{\gamma^5}{2c^2}\left[\frac{\dot{a}}{3\gamma^2} + \frac{v\gamma^2 a^2}{c^2}\left(\frac{5}{4} - \frac{1}{4} - \frac{v^2}{c^2}\right)\right]d^3 + (\;)\,d^4 + \cdots \\
&= \left(\frac{a\gamma^2}{2c}\right)d^2 + \frac{\gamma^3}{2c^2}\left(\frac{\dot{a}}{3} + \frac{v\gamma^2 a^2}{c^2}\right)d^3 + (\;)\,d^4 + \cdots \\
(c\mathscr{r} - lv)^{-3} &= \left[\frac{cd}{\gamma}\left(1 + \frac{\gamma^6 a^2}{8c^4}d^2\right)\right]^{-3} = \left(\frac{\gamma}{cd}\right)^3\left(1 - 3\frac{\gamma^6 a^2}{8c^4}d^2\right) + \cdots
\end{aligned}
$$

$$
\begin{aligned}
\mathbf{F}_{\text{self}} &= \frac{q^2}{8\pi\epsilon_0}\left(\frac{\gamma}{cd}\right)^3\left(1 - 3\frac{\gamma^6 a^2}{8c^4}d^2\right)\left\{\left[\left(\frac{a\gamma^2}{2c}\right)d^2 + \frac{\gamma^3}{2c^2}\left(\frac{\dot{a}}{3} + \frac{v\gamma^2 a^2}{c^2}\right)d^3\right]\frac{c^2}{\gamma^2} - acd^2\right\}\hat{\mathbf{x}} \\
&= \frac{q^2}{8\pi\epsilon_0}\frac{\gamma^3}{c^3 d}\left(1 - \frac{3}{8}\frac{\gamma^6 a^2}{c^4}d^2\right)\left[-\frac{ac}{2} + \frac{\gamma}{2}\left(\frac{\dot{a}}{3} + \frac{v\gamma^2 a^2}{c^2}\right)d\right]\hat{\mathbf{x}} \\
&= \frac{q^2}{8\pi\epsilon_0}\frac{\gamma^3}{c^3 d}\frac{1}{2}\left[-ac + \gamma\left(\frac{\dot{a}}{3} + \frac{v\gamma^2 a^2}{c^2}\right)d + (\;)\,d^2 + \cdots\right]\hat{\mathbf{x}} \\
&= \frac{q^2}{4\pi\epsilon_0}\left[-\gamma^3\frac{a}{4c^2 d} + \frac{\gamma^4}{4c^3}\left(\frac{\dot{a}}{3} + \frac{v\gamma^2 a^2}{c^2}\right) + (\;)\,d + \cdots\right]\hat{\mathbf{x}} \text{ (generalizing Eq. 11.95).}
\end{aligned}
$$

Switching to t: $v(t_r) = v(t) + \dot{v}(t)(t_r - t) + \cdots = v(t) - a(t)T = v(t) - a\gamma d/c$. (When multiplied by d, it doesn't matter—to this order—whether we evaluate at t or at t_r.)

$$1-\left[\frac{v(t_r)}{c}\right]^2 = 1 - \frac{[v(t)^2 - 2va\gamma d/c]}{c^2} = \left[1-\frac{v(t)^2}{c^2}\right]\left(1+\frac{2av\gamma^3 d}{c^3}\right), \text{ so}$$

$$\gamma = \left[1-\left(\frac{v(t_r)}{c}\right)^2\right]^{-1/2} = \gamma(t)\left(1-\frac{va\gamma^3}{c^3}d\right); \; a(t_r) = a(t) - T\dot{a} = a(t) - \frac{\dot{a}\gamma}{c}d.$$

Evaluating everything now at time t:

$$\begin{aligned}
\mathbf{F}_{\text{self}} &= \frac{q^2}{4\pi\epsilon_0}\left[-\gamma^3\frac{(1-3va\gamma^3 d/c^3)(a-\dot{a}\gamma d/c)}{4c^2 d} + \frac{\gamma^4}{4c^3}\left(\frac{\dot{a}}{3}+\frac{v\gamma^2 a^2}{c^2}\right) + (\,)\,d^2 + \cdots\right]\hat{\mathbf{x}} \\
&= \frac{q^2}{4\pi\epsilon_0}\left[-\frac{\gamma^3 a}{4c^2 d} + \frac{\gamma^3}{4c^2}\left(\frac{\dot{a}\gamma}{c} + 3\frac{va^2\gamma^2}{c^3}\right) + \frac{\gamma^4}{4c^3}\left(\frac{\dot{a}}{3}+\frac{v\gamma^2 a^2}{c^2}\right) + (\,)\,d + \cdots\right]\hat{\mathbf{x}} \\
&= \frac{q^2}{4\pi\epsilon_0}\left[-\frac{\gamma^3 a}{4c^2 d} + \frac{\gamma^4}{4c^3}\left(\dot{a} + \frac{\dot{a}}{3} + 3\frac{va^2\gamma^2}{c^2} + \frac{v\gamma^2 a^2}{c^2}\right) + (\,)\,d + \cdots\right]\hat{\mathbf{x}} \\
&= \frac{q^2}{4\pi\epsilon_0}\left[-\frac{\gamma^3 a}{4c^2 d} + \frac{\gamma^4}{3c^3}\left(\dot{a} + 3\frac{va^2\gamma^2}{c^2}\right) + (\,)\,d + \cdots\right]\hat{\mathbf{x}} \text{ (generalizing Eq. 11.96).}
\end{aligned}$$

The first term is the electromagnetic mass; the radiation reaction itself is the second term: $F_{\text{rad}}^{\text{int}} = \frac{\mu_0 q^2}{12\pi c}\gamma^4\left(\dot{a} + 3\frac{va^2\gamma^2}{c^2}\right)$ (generalizing Eq. 11.99), so the generalization of Eq. 11.100 is

$$\boxed{F_{\text{rad}} = \frac{\mu_0 q^2}{6\pi c}\gamma^4\left(\dot{a} + 3\frac{va^2\gamma^2}{c^2}\right).}$$

(b) $F_{\text{rad}} = A\gamma^4\left(\dot{a} + \frac{3\gamma^2 a^2 v}{c^2}\right)$, where $A \equiv \frac{\mu_0 q^2}{6\pi c}$. $P = Aa^2\gamma^6$ (Eq. 11.75). What we must show is that

$$\int_{t_1}^{t_2} F_{\text{rad}} v\,dt = -\int_{t_1}^{t_2} P\,dt, \quad \text{or} \quad \int_{t_1}^{t_2}\gamma^4\left(\dot{a}v + 3\frac{v^2a^2\gamma^2}{c^2}\right)dt = -\int_{t_1}^{t_2} a^2\gamma^6\,dt$$

(except for boundary terms—see Sect. 11.2.2).

Rewrite the first term: $\int_{t_1}^{t_2}\gamma^4\dot{a}v\,dt = \int_{t_1}^{t_2}(\gamma^4 v)\frac{da}{dt}dt = \gamma^4 va\Big|_{t_1}^{t_2} - \int_{t_1}^{t_2}\frac{d}{dt}(\gamma^4 v)a\,dt.$

Now $\frac{d}{dt}(\gamma^4 v) = 4\gamma^3\frac{d\gamma}{dt}v + \gamma^4 a$; $\frac{d\gamma}{dt} = \frac{d}{dt}\left(\frac{1}{\sqrt{1-v^2/c^2}}\right) = -\frac{1}{2}\frac{1}{(1-v^2/c^2)^{3/2}}\left(-\frac{2va}{c^2}\right) = \frac{va\gamma^3}{c^2}$. So

$$\frac{d}{dt}(\gamma^4 v) = 4\gamma^3 v\frac{va\gamma^3}{c^2} + \gamma^4 a = \gamma^6 a\left(1-\frac{v^2}{c^2}+4\frac{v^2}{c^2}\right) = \gamma^6 a\left(1+3\frac{v^2}{c^2}\right).$$

$\int_{t_1}^{t_2}\gamma^4\dot{a}v\,dt = \gamma^4 va\Big|_{t_1}^{t_2} - \int_{t_1}^{t_2}\gamma^6 a^2\left(1+3\frac{v^2}{c^2}\right)dt$, and hence

$$\int_{t_1}^{t_2}\gamma^4\left(\dot{a}v + \frac{3\gamma^2a^2v^2}{c^2}\right)dt = \gamma^4 va\Big|_{t_1}^{t_2} + \int_{t_1}^{t_2}\left[-\gamma^6a^2\left(1+3\frac{v^2}{c^2}\right) + 3\gamma^6\frac{a^2v^2}{c^2}\right]dt = \gamma^4 va\Big|_{t_1}^{t_2} - \int_{t_1}^{t_2}\gamma^6a^2\,dt. \quad \text{qed}$$

Problem 11.31

(a) $P = \frac{\mu_0 q^2 a^2\gamma^6}{6\pi c}$ (Eq. 11.75). $w = \sqrt{b^2+c^2t^2}$ (Eq. 10.45); $v = \dot{w} = \frac{c^2 t}{\sqrt{b^2+c^2t^2}}$;

$$a = \dot{v} = \frac{c^2}{\sqrt{b^2+c^2t^2}} - \frac{c^2t(c^2t)}{(b^2+c^2t^2)^{3/2}} = \frac{c^2}{(b^2+c^2t^2)^{3/2}}(b^2+c^2t^2-c^2t^2) = \frac{b^2c^2}{(b^2+c^2t^2)^{3/2}};$$

$$\gamma^2 = \frac{1}{1-v^2/c^2} = \frac{1}{1-[c^2t^2/(b^2+c^2t^2)]} = \frac{b^2+c^2t^2}{b^2+c^2t^2-c^2t^2} = \frac{1}{b^2}\left(b^2+c^2t^2\right).$$ So

$$P = \frac{\mu_0 q^2}{6\pi c}\frac{b^4c^4}{(b^2+c^2t^2)^3}\frac{(b^2+c^2t^2)^3}{b^6} = \boxed{\frac{q^2c}{6\pi\epsilon_0 b^2}}.$$ $\boxed{\text{Yes, it radiates}}$ (in fact, at a *constant rate*).

(b) $F_{\text{rad}} = \frac{\mu_0 q^2\gamma^4}{6\pi c}\left(\dot{a}+\frac{3\gamma^2a^2v}{c^2}\right)$; $\dot{a} = -\frac{3}{2}\frac{b^2c^2(2c^2t)}{(b^2+c^2t^2)^{5/2}} = -\frac{3b^2c^4t}{(b^2+c^2t^2)^{5/2}}$; $\left(\dot{a}+\frac{3\gamma^2a^2v}{c^2}\right) =$

$$-\frac{3b^2c^4t}{(b^2+c^2t^2)^{5/2}} + \frac{3}{c^2}\frac{(b^2+c^2t^2)}{b^2}\frac{b^4c^4}{(b^2+c^2t^2)^3}\frac{c^2t}{\sqrt{b^2+c^2t^2}} = 0.$$ $\boxed{F_{\text{rad}} = 0.}$ $\boxed{\text{No, the radiation reaction is zero.}}$

Chapter 12

Electrodynamics and Relativity

Problem 12.1

Let $\mathbf{u}$ be the velocity of a particle in $\mathcal{S}$, $\bar{\mathbf{u}}$ its velocity in $\bar{\mathcal{S}}$, and $\mathbf{v}$ the velocity of $\bar{\mathcal{S}}$ with respect to $\mathcal{S}$. Galileo's velocity addition rule says that $\mathbf{u} = \bar{\mathbf{u}} + \mathbf{v}$. For a free particle, $\mathbf{u}$ is constant (that's Newton's first law in $\mathcal{S}$).

(a) If $\mathbf{v}$ is constant, then $\bar{\mathbf{u}} = \bar{\mathbf{u}} - \mathbf{v}$ is also constant, so Newton's first law holds in $\bar{\mathcal{S}}$, and hence $\mathcal{S}$ is inertial.

(b) If $\bar{\mathcal{S}}$ is inertial, then $\bar{\mathbf{u}}$ is also constant, so $\mathbf{v} = \mathbf{u} - \bar{\mathbf{u}}$ is constant.

Problem 12.2

(a) $m_A\mathbf{u}_A + m_B\mathbf{u}_B = m_C\mathbf{u}_C + m_D\mathbf{u}_D; \quad \mathbf{u}_i = \bar{\mathbf{u}}_i + \mathbf{v}.$

$m_A(\bar{\mathbf{u}}_A + \mathbf{v}) + m_B(\bar{\mathbf{u}}_B + \mathbf{v}) = m_C(\bar{\mathbf{u}}_C + \mathbf{v}) + m_D(\bar{\mathbf{u}}_D + \mathbf{v}),$

$m_A\bar{\mathbf{u}}_A + m_B\bar{\mathbf{u}}_B + (m_A + m_B)\mathbf{v} = m_C\bar{\mathbf{u}}_C + m_D\bar{\mathbf{u}}_D + (m_C + m_D)\mathbf{v}.$

Assuming *mass* is conserved, $(m_A + m_B) = (m_C + m_D)$, it follows that

$m_A\bar{\mathbf{u}}_A + m_B\bar{\mathbf{u}}_B = m_C\bar{\mathbf{u}}_C + m_D\bar{\mathbf{u}}_D$, so momentum is conserved in $\bar{\mathcal{S}}$.

(b) $\frac{1}{2}m_Au_A^2 + \frac{1}{2}m_Bu_B^2 = \frac{1}{2}m_Cu_C^2 + \frac{1}{2}m_Du_D^2 \Rightarrow$

$$\tfrac{1}{2}m_A(\bar{u}_A^2 + 2\bar{\mathbf{u}}_A\cdot\mathbf{v} + v^2) + \tfrac{1}{2}m_B(\bar{u}_B^2 + 2\bar{\mathbf{u}}_B\cdot\mathbf{v} + v^2) = \tfrac{1}{2}m_C(\bar{u}_C^2 + 2\bar{\mathbf{u}}_C\cdot\mathbf{v} + v^2) + \tfrac{1}{2}m_D(\bar{u}_D^2 + 2\bar{\mathbf{u}}_D\cdot\mathbf{v} + v^2)$$

$$\begin{aligned}&\tfrac{1}{2}m_A\bar{u}_A^2 + \tfrac{1}{2}m_B\bar{u}_B^2 + 2\mathbf{v}\cdot(m_A\bar{\mathbf{u}}_A + m_B\bar{\mathbf{u}}_B) + \tfrac{1}{2}v^2(m_A + m_B)\\ &\quad = \tfrac{1}{2}m_C\bar{u}_C^2 + \tfrac{1}{2}m_D\bar{u}_D^2 + 2\mathbf{v}\cdot(m_C\bar{\mathbf{u}}_C + m_D\bar{\mathbf{u}}_D) + \tfrac{1}{2}v^2(m_C + m_D).\end{aligned}$$

But the middle terms are equal by conservation of momentum, and the last terms are equal by conservation of mass, so $\frac{1}{2}m_A\bar{u}_A^2 + \frac{1}{2}m_B\bar{u}_B^2 = \frac{1}{2}m_C\bar{u}_C^2 + \frac{1}{2}m_D\bar{u}_D^2$. qed

Problem 12.3

(a) $v_G = v_{AB} + v_{BC}$; $v_E = \frac{v_{AB}+v_{BC}}{1+v_{AB}v_{BC}/c^2} \approx v_G\left(1 - \frac{v_{AB}v_{BC}}{c^2}\right) \Rightarrow \frac{v_G - v_E}{v_G} = \frac{v_{AB}v_{BC}}{c^2}$.

In mi/h, $c = (186{,}000\,\text{mi/s}) \times (3600\,\text{sec/hr}) = 6.7 \times 10^8\,\text{mi/hr}$.

$\therefore \frac{v_G - v_E}{v_G} = \frac{(5)(60)}{(6.7\times10^8)^2} = 6.7 \times 10^{-16} \Rightarrow \boxed{6.7 \times 10^{-14}\%\text{ error,}}$ (pretty small!)

(b) $\left(\frac{1}{2}c + \frac{3}{4}c\right) / \left(1 + \frac{1}{2}\cdot\frac{3}{4}\right) = \left(\frac{5}{4}c\right) / \left(\frac{11}{8}\right) = \boxed{\frac{10}{11}c}$ (still *less* than c).

(c) To simplify the notation, let $\beta \equiv v_{AC}/c$, $\beta_1 \equiv v_{AB}/c$, $\beta_2 \equiv v_{BC}/c$. Then Eq. 12.3 says: $\beta = \frac{\beta_1+\beta_2}{1+\beta_1\beta_2}$, or:

$$\beta^2 = \frac{\beta_1^2 + 2\beta_1\beta_2 + \beta_2^2}{(1 + 2\beta_1\beta_2 + \beta_1^2\beta_2^2)} = \frac{1 + 2\beta_1\beta_2 + \beta_1^2\beta_2^2}{(1 + 2\beta_1\beta_2 + \beta_1^2\beta_2^2)} - \frac{(1 + \beta_1^2\beta_2^2 - \beta_1^2 - \beta_2^2)}{(1 + 2\beta_1\beta_2 + \beta_1^2\beta_2^2)} = 1 - \frac{(1-\beta_1^2)(1-\beta_2^2)}{(1+\beta_1\beta_2)^2} = 1 - \Delta,$$

where $\Delta \equiv (1-\beta_1^2)(1-\beta_2^2)/(1+\beta_1\beta_2)^2$ is clearly a *positive* number. So $\beta_2 < 1$, and hence $|v_{AC}| < c$. qed

Problem 12.4

(a) Velocity of bullet relative to ground: $\frac{1}{2}c + \frac{1}{3}c = \frac{5}{6}c = \frac{10}{12}c$.
Velocity of getaway car: $\frac{3}{4}c = \frac{9}{12}c$. Since $v_b > v_g$, $\boxed{\text{bullet } does \text{ reach target.}}$

(b) Velocity of bullet relative to ground: $\frac{\frac{1}{2}c+\frac{1}{3}c}{1+\frac{1}{2}\cdot\frac{1}{3}} = \frac{\frac{5}{6}c}{\frac{7}{6}} = \frac{5}{7}c = \frac{20}{28}c$.
Velocity of getaway car: $\frac{3}{4}c = \frac{21}{28}c$. Since $v_g > v_b$, $\boxed{\text{bullet does } not \text{ reach target.}}$

Problem 12.5

(a) Light from the 90th clock took $\frac{90\times10^9 \text{ m}}{3\times10^8 \text{ m/s}} = 300$ s $= 5$ min to reach me, so the time I *see* on the clock is $\boxed{\text{11:55 am.}}$

(b) I *observe* $\boxed{\text{12 noon.}}$

Problem 12.6

$$\begin{cases}\text{light signal leaves } a \text{ at time } t'_a\text{; arrives at earth at time } t_a = t'_a + d_a/c,\\ \text{light signal leaves } b \text{ at time } t'_b\text{; arrives at earth at time } t_b = t'_b + d_b/c.\end{cases}$$

$$\therefore \Delta t = t_b - t_a = t'_b - t'_a + \frac{(d_b - d_a)}{c} = \Delta t' + \frac{(-v\Delta t'\cos\theta)}{c} = \Delta t'\left[1 - \frac{v}{c}\cos\theta\right].$$

(Here d_a is the distance from a to earth, and d_b is the distance from b to earth.)

$$\Delta s = v\Delta t'\sin\theta = \frac{v\sin\theta\,\Delta t}{(1 - v/c\cos\theta)}; \quad \boxed{u = \frac{v\sin\theta}{(1-\frac{v}{c}\cos\theta)}} \text{ is the the apparent velocity.}$$

$$\frac{du}{d\theta} = \frac{v\left[(1-\frac{v}{c}\cos\theta)(\cos\theta) - \sin\theta(\frac{v}{c}\sin\theta)\right]}{(1-\frac{v}{c}\cos\theta)^2} = 0 \Rightarrow (1 - \frac{v}{c}\cos\theta)\cos\theta = \frac{v}{c}\sin^2\theta$$
$$\Rightarrow \cos\theta = \frac{v}{c}(\sin^2\theta + \cos^2\theta) = \frac{v}{c}$$

$\boxed{\theta_{\max} = \cos^{-1}(v/c).}$ At this maximal angle, $u = \frac{v\sqrt{1-v^2/c^2}}{1-v^2/c^2} = \frac{v}{\sqrt{1-v^2/c^2}}$.
As $v \to c$, $\boxed{u \to \infty,}$ because the denominator $\to 0$, even though $v < c$.

Problem 12.7

The student has not taken into account time dilation of the muon's "internal clock". In the *laboratory*, the muon lasts $\gamma\tau = \frac{\tau}{\sqrt{1-v^2/c^2}}$, where τ is the "proper" lifetime, 2×10^{-6} s. Thus

$$v = \frac{d}{\tau/\sqrt{1-v^2/c^2}} = \frac{d}{\tau}\sqrt{1-v^2/c^2}, \text{ where } d = 800 \text{ m}.$$

$$\left(\frac{\tau}{d}\right)^2 v^2 = 1 - \frac{v^2}{c^2}; \quad v^2\left[\left(\frac{\tau}{d}\right)^2 + \frac{1}{c^2}\right] = 1; \quad v^2 = \frac{1}{(\tau/d)^2 + (1/c)^2}.$$

$$\frac{v^2}{c^2} = \frac{1}{1+(\tau c/d)^2}; \quad \frac{\tau c}{d} = \frac{(2\times10^{-6})(3\times10^8)}{800} = \frac{6}{8} = \frac{3}{4}; \quad \frac{v^2}{c^2} = \frac{1}{1+9/16} = \frac{16}{25}; \quad \boxed{v = \frac{4}{5}c.}$$

Problem 12.8

(a) Rocket clock runs slow; so earth clock reads $\gamma t = \frac{1}{\sqrt{1-v^2/c^2}} \cdot 1$ hr. Here $\gamma = \frac{1}{\sqrt{1-v^2/c^2}} = \frac{1}{\sqrt{1-9/25}} = \frac{5}{4}$.

$\therefore$ According to earth clocks signal was sent $\boxed{\text{1 hr and 15 min}}$ after take-off.

(b) By earth observer, rocket is now a distance $\left(\frac{3}{5}c\right)\left(\frac{5}{4}\right)(1 \text{ hr}) = \frac{3}{4}c$ hr (three-quarters of a light hour) away. Light signal will therefore take $\frac{3}{4}$ hr to return to earth. Since it *left* 1 hr and 15 min after departure, light signal reaches earth $\boxed{\text{2 hrs after takeoff.}}$

(c) Earth clocks run slow: $t_{\text{rocket}} = \gamma \cdot (2 \text{ hrs}) = \frac{5}{4} \cdot (2 \text{ hrs}) = \boxed{2.5 \text{ hrs.}}$

Problem 12.9

$$L_c = 2L_v;\ \frac{L_c}{\gamma_c} = \frac{L_v}{\gamma_v};\ \text{so } \frac{2}{\gamma_c} = \frac{1}{\gamma_v} = \sqrt{1-\left(\tfrac{1}{2}\right)^2} = \sqrt{\tfrac{3}{4}};\ \frac{1}{\gamma_c^2} = 1 - \frac{v^2}{c^2} = \frac{3}{16};\ \frac{v^2}{c^2} = 1 - \frac{3}{16} = \frac{13}{16};\ \boxed{v = \frac{\sqrt{13}}{4}c.}$$

Problem 12.10

Say length of mast (at rest) is l. To an observer on the boat, height of mast is $l\sin\theta$, horizontal projection is $l\cos\theta$. To observer on dock, the former is unaffected, but the latter is Lorentz contracted to $\frac{1}{\gamma}l\cos\theta$. Therefore:

$$\tan\bar{\theta} = \frac{l\sin\theta}{\frac{1}{\gamma}l\cos\theta} = \gamma\tan\theta, \text{ or } \boxed{\tan\bar{\theta} = \frac{\tan\theta}{\sqrt{1-v^2/c^2}}.}$$

Problem 12.11

Naively, circumference/diameter $= \frac{1}{\gamma}(2\pi R)/(2R) = \pi/\gamma = \pi\sqrt{1-(\omega R/c)^2}$ — but this is nonsense. Point is: an accelerating object cannot remain rigid, in relativity. To decide what *actually* happens here, you need a specific model for the internal forces holding the disk together.

Problem 12.12

(iv) $\Rightarrow t = \frac{\bar{t}}{\gamma} + \frac{v\bar{x}}{c^2}$. Put this into (i), and solve for x:

$$\bar{x} = \gamma x - \gamma v\left(\frac{\bar{t}}{\gamma} + \frac{vx}{c^2}\right) = \gamma x\left(1 - \frac{v^2}{c^2}\right) - v\bar{t} = \gamma x\frac{1}{\gamma^2} - v\bar{t} = \frac{x}{\gamma} - v\bar{t};\ \boxed{x = \gamma(\bar{x} + v\bar{t}).}\ \checkmark$$

Similarly, (i) $\Rightarrow x = \frac{\bar{x}}{\gamma} + vt$. Put this into (iv) and solve for t:

$$\bar{t} = \gamma t - \frac{\gamma v}{c^2}\left(\frac{\bar{x}}{\gamma} + vt\right) = \gamma t\left(1 - \frac{v^2}{c^2}\right) - \frac{v}{c^2}\bar{x} = \frac{t}{\gamma} - \frac{v}{c^2}\bar{x};\ \boxed{t = \gamma\left(\bar{t} + \frac{v}{c^2}\bar{x}\right).}\ \checkmark$$

Problem 12.13

Let brother's accident occur at origin, time zero, in both frames. In system $\mathcal{S}$ (Sophie's), the coordinates of Sophie's cry are $x = 5 \times 10^5$ m, $t = 0$. In system $\bar{\mathcal{S}}$ (scientist's), $\bar{t} = \gamma(t - \frac{v}{c^2}x) = -\gamma vx/c^2$. Since this is *negative*, $\boxed{\text{Sophie's cry occurred } \textit{before} \text{ the accident,}}$ in $\bar{\mathcal{S}}$. $\gamma = \frac{1}{\sqrt{1-(12/13)^2}} = \frac{13}{\sqrt{169-144}} = \frac{13}{5}$. So $\bar{t} = -\left(\frac{13}{5}\right)\left(\frac{12}{13}c\right)(5 \times 10^5)/c^2 = -12 \times 10^5/3 \times 10^8 = -4 \times 10^{-3}$. $\boxed{4 \times 10^{-3}\text{ s earlier.}}$

Problem 12.14

(a) In $\mathcal{S}$ it moves a distance dy in time dt. In $\bar{\mathcal{S}}$, meanwhile, it moves a distance $d\bar{y} = dy$ in time $d\bar{t} = \gamma(dt - \frac{v}{c^2}dx)$.

$$\therefore \frac{d\bar{y}}{d\bar{t}} = \frac{dy}{\gamma(dt - \frac{v}{c^2}dx)} = \frac{(dy/dt)}{\gamma\left(1 - \frac{v}{c^2}\frac{dx}{dt}\right)};\ \text{or } \boxed{\bar{u}_y = \frac{u_y}{\gamma\left(1 - \frac{vu_x}{c^2}\right)};\ \bar{u}_z = \frac{u_z}{\gamma\left(1 - \frac{vu_x}{c^2}\right)}.}$$

(b) $\tan\bar{\theta} = -\dfrac{\bar{u}_y}{\bar{u}_x} = -\dfrac{u_y/\left[\gamma\left(1-\frac{vu_x}{c^2}\right)\right]}{(u_x - v)/\left(1-\frac{vu_x}{c^2}\right)} = \dfrac{1}{\gamma}\dfrac{(-u_y)}{(u_x - v)}$.

In this case $u_x = -c\cos\theta$; $u_y = c\sin\theta \Rightarrow \tan\bar{\theta} = \frac{1}{\gamma}\left(\frac{-c\sin\theta}{-c\cos\theta - v}\right)$.

$\boxed{\tan\bar{\theta} = \dfrac{1}{\gamma}\left(\dfrac{\sin\theta}{\cos\theta + v/c}\right).}$ [Compare $\tan\bar{\theta} = \gamma\frac{\sin\theta}{\cos\theta}$ in Prob. 12.10. The point is that *velocities* are sensitive not only to the transformation of *distances*, but also of *times*. That's why there is no *universal* rule for translating angles—you have to know whether it's an angle made by a *velocity* vector or a *position* vector.]

Problem 12.15

Bullet relative to ground: $\dfrac{5}{7}c$, Outlaws relative to police: $\dfrac{\frac{3}{4}c - \frac{1}{2}c}{1 - \frac{3}{4}\cdot\frac{1}{2}} = \dfrac{(1/4)c}{(5/8)} = \dfrac{2}{5}c.$

Bullet relative to outlaws: $\dfrac{\frac{5}{7}c - \frac{3}{4}c}{1 - \frac{5}{7}\cdot\frac{3}{4}} = \dfrac{-(1/28)c}{(13/28)} = -\dfrac{1}{13}c$. [Velocity of A relative to B is minus the velocity of B relative to A, so all entries below the diagonal are trivial. Note that in every case $v_{\text{bullet}} < v_{\text{outlaws}}$, so no matter how you look at it, the bad guys get away.]

speed of → relative to ↓	Ground	Police	Outlaws	Bullet	Do they escape?
Ground	0	$\frac{1}{2}c$	$\frac{3}{4}c$	$\frac{5}{7}c$	Yes
Police	$-\frac{1}{2}c$	0	$\frac{2}{5}c$	$\frac{1}{3}c$	Yes
Outlaws	$-\frac{3}{4}c$	$-\frac{2}{5}c$	0	$-\frac{1}{13}c$	Yes
Bullet	$-\frac{5}{7}c$	$-\frac{1}{3}c$	$\frac{1}{13}c$	0	Yes

Problem 12.16

(a) Moving clock runs slow, by a factor $\gamma = \frac{1}{\sqrt{1-(4/5)^2}} = \frac{5}{3}$. Since 18 years elapsed on the moving clock, $\frac{5}{3} \times 18 = 30$ years elapsed on the stationary clock. $\boxed{\text{51 years old.}}$

(b) By earth clock, it took 15 years to get there, at $\frac{4}{5}c$, so $d = \frac{4}{5}c \times 15$ years $= \boxed{12c \text{ years}}$ (12 light years).

(c) $\boxed{t = 15 \text{ years}, x = 12c \text{ years.}}$

(d) $\boxed{\bar{t} = 9 \text{ years}, \bar{x} = 0.}$ [She got *on* at the origin in $\bar{S}$, and rode along with $\bar{S}$, so she's *still* at the origin. If you doubt these values, use the Lorentz transformations, with x and t from (c).]

(e) Lorentz transformations: $\left\{\begin{array}{l}\tilde{x} = \gamma(x + vt)\\ \tilde{t} = \gamma(t + \frac{v}{c^2}x)\end{array}\right\}$ (note that v is *negative*, since $\tilde{S}$ is going to the *left*).

$\therefore \tilde{x} = \frac{5}{3}(12c \text{ yrs} + \frac{4}{5}c\cdot 15 \text{ yrs}) = \frac{5}{3}\cdot 24c \text{ yrs} = \boxed{40c \text{ years.}}$

$\tilde{t} = \frac{5}{3}(15 \text{ yrs} + \frac{4}{5}\frac{c}{c^2}\cdot 12c \text{ yrs}) = \frac{5}{3}\left(15 + \frac{48}{5}\right) \text{yrs} = (25 + 16) \text{ yrs} = \boxed{41 \text{ years.}}$

(f) Set her clock $\boxed{\text{ahead 32 years,}}$ from 9 to 41 ($\bar{t} \to \tilde{t}$). Return trip takes 9 years (moving time), so her clock will now read $\boxed{50}$ years at her arrival. Note that this is $\frac{5}{3}\cdot 30$ years—precisely what she would calculate if the *stay-at-home* had been the traveler, for 30 years of his own time.

(g) (i) $\bar{t} = 9$ yrs, $x = 0$. What is t? $t = \frac{v}{c^2}x + \frac{\bar{t}}{\gamma} = \frac{3}{5}\cdot 9 = \frac{27}{5} = 5.4$ years, and he started at age 21, so he's $\boxed{\text{26.4 years old.}}$ (*Younger* than the traveler (!) because to the traveler it's the stay-at-home who's moving.)

(ii) $\tilde{t} = 41$ yrs, $x = 0$. What is t? $t = \frac{\tilde{t}}{\gamma} = \frac{3}{5}\cdot 41 = \frac{123}{5} = 24.6$ years, and he started at 21, so he's $\boxed{\text{45.6 years old.}}$

(h) It will take another $\boxed{5.4 \text{ years}}$ of earth time for the return, so when she gets back, she will say her twin's age is $45.6 + 5.4 = \boxed{51}$ years—which is what we found in (a). But note that to make it work from traveler's point of view you *must* take into account the jump in *perceived* age of stay-at-home when she changes coordinates from $\bar{S}$ to $\bar{\bar{S}}$.

Problem 12.17

$$-\bar{a}^0\bar{b}^0 + \bar{a}^1\bar{b}^1 + \bar{a}^2\bar{b}^2 + \bar{a}^3\bar{b}^3 = -\gamma^2(a^0 - \beta a^1)(b^0 - \beta b^1) + \gamma^2(a^1 - \beta a^0)(b^1 - \beta a^0) + a^2b^2 + a^3b^3$$

$$= -\gamma^2(a^0b^0 - \beta \cancel{a^0} b^1 - \beta \cancel{a^1} b^0 + \beta^2 a^1 b^1 - a^1b^1 + \beta \cancel{a^1} b^0 + \beta \cancel{a^0} b^1 - \beta^2 a^0 b^0) + a^2b^2 + a^3b^3$$

$$= -\gamma^2 a^0 b^0 (1 - \beta^2) + \gamma^2 a^1 b^1 (1 - \beta^2) + a^2b^2 + a^3b^3$$

$$= -a^0b^0 + a^1b^1 + a^2b^2 + a^3b^3. \quad \text{qed} \quad [\textit{Note: } \gamma^2(1-\beta^2) = 1.]$$

Problem 12.18

(a) $\boxed{\begin{pmatrix} c\bar{t} \\ \bar{x} \\ \bar{y} \\ \bar{z} \end{pmatrix} = \begin{pmatrix} 1 & 0 & 0 & 0 \\ -\beta & 1 & 0 & 0 \\ 0 & 0 & 1 & 0 \\ 0 & 0 & 0 & 1 \end{pmatrix} \begin{pmatrix} ct \\ x \\ y \\ z \end{pmatrix}}$ (using the notation of Eq. 12.24, for best comparison).

(b) $\boxed{\Lambda = \begin{pmatrix} \gamma & 0 & -\gamma\beta & 0 \\ 0 & 1 & 0 & 0 \\ -\gamma\beta & 0 & \gamma & 0 \\ 0 & 0 & 0 & 1 \end{pmatrix}}.$

(c) Multiply the matrices: $\Lambda = \begin{pmatrix} \bar{\gamma} & 0 & -\bar{\gamma}\bar{\beta} & 0 \\ 0 & 1 & 0 & 0 \\ -\bar{\gamma}\bar{\beta} & 0 & \bar{\gamma} & 0 \\ 0 & 0 & 0 & 1 \end{pmatrix} \begin{pmatrix} \gamma & -\gamma\beta & 0 & 0 \\ -\gamma\beta & \gamma & 0 & 0 \\ 0 & 0 & 1 & 0 \\ 0 & 0 & 0 & 1 \end{pmatrix} = \boxed{\begin{pmatrix} \gamma\bar{\gamma} & -\gamma\bar{\gamma}\beta & -\bar{\gamma}\bar{\beta} & 0 \\ -\gamma\beta & \gamma & 0 & 0 \\ -\bar{\gamma}\gamma\bar{\beta} & \gamma\bar{\gamma}\beta\bar{\beta} & \bar{\gamma} & 0 \\ 0 & 0 & 0 & 1 \end{pmatrix}}.$

$\boxed{\text{Yes,}}$ the order *does* matter. In the other order, "bars" and "no-bars" would be switched, and this would give a *different matrix.*

Problem 12.19

(a) Since $\tanh\theta = \frac{\sinh\theta}{\cosh\theta}$, and $\cosh^2\theta - \sinh^2\theta = 1$, we have:

$$\gamma = \frac{1}{\sqrt{1 - v^2/c^2}} = \frac{1}{\sqrt{1 - \tanh^2\theta}} = \frac{\cosh\theta}{\sqrt{\cosh^2\theta - \sinh^2\theta}} = \cosh\theta\,; \ \gamma\beta = \cosh\theta\tanh\theta = \sinh\theta.$$

$$\therefore \boxed{\Lambda = \begin{pmatrix} \cosh\theta & -\sinh\theta & 0 & 0 \\ -\sinh\theta & \cosh\theta & 0 & 0 \\ 0 & 0 & 1 & 0 \\ 0 & 0 & 0 & 1 \end{pmatrix}}. \quad \text{Compare: } R = \begin{pmatrix} \cos\phi & \sin\phi & 0 \\ -\sin\phi & \cos\phi & 0 \\ 0 & 0 & 1 \end{pmatrix}.$$

(b) $\bar{u} = \dfrac{u - v}{1 - \frac{uv}{c^2}} \Rightarrow \dfrac{\bar{u}}{c} = \dfrac{(u/c) - (v/c)}{1 - \left(\frac{u}{c}\right)\left(\frac{v}{c}\right)} \Rightarrow \tanh\bar{\phi} = \dfrac{\tanh\phi - \tanh\theta}{1 - \tanh\phi\tanh\theta}$, where $\tanh\phi = u/c$, $\tanh\theta = v/c$; $\tanh\bar{\phi} = \bar{u}/c$. But a "trig" formula for hyperbolic functions (CRC Handbook, 18th Ed., p. 204) says:

$$\frac{\tanh\phi - \tanh\theta}{1 - \tanh\phi\tanh\theta} = \tanh(\phi - \theta). \quad \therefore \tanh\bar{\phi} = \tanh(\phi - \theta), \text{ or: } \boxed{\bar{\phi} = \phi - \theta.}$$

Problem 12.20

(a) (i) $I = -c^2\Delta t^2 + \Delta x^2 + \Delta y^2 + \Delta z^2 = -(5-15)^2 + (10-5)^2 + (8-3)^2 + (0-0)^2 = -100 + 25 + 25 = \boxed{-50.}$

(ii) $\boxed{\text{No.}}$ (In such a system $\Delta \bar{t} = 0$, so I would have to be ***positive***, which it ***isn't.***)

(iii) $\boxed{\text{Yes.}}$

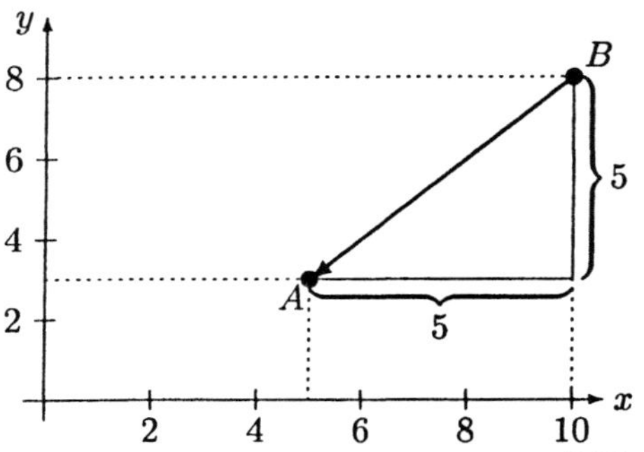

$\bar{S}$ travels in the direction from B toward A, making the trip in time $10/c$.

$$\therefore \mathbf{v} = \frac{-5\hat{\mathbf{x}} - 5\hat{\mathbf{y}}}{10/c} = \boxed{-\frac{c}{2}\hat{\mathbf{x}} - \frac{c}{2}\hat{\mathbf{y}}.}$$

Note that $\frac{v^2}{c^2} = \frac{1}{4} + \frac{1}{4} = \frac{1}{2}$, so $v = \frac{1}{\sqrt{2}}c$, safely less than c.

(b) (i) $I = -(3-1)^2 + (5-2)^2 + 0 + 0 = -4 + 9 = \boxed{5.}$

(ii) $\boxed{\text{Yes.}}$ By Lorentz transformation: $\Delta(c\bar{t}) = \gamma\left[\Delta(ct) - \beta(\Delta x)\right]$. We want $\Delta\bar{t} = 0$, so $\Delta(ct) = \beta(\Delta x)$; or $\frac{v}{c} = \frac{\Delta(ct)}{\Delta x} = \frac{(3-1)}{(5-2)} = \frac{2}{3}$. So $\boxed{v = \frac{2}{3}c,}$ in the $+x$ direction.

(iii) $\boxed{\text{No.}}$ (In such a system $\Delta x = \Delta y = \Delta z = 0$ so I would be negative, which it ***isn't.***)

Problem 12.21

Using Eq. 12.18 (iv): $\Delta\bar{t} = \gamma(\Delta t - \frac{v}{c^2}\Delta x) = 0 \Rightarrow \Delta t = \frac{v}{c^2}\Delta x$, or $v = \frac{\Delta t}{\Delta x}c^2 = \boxed{\frac{t_B - t_A}{x_B - x_A}c^2.}$

Problem 12.22

(a)

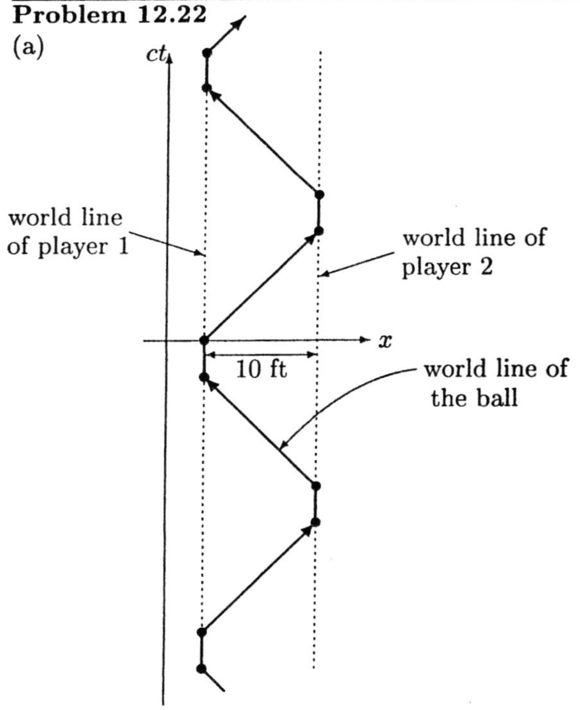

Truth is, you never *do* communicate with the other person *right now*—you communicate with the person he/she *will be* when the message gets there; and the response comes back to and older and wiser you.

(b) $\boxed{\textit{No way.}}$ It is true that a moving observer might say she arrived at B before she left A, but for the ***round trip*** everyone must agree that she arrives back after she set out.

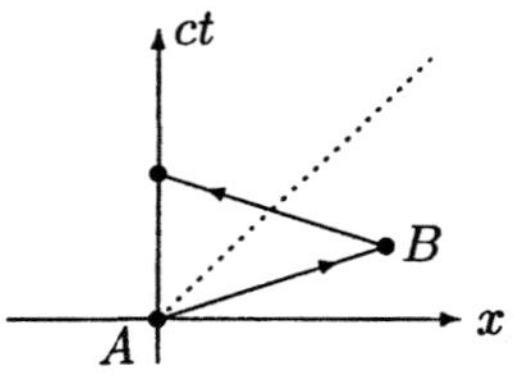

Problem 12.23

(a)

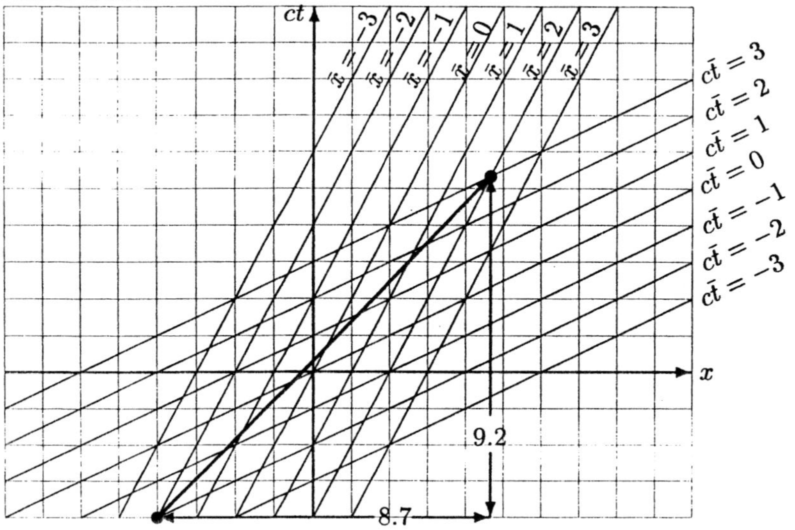

(b) $\frac{c}{v}$ = slope $= \frac{9.2}{8.7}$
$\Rightarrow v = \frac{8.7}{9.2}c = \boxed{0.95c.}$

(c) $v' = \frac{4}{5}c$, so $v = \frac{\frac{4}{5}c + \frac{3}{5}c}{1 + \frac{4}{5}\cdot\frac{3}{5}}$
$= \frac{(7/5)c}{(37/25)} = \boxed{\frac{35}{37}c} = 0.95c.$ ✓

Problem 12.24

(a) $\left(1 - \frac{u^2}{c^2}\right)\eta^2 = u^2$; $u^2\left(1 + \frac{\eta^2}{c^2}\right) = \eta^2$; $\boxed{\mathbf{u} = \frac{1}{\sqrt{1+\eta^2/c^2}}\boldsymbol{\eta}.}$

(b) $\frac{1}{\sqrt{1-u^2/c^2}} = \frac{1}{\sqrt{1-\tanh^2\theta}} = \frac{\cosh\theta}{\sqrt{\cosh^2\theta - \sinh^2\theta}} = \cosh\theta$; $\quad \eta = \frac{1}{\sqrt{1-u^2/c^2}}u = \cosh\theta\, c\tanh\theta = \boxed{c\sinh\theta.}$

Problem 12.25

(a) $u_x = u_y = u\cos 45° = \frac{1}{\sqrt{2}}\frac{2}{\sqrt{5}}c = \boxed{\sqrt{\frac{2}{5}}\,c.}$

(b) $\frac{1}{\sqrt{1-u^2/c^2}} = \frac{1}{\sqrt{1-4/5}} = \frac{\sqrt{5}}{\sqrt{5-4}} = \sqrt{5}$; $\quad \boldsymbol{\eta} = \frac{\mathbf{u}}{\sqrt{1-u^2/c^2}} \Rightarrow \boxed{\eta_x = \eta_y = \sqrt{2}\,c.}$

(c) $\eta^0 = \gamma c = \boxed{\sqrt{5}\,c.}$

(d) Eq. 12.45⇒
$$\begin{cases} \bar{u}_x = \frac{u_x - v}{1 - \frac{u_x v}{c^2}} = \frac{\sqrt{2/5}\,c - \sqrt{2/5}\,c}{1 - \frac{2}{5}} = \boxed{0.} \\ \bar{u}_y = \frac{1}{\gamma}\left(\frac{u_y}{1 - \frac{u_x v}{c^2}}\right) = \sqrt{1 - \frac{2}{5}}\,\frac{\sqrt{2/5}\,c}{1 - \frac{2}{5}} = \frac{\sqrt{2/5}}{\sqrt{3/5}}c = \boxed{\sqrt{\frac{2}{3}}\,c.} \end{cases}$$

(e) $\bar{\eta}_x = \gamma(\eta_x - \beta\eta^0) = \sqrt{1 - \frac{2}{5}}\left(\sqrt{2}\,c - \sqrt{\frac{2}{5}}\sqrt{5}\,c\right) = \boxed{0.}\quad \boxed{\bar{\eta}_y = \eta_y = \sqrt{2}\,c.}$

(f) $\frac{1}{\sqrt{1-\bar{u}^2/c^2}} = \frac{1}{\sqrt{1-(2/3)}} = \sqrt{3}$; $\quad \bar{\boldsymbol{\eta}} = \sqrt{3}\,\bar{\mathbf{u}} \Rightarrow \left\{ \begin{array}{l} \bar{\eta}_x = \sqrt{3}\,\bar{u}_x = 0. ✓ \\ \bar{\eta}_y = \sqrt{3}\,\bar{u}_y = \sqrt{2}\,c. ✓ \end{array} \right\}$

Problem 12.26

$$\eta^\mu\eta_\mu = -(\eta^0)^2 + \eta^2 = \frac{1}{(1 - u^2/c^2)}(-c^2 + u^2) = -c^2\frac{(1 - u^2/c^2)}{(1 - u^2/c^2)} = \boxed{-c^2.}$$

Problem 12.27

(a) From Prob. 11.31 we have $\gamma = \frac{1}{b}\sqrt{b^2+c^2t^2}$. $\therefore \tau = \int \frac{1}{\gamma}dt = b\int \frac{dt}{\sqrt{b^2+c^2t^2}} = \frac{b}{c}\ln(ct+\sqrt{b^2+c^2t^2})+k$; at $t=0$ we want $\tau=0$: $0=\frac{b}{c}\ln b+k$, so $k=-\frac{b}{c}\ln b$; $\boxed{\tau = \frac{b}{c}\ln\left[\frac{1}{b}(ct+\sqrt{b^2+c^2t^2})\right].}$

(b) $\sqrt{x^2-b^2}+x = be^{c\tau/b}$; $\sqrt{x^2-b^2} = be^{c\tau/b}-x$; $x^2-b^2 = b^2e^{2c\tau/b}-2xbe^{c\tau/b}+x^2$; $2xbe^{c\tau/b} = b^2(1+e^{2c\tau/b})$; $x = b\left(\frac{e^{c\tau/b}+e^{-c\tau/b}}{2}\right) = \boxed{b\cosh(c\tau/b).}$ Also from Prob. 11.31: $v = c^2t/\sqrt{b^2+c^2t^2}$.

$v = \frac{c}{x}\sqrt{x^2-b^2} = \frac{c}{b\cosh(c\tau/b)}\sqrt{b^2\cosh^2(c\tau/b)-b^2} = c\frac{\sqrt{\cosh^2(c\tau/b)-1}}{\cosh(c\tau/b)} = c\frac{\sinh(c\tau/b)}{\cosh(c\tau/b)} = \boxed{c\tanh\left(\frac{c\tau}{b}\right).}$

(c) $\eta^\mu = \gamma(c,v,0,0)$; $\gamma = \frac{x}{b} = \cosh\frac{c\tau}{b}$, so $\eta^\mu = \cosh\frac{c\tau}{b}\left(c, c\tanh\frac{c\tau}{b},0,0\right) = \boxed{c\left(\cosh\frac{c\tau}{b}, \sinh\frac{c\tau}{b},0,0\right).}$

Problem 12.28

(a) $m_Au_A+m_Bu_B = m_Cu_C+m_Du_D$; $\quad u_i = \dfrac{\bar{u}_i+v}{1+(\bar{u}_iv/c^2)}$.

$$m_A\frac{\bar{u}_A+v}{1+(\bar{u}_Av/c^2)} + m_B\frac{\bar{u}_B+v}{1+(\bar{u}_Bv/c^2)} = m_C\frac{\bar{u}_C+v}{1+(\bar{u}_Cv/c^2)} + m_D\frac{\bar{u}_D+v}{1+(\bar{u}_Dv/c^2)}.$$

This time, because the denominators are all different, we *cannot* conclude that

$m_A\bar{u}_A+m_B\bar{u}_B = m_C\bar{u}_C+m_D\bar{u}_D$.

As an explicit counterexample, suppose all the masses are equal, and $u_A=-u_B=v$; $u_C=u_D=0$. This is a symmetric "completely inelastic" collision in $\mathcal{S}$, and momentum is clearly conserved ($0=0$). But the Einstein velocity addition rule gives $\bar{u}_A=0$, $\bar{u}_B=-2u/(1+u^2/c^2)$, $\bar{u}_C=\bar{u}_D=-u$, so in $\bar{\mathcal{S}}$ the (incorrectly defined) momentum is *not* conserved:

$$m\left(\frac{-2u}{1+u^2/c^2}\right) \neq -2mu.$$

(b) $m_A\eta_A+m_B\eta_B = m_C\eta_C+m_D\eta_D$; $\quad \eta_i = \gamma(\bar{\eta}_i+\beta\bar{\eta}_i^0)$. (The inverse Lorentz transformation.)

$m_A\gamma(\bar{\eta}_A+\beta\bar{\eta}_A^0)+m_B\gamma(\bar{\eta}_B+\beta\bar{\eta}_B^0) = m_C\gamma(\bar{\eta}_C+\beta\bar{\eta}_C^0)+m_D\gamma(\bar{\eta}_D+\beta\bar{\eta}_D^0)$. The gamma's cancel:

$m_A\bar{\eta}_A+m_B\bar{\eta}_B+\beta(m_A\bar{\eta}_A^0+m_B\bar{\eta}_B^0) = m_C\bar{\eta}_C+m_D\bar{\eta}_D+\beta(m_C\bar{\eta}_C^0+m_D\bar{\eta}_D^0)$.

But $m_i\eta_i^0 = p_i^0 = E_i/c$, so if $\boxed{\textit{energy}\text{ is conserved}}$ in $\bar{\mathcal{S}}$ ($\bar{E}_A+\bar{E}_B=\bar{E}_C+\bar{E}_D$), then so too is the momentum (correctly defined):

$m_A\bar{\eta}_A+m_B\bar{\eta}_B = m_C\bar{\eta}_C+m_D\bar{\eta}_D$. qed

---

Problem 12.29

$\gamma mc^2 - mc^2 = nmc^2 \Rightarrow \gamma = n+1 = \frac{1}{\sqrt{1-u^2/c^2}} \Rightarrow 1-\frac{u^2}{c^2} = \frac{1}{(n+1)^2}$.

$\therefore \frac{u^2}{c^2} = 1-\frac{1}{(n+1)^2} = \frac{n^2+2n+1-1}{(n+1)^2} = \frac{n(n+2)}{(n+1)^2}$; $\boxed{u = \frac{\sqrt{n(n+2)}}{n+1}c.}$

Problem 12.30

$E_T = E_1+E_2+\cdots$; $\quad p_T = p_1+p_2+\cdots$; $\quad \bar{p}_T = \gamma(p_T-\beta E_T/c) = 0 \Rightarrow \beta = v/c = p_Tc/E_T$.

$v = c^2p_T/E_T = \boxed{c^2(p_1+p_2+\cdots)/(E_1+E_2+\cdots).}$

Problem 12.31

$$E_\mu = \frac{(m_\pi^2+m_\mu^2)}{2m_\pi}c^2 = \gamma m_\mu c^2 \Rightarrow \gamma = \frac{(m_\pi^2+m_\mu^2)}{2m_\pi m_\mu} = \frac{1}{\sqrt{1-v^2/c^2}};\ 1-\frac{v^2}{c^2} = \frac{1}{\gamma^2};$$

$$\frac{v^2}{c^2} = 1-\frac{1}{\gamma^2} = 1-\frac{4m_\pi^2m_\mu^2}{(m_\pi^2+m_\mu^2)^2} = \frac{m_\pi^4+2m_\pi^2m_\mu^2+m_\mu^4-4m_\pi^2m_\mu^2}{(m_\pi^2+m_\mu^2)^2} = \frac{(m_\pi^2-m_\mu^2)^2}{(m_\pi^2+m_\mu^2)^2};\ v = \boxed{\left(\frac{m_\pi^2-m_\mu^2}{m_\pi^2+m_\mu^2}\right)c.}$$

Problem 12.32

Initial momentum: $E^2 - p^2c^2 = m^2c^4 \Rightarrow p^2c^2 = (2mc^2)^2 - m^2c^4 = 3m^2c^4 \Rightarrow p = \sqrt{3}\,mc$.
Initial energy: $2mc^2 + mc^2 = 3mc^2$.

Each is conserved, so final energy is $3mc^2$, final momentum is $\sqrt{3}\,mc$.

$$E^2 - p^2c^2 = (3mc^2)^2 - (\sqrt{3}\,mc)^2c^2 = 6m^2c^4 = M^2c^4 \Rightarrow \boxed{M = \sqrt{6}\,m} \approx 2.5m.$$

(In this process some kinetic energy was converted into rest energy, so $M > 2m$.)

$$v = \frac{pc^2}{E} = \frac{\sqrt{3}\,mc\,c^2}{3mc^2} = \boxed{\frac{c}{\sqrt{3}}}.$$

Problem 12.33

First calculate pion's energy: $E^2 = p^2c^2 + m^2c^4 = \frac{9}{16}m^2c^4 + m^2c^4 = \frac{25}{16}m^2c^4 \Rightarrow E = \frac{5}{4}mc^2$.

$$\left.\begin{array}{ll}\text{Conservation of energy:} & \frac{5}{4}mc^2 = E_A + E_B \\ \text{Conservation of momentum:} & \frac{3}{4}mc^2 = p_A + p_B = \frac{E_A}{c} - \frac{E_B}{c} \Rightarrow \frac{3}{4}mc^2 = E_A - E_B\end{array}\right\} 2E_A = 2mc^2.$$

$$\Rightarrow \boxed{E_A = mc^2;} \quad \boxed{E_B = \frac{1}{4}mc^2.}$$

Problem 12.34

Classically, $E = \frac{1}{2}mv^2$. In a colliding beam experiment, the relative velocity (classically) is *twice* the velocity of either one, so the relative energy is $4E$.

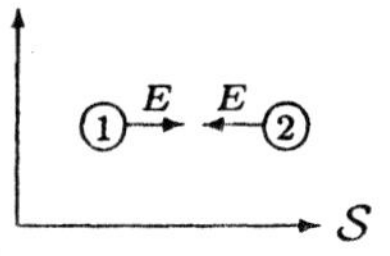

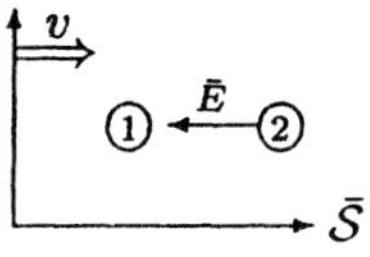

Let $\bar{S}$ be the system in which ① is at rest. Its speed v, relative to S, is just the speed of ① in S.

$\bar{p}^0 = \gamma(p^0 - \beta p^1) \Rightarrow \frac{\bar{E}}{c} = \gamma\left(\frac{E}{c} - \beta p\right)$, where p is the momentum of ② in S.

$E = \gamma Mc^2$, so $\gamma = \frac{E}{Mc^2}$; $p = -\gamma Mv = -\gamma M\beta c$; $\bar{E} = \gamma\left(\frac{E}{c} + \beta\gamma M\beta c\right)c = \gamma(E + \gamma Mc^2\beta^2)$.

$\gamma^2 = \frac{1}{1-\beta^2} \Rightarrow 1 - \beta^2 = \frac{1}{\gamma^2} \Rightarrow \beta^2 = 1 - \frac{1}{\gamma^2} = \frac{\gamma^2-1}{\gamma^2}$; $\bar{E} = \frac{E}{Mc^2}E + \left[\left(\frac{E}{Mc^2}\right)^2 - 1\right]Mc^2$.

$$\bar{E} = \frac{E^2}{Mc^2} + \frac{E^2}{Mc^2} - Mc^2; \quad \boxed{\bar{E} = \frac{2E^2}{Mc^2} - Mc^2.}$$

For $E = 30$ GeV and $Mc^2 = 1$ GeV, we have $\bar{E} = \frac{(2)(900)}{1} - 1 = 1800 - 1 = \boxed{1799\text{ GeV}} = \boxed{60E.}$

Problem 12.35

One photon is impossible, because in the "center of momentum" frame (Prob. 12.30) we'd be left with a photon *at rest*, whereas photons *have* to travel at speed c.

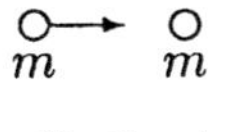

(before)

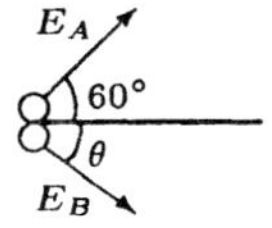

(after)

$$\left\{\begin{array}{ll}\text{Cons. of energy:} & \sqrt{p_0c^2 + m^2c^4} + mc^2 = E_A + E_B. \\ \text{Cons. of mom.:} & \left\{\begin{array}{ll}\text{horizontal:} & p_0 = \frac{E_A}{c}\cos 60° + \frac{E_B}{c}\cos\theta \Rightarrow E_B\cos\theta = p_0c - \frac{1}{2}E_A, \\ \text{vertical:} & 0 = \frac{E_A}{c}\sin 60° - \frac{E_B}{c}\sin\theta \Rightarrow E_B\sin\theta = \frac{\sqrt{3}}{2}E_A;\end{array}\right\} \text{square and add:}\end{array}\right.$$

$$E_B^2(\cos^2\theta + \sin^2\theta) = p_0c^2 - p_0cE_A + \frac{1}{4}E_A^2 + \frac{3}{4}E_A^2$$

$$\Rightarrow E_B^2 = p_0c^2 - p_0cE_A + E_A^2 = \left[\sqrt{p_0^2c^2 + m^2c^4} + mc^2 - E_A\right]^2$$

$$= p_0c^2 + m^2c^4 + 2\sqrt{p_0^2c^2 + m^2c^4}(mc^2 - E_A) + m^2c^4 - 2E_Amc^2 + E_A^2. \text{ Or:}$$

$$-p_0cE_A = 2m^2c^4 + 2mc^2\sqrt{p_0^2c^2 + m^2c^4} - 2E_A\sqrt{p_0^2c^2 + m^2c^4} - 2E_Amc^2;$$

$$E_A(mc^2 + \sqrt{p_0^2c^2 + m^2c^4} - p_0c/2) = m^2c^4 + mc^2\sqrt{p_0c^2 + m^2c^4};$$

$$E_A = mc^2\frac{(mc^2 + \sqrt{p_0^2c^2 + m^2c^4})}{(mc^2 + \sqrt{p_0^2c^2 + m^2c^4} - p_0c/2)} \cdot \frac{(mc^2 - \sqrt{p_0^2c^2 + m^2c^4} - p_0c/2)}{(mc^2 - \sqrt{p_0^2c^2 + m^2c^4} - p_0c/2)}$$

$$= mc^2\frac{(\cancel{m^2}c^4 - p_0^2c^2 - \cancel{m^2}c^4 - \frac{1}{2}p_0mc^3 - \frac{p_0c}{2}\sqrt{p_0^2c^2 + m^2c^4})}{(\cancel{m^2}c^4 - p_0mc^3 + \frac{p_0^2c^2}{4} - p_0^2c^2 - \cancel{m^2}c^4)} = \boxed{\frac{mc^2}{2}\frac{(mc + 2p_0 + \sqrt{p_0^2 + m^2c^2})}{(mc + \frac{3}{4}p_0)}}.$$

Problem 12.36

$$\mathbf{F} = \frac{d\mathbf{p}}{dt} = \frac{d}{dt}\frac{m\mathbf{u}}{\sqrt{1 - u^2/c^2}} = m\left\{\frac{\frac{d\mathbf{u}}{dt}}{\sqrt{1 - u^2/c^2}} + \mathbf{u}\left(-\frac{1}{2}\right)\frac{-\frac{1}{c^2}\,2\mathbf{u}\cdot\frac{d\mathbf{u}}{dt}}{(1 - u^2/c^2)^{3/2}}\right\}$$

$$= \frac{m}{\sqrt{1 - u^2/c^2}}\left\{\mathbf{a} + \frac{\mathbf{u}(\mathbf{u}\cdot\mathbf{a})}{(c^2 - u^2)}\right\}. \quad \text{qed}$$

Problem 12.37

At constant force you go in "hyperbolic" motion. Photon A, which left the origin at $t < 0$, catches up with you, but photon B, which passes the origin at $t > 0$, never does.

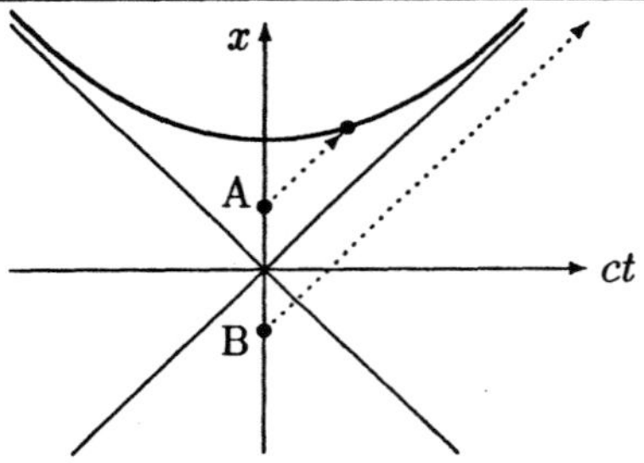

Problem 12.38

(a)

$$\alpha^0 = \frac{d\eta_0}{d\tau} = \frac{d\eta_0}{dt}\frac{dt}{d\tau} = \left[\frac{d}{dt}\left(\frac{c}{\sqrt{1 - u^2/c^2}}\right)\right]\frac{1}{\sqrt{1 - u^2/c^2}}$$

$$= \frac{c}{\sqrt{1 - u^2/c^2}}\left(-\frac{1}{2}\right)\frac{(-\frac{1}{c^2})\,2\mathbf{u}\cdot\mathbf{a}}{(1 - u^2/c^2)^{3/2}} = \boxed{\frac{1}{c}\frac{\mathbf{u}\cdot\mathbf{a}}{(1 - u^2/c^2)^2}}.$$

$$\boldsymbol{\alpha} = \frac{d\boldsymbol{\eta}}{d\tau} = \frac{dt}{d\tau}\frac{d\boldsymbol{\eta}}{dt} = \frac{1}{\sqrt{1 - u^2/c^2}}\frac{d}{dt}\left(\frac{\mathbf{u}}{\sqrt{1 - u^2/c^2}}\right) = \frac{1}{\sqrt{1 - u^2/c^2}}\left\{\frac{\mathbf{a}}{\sqrt{1 - u^2/c^2}} + \mathbf{u}\left(-\frac{1}{2}\right)\frac{-\frac{1}{c^2}2\mathbf{u}\cdot\mathbf{a}}{(1 - u^2/c^2)^{3/2}}\right\}$$

$$= \boxed{\frac{1}{(1 - u^2/c^2)}\left[\mathbf{a} + \frac{\mathbf{u}(\mathbf{u}\cdot\mathbf{a})}{(c^2 - u^2)}\right]}.$$

(b) $$\alpha_\mu\alpha^\mu = -(\alpha^0)^2 + \boldsymbol{\alpha}\cdot\boldsymbol{\alpha} = -\frac{1}{c^2}\frac{(\mathbf{u}\cdot\mathbf{a})^2}{(1-u^2/c^2)^4} + \frac{1}{(1-u^2/c^2)^4}\left[\mathbf{a}\left(1-\frac{u^2}{c^2}\right)+\frac{1}{c^2}\mathbf{u}(\mathbf{u}\cdot\mathbf{a})\right]^2$$

$$= \frac{1}{(1-u^2/c^2)^4}\left\{-\frac{1}{c^2}(\mathbf{u}\cdot\mathbf{a})^2 + a^2\left(1-\frac{u^2}{c^2}\right)^2 + \frac{2}{c^2}\left(1-\frac{u^2}{c^2}\right)(\mathbf{u}\cdot\mathbf{a})^2 + \frac{1}{c^4}u^2(\mathbf{u}\cdot\mathbf{a})^2\right\}$$

$$= \frac{1}{(1-u^2/c^2)^4}\left\{a^2\left(1-\frac{u^2}{c^2}\right)^2 + \frac{(\mathbf{u}\cdot\mathbf{a})^2}{c^2}\underbrace{\left(-1+2-2\frac{u^2}{c^2}+\frac{u^2}{c^2}\right)}_{(1-\frac{u^2}{c^2})}\right\}$$

$$= \boxed{\frac{1}{(1-u^2/c^2)^2}\left[a^2+\frac{(\mathbf{u}\cdot\mathbf{a})^2}{(c^2-u^2)}\right].}$$

(c) $\eta^\mu\eta_\mu = -c^2$, so $\frac{d}{d\tau}(\eta^\mu\eta_\mu) = \alpha^\mu\eta_\mu + \eta^\mu\alpha_\mu = 2\alpha^\mu\eta_\mu = 0$, so $\boxed{\alpha^\mu\eta_\mu = 0.}$

(d) $K^\mu = \frac{dp^\mu}{d\tau} = \frac{d}{d\tau}(m\eta^\mu) = \boxed{m\alpha^\mu.}$ $\boxed{K^\mu\eta_\mu = m\alpha^\mu\eta_\mu = 0.}$

Problem 12.39

$K_\mu K^\mu = -(K^0)^2 + \mathbf{K}\cdot\mathbf{K}$. From Eq. 12.70, $\mathbf{K}\cdot\mathbf{K} = \frac{F^2}{(1-u^2/c^2)}$. From Eq. 12.71:

$$K^0 = \frac{1}{c}\frac{dE}{d\tau} = \frac{1}{c\sqrt{1-u^2/c^2}}\frac{d}{dt}\left(\frac{mc^2}{\sqrt{1-u^2/c^2}}\right) = \frac{mc}{\sqrt{1-u^2/c^2}}\left[-\frac{1}{2}\frac{(-1/c^2)}{(1-u^2/c^2)^{3/2}}2\mathbf{u}\cdot\mathbf{a}\right] = \frac{m}{c}\frac{(\mathbf{u}\cdot\mathbf{a})}{(1-u^2/c^2)^2}.$$

But (Eq. 12.73): $\mathbf{u}\cdot\mathbf{F} = uF\cos\theta = \frac{m}{\sqrt{1-u^2/c^2}}\left[(\mathbf{u}\cdot\mathbf{a}) + \frac{u^2(\mathbf{u}\cdot\mathbf{a})}{c^2(1-u^2/c^2)}\right] = \frac{m(\mathbf{u}\cdot\mathbf{a})}{(1-u^2/c^2)^{3/2}}$, so

$$K^0 = \frac{uF\cos\theta}{c\sqrt{1-u^2/c^2}};\quad K_\mu K^\mu = \frac{F^2}{(1-u^2/c^2)} - \frac{u^2F^2\cos^2\theta}{c^2(1-u^2/c^2)} = \left[\frac{1-(u^2/c^2)\cos^2\theta}{(1-u^2/c^2)}\right]F^2.\quad \text{qed}$$

Problem 12.40

$$\mathbf{F} = \frac{m}{\sqrt{1-u^2/c^2}}\left[\mathbf{a}+\frac{\mathbf{u}(\mathbf{u}\cdot\mathbf{a})}{c^2-u^2}\right] = q(\mathbf{E}+\mathbf{u}\times\mathbf{B}) \Rightarrow \mathbf{a}+\frac{\mathbf{u}(\mathbf{u}\cdot\mathbf{a})}{(c^2-u^2)} = \frac{q}{m}\sqrt{1-u^2/c^2}(\mathbf{E}+\mathbf{u}\times\mathbf{B}).$$

$$\text{Dot in } \mathbf{u}\text{: } (\mathbf{u}\cdot\mathbf{a}) + \frac{u^2(\mathbf{u}\cdot\mathbf{a})}{c^2(1-u^2/c^2)} = \frac{\mathbf{u}\cdot\mathbf{a}}{(1-u^2/c^2)} = \frac{q}{m}\sqrt{1-u^2/c^2}\,[\mathbf{u}\cdot\mathbf{E} + \underbrace{\mathbf{u}\cdot(\mathbf{u}\times\mathbf{B})}_{=0}];$$

$$\therefore \frac{\mathbf{u}(\mathbf{u}\cdot\mathbf{a})}{(c^2-u^2)} = \frac{q}{m}\sqrt{1-u^2/c^2}\frac{\mathbf{u}(\mathbf{u}\cdot\mathbf{E})}{c^2}.\quad \text{So } \mathbf{a} = \frac{q}{m}\sqrt{1-u^2/c^2}\left[\mathbf{E}+\mathbf{u}\times\mathbf{B}-\frac{1}{c^2}\mathbf{u}(\mathbf{u}\cdot\mathbf{E})\right].\quad \text{qed}$$

Problem 12.41

One way to see it is to look back at the general formula for $\mathbf{E}$ (Eq. 10.29). For a uniform infinite plane of charge, moving at constant velocity in the plane, $\dot{\mathbf{J}} = 0$ and $\dot{\rho} = 0$, while ρ (or rather, σ) is independent of t (so retardation does nothing). Therefore the field is exactly the same as it would be for a plane at rest (except that σ itself is altered by Lorentz contraction).

A more elegant argument exploits the fact that $\mathbf{E}$ is a *vector* (whereas $\mathbf{B}$ is a *pseudo*vector). This means that any given component changes sign if the configuration is reflected in a plane perpendicular to that direction. But in Fig. 12.35(b), if we reflect in the xy plane the configuration is unaltered, so the z component of $\mathbf{E}$ would have to stay the *same*. Therefore it must in fact be zero. (By contrast, if you reflect in a plane perpendicular to the y direction the charges trade places, so it is perfectly appropriate that the y component of $\mathbf{E}$ should reverse its sign.)

Problem 12.42

(a) Field is σ_0/ϵ_0, and it points perpendicular to the positive plate, so:

$$\mathbf{E}_0 = \frac{\sigma_0}{\epsilon_0}(\cos 45^\circ\,\hat{\mathbf{x}} + \sin 45^\circ\,\hat{\mathbf{y}}) = \boxed{\frac{\sigma_0}{\sqrt{2}\,\epsilon_0}(-\hat{\mathbf{x}}+\hat{\mathbf{y}}).}$$

(b) From Eq. 12.108, $E_x = E_{x_0} = -\frac{\sigma_0}{\sqrt{2}\,\epsilon_0}$; $E_y = \gamma E_{y_0} = \gamma\frac{\sigma_0}{\sqrt{2}\,E_0}$. So $\boxed{\mathbf{E} = \frac{\sigma_0}{\sqrt{2}\,\epsilon_0}(-\hat{\mathbf{x}}+\gamma\,\hat{\mathbf{y}}).}$

(c) From Prob. 12.10: $\tan\theta = \gamma$, so $\boxed{\theta = \tan^{-1}\gamma.}$

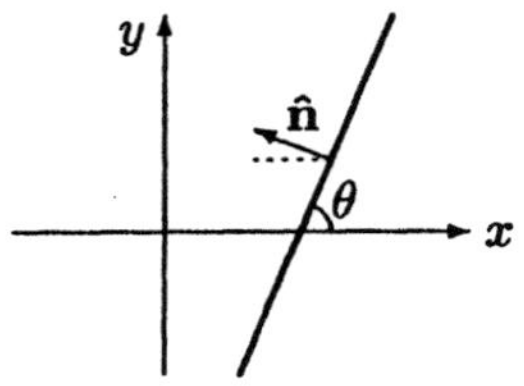

(d) Let $\hat{\mathbf{n}}$ be a unit vector perpendicular to the plates in $\mathcal{S}$—evidently $\hat{\mathbf{n}} = -\sin\theta\,\hat{\mathbf{x}} + \cos\theta\,\hat{\mathbf{y}}$; $|E| = \frac{\sigma_0}{\sqrt{2}\,\epsilon_0}\sqrt{1+\gamma^2}$.

So the angle ϕ between $\hat{\mathbf{n}}$ and $\mathbf{E}$ is:

$$\frac{\mathbf{E}\cdot\hat{\mathbf{n}}}{|E|} = \cos\phi = \frac{1}{\sqrt{1+\gamma^2}}(\sin\theta + \gamma\cos\theta) = \frac{\cos\theta}{\sqrt{1+\gamma^2}}(\tan\theta+\gamma) = \frac{2\gamma}{\sqrt{1+\gamma^2}}\cos\theta$$

But $\gamma = \tan\theta = \frac{\sin\theta}{\cos\theta} = \frac{\sqrt{1-\cos^2\theta}}{\cos\theta} = \sqrt{\frac{1}{\cos^2\theta}-1} \Rightarrow \frac{1}{\cos}{}^2\theta = \gamma^2+1 \Rightarrow \cos\theta = \frac{1}{\sqrt{1+\gamma^2}}$. So $\boxed{\cos\phi = \left(\frac{2\gamma}{1+\gamma^2}\right).}$

Evidently the field is $\boxed{\textit{not}}$ perpendicular to the plates in $\mathcal{S}$.

Problem 12.43

(a) $\mathbf{E} = \dfrac{1}{4\pi\epsilon_0}\dfrac{q(1-v^2/c^2)}{(1-\frac{v^2}{c^2}\sin^2\theta)^{3/2}}\dfrac{\hat{\mathbf{R}}}{R^2}$ (Eq. 12.92) $\Rightarrow$

$$\int\mathbf{E}\cdot d\mathbf{a} = \frac{q(1-v^2/c^2)}{4\pi\epsilon_0}\int\frac{R^2\sin\theta\,d\theta\,d\phi}{R^2(1-\frac{v^2}{c^2}\sin^2\theta)^{3/2}}$$

$$= \frac{q(1-v^2/c^2)}{4\pi\epsilon_0}2\pi\int_0^\pi\frac{\sin\theta\,d\theta}{(1-\frac{v^2}{c^2}\sin^2\theta)^{3/2}}.$$ Let $u=\cos\theta$, so $du = -\sin\theta\,d\theta$, $\sin^2\theta = 1-u^2$.

$$= \frac{q(1-v^2/c^2)}{2\epsilon_0}\int_{-1}^{1}\frac{du}{[1-\frac{v^2}{c^2}+\frac{v^2}{c^2}u^2]^{3/2}} = \frac{q(1-v^2/c^2)}{2\epsilon_0}\left(\frac{c}{v}\right)^3\int_{-1}^{1}\frac{du}{\left(\frac{c^2}{v^2}-1+u^2\right)^{3/2}}.$$

The integral is: $$\left.\frac{u}{\left(\frac{c^2}{v^2}-1\right)\sqrt{\frac{c^2}{v^2}-1+u^2}}\right|_{-1}^{+1} = \frac{2}{\left(\frac{c^2}{v^2}-1\right)\frac{c}{v}} = \left(\frac{v}{c}\right)^3\frac{2}{(1-v^2/c^2)}.$$

So $$\int\mathbf{E}\cdot d\mathbf{a} = \frac{q(1-v^2/c^2)}{2\epsilon_0}\left(\frac{c}{v}\right)^3\left(\frac{v}{c}\right)^3\frac{2}{(1-v^2/c^2)} = q.\ \checkmark$$

(b) Using Eq. 12.111 and Eq. 12.92, $$\mathbf{S} = \frac{1}{\mu_0}(\mathbf{E}\times\mathbf{B}) = \frac{1}{\mu_0}\frac{1}{4\pi\epsilon_0}\frac{1}{4\pi}\frac{\mu_0\,q^2(1-v^2/c^2)^2v\sin\theta}{R^4(1-\frac{v^2}{c^2}\sin^2\theta)^3}\underbrace{(\hat{\mathbf{R}}\times\hat{\boldsymbol{\phi}})}_{-\hat{\boldsymbol{\theta}}};$$

$$\mathbf{S} = \boxed{-\frac{q^2}{16\pi^2\epsilon_0}\frac{(1-v^2/c^2)^2 v\sin\theta}{R^4(1-\frac{v^2}{c^2}\sin^2\theta)^3}\hat{\boldsymbol{\theta}}.}$$

Problem 12.44

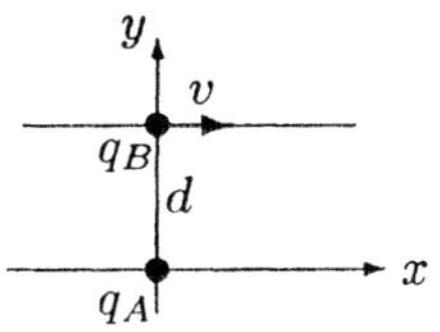

(a) Fields of A at B: $\mathbf{E} = \frac{1}{4\pi\epsilon_0}\frac{q_A}{d^2}\hat{\mathbf{y}}$; $\mathbf{B} = 0$. So force on q_B is $\boxed{\mathbf{F} = \frac{1}{4\pi\epsilon_0}\frac{q_A q_B}{d^2}\hat{\mathbf{y}}.}$

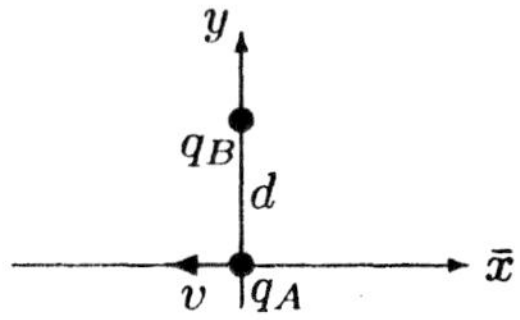

(b) (i) From Eq. 12.68: $\boxed{\bar{\mathbf{F}} = \frac{\gamma}{4\pi\epsilon_0}\frac{q_A q_B}{d^2}\hat{\mathbf{y}}.}$ (*Note:* here the particle is at rest in $\bar{\mathcal{S}}$.)

(ii) From Eq. 12.92, with $\theta = 90°$: $\bar{E} = \frac{1}{4\pi\epsilon_0}\frac{q_A(1-v^2/c^2)}{(1-v^2/c^2)^{3/2}}\frac{1}{d^2}\hat{\mathbf{y}} = \frac{\gamma}{4\pi\epsilon_0}\frac{q_A}{d^2}\hat{\mathbf{y}}$ (this also follows from Eq. 12.108).

$\bar{\mathbf{B}} \neq 0$, but since $v_B = 0$ in $\bar{\mathcal{S}}$, there is no magnetic force anyway, and $\boxed{\bar{\mathbf{F}} = \frac{\gamma}{4\pi\epsilon_0}\frac{q_A q_B}{d^2}\hat{\mathbf{y}}}$ (as before).

Problem 12.45

Here $\theta = 90°$, $\hat{\boldsymbol{\imath}} = \hat{\mathbf{y}}$, $\hat{\boldsymbol{\phi}} = \hat{\mathbf{z}}$, $\imath = r$, so (using $c^2 = 1/\mu_0\epsilon_0$):

$$\mathbf{E} = -\frac{q}{4\pi\epsilon_0}\frac{\gamma}{r^2}\hat{\mathbf{y}}, \quad \mathbf{B} = -\frac{q}{4\pi\epsilon_0}\frac{v}{c^2}\frac{\gamma}{r^2}\hat{\mathbf{z}}, \quad \text{where } \gamma = \frac{1}{\sqrt{1-v^2/c^2}}.$$

Note that $(E^2 - B^2c^2) = \left(\frac{q}{4\pi\epsilon_0 r}\right)^2\gamma^2\left(1-\frac{v^2}{c^2}\right) = \left(\frac{q}{4\pi\epsilon_0 r}\right)^2$ is invariant, because it doesn't depend on v. We can use this as a check.

System A: $v_A = v$, so $\mathbf{E} = -\frac{q}{4\pi\epsilon_0}\frac{\bar{\gamma}}{r^2}\hat{\mathbf{y}}$, $\mathbf{B} = -\frac{q}{4\pi\epsilon_0}\frac{v}{c^2}\frac{\bar{\gamma}}{r^2}\hat{\mathbf{z}}$, where $\bar{\gamma} = \frac{1}{\sqrt{1-v^2/c^2}}$.

$$\mathbf{F} = q[\mathbf{E} + (-v\hat{\mathbf{x}})\times\mathbf{B}] = -\frac{q^2}{4\pi\epsilon_0}\frac{\bar{\gamma}}{r^2}[\hat{\mathbf{y}} - \frac{v^2}{c^2}(\hat{\mathbf{x}}\times\hat{\mathbf{z}})] = -\frac{q^2}{4\pi\epsilon_0}\frac{\bar{\gamma}}{r^2}(1+\frac{v^2}{c^2})\hat{\mathbf{y}}.$$

System B: $v_B = \frac{v+v}{1+v^2/c^2} = \frac{2v}{(1+v^2/c^2)}$

$$\gamma_B = \frac{1}{\sqrt{1-\frac{4v^2/c^2}{(1+v^2/c^2)^2}}} = \frac{(1+v^2/c^2)}{\sqrt{1-2\frac{v^2}{c^2}+\frac{v^4}{c^4}}} = \frac{(1+v^2/c^2)}{(1-v^2/c^2)} = \bar{\gamma}^2(1+\frac{v^2}{c^2})\,;\; v_b\gamma_B = 2v\bar{\gamma}^2.$$

$$\therefore \mathbf{E} = -\frac{q}{4\pi\epsilon_0}\frac{1}{r^2}\bar{\gamma}^2(1+\frac{v^2}{c^2})\hat{\mathbf{y}}\,;\; \mathbf{B} = -\frac{q}{4\pi\epsilon_0}\frac{2v}{c^2}\frac{\bar{\gamma}^2}{r^2}\hat{\mathbf{z}}.$$

[*Check*: $E^2 - B^2c^2 = \left(\frac{q}{4\pi\epsilon_0 r}\right)^2\bar{\gamma}^4\left(1+\frac{2v^2}{c^2}+\frac{v^4}{c^4}-\frac{4v^2}{c^2}\right) = \left(\frac{q}{4\pi\epsilon_0 r}\right)^2\bar{\gamma}^4\frac{1}{\bar{\gamma}^4} = \left(\frac{q}{4\pi\epsilon_0 r}\right)^2$. ✓]

$\mathbf{F} = q\mathbf{E} = -\frac{q^2}{4\pi\epsilon_0}\frac{\bar{\gamma}^2}{r^2}(1+\frac{v^2}{c^2})\hat{\mathbf{y}}$. ($+q$ at rest $\Rightarrow$ no magnetic force). [*Check*: Eq. 12.68 $\Rightarrow F_A = \frac{1}{\bar{\gamma}}F_B$. ✓]

System C: $v_C = 0$. $\mathbf{E} = -\frac{q}{4\pi\epsilon_0}\frac{1}{r^2}\hat{\mathbf{y}}$; $\mathbf{B} = 0$; $\mathbf{F} = q\mathbf{E} = -\frac{q^2}{4\pi\epsilon_0}\frac{1}{r^2}\hat{\mathbf{y}}$.

[The relative velocity of B and C is $2v/(1+v^2/c^2)$, and the corresponding γ is $\bar{\gamma}^2(1+v^2/c^2)$. So Eq. 12.68 $\Rightarrow F_C = \frac{1}{\bar{\gamma}^2(1+v^2/c^2)}F_B$. ✓]

Summary:

$(-\frac{q}{4\pi\epsilon_0 r^2})\gamma\,\hat{\mathbf{y}}$	$(-\frac{q}{4\pi\epsilon_0 r^2})\gamma^2(1+\frac{v^2}{c^2})\hat{\mathbf{y}}$	$(-\frac{q}{4\pi\epsilon_0 r^2})\hat{\mathbf{y}}$
$(-\frac{q}{4\pi\epsilon_0 r^2})\frac{v}{c^2}\gamma\,\hat{\mathbf{z}}$	$(-\frac{q}{4\pi\epsilon_0 r^2})\frac{2v}{c^2}\gamma^2\hat{\mathbf{z}}$	0
$(-\frac{q^2}{4\pi\epsilon_0 r^2})\gamma(1+\frac{v^2}{c^2})\hat{\mathbf{y}}$	$(-\frac{q^2}{4\pi\epsilon_0 r^2})\gamma^2(1+\frac{v^2}{c^2})\hat{\mathbf{y}}$	$(-\frac{q}{4\pi\epsilon_0 r^2})\hat{\mathbf{y}}$

$\gamma = \frac{1}{\sqrt{1-v^2/c^2}}$.

Problem 12.46

(a) From Eq. 12.108:

$$\bar{\mathbf{E}}\cdot\bar{\mathbf{B}} = \bar{E}_x\bar{B}_x + \bar{E}_y\bar{B}_y + \bar{E}_z\bar{B}_z = E_xB_x + \gamma^2(E_y - vB_z)(B_y + \frac{v}{c^2}E_z) + \gamma(E_z + vB_y)(B_z - \frac{v}{c^2}E_y)$$

$$= E_xB_x + \gamma^2\{E_yB_y + \frac{v}{c^2}\cancel{E_y}E_z - v\cancel{B_y}B_z - \frac{v^2}{c^2}E_zB_z + E_zB_z - \frac{v}{c^2}\cancel{E_y}E_z + v\cancel{B_y}B_z - \frac{v^2}{c^2}E_yB_y\}$$

$$= E_xB_x + \gamma^2\left[E_yB_y\left(1-\frac{v^2}{c^2}\right) + E_zB_z\left(1-\frac{v^2}{c^2}\right)\right] = E_xB_x + E_yB_y + E_zB_z = \mathbf{E}\cdot\mathbf{B}.\quad \text{qed}$$

(b) $$\bar{E}^2 - c^2\bar{B}^2 = [E_x^2 + \gamma^2(E_y - vB_z)^2 + \gamma^2(E_z + vB_y)^2] - c^2[B_x^2 + \gamma^2(B_y + \frac{v}{c^2}E_z)^2 + \gamma^2(B_z - \frac{v}{c^2}E_y)^2]$$

$$= E_x^2 + \gamma^2(E_y^2 - 2\cancel{E_yv}B_z + v^2B_z^2 + E_z^2 + 2\cancel{E_zv}B_y + v^2B_y^2 - c^2B_y^2 - c^2 2\frac{v}{\cancel{c^2}}B_yE_z$$

$$- c^2\frac{v^2}{c^4}E_z^2 - c^2B_z^2 + c^2 2\frac{v}{\cancel{c^2}}B_zE_y \;\; - c^2\frac{v^2}{c^4}E_y^2) - c^2B_x^2$$

$$= E_x^2 - c^2B_x^2 + \gamma^2\left[E_y^2\left(1-\frac{v^2}{c^2}\right) + E_z^2\left(1-\frac{v^2}{c^2}\right) - c^2B_y^2\left(1-\frac{v^2}{c^2}\right) - c^2B_z^2\left(1-\frac{v^2}{c^2}\right)\right]$$

$$= (E_x^2 + E_y^2 + E_z^2) - c^2(B_x^2 + B_y^2 + B_z^2) = E^2 - B^2c^2.\quad \text{qed}$$

(c) **No.** For if $\mathbf{B} = 0$ in one system, then $(E^2 - c^2B^2)$ is ***positive***. Since it is invariant, it must be positive in *any* system. Therefore $\mathbf{E} \neq 0$ in *all* systems.

Problem 12.47

(a) Making the appropriate modifications in Eq. 9.48 (and picking $\delta = 0$ for convenience),

$$\boxed{\mathbf{E}(x,y,z,t) = E_0\cos(kx-\omega t)\,\hat{\mathbf{y}},\quad \mathbf{B}(x,y,z,t) = \frac{E_0}{c}\cos(kx-\omega t)\,\hat{\mathbf{z}},\quad \text{where } k \equiv \frac{\omega}{c}.}$$

(b) Using Eq. 12.108 to transform the fields:

$$\bar{E}_x = \bar{E}_z = 0,\quad \bar{E}_y = \gamma(E_y - vB_z) = \gamma E_0\left[\cos(kx-\omega t) - \frac{v}{c}\cos(kx-\omega t)\right] = \alpha E_0\cos(kx-\omega t),$$

$$\bar{B}_x = \bar{B}_y = 0,\quad \bar{B}_z = \gamma(B_z - \frac{v}{c^2}E_y) = \gamma E_0\left[\frac{1}{c}\cos(kx-\omega t) - \frac{v}{c^2}\cos(kx-\omega t)\right] = \alpha\frac{E_0}{c}\cos(kx-\omega t),$$

where $$\boxed{\alpha \equiv \gamma\left(1-\frac{v}{c}\right) = \sqrt{\frac{1-v/c}{1+v/c}}.}$$

Now the inverse Lorentz transformations (Eq. 12.19) $\Rightarrow x = \gamma(\bar{x} + v\bar{t})$ and $t = \gamma\left(\bar{t} + \frac{v}{c^2}\bar{x}\right)$, so

$$kx - \omega t = \gamma\left[k(\bar{x} + v\bar{t}) - \omega\left(\bar{t} + \frac{v}{c^2}\bar{x}\right)\right] = \gamma\left[\left(k - \frac{\omega v}{c^2}\right)\bar{x} - (\omega - kv)\bar{t}\right] = \bar{k}\bar{x} - \bar{\omega}\bar{t},$$

where (recalling that $k = \omega/c$): $\bar{k} \equiv \gamma\left(k - \frac{\omega v}{c^2}\right) = \gamma k(1 - v/c) = \alpha k$ and $\bar{\omega} \equiv \gamma\omega(1 - v/c) = \alpha\omega$.

Conclusion:

$$\boxed{\begin{gathered}\bar{\mathbf{E}}(\bar{x}, \bar{y}, \bar{z}, \bar{t}) = \bar{E}_0 \cos(\bar{k}\bar{x} - \bar{\omega}\bar{t})\,\hat{\mathbf{y}}, \quad \bar{\mathbf{B}}(\bar{x}, \bar{y}, \bar{z}, \bar{t}) = \frac{\bar{E}_0}{c}\cos(\bar{k}\bar{x} - \bar{\omega}\bar{t})\,\hat{\mathbf{z}}, \\ \text{where } \bar{E}_0 = \alpha E_0, \quad \bar{k} = \alpha k, \quad \bar{\omega} = \alpha\omega, \quad \text{and } \alpha \equiv \sqrt{\frac{1 - v/c}{1 + v/c}}.\end{gathered}}$$

(c) $\boxed{\bar{\omega} = \omega\sqrt{\frac{1 - v/c}{1 + v/c}}.}$ This is the $\boxed{\text{Doppler shift}}$ for light. $\bar{\lambda} = \frac{2\pi}{\bar{k}} = \frac{2\pi}{\alpha k} = \frac{\lambda}{\alpha}$. The velocity of the wave in $\bar{S}$ is $\bar{v} = \frac{\bar{\omega}}{2\pi}\bar{\lambda} = \frac{\omega}{\lambda} = \boxed{c.}$ $\boxed{\text{Yup,}}$ this is exactly what I expected (the velocity of a light wave is the same in any inertial system).

(d) Since intensity goes like E^2, the ratio is $\frac{\bar{I}}{I} = \frac{\bar{E}_0^2}{E_0^2} = \alpha^2 = \boxed{\frac{1 - v/c}{1 + v/c}.}$

Dear Al,

The amplitude, frequency, and intensity of the light wave will all $\boxed{\text{decrease to zero}}$ as you run faster and faster. It'll get so faint you won't be able to see it, and so red-shifted even your night-vision goggles won't help. But it'll still be going 3×10^8 m/s relative to you. Sorry about that.

Sincerely,

David

Problem 12.48

$\bar{t}^{02} = \Lambda^0_\lambda\Lambda^2_\sigma t^{\lambda\sigma} = \Lambda^0_0\Lambda^2_2 t^{02} + \Lambda^0_1\Lambda^2_2 t^{12} = \gamma t^{02} + (-\gamma\beta)t^{12} = \gamma(t^{02} - \beta t^{12})$.

$\bar{t}^{03} = \Lambda^0_\lambda\Lambda^3_\sigma t^{\lambda\sigma} = \Lambda^0_0\Lambda^3_3 t^{03} + \Lambda^0_1\Lambda^3_3 t^{13} = \gamma t^{03} + (-\gamma\beta)t^{13} = \gamma(t^{03} - \beta t^{13}) = \gamma(t^{03} + \beta t^{31})$.

$\bar{t}^{23} = \Lambda^2_\lambda\Lambda^3_\sigma t^{\lambda\sigma} = \Lambda^2_2\Lambda^3_3 t^{23} = t^{23}$.

$\bar{t}^{31} = \Lambda^3_\lambda\Lambda^1_\sigma t^{\lambda\sigma} = \Lambda^3_3\Lambda^1_0 t^{30} + \Lambda^3_3\Lambda^1_1 t^{31} = (-\gamma\beta)t^{30} + \gamma t^{31} = \gamma(t^{31} + \beta t^{03})$.

$\bar{t}^{12} = \Lambda^1_\lambda\Lambda^2_\sigma t^{\lambda\sigma} = \Lambda^1_0\Lambda^2_2 t^{02} + \Lambda^1_1\Lambda^2_2 t^{12} = (-\gamma\beta)t^{02} + \gamma t^{12} = \gamma(t^{12} - \beta t^{02})$.

Problem 12.49

Suppose $t^{\nu\mu} = \pm t^{\mu\nu}$ (+ for symmetric, − for antisymmetric).

$\bar{t}^{\kappa\lambda} = \Lambda^\kappa_\mu\Lambda^\lambda_\nu t^{\mu\nu}$

$\bar{t}^{\lambda\kappa} = \Lambda^\lambda_\mu\Lambda^\kappa_\nu t^{\mu\nu} = \Lambda^\lambda_\nu\Lambda^\kappa_\mu t^{\nu\mu}$ [Because μ and ν are both summed from $0 \to 3$, it doesn't matter which we call μ and and which call ν.]

$= \Lambda^\kappa_\mu\Lambda^\lambda_\mu(\pm t^{\mu\nu})$ [I used the symmetry of $t^{\mu\nu}$, and wrote the Λ's in the other order.]

$= \pm\bar{t}^{\kappa\lambda}$. qed

Problem 12.50

$$\begin{aligned}F^{\mu\nu}F_{\mu\nu} &= F^{00}F^{00} - F^{01}F^{01} - F^{02}F^{02} - F^{03}F^{03} - F^{10}F^{10} - F^{20}F^{20} - F^{30}F^{30}\\ &\quad + F^{11}F^{11} + F^{12}F^{12} + F^{13}F^{13} + F^{21}F^{21} + F^{22}F^{22} + F^{23}F^{23} + F^{31}F^{31} + F^{32}F^{32} + F^{33}F^{33}\\ &= -(E_x/c)^2 - (E_y/c)^2 - (E_z/c)^2 - (E_x/c)^2 - (E_y/c)^2 - (E_z/c)^2 + B_z^2 + B_y^2 + B_z^2 + B_x^2 + B_y^2 + B_x^2\\ &= 2B^2 - 2E^2/c^2 = \boxed{2\left(B^2 - \frac{E^2}{c^2}\right)},\end{aligned}$$

which, apart from the constant factor $-\frac{2}{c^2}$, is the invariant we found in Prob. 12.46(b).

$\boxed{G^{\mu\nu}G_{\mu\nu} = 2(E^2/c^2 - B^2)}$ (the same invariant).

$$\begin{aligned}F^{\mu\nu}G_{\mu\nu} &= -2\left(F^{01}G^{01} + F^{02}G^{02} + F^{03}G^{03}\right) + 2\left(F^{12}G^{12} + F^{13}G^{13} + F^{23}G^{23}\right)\\ &= -2\left(\frac{1}{c}E_xB_x + \frac{1}{c}E_yB_y + \frac{1}{c}E_zB_z\right) 2\left[B_z(-E_z/c) + (-B_y)(E_y/c) + B_x(-E_x/c)\right]\\ &= -\frac{2}{c}(\mathbf{E}\cdot\mathbf{B}) - \frac{2}{c}(\mathbf{E}\cdot\mathbf{B}) = \boxed{-\frac{4}{c}(\mathbf{E}\cdot\mathbf{B}),}\end{aligned}$$

which, apart from the factor $-4/c$, is the invariant of Prob. 12.46(a). [These are, incidentally, the *only* fundamental invariants you can construct from **E** and **B**.]

Problem 12.51

$$\left.\begin{aligned}\mathbf{E} &= \frac{1}{4\pi\epsilon_0}\frac{2\lambda}{x}\hat{\mathbf{x}} = \frac{\mu_0}{2\pi}\frac{\lambda c^2}{x}\hat{\mathbf{x}}\\ \mathbf{B} &= \frac{\mu_0}{4\pi}\frac{2\lambda v}{x}\hat{\mathbf{y}} = \frac{\mu_0}{2\pi}\frac{\lambda v}{x}\hat{\mathbf{y}}\end{aligned}\right\} \boxed{F^{\mu\nu} = \frac{\mu_0\lambda}{2\pi x}\begin{pmatrix}0 & c & 0 & 0\\ -c & 0 & 0 & -v\\ 0 & 0 & 0 & 0\\ 0 & v & 0 & 0\end{pmatrix}.}$$

Problem 12.52

$\partial_\nu F^{\mu\nu} = \mu_0 J^\mu$. Differentiate: $\partial_\mu\partial_\nu F^{\mu\nu} = \mu_0\partial_\mu J^\mu$.

But $\partial_\mu\partial_\nu = \partial_\nu\partial_\mu$ (the combination is *symmetric*) while $F^{\nu\mu} = -F^{\mu\nu}$ (*anti*symmetric).

$\therefore\ \partial_\mu\partial_\nu F^{\mu\nu} = 0$. [Why? Well, these indices are both summed from $0 \to 3$, so it doesn't matter which we call μ, which ν: $\partial_\mu\partial_\nu F^{\mu\nu} = \partial_\nu\partial_\mu F^{\nu\mu} = \partial_\mu\partial_\nu(-F^{\mu\nu}) = -\partial_\mu\partial_\nu F^{\mu\nu}$. But if a quantity is equal to minus itself, it must be zero.] *Conclusion:* $\partial_\mu J^\mu = 0$. qed

Problem 12.53

We know that $\partial_\nu G^{\mu\nu} = 0$ is equivalent to the two homogeneous Maxwell equations, $\nabla\cdot\mathbf{B} = 0$ and $\nabla\times\mathbf{E} = -\frac{\partial\mathbf{B}}{\partial t}$. All we have to show, then, is that $\partial_\lambda F_{\mu\nu} + \partial_\mu F_{\nu\lambda} + \partial_\nu F_{\lambda\mu} = 0$ is *also* equivalent to them. Now this equation stands for 64 separate equations ($\mu = 0 \to 3$, $\nu = 0 \to 3$, $\lambda = 0 \to 3$, and $4\times4\times4 = 64$). But many of them are redundant, or trivial.

Suppose two indices are the same (say, $\mu = \nu$). Then $\partial_\lambda F_{\mu\mu} + \partial_\mu F_{\mu\lambda} + \partial_\mu F_{\lambda\mu} = 0$. But $F_{\mu\mu} = 0$ and $F_{\mu\lambda} = -F_{\lambda\mu}$, so this is trivial: $0 = 0$. To get anything significant, then, μ, ν, λ must all be ***different***. They could be ***all spatial*** ($\mu,\nu,\lambda = 1,2,3 = x,y,z$ — or some permutation thereof), or ***one temporal*** and ***two spatial*** ($\mu = 0$, $\nu,\lambda = 1,2$ or $2,3$, or $1,3$ — or some permutation). Let's examine these two cases separately.

All spatial: say, $\mu = 1$, $\nu = 2$, $\lambda = 3$ (other permutations yield the same equation, or minus it).

$$\partial_3 F_{12} + \partial_1 F_{23} + \partial_2 F_{31} = 0 \Rightarrow \frac{\partial}{\partial z}(B_z) + \frac{\partial}{\partial x}(B_x) + \frac{\partial}{\partial y}(B_y) = 0 \Rightarrow \nabla\cdot\mathbf{B} = 0.$$

One temporal: say, $\mu = 0$, $\nu = 1$, $\lambda = 2$ (other permutations of these indices yield the same result, or minus it).

$$\partial_2 F_{01} + \partial_0 F_{12} + \partial_1 F_{20} = 0 \Rightarrow \frac{\partial}{\partial y}\left(-\frac{E_x}{c}\right) + \frac{\partial}{\partial (ct)}(B_z) + \frac{\partial}{\partial x}\left(\frac{E_y}{c}\right) = 0,$$

or $-\frac{\partial B_z}{\partial t} + \left(\frac{\partial E_x}{\partial y} - \frac{\partial E_y}{\partial x}\right) = 0$, which is the z component of $-\frac{\partial \mathbf{B}}{\partial t} = \boldsymbol{\nabla}\times\mathbf{E}$. (If $\mu = 0$, $\nu = 1$, $\lambda = 2$, we get the y component; for $\nu = 2$, $\lambda = 3$ we get the x component.)

Conclusion: $\partial_\lambda F_{\mu\nu} + \partial_\mu F_{\nu\lambda} + \partial_\nu F_{\lambda\mu} = 0$ is equivalent to $\boldsymbol{\nabla}\cdot\mathbf{B} = 0$ and $\frac{\partial \mathbf{B}}{\partial t} = -\boldsymbol{\nabla}\times\mathbf{E}$, and hence to $\partial_\nu G^{\mu\nu} = 0$. qed

Problem 12.54

$K^0 = q\eta_\nu F^{0\nu} = q(\eta_1 F^{01} + \eta_2 F^{02} + \eta_3 F^{03}) = q(\boldsymbol{\eta}\cdot\mathbf{E})/c = \boxed{\frac{q}{c}\gamma\mathbf{u}\cdot\mathbf{E}.}$ Now from Eq. 12.71 we know that $K^0 = \frac{1}{c}\frac{dW}{d\tau}$, where W is the energy of the particle. Since $d\tau = \frac{1}{\gamma}dt$, we have:

$$\frac{1}{c}\gamma\frac{dW}{dt} = \frac{q}{c}\gamma(\mathbf{u}\cdot\mathbf{E}) \Rightarrow \boxed{\frac{dW}{dt} = q(\mathbf{u}\cdot\mathbf{E}).}$$

This says *the power delivered to the particle is force* ($q\mathbf{E}$) *times velocity* ($\mathbf{u}$) — which is as it *should* be.

Problem 12.55

$$\overline{\partial^0\phi} = \frac{\partial}{\partial \bar{x}_0}\phi = -\frac{1}{c}\frac{\partial}{\partial \bar{t}}\phi = -\frac{1}{c}\left(\frac{\partial\phi}{\partial t}\frac{\partial t}{\partial \bar{t}} + \frac{\partial\phi}{\partial x}\frac{\partial x}{\partial \bar{t}} + \frac{\partial\phi}{\partial y}\frac{\partial y}{\partial \bar{t}} + \frac{\partial\phi}{\partial z}\frac{\partial z}{\partial \bar{t}}\right).$$

From Eq. 12.19, we have: $\frac{\partial t}{\partial \bar{t}} = \gamma$, $\quad\frac{\partial x}{\partial \bar{t}} = \gamma v$, $\quad\frac{\partial y}{\partial \bar{t}} = \frac{\partial z}{\partial \bar{t}} = 0.$

So $\overline{\partial^0\phi} = -\frac{1}{c}\gamma\left(\frac{\partial\phi}{\partial t} + v\frac{\partial\phi}{\partial x}\right)$ or (since $ct = x^0 = -x_0$): $\overline{\partial^0\phi} = \gamma\left(\frac{\partial\phi}{\partial x_0} - \frac{v}{c}\frac{\partial\phi}{\partial x^1}\right) = \gamma\left[(\partial^0\phi) - \beta(\partial^1\phi)\right].$

$$\overline{\partial^1\phi} = \frac{\partial}{\partial \bar{x}}\phi = \frac{\partial\phi}{\partial t}\frac{\partial t}{\partial x^1} + \frac{\partial\phi}{\partial x}\frac{\partial x}{\partial \bar{x}} + \frac{\partial\phi}{\partial y}\frac{\partial y}{\partial \bar{x}} + \frac{\partial\phi}{\partial z}\frac{\partial z}{\partial \bar{x}} = \gamma\frac{v}{c^2}\frac{\partial\phi}{\partial t} + \gamma\frac{\partial\phi}{\partial x} = \gamma\left(\frac{\partial\phi}{\partial x_1} - \frac{v}{c}\frac{\partial\phi}{\partial x_0}\right) = \gamma\left[(\partial^1\phi) - \beta(\partial^0\phi)\right].$$

$$\overline{\partial^2\phi} = \frac{\partial\phi}{\partial \bar{y}} = \frac{\partial\phi}{\partial t}\frac{\partial t}{\partial \bar{y}} + \frac{\partial\phi}{\partial x}\frac{\partial x}{\partial \bar{y}} + \frac{\partial\phi}{\partial y}\frac{\partial y}{\partial \bar{y}} + \frac{\partial\phi}{\partial z}\frac{\partial z}{\partial \bar{y}} = \frac{\partial\phi}{\partial y} = \partial^2\phi.$$

$$\overline{\partial^3\phi} = \frac{\partial\phi}{\partial \bar{z}} = \frac{\partial\phi}{\partial t}\frac{\partial t}{\partial \bar{z}} + \frac{\partial\phi}{\partial x}\frac{\partial x}{\partial \bar{z}} + \frac{\partial\phi}{\partial y}\frac{\partial y}{\partial \bar{z}} + \frac{\partial\phi}{\partial z}\frac{\partial z}{\partial \bar{z}} = \frac{\partial\phi}{\partial z} = \partial^3\phi.$$

Conclusion: $\partial^\mu\phi$ transforms in the same way as a^μ (Eq. 12.27)—and hence is a contravariant 4-vector. qed

Problem 12.56

According to Prob. 12.53, $\frac{\partial G^{\mu\nu}}{\partial x^\nu} = 0$ is equivalent to Eq. 12.129. Using Eq. 12.132, we find (in the notation of Prob. 12.55):

$$\begin{aligned}
\frac{\partial F_{\mu\nu}}{\partial x^\lambda} + \frac{\partial F_{\nu\lambda}}{\partial x^\mu} + \frac{\partial F_{\lambda\mu}}{\partial x^\nu} &= \partial_\lambda F_{\mu\nu} + \partial_\mu F_{\nu\lambda} + \partial_\nu F_{\lambda\mu} \\
&= \partial_\lambda(\partial_\mu A_\nu - \partial_\nu A_\mu) + \partial_\mu(\partial_\nu A_\lambda - \partial_\lambda A_\nu) + \partial_\nu(\partial_\lambda A_\mu - \partial_\mu A_\lambda) \\
&= (\partial_\lambda\partial_\mu A_\nu - \partial_\mu\partial_\lambda A_\nu) + (\partial_\mu\partial_\nu A_\lambda - \partial_\nu\partial_\mu A_\lambda) + (\partial_\nu\partial_\lambda A_\mu - \partial_\lambda\partial_\nu A_\mu) = 0. \quad \text{qed}
\end{aligned}$$

[Note that $\partial_\lambda\partial_\mu A_\nu = \frac{\partial^2 A_\nu}{\partial x^\lambda \partial x^\mu} = \frac{\partial^2 A_\nu}{\partial x^\mu \partial x^\lambda} = \partial_\mu\partial_\lambda A_\nu$, by equality of cross-derivatives.]

Problem 12.57

Step 1: rotate from xy to XY, using Eq. 1.29:

$$X = \cos\phi\, x + \sin\phi\, y$$
$$Y = -\sin\phi\, x + \cos\phi\, y$$

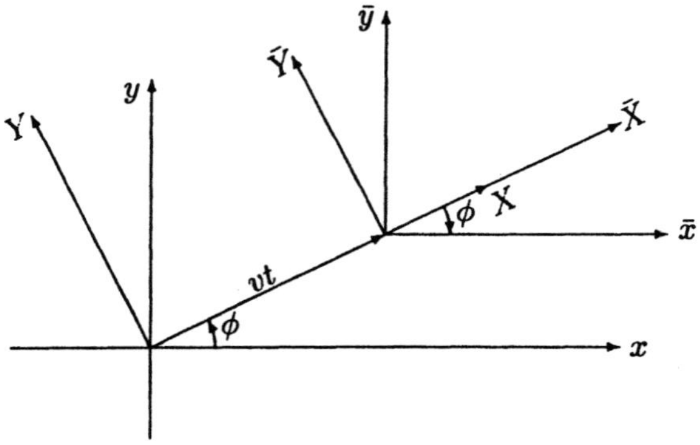

Step 2: Lorentz-transform from XY to $\bar{X}\bar{Y}$, using Eq. 12.18:

$$\bar{X} = \gamma(X - vt) = \gamma[\cos\phi\, x + \sin\phi\, y - \beta ct]$$
$$\bar{Y} = Y = -\sin\phi\, x + \cos\phi\, y$$
$$\bar{Z} = Z = z$$
$$c\bar{t} = \gamma(ct - \beta X) = \gamma\big[ct - \beta(\cos\phi\, x + \sin\phi\, y)\big]$$

Step 3: Rotate from $\bar{X}\bar{Y}$ to $\bar{x}\bar{y}$, using Eq. 1.29 with negative ϕ:

$$\bar{x} = \cos\phi\,\bar{X} - \sin\phi\,\bar{Y} = \gamma\cos\phi[\cos\phi\, x + \sin\phi\, y - \beta ct] - \sin\phi[-\sin\phi\, x + \cos\phi\, y]$$
$$= (\gamma\cos^2\phi + \sin^2\phi)x + (\gamma - 1)\sin\phi\cos\phi\, y - \gamma\beta\cos\phi\,(ct)$$
$$\bar{y} = \sin\phi\,\bar{X} + \cos\phi\,\bar{Y} = \gamma\sin\phi(\cos\phi\, x + \sin\phi\, y - \beta ct) + \cos\phi(-\sin\phi\, c + \cos\phi\, y)$$
$$= (\gamma - 1)\sin\phi\cos\phi\, x + (\gamma\sin^2\phi + \cos^2\phi)y - \gamma\beta\sin\phi\,(ct)$$

In matrix form:
$$\begin{pmatrix} c\bar{t} \\ \bar{x} \\ \bar{y} \\ \bar{z} \end{pmatrix} = \boxed{\begin{pmatrix} \gamma & -\gamma\beta\cos\phi & -\gamma\beta\sin\phi & 0 \\ -\gamma\beta\cos\phi & (\gamma\cos^2\phi + \sin^2\phi) & (\gamma - 1)\sin\phi\cos\phi & 0 \\ -\gamma\beta\sin\phi & (\gamma - 1)\sin\phi\cos\phi & (\gamma\sin^2\phi + \cos^2\phi) & 0 \\ 0 & 0 & 0 & 1 \end{pmatrix}} \begin{pmatrix} ct \\ x \\ y \\ z \end{pmatrix}.$$

Problem 12.58

In center-of-momentum system, threshold occurs when incident energy is *just* sufficient to cover the *rest* energy of the resulting particles, with none "wasted" as kinetic energy. Thus, in lab system, we want the outgoing K and Σ to have the *same velocity*, at threshold:

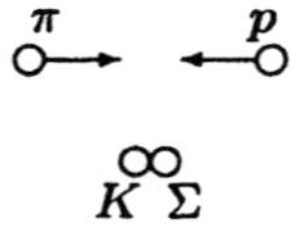

Initial momentum: p_π; initial energy of π: $E^2 - p^2c^2 = m^2c^4 \Rightarrow E_\pi^2 = m_\pi^2c^4 + p_\pi^2c^2$.

Total initial energy: $m_pc^2 + \sqrt{m_\pi^2c^4 + p_\pi^2c^2}$. These are also the final energy and momentum: $E^2 - p^2c^2 = (m_K + m_\Sigma)^2c^4$.

$$\left(m_pc^2 + \sqrt{m_\pi^2c^4 + p_\pi^2c^2}\right)^2 - p_\pi^2c^2 = (m_K + m_\Sigma)^2c^4$$

$$m_p^2\cancel{c^4} + \frac{2m_pc^2}{c^4}\sqrt{m_\pi^2c^2 + p_\pi^2}\,c + m_\pi^2\cancel{c^4} + \cancel{p_\pi^2c^2} - \cancel{p_\pi^2c^2} = (m_K + m_\Sigma)^2\cancel{c^4}$$

$$\frac{2m_p}{c}\sqrt{m_\pi^2c^2 + p_\pi^2} = (m_K + m_\Sigma)^2 - m_p^2 - m_\pi^2$$

$$(m_\pi^2 c^2 + p_\pi^2)\frac{4m_p^2}{c^2} = (m_K + m_\Sigma)^4 - 2(m_p^2 + m_\pi^2)(m_K + m_\Sigma)^2 + m_p^4 + m_\pi^4 + 2m_p^2 m_\pi^2$$

$$\frac{4m_p^2}{c^2}p_\pi^2 = (m_K + m_\Sigma)^4 - 2(m_p^2 + m_\pi^2)(m_K + m_\Sigma)^2 + (m_p^2 - m_\pi^2)^2$$

$$\boxed{p_\pi = \frac{c}{2m_p}\sqrt{(m_K + m_\Sigma)^4 - 2(m_p^2 + m_\pi^2)(m_K + m_\Sigma)^2 + (m_p^2 - m_\pi^2)^2}}$$

$$= \frac{1}{(2m_p c^2)c}\sqrt{(m_K c^2 + m_\Sigma c^2)^4 - 2\left[(m_p c^2)^2 + (m_\pi c^2)^2\right](m_K c^2 + m_\Sigma c^2)^2 + \left[(m_p c^2)^2 - (m_\pi c^2)^2\right]^2}$$

$$= \frac{1}{2c(900)}\sqrt{(1700)^4 - 2\left[(900)^2 + (150)^2\right](1700)^2 + \left[(900)^2 - (150)^2\right]^2}$$

$$= \frac{1}{1800c}\sqrt{(8.35 \times 10^{12}) - (4.81 \times 10^{12}) + (0.62 \times 10^{12})} = \frac{1}{1800c}(2.04 \times 10^6) = \boxed{1133 \text{ MeV}/c.}$$

Problem 12.59

In CM:

Before

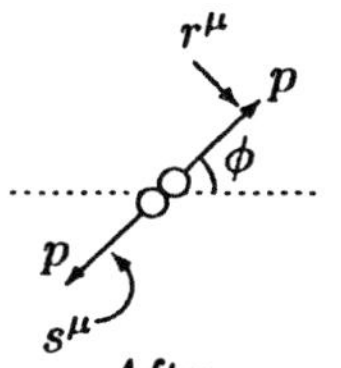

After

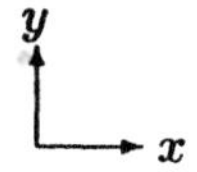

(p = magnitude of 3-momentum in CM, ϕ = CM scattering angle)

Outgoing 4-momenta: $r^\mu = \left(\frac{E}{c}, p\cos\phi, p\sin\phi, 0\right)$; $\quad s^\mu = \left(\frac{E}{c}, -p\cos\phi, -p\sin\phi, 0\right)$.

In Lab:

Before

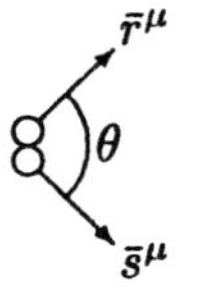

Problem: calculate θ, in terms of p, ϕ.

Lorentz transformation: $\bar{r}_x = \gamma(r_x - \beta r^0)$; $\bar{r}_y = r_y$; $\bar{s}_x = \gamma(s_x - \beta s^0)$; $\bar{s}_y = s_y$.

Now $E = \gamma mc^2$; $p = -\gamma mv$ (v here is to the *left*); $E^2 - p^2c^2 = m^2c^4$, so $\beta = -\frac{pc}{E}$.
$\therefore \bar{r}_x = \gamma\left(p\cos\phi + \frac{pc}{E}\frac{E}{c}\right) = \gamma p(1 + \cos\phi)$; $\bar{r}_y = p\sin\phi$; $\bar{s}_x = \gamma p(1 - \cos\phi)$; $\bar{s}_y = -p\sin\phi$.

$$\cos\theta = \frac{\bar{\mathbf{r}} \cdot \bar{\mathbf{s}}}{\bar{r}\bar{s}} = \frac{\gamma^2 p^2(1 - \cos^2\phi) - p^2\sin^2\phi}{\sqrt{\left[\gamma^2p^2(1+\cos\phi)^2 + p^2\sin^2\phi\right]\left[\gamma^2p^2(1-\cos\phi)^2 + p^2\sin^2\phi\right]}}$$

$$= \frac{(\gamma^2 - 1)\sin^2\phi}{\sqrt{\left[\gamma^2(1+\cos\phi)^2 + \sin^2\phi\right]\left[\gamma^2(1-\cos\phi)^2 + \sin^2\phi\right]}}$$

$$= \frac{(\gamma^2 - 1)}{\sqrt{\left[\gamma^2\left(\frac{1+\cos\phi}{\sin\phi}\right)^2 + 1\right]\left[\gamma^2\left(\frac{1-\cos\phi}{\sin\phi}\right)^2 + 1\right]}} = \frac{(\gamma^2 - 1)}{\sqrt{\left(\gamma^2\cot^2\frac{\phi}{2} + 1\right)\left(\gamma^2\tan^2\frac{\phi}{2} + 1\right)}}$$

$$\cos\theta = \frac{\omega}{\sqrt{\left(1+\cot^2\frac{\phi}{2}+\omega\cot^2\frac{\phi}{2}\right)\left(1+\tan^2\frac{\phi}{2}+\omega\tan^2\frac{\phi}{2}\right)}} \quad (\text{where } \omega \equiv \gamma^2-1)$$

$$= \frac{\omega}{\sqrt{\left(\csc^2\frac{\phi}{2}+\omega\cot^2\frac{\phi}{2}\right)\left(\sec^2\frac{\phi}{2}+\omega\tan^2\frac{\phi}{2}\right)}} = \frac{\omega\sin\frac{\phi}{2}\cos\frac{\phi}{2}}{\sqrt{\left(1+\omega\cos^2\frac{\phi}{2}\right)\left(1+\omega\sin^2\frac{\phi}{2}\right)}}$$

$$= \frac{\frac{1}{2}\omega\sin\phi}{\sqrt{\left[1+\frac{1}{2}\omega(1+\cos\phi)\right]\left[1+\frac{1}{2}\omega(1-\cos\phi)\right]}} = \frac{\sin\phi}{\sqrt{\left[\left(\frac{2}{\omega}+1\right)+\cos\phi\right]\left[\left(\frac{2}{\omega}+1\right)-\cos\phi\right]}}$$

$$= \frac{\sin\phi}{\sqrt{\left(\frac{2}{\omega}+1\right)^2-\cos^2\phi}} = \frac{\sin\phi}{\sqrt{\frac{4}{\omega^2}+\frac{4}{\omega}+\sin^2\phi}} = \frac{1}{\sqrt{1+(\tau/\sin\phi)^2}}, \text{ where } \tau^2 = \frac{4}{\omega^2}+\frac{4}{\omega}.$$

$\sin\theta = \frac{\tau}{\sin\phi}$. $\tau^2 = \frac{4}{\omega^2}(1+\omega) = \frac{4}{(\gamma^2-1)^2}\gamma^2$, so $\tan\theta = \frac{2\gamma}{(\gamma^2-1)\sin\phi}$.

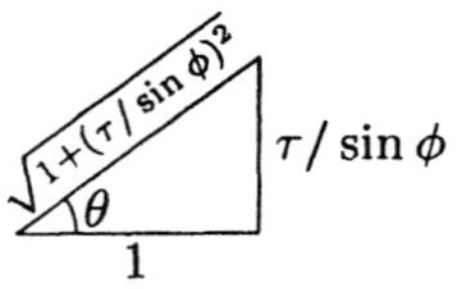

Or, since $(\gamma^2-1) = \gamma^2\left(1-\frac{1}{\gamma^2}\right) = \gamma^2\frac{v^2}{c^2}$, $\boxed{\tan\theta = \frac{2c^2}{\gamma v^2\sin\phi}}$.

Problem 12.60

$\frac{dp}{d\tau} = K$ (a constant) $\Rightarrow \frac{dp}{dt}\frac{dt}{d\tau} = K$. But $\frac{dt}{d\tau} = \frac{1}{\sqrt{1-u^2/c^2}}$; $p = \frac{mu}{\sqrt{1-u^2/c^2}}$.

$\therefore \frac{d}{dt}\left(\frac{u}{\sqrt{1-u^2/c^2}}\right) = \frac{K}{m}\sqrt{1-u^2/c^2}$. Multiply by $\frac{dt}{dx} = \frac{1}{u}$:

$$\frac{dt}{dx}\frac{d}{dt}\left(\frac{u}{\sqrt{1-u^2/c^2}}\right) = \frac{d}{dx}\left(\frac{u}{\sqrt{1-u^2/c^2}}\right) = \frac{K}{m}\frac{\sqrt{1-u^2/c^2}}{u}. \text{ Let } w = \frac{u}{\sqrt{1-u^2/c^2}}.$$

$$\frac{dw}{dx} = \frac{K}{m}\frac{1}{w}; \quad w\frac{dw}{dx} = \frac{1}{2}\frac{d}{dx}w^2 = \frac{k}{m}; \quad \frac{d(w^2)}{dx} = \frac{2K}{m} \Rightarrow d(w^2) = \frac{2K}{m}(dx).$$

$\therefore w^2 = \frac{2K}{m}x +$ constant. But at $t=0$, $x=0$ and $u=0$ (so $w=0$), and hence the constant is 0.

$$w^2 = \frac{2K}{m}x = \frac{u^2}{1-u^2/c^2}; \quad u^2 = \frac{2Kx}{m} - \frac{2Kx}{mc^2}u^2; \quad u^2\left(1+\frac{2Kx}{mc^2}\right) = \frac{2Kx}{m}.$$

$$u^2 = \frac{2Kx/m}{1+\frac{2Kx}{mc^2}} = \frac{c^2}{1+\left(\frac{mc^2}{2Kx}\right)}; \quad \frac{dx}{dt} = \frac{c}{\sqrt{1+\left(\frac{mc^2}{2Kx}\right)}}; \quad ct = \int\sqrt{1+\left(\frac{mc^2}{2Kx}\right)}\,dx.$$

Let $\frac{mc^2}{2K} \equiv a^2$; $ct = \int\frac{\sqrt{x+a^2}}{\sqrt{x}}dx$. Let $x \equiv y^2$; $dx = 2y\,dy$; $\sqrt{x} = y$.

$$ct = \int\frac{\sqrt{y^2+a^2}}{y}2y\,dy = 2\int\sqrt{y^2+a^2}\,dy = \left[y\sqrt{y^2+a^2}+a^2\ln(y+\sqrt{y^2+a^2})\right] + \text{constant.}$$

At $t=0$, $x=0 \Rightarrow y=0$, so $0 = a^2\ln a +$ constant $\Rightarrow$ constant $= -a^2\ln a$.

$$\therefore ct = y\sqrt{y^2+a^2}+a^2\ln\left(y/a+\sqrt{(y/a)^2+1}\right) = a^2\left[\left(\frac{y}{a}\right)\sqrt{\left(\frac{y}{a}\right)^2+1}+\ln\left(\frac{y}{a}+\sqrt{\left(\frac{y}{a}\right)^2+1}\right)\right].$$

Let: $z \equiv y/a = \sqrt{x}\sqrt{\frac{2K}{mc^2}} = \sqrt{\frac{2Kx}{mc^2}}$. Then $\boxed{\frac{2Kt}{mc} = z\sqrt{1+z^2}+\ln(z+\sqrt{1+z^2})}$.

Problem 12.61

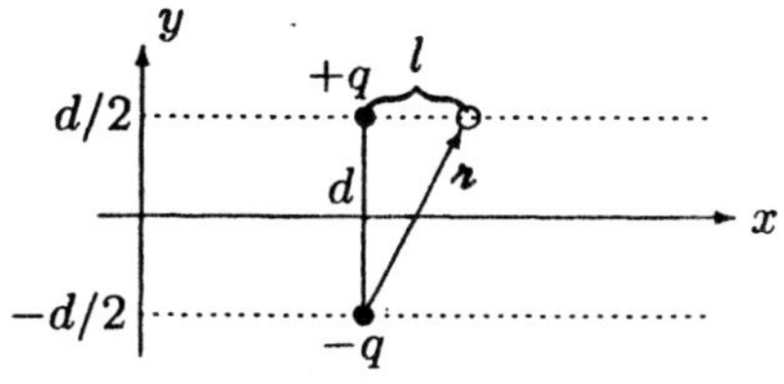

(a) $x(t) = \frac{c}{\alpha}\left[\sqrt{1+(\alpha t)^2} - 1\right]$, where $\alpha = \frac{F}{mc}$. The force of $+q$ on $-q$ will be the mirror image of the force of $-q$ on $+q$ (in the x axis), so the *net* force is in the x direction (the net *magnetic* force is zero). All we need is the x component of $\mathbf{E}$.

The field at $+q$ due to $-q$ is: (Eq. 10.65)

$$\mathbf{E} = -\frac{q}{4\pi\epsilon_0}\frac{\mathscr{r}}{(\boldsymbol{\mathscr{r}}\cdot\mathbf{u})^3}\left[\mathbf{u}(c^2 - v^2) + \mathbf{u}(\boldsymbol{\mathscr{r}}\cdot\mathbf{a}) - \mathbf{a}(\boldsymbol{\mathscr{r}}\cdot\mathbf{u})\right].$$

$\mathbf{u} = c\hat{\boldsymbol{\mathscr{r}}} - \mathbf{v} \Rightarrow u_x = c\frac{l}{\mathscr{r}} - v = \frac{1}{\mathscr{r}}(cl - v\mathscr{r})$; $\boldsymbol{\mathscr{r}}\cdot\mathbf{u} = c\mathscr{r} - \boldsymbol{\mathscr{r}}\cdot\mathbf{v} = (c\mathscr{r} - lv)$; $\boldsymbol{\mathscr{r}}\cdot\mathbf{a} = la$. So:

$$E_x = -\frac{q}{4\pi\epsilon_0}\frac{\mathscr{r}}{(c\mathscr{r} - vl)^3}\Big[\frac{1}{\mathscr{r}}(cl - v\mathscr{r})(c^2 - v^2) + \underbrace{\frac{1}{\mathscr{r}}(cl - \cancel{v\mathscr{r}})la - a(c\mathscr{r} - \cancel{lv})}_{\frac{1}{\mathscr{r}}ca(l^2 - \mathscr{r}^2) = -cad^2/\mathscr{r}}\Big]$$

$$= -\frac{q}{4\pi\epsilon_0}\frac{1}{(c\mathscr{r} - vl)^3}\left[(cl - v\mathscr{r})(c^2 - v^2) - cad^2.\right]$$

The *force* on $+q$ is qE_x, and there is an equal force on $-q$, so the net force on the dipole is:

$$\boxed{\mathbf{F} = -\frac{2q^2}{4\pi\epsilon_0}\frac{1}{(c\mathscr{r} - lv)^3}\left[(cl - v\mathscr{r})(c^2 - v^2) - cad^2\right]\hat{\mathbf{x}}.}$$

It remains to determine $\mathscr{r}$, l, v, and a, and plug these in.

$$v(t) = \frac{dx}{dt} = \frac{c}{\alpha}\frac{1}{2}\frac{1}{\sqrt{1+(\alpha t)^2}}2\alpha^2 t = \frac{c\alpha t}{\sqrt{1-(\alpha t)^2}}\,;\; v = v(t_r) = \frac{c\alpha t_r}{T},\ \text{where } T \equiv \sqrt{1+(\alpha t_r)^2}.$$

$$a(t_r) = \frac{dv}{dt_r} = \frac{c\alpha}{T} + c\alpha t_r\left(-\frac{1}{2}\right)\frac{2\alpha^2 t_r}{T^3} = \frac{c\alpha}{T^3}\left[1 + (\alpha t_r)^2 - (\alpha t_r)^2\right] = \frac{c\alpha}{T^3}.$$

Now calculate t_r: $c^2(t - t_r)^2 = \mathscr{r}^2 = l^2 + d^2$; $l = x(t) - x(t_r) = \frac{c}{\alpha}\left[\sqrt{1+(\alpha t)^2} - \sqrt{1+(\alpha t_r)^2}\right]$, so
$\cancel{t^2} - 2tt_r + \cancel{t_r^2} = \frac{1}{\alpha^2}\left[1 + (\cancel{\alpha t})^2 + 1 + (\cancel{\alpha t_r})^2 - 2\sqrt{1+(\alpha t)^2}\sqrt{1+(\alpha t_r)^2}\right] + (d/c)^2$
($\bigstar$) $\sqrt{1+(\alpha t)^2}\sqrt{1+(\alpha t_r)^2} = 1 + \alpha^2 tt_r + \frac{1}{2}\left(\frac{\alpha d}{c}\right)^2$. Square both sides:

$$\cancel{1} + (\alpha t)^2 + (\alpha t_r)^2 + \cancel{\alpha^4 t^2 t_r^2} = \cancel{1} + \cancel{\alpha^4 t^2 t_r^2} + \frac{1}{4}\left(\frac{\alpha d}{c}\right)^4 + 2\alpha^2 tt_r + \left(\frac{\alpha d}{c}\right)^2 + \alpha^2 tt_r\left(\frac{\alpha d}{c}\right)^2$$

$$t^2 + t_r^2 - 2tt_r - tt_r\left(\frac{\alpha d}{c}\right)^2 - \left(\frac{d}{c}\right)^2 - \frac{\alpha^2}{4}\left(\frac{d}{c}\right)^4 = 0.$$

At this point we *could* solve for t_r in terms of t, but since v and a are already expressed in terms of t_r it is simpler to solve for t (in terms of t_r), and express everything in terms of t_r:

$$t^2 - tt_r\left[2 + \left(\frac{\alpha d}{c}\right)^2\right] + \left[t_r^2 - \left(\frac{d}{c}\right)^2 - \frac{\alpha^2}{4}\left(\frac{d}{c}\right)^4\right] = 0 \Longrightarrow$$

$$t = \frac{1}{2}\left\{t_r\left[2 + \left(\frac{\alpha d}{c}\right)^2\right] \pm \sqrt{t_r^2\left[\cancel{4} + 4\left(\frac{\alpha d}{c}\right)^2 + \left(\frac{\alpha d}{c}\right)^4\right] - \cancel{4t_r^2} + 4\left(\frac{d}{c}\right)^2 + \alpha^2\left(\frac{d}{c}\right)^4}\right\}$$

$$= t_r\left[1 + \frac{1}{2}\left(\frac{\alpha d}{c}\right)^2\right] \pm \sqrt{\left[1 + (\alpha t_r)^2\right]\left(\frac{d}{c}\right)^2\left[1 + \left(\frac{\alpha d}{2c}\right)^2\right]}$$

Which sign? For small α we want $t \approx t_r + d/c$, so we need the + sign:

$$t = t_r\left[1+\frac{1}{2}\left(\frac{\alpha d}{c}\right)^2\right]+\frac{d}{c}TD,\quad\text{where } D\equiv\sqrt{1+\left(\frac{\alpha d}{2c}\right)^2}.$$

So $\mathscr{r} = c(t-t_r) \Rightarrow \mathscr{r} = \frac{ct_r}{2}\left(\frac{\alpha d}{c}\right)^2 + dTD$. Now go back to Eq. (★) and solve for $\sqrt{1+(\alpha t)^2}$:

$$\begin{aligned}\sqrt{1+(\alpha t)^2} &= \frac{1}{T}\left\{1+\frac{1}{2}\left(\frac{\alpha d}{c}\right)^2+\alpha^2 t_r\left[t_r\left(1+\frac{1}{2}\left(\frac{\alpha d}{c}\right)^2\right)+\frac{d}{c}TD\right]\right\}\\ &= \frac{1}{T}\left\{\underbrace{[1+(\alpha t_r)^2]}_{T^2}\left[1+\frac{1}{2}\left(\frac{\alpha d}{c}\right)^2\right]+\frac{\alpha^2 t_r d}{c}TD\right\} = \left[1+\frac{1}{2}\left(\frac{\alpha d}{c}\right)^2\right]T+\frac{\alpha^2 t_r d}{c}D.\end{aligned}$$

$$l = \frac{c}{\alpha}\left[\sqrt{1+(\alpha t^2)} - \sqrt{1+(\alpha t_r)^2}\right] = \frac{c}{\alpha}\left\{\left[\cancel{1}+\frac{1}{2}\left(\frac{\alpha d}{c}\right)^2\right]T+\frac{\alpha^2 t_r d}{c}D - \cancel{T}\right\} = \alpha d\left(\frac{d}{2c}T + t_r D\right).$$

Putting all this in, the numerator in square brackets in $\mathbf{F}$ becomes:

$$\begin{aligned}[\ \ ] &= \left\{c\alpha d\left(\frac{d}{2c}T+t_rD\right) - \frac{c\alpha t_r}{T}\left[\frac{ct_r}{2}\left(\frac{\alpha d}{c}\right)^2+dTD\right]\right\}\left(c^2-\frac{c^2\alpha^2t_r^2}{T^2}\right) - c\frac{c\alpha}{T^3}d^2\\ &= c\alpha d\left[\frac{d}{2c}T+\cancel{t_rD}-\frac{d(\alpha t_r)^2}{2cT}-\cancel{t_rD}\right]\frac{c^2}{T^2}\left[1+(\cancel{\alpha t_r})^2-(\cancel{\alpha t_r})^2\right] - \frac{c^2\alpha d^2}{T^3}\\ &= \frac{c^2\alpha d^2}{T^3}\left[\frac{1}{2}T^2-\frac{1}{2}(\alpha t_r)^2-1\right] = \frac{c^2\alpha d^2}{2T^3}\left[1+(\alpha t_r)^2-(\alpha t_r)^2-2\right] = -\frac{c^2\alpha d^2}{2T^3}.\end{aligned}$$

$\therefore \mathbf{F} = \dfrac{q^2}{4\pi\epsilon_0}\dfrac{c^2\alpha d^2}{[(c\mathscr{r}-lv)T]^3}\hat{\mathbf{x}}.$ It remains to compute the denominator:

$$\begin{aligned}(c\mathscr{r}-lv)T &= \left\{c\left[\frac{ct_r}{2}\left(\frac{\alpha d}{c}\right)^2+dTD\right] - \alpha d\left(\frac{d}{2c}T+t_rD\right)\frac{c\alpha t_r}{T}\right\}T\\ &= \left[\frac{1}{2}\alpha^2\cancel{t_r}d^2 + cdTD - \frac{1}{2}\alpha^2\cancel{t_r}d^2 - \frac{cd(\alpha t_r)^2}{T}D\right]T = cdD[\underbrace{T^2-(\alpha t_r)^2}_{1+(\cancel{\alpha t_r})^2-(\cancel{\alpha t_r})^2}] = dcD.\end{aligned}$$

$$\therefore \mathbf{F} = \frac{q^2}{4\pi\epsilon_0}\frac{c^2d^2\alpha}{c^3d^3D^3}\hat{\mathbf{x}} = \boxed{\frac{q^2}{4\pi\epsilon_0}\frac{\alpha}{cd[1+(\alpha d/2c)^2]^{3/2}}\hat{\mathbf{x}}}\quad\left(\alpha=\frac{F}{mc}\right).$$

Energy must come from the "reservoir" of energy stored in the electromagnetic fields.

(b) $F = mc\alpha = \underset{\uparrow}{\frac{1}{2}}\dfrac{q^2}{4\pi\epsilon_0}\dfrac{\alpha}{cd[1+(\alpha d/2c)^2]^{3/2}} \Rightarrow \left[1+\left(\frac{\alpha d}{2c}\right)^2\right]^{3/2} = \dfrac{q^2}{8\pi\epsilon_0 mc^2 d} = \dfrac{\mu_0 q^2}{8\pi m d}.$

(force on one end only)

$\therefore \alpha = \dfrac{2c}{d}\sqrt{\left(\dfrac{\mu_0 q^2}{8\pi m d}\right)^{2/3} - 1}$, so $\boxed{F = \dfrac{2mc^2}{d}\sqrt{\left(\dfrac{\mu_0 q^2}{8\pi m d}\right)^{2/3} - 1}.}$

Problem 12.62

(a) $A^\mu = (V/c, A_x, A_y, A_z)$ is a 4-vector (like $x^\mu = (ct, x, y, z)$), so (using Eq. 12.19): $V = \gamma(\bar{V} + v\bar{A}_x)$. But $\bar{V} = 0$, and

$$\bar{A}_x = \frac{\mu_0}{4\pi}\frac{(\mathbf{m}\times\bar{\mathbf{r}})_x}{\bar{r}^3}.$$

Now $(\mathbf{m}\times\bar{\mathbf{r}})_x = m_y\bar{z} - m_z\bar{y} = m_y z - m_z y$. So

$$V = \gamma v\frac{\mu_0}{4\pi}\frac{(m_y z - m_z y)}{\bar{r}^3}.$$

Now $\bar{x} = \gamma(x - vt) = \gamma R_x$, $\bar{y} = y = R_y$, $\bar{z} = z = R_z$, where $\mathbf{R}$ is the vector (in $\mathcal{S}$) from the (instantaneous) location of the dipole to the point of observation. Thus

$$\bar{r}^2 = \gamma^2 R_x^2 + R_y^2 + R_z^2 = \gamma^2(R_x^2 + R_y^2 + R_z^2) + (1-\gamma^2)(R_y^2 + R_z^2) = \gamma^2\left(R^2 - \frac{v^2}{c^2}R^2\sin^2\theta\right)$$

(where θ is the angle between $\mathbf{R}$ and the x axis, so that $R_y^2 + R_z^2 = R^2\sin^2\theta$).

$$\therefore V = \frac{\mu_0}{4\pi}\frac{v\gamma(m_y R_z - m_z R_y)}{\gamma^3 R^3\left(1 - \frac{v^2}{c^2}\sin^2\theta\right)^{3/2}};\quad \text{but } \mathbf{v}\cdot(\mathbf{m}\times\mathbf{R}) = v(\mathbf{m}\times\mathbf{R})_x = v(m_y R_z - m_z R_y),\quad \text{so}$$

$$\boxed{V = \frac{\mu_0}{4\pi}\frac{\mathbf{v}\cdot(\mathbf{m}\times\mathbf{R})\left(1-\frac{v^2}{c^2}\right)}{R^3\left(1 - \frac{v^2}{c^2}\sin^2\theta\right)^{3/2}},}$$

or, using $\mu_0 = \frac{1}{\epsilon_0 c^2}$ and $\mathbf{v}\cdot(\mathbf{m}\times\mathbf{R}) = \mathbf{R}\cdot(\mathbf{v}\times\mathbf{m})$: $V = \dfrac{1}{4\pi\epsilon_0}\dfrac{\hat{\mathbf{R}}\cdot(\mathbf{v}\times\mathbf{m})\left(1-\frac{v^2}{c^2}\right)}{c^2R^2\left(1 - \frac{v^2}{c^2}\sin^2\theta\right)^{3/2}}.$

(b) In the nonrelativistic limit ($v^2 \ll c^2$):

$$V = \frac{1}{4\pi\epsilon_0}\frac{\hat{\mathbf{R}}\cdot(\mathbf{v}\times\mathbf{m})}{c^2R^2} = \frac{1}{4\pi\epsilon_0}\frac{\hat{\mathbf{R}}\cdot\mathbf{p}}{R^2},\quad \text{with } \mathbf{p} = \frac{\mathbf{v}\times\mathbf{m}}{c^2},$$

which is the potential of an *electric* dipole.

Problem 12.63

(a) $\mathbf{B} = -\frac{\mu_0}{2}K\hat{\mathbf{y}}$ (Eq. 5.56); $\mathbf{N} = \mathbf{m}\times\mathbf{B}$ (Eq. 6.1), so $\mathbf{N} = -\frac{\mu_0}{2}mK(\hat{\mathbf{z}}\times\hat{\mathbf{y}})$.

$$\boxed{\mathbf{N} = \frac{\mu_0}{2}mK\hat{\mathbf{x}}} = \frac{\mu_0}{2}(\lambda v l^2)(\sigma v)\hat{\mathbf{x}} = \frac{\mu_0}{2}\lambda\sigma v^2 l^2\hat{\mathbf{x}}.$$

(b) 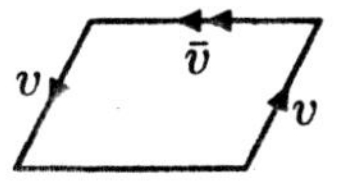

Charge density on the front side: λ_0 ($\lambda = \gamma\lambda_0$);
Charge density on the back side: $\bar{\lambda} = \bar{\gamma}\lambda_0$, where $\bar{v} = \frac{2v}{1+v^2/c^2} \Rightarrow$

$$\bar{\gamma} = \frac{1}{\sqrt{1 - \frac{4v^2/c^2}{(1+v^2/c^2)^2}}} = \frac{(1+v^2/c^2)}{\sqrt{1 + 2\frac{v^2}{c^2} + \frac{v^4}{c^4} - 4\frac{v^2}{c^2}}} = \frac{1+v^2/c^2}{\sqrt{1 - 2\frac{v^2}{c^2} + \frac{v^4}{c^4}}} = \frac{(1+v^2/c^2)}{(1-v^2/c^2)} = \gamma^2\left(1+\frac{v^2}{c^2}\right).$$

Length of front and back sides in this frame: l/γ. So the net charge on the back side is:

$$q_+ = \bar{\lambda}\frac{l}{\gamma} = \gamma^2\left(1+\frac{v^2}{c^2}\right)\frac{\lambda}{\gamma}\frac{l}{\gamma} = \left(1+\frac{v^2}{c^2}\right)\lambda l.$$

Net charge on front side is:

$$q_- = \lambda_0 \frac{l}{\gamma} = \frac{\lambda}{\gamma}\frac{l}{\gamma} = \frac{1}{\gamma^2}\lambda l.$$

So the dipole moment (note: charges on *sides* are equal):

$$\mathbf{p} = (q_+)\frac{l}{2}\,\hat{\mathbf{y}} - (q_-)\frac{l}{2}\,\hat{\mathbf{y}} = \left[\left(1+\frac{v^2}{c^2}\right)\lambda l\frac{l}{2} - \frac{1}{\gamma^2}\lambda l\frac{l}{2}\right]\hat{\mathbf{y}} = \frac{\lambda l^2}{2}\left(1+\frac{v^2}{c^2}-1+\frac{v^2}{c^2}\right)\hat{\mathbf{y}} = \boxed{\frac{\lambda l^2 v^2}{c^2}\hat{\mathbf{y}}}.$$

$\mathbf{E} = \frac{\sigma_0}{2\epsilon_0}\hat{\mathbf{z}}$, where $\sigma = \gamma\sigma_0$, so $\mathbf{N} = \mathbf{p}\times\mathbf{E} = \dfrac{\lambda l^2 v^2}{c^2}\dfrac{\sigma}{2\epsilon_0\gamma}(\hat{\mathbf{y}}\times\hat{\mathbf{z}}) = \boxed{\dfrac{1}{\gamma}\dfrac{\mu_0}{2}\lambda\sigma l^2 v^2\hat{\mathbf{x}}.}$

So apart from the relativistic factor of γ the torque is the same in both systems—but in $\mathcal{S}$ it is the torque exerted by a *magnetic* field on a *magnetic* dipole, whereas in $\bar{\mathcal{S}}$ it is the torque exerted by an *electric* field on an *electric* dipole.

Problem 12.64

Choose axes so that $\mathbf{E}$ points in the z direction and $\mathbf{B}$ in the yz plane: $\mathbf{E} = (0,0,E)$; $\mathbf{B} = (0, B\cos\phi, B\sin\phi)$. Go to a frame moving at speed v in the x direction:

$$\bar{\mathbf{E}} = \left(0, -\gamma v B\sin\phi, \gamma(E + vB\cos\phi)\right); \quad \bar{\mathbf{B}} = \left(0, \gamma(B\cos\phi + \frac{v}{c^2}E), \gamma B\sin\phi\right).$$

(I used Eq. 12.108.) Parallel provided $\dfrac{-\gamma v B\sin\phi}{\gamma(B\cos\phi + \frac{v}{c^2}E)} = \dfrac{\gamma(E+vB\cos\phi)}{\gamma B\sin\phi}$, or

$$-vB^2\sin^2\phi = \left(B\cos\phi + \frac{v}{c^2}E\right)(E + vB\cos\phi) = EB\cos\phi + vB^2\cos^2\phi + \frac{v}{c^2}E^2 + \frac{v^2}{c^2}EB\cos\phi,$$

$$0 = vB^2 + \frac{v}{c^2}E^2 + EB\cos\phi\left(1+\frac{v^2}{c^2}\right); \quad \frac{v}{1+v^2/c^2} = -\frac{EB\cos\phi}{B^2+E^2/c^2}.$$

Now $\mathbf{E}\times\mathbf{B} = \begin{vmatrix} \hat{\mathbf{x}} & \hat{\mathbf{y}} & \hat{\mathbf{z}} \\ 0 & 0 & E \\ 0 & B\cos\phi & B\sin\phi \end{vmatrix} = -EB\cos\phi\,\hat{\mathbf{x}}$. So $\dfrac{\mathbf{v}}{1+v^2/c^2} = \dfrac{\mathbf{E}\times\mathbf{B}}{B^2+E^2/c^2}$. qed

$\boxed{\text{No,}}$ there can be no frame in which $\mathbf{E}\perp\mathbf{B}$, for $(\mathbf{E}\cdot\mathbf{B})$ is invariant, and since it is not zero in $\mathcal{S}$ it can't be zero in $\bar{\mathcal{S}}$.

Problem 12.65

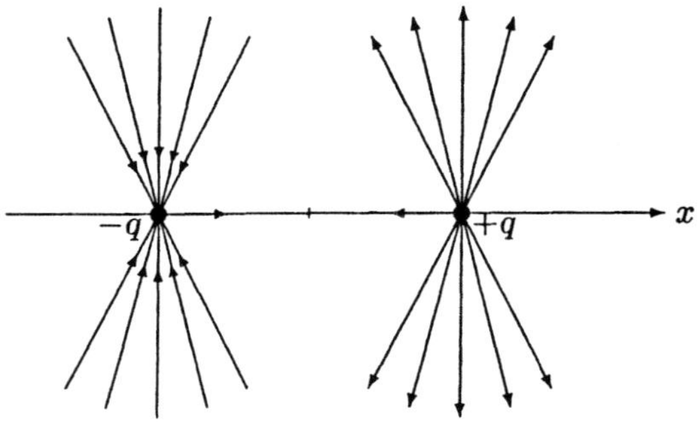

Just before:
Field lines emanate from *present* position of particle.

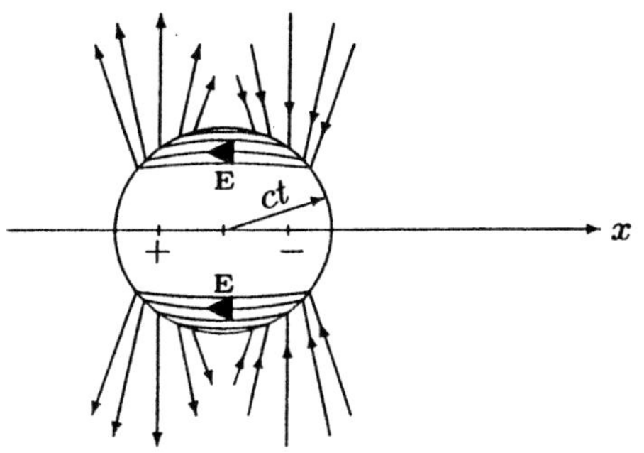

Just after: Field lines *outside* sphere of radius ct emanate from position particle *would* have reached, had it kept going on its original "flight plan". *Inside* the sphere $E = 0$. On the surface the lines connect up (since they cannot simply *terminate* in empty space), as suggested in the figure.

This produces a dense cluster of tangentially-directed field lines, which expand with the spherical shell. This is a pictorial way of understanding the generation of *electromagnetic radiation.*

Problem 12.66

Equation 12.68 assumes the particle is (instantaneously) at rest in $\mathcal{S}$. Here the particle is at rest in $\bar{\mathcal{S}}$. So $\mathbf{F}_\perp = \frac{1}{\gamma}\bar{\mathbf{F}}_\perp$, $F_\parallel = \bar{F}_\parallel$. Using $\bar{\mathbf{F}} = q\bar{\mathbf{E}}$, then,

$$F_x = \bar{F}_x = q\bar{E}_x, \quad F_y = \frac{1}{\gamma}\bar{F}_y = \frac{1}{\gamma}q\bar{E}_y, \quad F_z = \frac{1}{\gamma}\bar{F}_z = \frac{1}{\gamma}q\bar{E}_z.$$

Invoking Eq. 12.108:

$$F_x = qE_x, \quad F_y = \frac{1}{\gamma}q\gamma(E_y - vB_z) = q(E_y - vB_z), \quad F_z = \frac{1}{\gamma}q\gamma(E_z + vB_y) = q(E_z + vB_y).$$

But $\mathbf{v}\times\mathbf{B} = -vB_z\,\hat{\mathbf{x}} + vB_y\,\hat{\mathbf{z}}$, so $\mathbf{F} = q(\mathbf{E} + \mathbf{v}\times\mathbf{B})$. qed

---

Problem 12.67

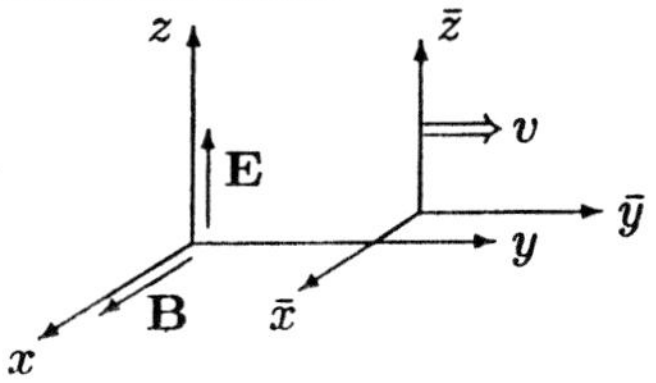

Rewrite Eq. 12.108 with $x \to y$, $y \to z$, $z \to x$:

$$\boxed{\begin{array}{lll} \bar{E}_y = E_y & \bar{E}_z = \gamma(E_z - vB_x) & \bar{E}_x = \gamma(E_x + vB_z) \\ \bar{B}_y = B_y & \bar{B}_z = \gamma\left(B_z + \frac{v}{c^2}E_x\right) & \bar{B}_x = \gamma\left(B_x - \frac{v}{c^2}E_z\right) \end{array}}$$

This gives the fields in system $\bar{\mathcal{S}}$ moving in the y direction at speed v.

Now $\mathbf{E} = (0, 0, E_0)$; $\mathbf{B} = (B_0, 0, 0)$, so $\bar{E}_y = 0$, $\bar{E}_z = \gamma(E_0 - vB_0)$, $\bar{E}_x = 0$.
If we want $\bar{\mathbf{E}} = 0$, we must pick v so that $E_0 - vB_0 = 0$; i.e. $\boxed{v = E_0/B_0.}$
(The condition $E_0/B_0 < c$ guarantees that there is no problem *getting* to such a system.)
With this, $\bar{B}_y = 0$, $\bar{B}_z = 0$, $\bar{B}_x = \gamma(B_0 - \frac{v}{c^2}E_0) = \gamma B_0(1 - \frac{v^2}{c^2}) = \gamma B_0\frac{1}{\gamma^2} = \frac{1}{\gamma}B_0$; $\boxed{\bar{\mathbf{B}} = \frac{1}{\gamma}B_0\,\hat{\mathbf{x}}.}$

The trajectory in $\bar{\mathcal{S}}$: Since the particle started out at rest at the origin in $\mathcal{S}$, it started out with velocity $-v\hat{\mathbf{y}}$ in $\bar{\mathcal{S}}$. According to Eq. 12.72 it will move in a circle of radius R, given by

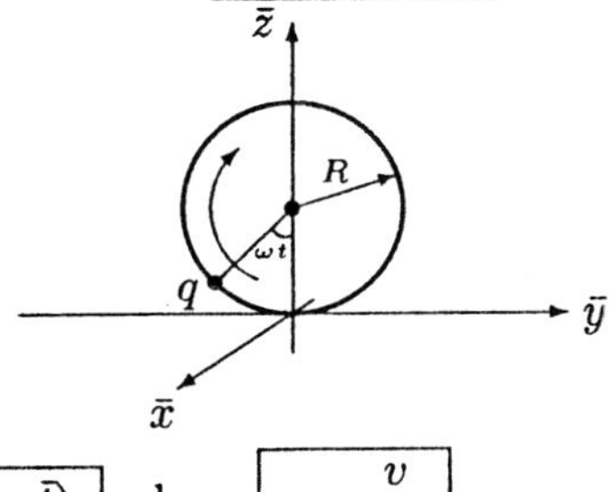

$$p = qBR, \text{ or } \gamma mv = q\left(\frac{1}{\gamma}B_0\right)R \Rightarrow \boxed{R = \frac{m\gamma^2 v}{qB_0}.}$$

The actual trajectory is given by $\boxed{\bar{x} = 0;\ \bar{y} = -R\sin\omega\bar{t};\ \bar{z} = R(1 - \cos\omega\bar{t});}$ where $\boxed{\omega = \frac{v}{R}.}$

The trajectory in $\mathcal{S}$: The Lorentz transformations Eqs. 12.18 and 12.19, for the case of relative motion in the y direction, read:

$$\begin{array}{ll} \bar{x} = x & x = \bar{x} \\ \bar{y} = \gamma(y - vt) & y = \gamma(\bar{y} + v\bar{t}) \\ \bar{z} = z & z = \bar{z} \\ \bar{t} = \gamma\left(t - \frac{v}{c^2}y\right) & t = \gamma\left(\bar{t} + \frac{v}{c^2}\bar{y}\right) \end{array}$$

So the trajectory in $\mathcal{S}$ is given by:

$$x = 0; \quad y = \gamma(-R\sin\omega\bar{t} + v\bar{t}) = \gamma\left\{-R\sin\left[\omega\gamma\left(t - \frac{v}{c^2}y\right)\right] + v\gamma\left(t - \frac{v}{c^2}y\right)\right\}, \text{ or}$$

$$\left.\underbrace{y\left(1 + \gamma^2\frac{v^2}{c^2}\right)}_{\gamma^2 y\left(1 - \frac{v^2}{c^2} + \frac{v^2}{c^2}\right) = \gamma^2 y} = \gamma^2 vt - \gamma R\sin\left[\omega\gamma\left(t - \frac{v}{c^2}y\right)\right]\right\} (y - vt)\gamma = -R\sin\left[\omega\gamma\left(t - \frac{v}{c^2}y\right)\right];$$

$$z = R(1 - \cos^2\omega\bar{t}) = R\left[1 - \cos\omega\gamma\left(t - \frac{v}{c^2}y\right)\right].$$

So: $\boxed{x = 0;\ y = vt - \frac{R}{\gamma}\sin\left[\omega\gamma\left(t - \frac{v}{c^2}y\right)\right];\ z = R - R\cos\left[\omega\gamma\left(t - \frac{v}{c^2}y\right)\right].}$

We can get rid of the trigonometric terms by the usual trick:

$$\left.\begin{array}{l} \gamma(y - vt) = -R\sin\left[\omega\gamma(t - \frac{v}{c^2}y)\right] \\ z - R = -R\cos\left[\omega\gamma(t - \frac{v}{c^2}y)\right] \end{array}\right\} \Rightarrow \boxed{\gamma^2(y - vt)^2 + (z - R)^2 = R^2.}$$

Absent the γ^2, this would be the cycloid we found back in Ch. 5 (Eq. 5.9). The γ^2 makes it, as it were, an *elliptical* cycloid — same picture as p. 206, but with the horizontal axis stretched out.

Problem 12.68

(a) $\left.\begin{array}{l}\mathbf{D} = \epsilon_0\mathbf{E} + \mathbf{P} \text{ suggests } \mathbf{E} \to \frac{1}{\epsilon_0}\mathbf{D} \\ \mathbf{H} = \frac{1}{\mu_0}\mathbf{B} - \mathbf{M} \text{ suggests } \mathbf{B} \to \mu_0\mathbf{H}\end{array}\right\}$ but it's a little cleaner if we divide by μ_0 while we're at it, so that

$\mathbf{E} \to \frac{1}{\mu_0\epsilon_0}\mathbf{D} = c^2\mathbf{D}$, $\mathbf{B} \to \mathbf{H}$. Then:

$$\boxed{D^{\mu\nu} = \left\{\begin{array}{cccc} 0 & cD_x & cD_y & cD_z \\ -cD_x & 0 & H_z & -H_y \\ -cD_y & -H_z & 0 & H_x \\ -cD_z & H_y & -H_x & 0 \end{array}\right\}}$$

Then (following the derivation on p. 539):

$$\frac{\partial}{\partial x^\nu}D^{0\nu} = c\,\boldsymbol{\nabla}\cdot\mathbf{D} = c\rho_f = J_f^0\ ; \quad \frac{\partial}{\partial x^\nu}D^{1\nu} = \frac{1}{c}\frac{\partial}{\partial t}(-cD_x) + (\boldsymbol{\nabla}\times\mathbf{H})_x = (J_f)_x\ ; \text{ so } \boxed{\frac{\partial D^{\mu\nu}}{\partial x^\nu} = J_f^\mu,}$$

where $\boxed{J_f^\mu = (c\rho_f, \mathbf{J}_f).}$ Meanwhile, the homogeneous Maxwell equations ($\boldsymbol{\nabla}\cdot\mathbf{B} = 0$, $\mathbf{E} = -\frac{\partial\mathbf{B}}{\partial t}$) are unchanged,

and hence $\boxed{\frac{\partial G^{\mu\nu}}{\partial x^\nu} = 0.}$

(b) $$\boxed{H^{\mu\nu} = \left\{\begin{array}{cccc} 0 & H_x & H_y & H_z \\ -H_x & 0 & -cD_z & cD_y \\ -H_y & cD_z & 0 & -cD_x \\ -H_z & -cD_y & cD_x & 0 \end{array}\right\}}$$

(c) If the material is at rest, $\eta_\nu = (-c, 0, 0, 0)$, and the sum over ν collapses to a single term:

$$D^{\mu 0}\eta_0 = c^2\epsilon F^{\mu 0}\eta_0 \Rightarrow D^{\mu 0} = c^2\epsilon F^{\mu 0} \Rightarrow -c\mathbf{D} = -c^2\epsilon\frac{\mathbf{E}}{c} \Rightarrow \mathbf{D} = \epsilon\mathbf{E} \text{ (Eq. 4.32)}, \checkmark$$

$$H^{\mu 0}\eta_0 = \frac{1}{\mu}G^{\mu 0}\eta_0 \Rightarrow H^{\mu 0} = \frac{1}{\mu}G^{\mu 0} \Rightarrow -\mathbf{H} = -\frac{1}{\mu}\mathbf{B} \Rightarrow \mathbf{H} = \frac{1}{\mu}\mathbf{B} \text{ (Eq. 6.31)}. \checkmark$$

(d) In general, $\eta_\nu = \gamma(-c, \mathbf{u})$, so, for $\mu = 0$:

$$D^{0\nu}\eta_\nu = D^{01}\eta_1 + D^{02}\eta_2 + D^{03}\eta_3 = cD_x(\gamma u_x) + cD_y(\gamma u_y) + cD_z(\gamma u_z) = \gamma c(\mathbf{D}\cdot\mathbf{u}),$$

$$F^{0\nu}\eta_\nu = F^{01}\eta_1 + F^{02}\eta_2 + F^{03}\eta_3 = \frac{E_x}{c}(\gamma u_x) + \frac{E_y}{c}(\gamma u_y) + \frac{E_z}{c}(\gamma u_z) = \frac{\gamma}{c}(\mathbf{E}\cdot\mathbf{u}), \text{ so}$$

$$D^{0\nu}\eta_\nu = c^2\epsilon F^{0\nu}\eta_\nu \Rightarrow \gamma c(\mathbf{D}\cdot\mathbf{u}) = c^2\epsilon\left(\frac{\gamma}{c}\right)(\mathbf{E}\cdot\mathbf{u}) \Rightarrow \mathbf{D}\cdot\mathbf{u} = \epsilon(\mathbf{E}\cdot\mathbf{u}). \qquad [1]$$

$$H^{0\nu}\eta_\nu = H^{01}\eta_1 + H^{02}\eta_2 + H^{03}\eta_3 = H_x(\gamma u_x) + H_y(\gamma u_y) + H_z(\gamma u_z) = \gamma(\mathbf{H}\cdot\mathbf{u}),$$

$$G^{0\nu}\eta_\nu = G^{01}\eta_1 + G^{02}\eta_2 + G^{03}\eta_3 = B_x(\gamma u_x) + B_y(\gamma u_y) + B_z(\gamma u_z) = \gamma(\mathbf{B}\cdot\mathbf{u}), \text{ so}$$

$$H^{0\nu}\eta_\nu = \frac{1}{\mu}G^{0\nu}\eta_\nu \Rightarrow \gamma(\mathbf{H}\cdot\mathbf{u}) = \frac{1}{\mu}(\gamma)(\mathbf{B}\cdot\mathbf{u}) \Rightarrow \mathbf{H}\cdot\mathbf{u} = \frac{1}{\mu}(\mathbf{B}\cdot\mathbf{u}). \qquad [2]$$

Similarly, for $\mu = 1$:

$$\begin{aligned} D^{1\nu}\eta_\nu &= D^{10}\eta_0 + D^{12}\eta_2 + D^{13}\eta_3 = (-cD_x)(-\gamma c) + H_z(\gamma u_y) + (-H_y)(\gamma u_z) = \gamma(c^2D_x + u_yH_z - u_zH_y) \\ &= \gamma\left[c^2\mathbf{D} + (\mathbf{u}\times\mathbf{H})\right]_x, \\ F^{1\nu}\eta_\nu &= F^{10}\eta_0 + F^{12}\eta_2 + F^{13}\eta_3 = \frac{-E_x}{c}(-\gamma c) + B_z(\gamma u_y) + (-B_y)(\gamma u_z) = \gamma(E_x + u_yB_z - u_zB_y) \\ &= \gamma\left[\mathbf{E} + (\mathbf{u}\times\mathbf{B})\right]_x, \text{ so } D^{1\nu}\eta_\nu = c^2\epsilon F^{1\nu}\eta_\nu \Rightarrow \end{aligned}$$

$$\gamma\left[c^2\mathbf{D} + (\mathbf{u}\times\mathbf{H})\right]_x = c^2\epsilon(\gamma)\left[\mathbf{E} + (\mathbf{u}\times\mathbf{B})\right]_x \Rightarrow \mathbf{D} + \frac{1}{c^2}(\mathbf{u}\times\mathbf{H}) = \epsilon\left[\mathbf{E} + (\mathbf{u}\times\mathbf{B})\right]. \qquad [3]$$

$$\begin{aligned} H^{1\nu}\eta_\nu &= H^{10}\eta_0 + H^{12}\eta_2 + H^{13}\eta_3 = (-H_x)(-\gamma c) + (-cD_z)(\gamma u_y) + (cD_y)(\gamma u_z) \\ &= \gamma c(H_x - u_yD_z + u_zD_y) = \gamma c\left[\mathbf{H} - (\mathbf{u}\times\mathbf{D})\right]_x, \\ G^{1\nu}\eta_\nu &= G^{10}\eta_0 + G^{12}\eta_2 + G^{13}\eta_3 = (-B_x)(-\gamma c) + \left(-\frac{E_z}{c}\right)(\gamma u_y) + \left(\frac{E_y}{c}\right)(\gamma u_z) \\ &= \frac{\gamma}{c}(c^2B_x - u_yE_z + u_zE_y) = \frac{\gamma}{c}\left[c^2\mathbf{B} - (\mathbf{u}\times\mathbf{E})\right]_x, \text{ so } H^{1\nu}\eta_\nu = \frac{1}{\mu}G^{1\nu}\eta_\nu \Rightarrow \end{aligned}$$

$$\gamma c\left[\mathbf{H} - (\mathbf{u}\times\mathbf{D})\right]_x = \frac{1}{\mu}\frac{\gamma}{c}\left[c^2\mathbf{B} - (\mathbf{u}\times\mathbf{E})\right]_x \Rightarrow \mathbf{H} - (\mathbf{u}\times\mathbf{D}) = \frac{1}{\mu}\left[\mathbf{B} - \frac{1}{c^2}(\mathbf{u}\times\mathbf{E})\right]. \qquad [4]$$

Use Eq. [4] as an expression for $\mathbf{H}$, plug this into Eq. [3], and solve for $\mathbf{D}$:

$$\mathbf{D} + \frac{1}{c^2}\mathbf{u}\times\left\{(\mathbf{u}\times\mathbf{D}) + \frac{1}{\mu}\left[\mathbf{B} - \frac{1}{c^2}(\mathbf{u}\times\mathbf{E})\right]\right\} = \epsilon\left[\mathbf{E} + (\mathbf{u}\times\mathbf{B})\right];$$

$$\mathbf{D} + \frac{1}{c^2}\left[(\mathbf{u}\cdot\mathbf{D})\mathbf{u} - u^2\mathbf{D}\right] = \epsilon\left[\mathbf{E} + (\mathbf{u}\times\mathbf{B})\right] - \frac{1}{\mu c^2}(\mathbf{u}\times\mathbf{B}) + \frac{1}{\mu c^4}\left[\mathbf{u}\times(\mathbf{u}\times\mathbf{E})\right].$$

Using Eq. [1] to rewrite $\mathbf{u}\cdot\mathbf{D}$:

$$\mathbf{D}\left(1-\frac{u^2}{c^2}\right) = -\frac{\epsilon}{c^2}(\mathbf{E}\cdot\mathbf{u})\mathbf{u}+\epsilon\left[\mathbf{E}+(\mathbf{u}\times\mathbf{B})\right]-\frac{1}{\mu c^2}(\mathbf{u}\times\mathbf{B})+\frac{1}{\mu c^4}\left[(\mathbf{E}\cdot\mathbf{u})\mathbf{u}-u^2\mathbf{E}\right]$$

$$= \epsilon\left\{\left[1-\frac{u^2}{\epsilon\mu c^4}\right]\mathbf{E}-\frac{1}{c^2}\left[1-\frac{1}{\epsilon\mu c^2}\right](\mathbf{E}\cdot\mathbf{u})\mathbf{u}+(\mathbf{u}\times\mathbf{B})\left[1-\frac{1}{\epsilon\mu c^2}\right]\right\}.$$

Let $\boxed{\gamma\equiv\frac{1}{\sqrt{1-u^2/c^2}},\quad v\equiv\frac{1}{\sqrt{\epsilon\mu}}.}$ Then

$$\boxed{\mathbf{D}=\gamma^2\epsilon\left\{\left(1-\frac{u^2v^2}{c^4}\right)\mathbf{E}+\left(1-\frac{v^2}{c^2}\right)\left[(\mathbf{u}\times\mathbf{B})-\frac{1}{c^2}(\mathbf{E}\cdot\mathbf{u})\mathbf{u}\right]\right\}.}$$

Now use Eq. [3] as an expression for $\mathbf{D}$, plug this into Eq. [4], and solve for $\mathbf{H}$:

$$\mathbf{H}-\mathbf{u}\times\left\{-\frac{1}{c^2}(\mathbf{u}\times\mathbf{H})+\epsilon\left[\mathbf{E}+(\mathbf{u}\times\mathbf{B})\right]\right\}=\frac{1}{\mu}\left[\mathbf{B}-\frac{1}{c^2}(\mathbf{u}\times\mathbf{E})\right];$$

$$\mathbf{H}+\frac{1}{c^2}\left[(\mathbf{u}\cdot\mathbf{H})\mathbf{u}-u^2\mathbf{H}\right]=\frac{1}{\mu}\left[\mathbf{B}-\frac{1}{c^2}(\mathbf{u}\times\mathbf{E})\right]+\epsilon(\mathbf{u}\times\mathbf{E})+\epsilon\left[\mathbf{u}\times(\mathbf{u}\times\mathbf{B})\right].$$

Using Eq. [2] to rewrite $\mathbf{u}\cdot\mathbf{H}$:

$$\mathbf{H}\left(1-\frac{u^2}{c^2}\right) = -\frac{1}{\mu c^2}(\mathbf{B}\cdot\mathbf{u})\mathbf{u}+\frac{1}{\mu}\left[\mathbf{B}-\frac{1}{c^2}(\mathbf{u}\times\mathbf{E})\right]+\epsilon(\mathbf{u}\times\mathbf{E})+\epsilon\left[(\mathbf{B}\cdot\mathbf{u})\mathbf{u}-u^2\mathbf{B}\right]$$

$$= \frac{1}{\mu}\left\{\left[1-\mu\epsilon u^2\right]\mathbf{B}+\left(\epsilon\mu-\frac{1}{c^2}\right)\left[(\mathbf{u}\times\mathbf{E})+(\mathbf{B}\cdot\mathbf{u})\mathbf{u}\right]\right\}.$$

$$\boxed{\mathbf{H}=\frac{\gamma^2}{\mu}\left\{\left(1-\frac{u^2}{v^2}\right)\mathbf{B}+\left(\frac{1}{v^2}-\frac{1}{c^2}\right)\left[(\mathbf{u}\times\mathbf{E})+(\mathbf{B}\cdot\mathbf{u})\mathbf{u}\right]\right\}.}$$

Problem 12.69

We know that (proper) power transforms as the zeroth component of a 4-vector: $K^0=\frac{1}{c}\frac{dW}{d\tau}$. The Larmor formula says that for $v=0$, $\frac{dW}{dt}=\frac{\mu_0 q^2a^2}{6\pi c}$ (Eq. 11.70). Can we think of a 4-vector whose zeroth component reduces to this when the velocity is zero?

Well, a^2 smells like $(\alpha^\nu\alpha_\nu)$, but how do we get a 4-vector in here? How about η^μ, whose zeroth component is just c, when $v=0$? Try, then:

$$K^\mu=\frac{\mu_0q^2}{6\pi c^3}(\alpha^\nu\alpha_\nu)\eta^\mu.$$

This has the right transformation properties, but we must check that it does reduce to the Larmor formula when $v\to 0$:

$\frac{dW}{dt}=\frac{1}{\gamma}\frac{dW}{d\tau}=\frac{1}{\gamma}cK^0=\frac{1}{\gamma}c\frac{\mu_0q^2}{6\pi c^3}(\alpha^\nu\alpha_\nu)\eta^0$, but $\eta^0=c\gamma$, so $\boxed{\frac{dW}{dt}=\frac{\mu_0q^2}{6\pi c}(\alpha^\nu\alpha_\nu).}$ [Incidentally, this tells us that the power itself (as opposed to *proper* power) is a *scalar*. If this had been obvious from the start, we could simply have looked for a Lorentz *scalar* that generalizes the Larmor formula.]

In Prob. 12.38(b) we calculated $(\alpha^\nu\alpha_\nu)$ in terms of the *ordinary* velocity and acceleration:

$$\alpha^\nu\alpha_\nu = \gamma^4\left[a^2 + \frac{(\mathbf{v}\cdot\mathbf{a})^2}{(c^2-v^2)}\right] = \gamma^6\left[a^2\gamma^{-2} + \frac{1}{c^2}(\mathbf{v}\cdot\mathbf{a})^2\right]$$
$$= \gamma^6\left[a^2\left(1-\frac{v^2}{c^2}\right) + \frac{1}{c^2}(\mathbf{v}\cdot\mathbf{a})^2\right] = \gamma^6\left\{a^2 - \frac{1}{c^2}\left[v^2a^2 - (\mathbf{v}\cdot\mathbf{a})^2\right]\right\}.$$

Now $\mathbf{v}\cdot\mathbf{a} = va\cos\theta$, where θ is the angle between $\mathbf{v}$ and $\mathbf{a}$, so:

$$v^2a^2 - (\mathbf{v}\cdot\mathbf{a})^2 = v^2a^2(1-\cos^2\theta) = v^2a^2\sin^2\theta = |\mathbf{v}\times\mathbf{a}|^2.$$
$$\alpha^\nu\alpha_\nu = \gamma^6\left(a^2 - \left|\frac{\mathbf{v}\times\mathbf{a}}{c}\right|^2\right).$$

$$\boxed{\frac{dW}{dt} = \frac{\mu_0 q^2}{6\pi c}\gamma^6\left(a^2 - \left|\frac{\mathbf{v}\times\mathbf{a}}{c}\right|^2\right),}$$ which is Liénard's formula (Eq. 11.73).

Problem 12.70

(a) It's inconsistent with the constraint $\eta_\mu K^\mu = 0$ (Prob. 12.38(d)).

(b) We want to find a 4-vector b^μ with the property that $(\frac{d\alpha^\mu}{d\tau} + b^\mu)\eta_\mu = 0$. How about $b^\mu = \kappa(\frac{d\alpha^\nu}{d\tau}\eta_\nu)\eta^\mu$? Then $(\frac{d\alpha^\mu}{d\tau} + b^\mu)\eta_\mu = \frac{d\alpha^\mu}{d\tau}\eta_\mu + \kappa\frac{d\alpha^\nu}{d\tau}\eta_\nu(\eta^\mu\eta_\mu)$. But $\eta^\mu\eta_\mu = -c^2$, so this becomes $(\frac{d\alpha^\mu}{d\tau}\eta_\mu) - c^2\kappa(\frac{d\alpha^\nu}{d\tau}\eta_\nu)$, which *is* zero, if we pick $\kappa = 1/c^2$. This suggests $\boxed{K^\mu_{\text{rad}} = \frac{\mu_0 q^2}{6\pi c}\left(\frac{d\alpha^\mu}{d\tau} + \frac{1}{c^2}\frac{d\alpha^\nu}{d\tau}\eta_\nu\eta^\mu\right).}$ Note that $\eta^\mu = (c, \mathbf{v})\gamma$, so the spatial components of b^μ vanish in the nonrelativistic limit $v \ll c$, and hence this still reduces to the Abraham-Lorentz formula. [Incidentally, $\alpha^\nu\eta_\nu = 0 \Rightarrow \frac{d}{d\tau}(\alpha^\nu\eta_\nu) = 0 \Rightarrow \frac{d\alpha^\nu}{d\tau}\eta_\nu + \alpha^\nu\frac{d\eta_\nu}{d\tau} = 0$, so $\frac{d\alpha^\nu}{d\tau}\eta_\nu = -\alpha^\nu\alpha_\nu$, and hence b^μ can just as well be written $-\frac{1}{c^2}(\alpha^\nu\alpha_\nu)\eta^\mu$.]

Problem 12.71

Define the electric current 4-vector as before: $J^\mu_e = (c\rho_e, \mathbf{J}_e)$, and the magnetic current the same way: $J^\mu_m = (c\rho_m, \mathbf{J}_m)$. The fundamental laws are then

$$\boxed{\partial_\nu F^{\mu\nu} = \mu_0 J^\mu_e, \quad \partial_\nu G^{\mu\nu} = \frac{\mu_0}{c}J^\mu_m, \quad K^\mu = \left(q_e F^{\mu\nu} + \frac{q_m}{c}G^{\mu\nu}\right)\eta_\nu.}$$

The first of these reproduces $\nabla\cdot\mathbf{E} = (1/\epsilon_0)\rho_e$ and $\nabla\times\mathbf{B} = \mu_0\mathbf{J}_e + \mu_0\epsilon_0\partial\mathbf{E}/\partial t$, just as before (p. 539); the second yields $\nabla\cdot\mathbf{B} = (\mu_0/c)(c\rho_m) = \mu_0\rho_m$ and $-(1/c)(\partial\mathbf{B}/\partial t + \nabla\times\mathbf{E}) = (\mu_0/c)\mathbf{J}_m$, or $\nabla\times\mathbf{E} = -\mu_0\mathbf{J}_m - \partial\mathbf{B}/\partial t$ (generalizing page 540). These are Maxwell's equations with magnetic charge (Eq. 7.43). The third (following the argument on p. 540) says

$$K^1 = \frac{q_e}{\sqrt{1-u^2/c^2}}\left[\mathbf{E} + (\mathbf{u}\times\mathbf{B})\right]_x + \frac{q_m}{c}\left[\frac{-c}{\sqrt{1-u^2/c^2}}(-B_x) + \frac{u_y}{\sqrt{1-u^2/c^2}}\left(-\frac{E_z}{c}\right) + \frac{u_z}{\sqrt{1-u^2/c^2}}\left(\frac{E_y}{c}\right)\right],$$
$$\mathbf{K} = \frac{1}{\sqrt{1-u^2/c^2}}\left\{q_e\left[\mathbf{E} + (\mathbf{u}\times\mathbf{B})\right] + q_m\left[\mathbf{B} - \frac{1}{c^2}(\mathbf{u}\times\mathbf{E})\right]\right\}, \text{ or}$$
$$\mathbf{F} = q_e\left[\mathbf{E} + (\mathbf{u}\times\mathbf{B})\right] + q_m\left[\mathbf{B} - \frac{1}{c^2}(\mathbf{u}\times\mathbf{E})\right],$$

which is the generalized Lorentz force law (Eq. 7.69).